THE LOGICO-ALGEBRAIC APPROACH TO QUANTUM MECHANICS

VOLUME I

HISTORICAL EVOLUTION

THE UNIVERSITY OF WESTERN ONTARIO SERIES IN PHILOSOPHY OF SCIENCE

A SERIES OF BOOKS

ON PHILOSOPHY OF SCIENCE, METHODOLOGY,

AND EPISTEMOLOGY

PUBLISHED IN CONNECTION WITH

THE UNIVERSITY OF WESTERN ONTARIO

PHILOSOPHY OF SCIENCE PROGRAMME

VOLUME 5

THE LOGICO-ALGEBRAIC APPROACH TO QUANTUM MECHANICS

VOLUME I

Historical Evolution

Edited by

C. A. HOOKER

University of Western Ontario, Ontario, Canada

D. REIDEL PUBLISHING COMPANY

DORDRECHT-HOLLAND / BOSTON-U.S.A.

Library of Congress Cataloging in Publication Data

Main entry under title:

The Logico-algebraic approach to quantum mechanics.

(The University of Western Ontario series in philosophy of science ; v. 5)
Includes bibliographical references.
CONTENTS: v. 1. Historical evolution.
1. Quantum theory—Addresses, essays, lectures. 2. Logic, Symbolic and mathematical—Addresses, essays, lectures. I. Hooker, Clifford Alan. II. Series: London, Ont. University of Western Ontario. Series in philosophy of science ; v. 5.
QC174.125.L63 530.1'2 75-8737

ISBN-13: 978-90-277-0613-3 e-ISBN-13: 978-94-010-1795-4
DOI: 10.1007/978-94-010-1795-4

Published by D. Reidel Publishing Company,
P.O. Box 17, Dordrecht, Holland

Sold and distributed in the U.S.A., Canada, and Mexico
by D. Reidel Publishing Company, Inc.
306 Dartmouth Street, Boston,
Mass. 02116, U.S.A.

TABLE OF CONTENTS

PREFACE

The twentieth century has witnessed a striking transformation in the understanding of the theories of mathematical physics. There has emerged clearly the idea that physical theories are significantly characterized by their abstract mathematical structure. This is in opposition to the traditional opinion that one should look to the specific applications of a theory in order to understand it. One might with reason now espouse the view that to understand the deeper character of a theory one must know its abstract structure and understand the significance of that structure, while to understand how a theory might be modified in light of its experimental inadequacies one must be intimately acquainted with how it is applied.

Quantum theory itself has gone through a development this century which illustrates strikingly the shifting perspective. From a collection of intuitive physical maneuvers under Bohr, through a formative stage in which the mathematical framework was bifurcated (between Schrödinger and Heisenberg) to an elegant culmination in von Neumann's Hilbert space formulation the elementary theory moved, flanked even at the later stage by the ill-understood formalisms for the relativistic version and for the field-theoretic alternative; after that we have a gradual, but constant, elaboration of all these quantal theories as abstract mathematical structures (their point of departure being von Neumann's formalism) until at the present time theoretical work is heavily preoccupied with the manipulation of purely abstract structures. (The earlier history is set out in some detail in e.g. M. Jammer, *The Conceptual Development of the Quantum Theory*, McGraw-Hill, 1966 and a subsequent companion volume now in press.) A similar story holds for the development of relativistic theories, and of course for the recent attempts to consistently combine them.

Despite this evolution of pure mathematical sophistication and self-consciousness we are at the present time, so it seems, neither any closer to an adequate atomic theory nor yet to a satisfactory physical interpretation of even elementary quantum theory. At least in regard to the latter problem, in my opinion, the situation stems directly from the fact that our conceptual understanding of physical theory has been even more slowly evolving than has our understanding of the mathematical structure of those theories themselves. In this respect the philosophy of science has itself undergone its own twentieth century revolution. From the 'everyday' intuitions and debates about determinism, mechanism and the like there has emerged slowly a tradition of conceptual analysis in which the ideal is to characterize conceptually interesting questions so sharply that they admit of formally precise answers. (Thus "Is the theory deterministic?" becomes "Does the theory admit such and such mathematical construction?" and an analysis of the predicate '-is deterministic' becomes '-has structure *S*'.) The examination of physical theory from this point of view has led to many rich and fruitful insights.

It is precisely the combination of the emerging mathematical sophistication and self-consciousness with the emerging formal sophistication and insight into the structure of conceptual schemes that is the foundation for the claim that the deep understanding of physical theory lies in the understanding of its abstract structures.

(In respect of these revolutions we may note: (i) That each had its origins in preceding centuries, though not there the dominant tradition, one thinks of Hamilton and Jacobi or of Boole – even so most of the development lies in the twentieth century, mathematical logic, formal syntax and semantics, the use in physics of Hilbert space, group theory, lattice theory and so on all essentially belong to the last 60 years. (ii) In respect of contributions to the various developments the departmental origins of salary were not closely correlated with type of contribution – indeed the various disciplines involved have never coped with the blurring of their separateness and are still in the early stages of adjusting to the changing intellectual perspective.)

The papers in this volume all belong to one strand of this complex development – the understanding of elementary quantum theory through examination of its formal, or abstract, structure. Remarkably, this is the first time (to my knowledge) that anyone has attempted to collect them together, though the body of literature has been well known to anyone approaching the subject in the last decade. (This is in itself witness to the newness of the perspective.) Considering the span of time covered and the diversity of authorship, the collection is satisfyingly complete – a tribute to the eagerness of all concerned to see their work set in the wider perspective of the emerging field of study. Even casual examination of the dates of the papers – they are arranged nearly chronologically, not alphabetically or by theme – reveals an explosion of interest and fruitful work in the late 1960's. (Occasionally I have foregone strict chronological order so that related articles might be grouped together. This is so in the case of Reichenbach's 3-valued approach and responses, and the work of Kochen and Specker, to which should be attached the earlier paper by Specker.)

The present volume barely reaches the beginnings of the recent explosion of interest and productivity, some papers with a broader perspective are included for the reader's benefit (principally those by Holland, Piron, Gudder and Greechie and van Fraassen), but by and large this volume concentrates on the foundational work laid down in the 'long period of initial struggle', 1935–1965. It is my intention to devote a second volume (to appear shortly) to recent research.

These articles do not belong to a single tradition concerning quantum theory, nothing like that could have emerged until the 1970's and the major points of view are only just now emerging clearly. Nor are they written from the same perspective – some are written by mathematicians for mathematicians, others by logicians for philosophers. These articles represent the 'raw material' for study. This volume is designed to be a basic reference text, not the presentation of a particular doctrine. I have my own understanding of the significance of these papers, but that will appear as part of Volume II which will be more devoted to works that analyse and interpret the mathematical material than is this volume.

In keeping with this referential role for the volume I shall refrain here from taking issue with the various interpretive battles now raging and from any attempt to evaluate the relative significances of the various mathematical contributions. Where this text is used – senior undergraduate or graduate course – one assumes a competent leader who will set the material in some preferred order. Some passing remarks may be in order for the benefit of the disciplinarily one-sided reader. The approach of Birkhoff and von Neumann is connected to the structure of logical systems clearly for the first time in the work of Kochen and Specker, who also make clear the significance of Gleason's theorem in this context. The connection between this approach and lattice and POset theory is discussed at length in Holland and reviewed in Piron and Gudder and Greechie, while its corresponding and close association with group theory is developed by Poole upon the basis of the fundamental paper by Foulis. The important paper by Varadarajan ought to be read in conjunction with the discussion by Kochen and Specker concerning the probabilistic constraints on acceptable boolean embeddings (cf. Zieler and Schlessinger) and compared with the discussions by Suppes. Finally, it is well known to those in philosophy that these researches have generated a heated discussion concerning the nature of logic, a debate of great profundity but one barely hinted at here (cf. Kochen and Specker, Suppes, van Fraassen); rather these papers serve as the background to the debate, the debate itself being taken up in Volume II.

As I remarked earlier, the material spreads across the boundaries between mathematics, physics and philosophy. From a mathematician's point of view this volume is designed to offer some of the basic source material for a study of the kind of axiomatic approach to quantum theory followed by George Mackey, Josef Jauch, and others, to connect it to the physical and conceptual (chiefly logical here) issues and to introduce several areas of mathematical enquiry delightful in their own rights. From the philosopher's point of view it is designed as a basic reference text to educate in the formalism and results he (she!) must know in order to competently follow the current debates and to contribute thereto. For the physicist the book offers an introduction to that complex of mathematical and philosophical argument which constitutes a first example in the new way of studying physical theories.

From the point of view of an adequately complete reference source it must be admitted that several other entire volumes ought also to be included! Of course this is not possible. I shall mention now several other volumes which, if the reader will employ them to complement the present selection of journal articles, will provide him/her with a well rounded reference library. Among the more important volumes are:

Bub, J., *The Interpretation of Quantum Mechanics*, D. Reidel Publishing Co., Dordrecht, 1974.

Hooker, C. A., *Contemporary Research in the Foundations and Philosophy of Quantum Theory*, D. Reidel Publishing Co., Dordrecht, 1974.

Jauch, J., *The Foundations of Quantum Mechanics*, Addison-Wesley, New York, 1968.
Mackey, G. W., *The Mathematical Foundations of Quantum Mechanics*, W. A. Benjamin, New York, 1963.
Varadarajan, V. S., *The Geometry of Quantum Mechanics*, 2 vols., Van Nostrand, Princeton, N.J., 1968.

London, Ontario, 1974.

ACKNOWLEDGEMENTS

I should like to acknowledge the generous co-operation of all of those journal editors and publishing houses which granted copyright permission so readily and so generously, thereby making this volume possible. I should like particularly to acknowledge the work of Mrs A. Smith, without whose indefatigable efforts and intelligence I should not have been able to assemble the volume in anything but a semi-infinite time.

'The Logic of Quantum Mechanics' by G. Birkhoff and J. von Neumann in *Annals of Mathematics* **37** (1936), 823–43. Reprinted by permission of the surviving author (Birkhoff) and the Estate of von Neumann and *Annals of Mathematics*.

'The Logic of Complementarity and the Foundation of Quantum Theory' by M. Strauss in *Modern Physics and Its Philosophy* (ed. by M. Strauss), D. Reidel Publishing Co., Dordrecht-Holland, 1972. (Note: This is a translation of an article appearing in 1936 together wish a *postscript* added in 1971.) Reprinted by permission of the author and D. Reidel Publishing Co.

'Mathematics as Logical Syntax – A Method to Formalize the Language of a Physical Theory' by M. Strauss in *Erkenntnis* **7** (1937–38), 147–153. Reprinted by permission of the author and D. Reidel Publ. Co.

'Three-Valued Logic and the Interpretation of Quantum Mechanics' by H. Reichenbach. An edited portion of sections 29–37 of *Philosophic Foundations of Quantum Mechanics* by H. Reichenbach, University of California Press, Los Angeles, 1944. Reprinted by permission of Maria Reichenbach. Originally published by the University of California Press; reprinted by permission of the Regents of the University of California.

'Three-Valued Logic', by H. Putnam in *Philosophical Studies* **VIII** (1957), 73–80. Reprinted by permission of the author and *Philosophical Studies*.

'Reichenbach's Interpretation of Quantum Mechanics' by P. Feyerabend in *Philosophical Studies* **IX** (1958), 49–59. Reprinted by permission of the author and *Philosophical Studies*.

'Measures on the Closed Subspaces of a Hilbert Space' by A. M. Gleason in *Journal of Mathematics and Mechanics* **6** (1957), 885–93. Reprinted by permission of the author and *Indiana University Mathematics Journal*. (Originally *Journal of Mathematics and Mechanics*.)

'The Logic of Propositions which are not Simultaneously Decidable' by E. P. Specker in *Dialectica* **14** (1960), 239–46. (A translation by Mr. A. Stairs.) Reprinted by permission of the author and *Dialectica*.

'Baer *-Semigroups' by D. J. Foulis in *Proceedings of the American Mathematical Society* **11** (1960), 648–54. Reprinted by permission of the author and American Mathematical Society.

'Axioms for Non-Relativistic Quantum Mechanics' by N. Zierler in *Pacific Journal of*

Mathematics **11** (1961), 1151–69. Reproduction is by permission of the author and the copyright owners, the *Pacific Journal of Mathematics*.
'Probability in Physics and a Theorem on Simultaneous Observability' by V. S. Varadarajan in *Communications in Pure and Applied Mathematics* **15** (1962), 189–217. (As corrected *Loc. cit.* **18** (1965).) Reprinted by permission of John Wiley & Sons, Inc.
'Semantic Representation of the Probability of Formulas in Formalized Theories' by J. Łoś in *Studia Logico* **14** (1963), 183–96. Reprinted by permission of the author and *Studia Logica*.
'The Structure of the Propositional Calculus of a Physical Theory' by F. Kamber in *Nachrichten der Akademie der Wissenschaften Mathematisch-Physikalische Klasse* **10** (1964) 103–124. Reprinted by permission of the author and *Nachrichten der Akademie der Wissenschaften Mathematisch-Physikalische Klasse*.
'Boolean Embeddings of Orthomodular Sets and Quantum Logic' by N. Zierler and M. Schlessinger in *Duke Mathematical Journal* **32** (1965), 251–62. Reprinted by permission of the author and publisher. © Copyright 1961, Duke University Press, Durham, North Carolina.
'Logical Structures Arising in Quantum Theory' by S. Kochen and E. P. Specker in *The Theory of Models* (ed. by J. Addison, L. Henkin, and A. Tarski), North-Holland Publishing Co., Amsterdam, 1965. Reprinted by permission of the authors and North-Holland Publishing Co.
'The Calculus of Partial Propositional Functions' by S. Kochen and E. P. Specker in *Logic, Methodology and the Philosophy of Science* (ed. by U. Bar-Hillel), North-Holland Publishing Co., Amsterdam, 1965. Reprinted by permission of the authors and North-Holland Publishing Co.
'The Problem of Hidden Variables in Quantum Mechanics' by S. Kochen and E. P. Specker in *Journal of Mathematics and Mechanics* **17** (1967), 59–67. Reprinted by permission of the authors and *Indiana University Mathematics Journal*. (Originally *Journal of Mathematics and Mechanics*.)
'Logics Appropriate to Empirical Theories' by P. Suppes in *The Theory of Models* (ed. by J. W. Addison, L. Henkin, and A. Tarski), North-Holland Publishing Co., Amsterdam, 1965. Reprinted by permission of the authors and North-Holland Publishing Co.
'The Probabilistic Argument for a Non-Classical Logic of Quantum Mechanics' by P. Suppes in *Philosophy of Science* **33** (1966), 14–21. Reprinted by permission of the author and *Philosophy of Science*.
'Foundations of Quantum Mechanics' by M. Strauss in *Modern Physics and Its Philosophy*, D. Reidel Publishing Co., Dordrecht-Holland, 1972. (Note: This is a translation of an article appearing in German in *Mikrokosmos-makrokosmos*, vol. II (ed. by H. Ley and R. Lother), Berlin, 1967.) Reprinted by permission of the author and D. Reidel Publishing Co.
'Baer *-Semigroups and the Logic of Quantum Mechanics' by J. C. T. Pool in *Communications on Mathematical Physics* **9** (1948), 118–41. Reprinted by permission of the author and *Communications on Mathematical Physics*.
'Semimodularity and the Logic of Quantum Mechanics' by J. C. T. Pool in *Communications on Mathematical Physics* **9** (1968), 212–28. Reprinted by permission of the author and *Communications on Mathematical Physics*.
'On the Structure of Quantum Logic' by P.D. Finch in *The Journal of Symbolic Logic* **34** (1969), 275–82. Reprinted by permission of the author and the publisher, The Association for Symbolic Logic, from *The Journal of Symbolic Logic*.
'On the Structure of Quantal Proposition Systems' by J. M. Jauch and C. Piron in *Helvetica*

Physica Acta **43** (1969), 842–8. Reprinted by permission of the author and *Helvetica Physica Acta*.

'The Current Interest in Orthomodular Lattices' by S. S. Holland Jr. in *Trends in Lattice Theory*, Van Nostrand, New York, 1970. Reprinted by permission of the author and the publisher. © Copyright 1970 by Litton Educational Publishing, Inc.

'Integration Theory of Observables' by N. S. Kronfli in *International Journal of Theoretical Physics* **3** (1970), 199–204. Reprinted by permission of the Estate of the author and *International Journal of Theoretical Physics*.

'Probabilistic Formulation of Classical Mechanics' by N. S. Kronfli in *International Journal of Theoretical Physics* **3** (1970), 395–400. Reprinted by permission of the author and *International Journal of Theoretical Physics*.

'Atomicity and Determinism in Boolean Systems' by N. S. Kronfli in *International Journal of Theoretical Physics* **4** (1971), 141–3. Reprinted by permission of the Estate of the author and *International Journal of Theoretical Physics*.

'Survey of General Quantum Physics' by C. Piron in *Foundations of Physics* **2** (1972), 287–314. Reprinted with permission of the author and *Foundations of Physics*. Copyright held by the publisher, Plenum Publishing Corp.

'Quantum Logics' by S. P. Gudder and R. J. Greechie in *Contemporary Research in the Foundations and Philosophy of Quantum Theory* (ed. by C. A. Hooker), D. Reidel Publishing Co., Dordrecht-Holland, 1974. Reprinted by permission of the authors and D. Reidel Publishing Co.

'The Labyrinth of Quantum Logics' by B. C. van Fraassen in *Boston Studies in the Philosophy of Science*, vol. XIII, D. Reidel Publishing Co., Dordrecht-Holland, 1974. Reprinted by permission of the author and D. Reidel Publishing Co.

GARRETT BIRKHOFF AND JOHN VON NEUMANN

THE LOGIC OF QUANTUM MECHANICS

1. *Introduction*

One of the aspects of quantum theory which has attracted the most general attention, is the novelty of the logical notions which it presupposes. It asserts that even a complete mathematical description of a physical system $\mathfrak{S}$ does not in general enable one to predict with certainty the result of an experiment on $\mathfrak{S}$, and that in particular one can never predict with certainty both the position and the momentum of $\mathfrak{S}$ (Heisenberg's Uncertainty Principle). It further asserts that most pairs of observations are incompatible, and cannot be made on $\mathfrak{S}$ simultaneously (Principle of Non-commutativity of Observations).

The object of the present paper is to discover what logical structure one may hope to find in physical theories which, like quantum mechanics, do not conform to classical logic. Our main conclusion, based on admittedly heuristic arguments, is that one can reasonably expect to find a calculus of propositions which is formally indistinguishable from the calculus of linear subspaces with respect to *set products*, *linear sums*, and *orthogonal complements* – and resembles the usual calculus of propositions with respect to *and*, *or*, and *not*.

In order to avoid being committed to quantum theory in its present form, we have first (in Sections 2–6) stated the heuristic arguments which suggest that such a calculus is the proper one in quantum mechanics, and then (in Sections 7–14) reconstructed this calculus from the axiomatic standpoint. In both parts an attempt has been made to clarify the discussion by continual comparison with classical mechanics and its propositional calculi. The paper ends with a few tentative conclusions which may be drawn from the material just summarized.

I. PHYSICAL BACKGROUND

2. *Observations on Physical Systems*

The concept of a physically observable "physical system" is present in all branches of physics, and we shall assume it.

C. A. Hooker (ed.), The Logico-Algebraic Approach to Quantum Mechanics, 1–26.

It is clear that an "observation" of a physical system $\mathfrak{S}$ can be described generally as a writing down of the readings from various[1] compatible measurements. Thus if the measurements are denoted by the symbols $\mu_1, \ldots, \mu_n$, then an observation of $\mathfrak{S}$ amounts to specifying numbers $x_1, \ldots, x_n$ corresponding to the different μ_k.

It follows that the most general form of a prediction concerning $\mathfrak{S}$ is that the point $(x_1, \ldots, x_n)$ determined by actually measuring $\mu_1, \ldots, \mu_n$, will lie in a subset S of $(x_1, \ldots, x_n)$-space. Hence if we call the $(x_1, \ldots, x_n)$-spaces associated with $\mathfrak{S}$, its "observation-spaces," we may call the subsets of the observation-spaces associated with any physical system $\mathfrak{S}$, the "experimental propositions" concerning $\mathfrak{S}$.

3. *Phase-Spaces*

There is one concept which quantum theory shares alike with classical mechanics and classical electrodynamics. This is the concept of a mathematical "phase-space."

According to this concept, any physical system $\mathfrak{S}$ is at each instant hypothetically associated with a "point" p in a fixed phase-space Σ; this point is supposed to represent mathematically the "state" of $\mathfrak{S}$, and the "state" of $\mathfrak{S}$ is supposed to be ascertainable by "maximal"[2] observations.

Furthermore, the point p_0 associated with $\mathfrak{S}$ at a time t_0, together with a prescribed mathematical "law of propagation," fix the point p_t associated with $\mathfrak{S}$ at any later time t; this assumption evidently embodies the principle of *mathematical causation.*[3]

Thus in classical mechanics, each point of Σ corresponds to a choice of n position and n conjugate momentum coordinates – and the law of propagation may be Newton's inverse-square law of attraction. Hence in this case Σ is a region of ordinary $2n$-dimensional space. In electrodynamics, the points of Σ can only be specified after certain *functions* – such as the electromagnetic and electrostatic potential – are known; hence Σ is a function-space of infinitely many dimensions. Similarly, in quantum theory the points of Σ correspond to so-called "wave-functions," and hence Σ is again a function-space – usually[4] assumed to be Hilbert space.

In electrodynamics, the law of propagation is contained in Maxwell's equations, and in quantum theory, in equations due to Schrödinger. In any case, the law of propagation may be imagined as inducing a steady fluid motion in the phase-space.

It has proved to be a fruitful observation that in many important cases of classical dynamics, this flow conserves volumes. It may be noted that in quantum mechanics, the flow conserves distances (i.e., the equations are "unitary").

4. *Propositions as Subsets of Phase-Space*

Now before a phase-space can become imbued with reality, its elements and subsets must be correlated in some way with "experimental propositions" (which are subsets of different observation-spaces). Moreover, this must be so done that set-theoretical inclusion (which is the analogue of logical implication) is preserved.

There is an obvious way to do this in dynamical systems of the classical type.[5] One can measure position and its first time-derivative velocity – and hence momentum – explicitly, and so establish a one-one correspondence which preserves inclusion between subsets of phase-space and subsets of a suitable observation-space.

In the cases of the kinetic theory of gases and of electromagnetic waves no such simple procedure is possible, but it was imagined for a long time that "demons" of small enough size could by tracing the motion of each particle, or by a dynamometer and infinitesimal point-charges and magnets, measure quantities corresponding to every coordinate of the phase-space involved.

In quantum theory not even this is imagined, and the possibility of predicting in general the readings from measurements on a physical system $\mathfrak{S}$ from a knowledge of its "state" is denied; only statistical predictions are always possible.

This has been interpreted as a renunciation of the doctrine of predetermination; a thoughtful analysis shows that another and more subtle idea is involved. The central idea is that physical quantities are *related*, but are not all computable from a number of *independent basic* quantities (such as position and velocity).[6]

We shall show in Section 12 that this situation has an exact algebraic analogue in the calculus of propositions.

5. *Propositional Calculi in Classical Dynamics*

Thus we see that an uncritical acceptance of the ideas of classical dynamics (particularly as they involve *n*-body problems) leads one to identify each subset of phase-space with an experimental proposition (the proposition that the system considered has position and momentum coordinates satisfying certain conditions) and conversely.

This is easily seen to be unrealistic; for example, how absurd it would be to call an "experimental proposition," the assertion that the angular momentum (in radians per second) of the earth around the sun was at a particular instant a rational number!

Actually, at least in statistics, it seems best to assume that it is the *Lebesgue-measurable* subsets of a phase-space which correspond to experimental propositions, two subsets being identified, if their difference has *Lebesgue-measure* 0.[7]

But in either case, the set-theoretical sum and product of any two subsets, and the complement of any one subset of phase-space corresponding to experimental propositions, has the same property. That is, by definition[8]

The experimental propositions concerning any system in classical mechanics, correspond to a "field" of subsets of its phase-space. More precisely: To the "quotient" of such a field by an ideal in it. At any rate they form a "Boolean Algebra."[9]

In the axiomatic discussion of propositional calculi which follows, it will be shown that this is inevitable when one is dealing with exclusively compatible measurements, and also that it is logically immaterial which particular field of sets is used.

6. *A Propositional Calculus for Quantum Mechanics*

The question of the connection in quantum mechanics between subsets of observation-spaces (or "experimental propositions") and subsets of the phase-space of a system $\mathfrak{S}$, has not been touched. The present section will be devoted to defining such a connection, proving some facts about it, and obtaining from it heuristically by introducing a plausible postulate, a propositional calculus for quantum mechanics.

Accordingly, let us observe that if $\alpha_1, \dots, \alpha_n$ are any compatible observations on a quantum-mechanical system $\mathfrak{S}$ with phase-space Σ, then[10] there exists a set of mutually orthogonal closed linear subspaces Ω_i of Σ (which correspond to the families of proper functions satisfying $\alpha_1 f = \lambda_{i,1} f, \dots, \alpha_n f = \chi_{i,n} f$) such that *every* point (or function) $f \in \Sigma$ can be uniquely written in the form

$$f = c_1 f_1 + c_2 f_2 + c_3 f_3 + \cdots [f_i \in \Omega_i]$$

Hence if we state the

DEFINITION. By the "mathematical representative" of a subset S of any observation-space (determined by compatible observations $\alpha_1, \ldots, \alpha_n$) for a quantum-mechanical system $\mathfrak{S}$, will be meant the set of all points f of the phase-space of $\mathfrak{S}$, which are linearly determined by proper functions f_k satisfying $\alpha_1 f_k = \lambda_1 f_k, \ldots, \alpha_n f_k = \lambda_n f_k$, where $(\lambda_1, \ldots, \lambda_n) \in S$.

Then it follows immediately: (1) that the mathematical representative of any experimental proposition is a closed linear subspace of Hilbert space (2) since all operators of quantum mechanics are Hermitian, that the mathematical representative of the *negative*[11] of any experimental proposition is the *orthogonal complement* of the mathematical representative of the proposition itself (3) the following three conditions on two experimental propositions P and Q concerning a given type of physical system are equivalent:

(3a) The mathematical representative of P is a subset of the mathematical representative of Q.

(3b) P implies Q – that is, whenever one can predict P with certainty, one can predict Q with certainty.

(3c) For any statistical ensemble of systems, the probability of P is at most the probability of Q.

The equivalence of (3a)–(3c) leads one to regard the aggregate of the mathematical representatives of the experimental propositions concerning any physical system $\mathfrak{S}$, as representing mathematically the propositional calculus for $\mathfrak{S}$.

We now introduce the

POSTULATE. *The set-theoretical product of any two mathematical representatives of experimental propositions concerning a quantum-mechanical system, is itself the mathematical representative of an experimental proposition.*

Remarks. This postulate would clearly be implied by the not unnatural conjecture that all Hermitian-symmetric operators in Hilbert space (phase-space) correspond to observables;[12] it would even be implied by the conjecture that those operators which correspond to observables coincide with the Hermitian-symmetric elements of a suitable operator-ring M.[13]

Now the closed linear sum $\Omega_1 + \Omega_2$ of any two closed linear subspaces Ω_i of Hilbert space, is the orthogonal complement of the set-product $\Omega_1' \cdot \Omega_2'$ of the orthogonal complements Ω_i' of the Ω_i; hence if one adds the above postulate to the usual postulates of quantum theory, then one can deduce that

The set-product and closed linear sum of any two, and the orthogonal complement of any one closed linear subspace of Hilbert space representing mathematically an experimental proposition concerning a quantum-mechanical system $\mathfrak{S}$, itself represents an experimental proposition concerning $\mathfrak{S}$.

This defines the calculus of experimental propositions concerning $\mathfrak{S}$, as a calculus with three operations and a relation of implication, which closely resembles the systems defined in Section 5. We shall now turn to the analysis and comparison of all three calculi from an axiomatic-algebraic standpoint.

II. ALGEBRAIC ANALYSIS

7. *Implication as Partial Ordering*

It was suggested above that in any physical theory involving a phase-space, the experimental propositions concerning a system $\mathfrak{S}$ correspond to a family of subsets of its phase-space Σ, in such a way that "x implies y" (x and y being any two experimental propositions) means that the subset of Σ corresponding to x is contained set-theoretically in the subset corresponding to y. This hypothesis clearly is important in proportion as relationships of implication exist between experimental propositions corresponding to subsets of different observation-spaces.

The present section will be devoted to corroborating this hypothesis by identifying the algebraic-axiomatic properties of logical implication with those of set-inclusion.

It is customary to admit as relations of "implication," only relations satisfying

S1: x implies x.

S2: If x implies y and y implies z, then x implies z.

S3: If x implies y and y implies x, then x and y are logically equivalent.

In fact, S3 need not be stated as a postulate at all, but can be regarded as a definition of logical equivalence. Pursuing this line of thought, one can interpret as a "physical quality," the set of all experimental propositions logically equivalent to a given experimental proposition.[14]

Now if one regards the set S_x of propositions implying a given proposition x as a "mathematical representative" of x, then by S3 the correspondence between the x and the S_x is one-one, and x implies y if and only if $S_x \subset S_y$. While conversely, if L is any system of subsets X of a fixed class Γ, then there is an isomorphism which carries inclusion into logical implication between L and the system L^* of propositions "x is a point of X," $X \in L$.

Thus we see that the properties of logical implication are indistinguishable from those of set-inclusion, and that therefore it is *algebraically* reasonable to try to correlate physical qualities with subsets of phase-space.

A system satisfying S1–S3, and in which the relation "x implies y" is written $x \subset y$, is usually[15] called a "partially ordered system," and thus our first postulate concerning propositional calculi is that *the physical qualities attributable to any physical system form a partially ordered system.*

It does not seem excessive to require that in addition any such calculus contain two special propositions: the proposition $\square$ that the system considered *exists*, and the proposition $\circledcirc$ that it does *not exist*. Clearly

S4: $\circledcirc \subset x \subset \square$ for any x.

$\circledcirc$ is, from a logical standpoint, the "identically false" or "absurd" proposition; $\square$ is the "identically true" or "self-evident" proposition.

8. *Lattices*

In any calculus of propositions, it is natural to imagine that there is a weakest proposition implying, and a strongest proposition implied by, a given pair of propositions. In fact, investigations of partially ordered systems from different angles all indicate that the first property which they are likely to possess, is the existence of greatest lower bounds and least upper bounds to subsets of their elements. Accordingly, we state

DEFINITION. A partially ordered system L will be called a "lattice" if and only if to any pair x and y of its elements there correspond

S5: A "meet" or "greatest lower bound" $x \cap y$ such that (5a) $x \cap y \subset x$, (5b) $x \cap y \subset y$, (5c) $z \subset x$ and $z \subset y$ imply $z \subset x \cap y$.

S6: A "join" or "least upper bound" $x \cup y$ satisfying (6a) $x \cup y \supset x$, (6b) $x \cup y \supset y$, (6c) $w \supset x$ and $w \supset y$ imply $w \supset x \cup y$.

The relation between meets and joins and abstract inclusion can be summarized as follows,[16]

(8.1) In any lattice L, the following formal identities are true,

L1: $a \cap a = a$ and $a \cup a = a$.
L2: $a \cap b = b \cap a$ and $a \cup b = b \cup a$.
L3: $a \cap (b \cap c) = (a \cap b) \cap c$ and $a \cup (b \cup c) = (a \cup b) \cup c$.
L4: $a \cup (a \cap b) = a \cap (a \cup b) = a$.

Moreover, the relations $a \supset b$, $a \cap b = b$, and $a \cup b = a$ are equivalent – each implies both of the others.

(8.2) Conversely, in any set of elements satisfying L2–L4 (L1 is redundant), $a \cap b = b$ and $a \cup b = a$ are equivalent. And if one defines them to mean $a \supset b$, then one reveals L as a lattice.

Clearly L1–L4 are well-known formal properties of *and* and *or* in ordinary logic. This gives an algebraic reason for admitting as a *postulate* (if necessary) the statement that a given calculus of propositions is a lattice. There are other reasons[17] which impel one to admit as a postulate the stronger statement that the set-product of any two subsets of a phase-space which correspond to physical qualities, itself represents a physical quality – this is, of course, the Postulate of Section 6.

It is worth remarking that in classical mechanics, one can easily define the meet or join of any two experimental propositions as an *experimental proposition* – simply by having independent observers read off the measurements which either proposition involves, and combining the results logically. This is true in quantum mechanics only exceptionally – only when all the measurements involved commute (are compatible); in general, one can only express the join or meet of two given experimental propositions as a class of logically equivalent experimental propositions – i.e., as a *physical quality.*[18]

9. *Complemented Lattices*

Besides the (binary) operations of meet- and join-formation, there is a third (unary) operation which may be defined in partially ordered systems. This is the operation of *complementation.*

In the case of lattices isomorphic with "fields" of sets, complementation corresponds to passage to the set-complement. In the case of closed linear subspaces of Hilbert space (or of Cartesian n-space), it corresponds to passage to the orthogonal complement. In either case, denoting the "complement" of an element a by a', one has the formal identities,

L71: $(a')' = a$.
L72: $a \cap a' = \circledcirc$ and $a \cup a' = \square$.
L73: $a \subset b$ implies $a' \supset b'$.

By definition, L71 and L73 amount to asserting that complementation is a "dual automorphism" of period two. It is an immediate corollary of this and the duality between the definitions (in terms of inclusion) of meet and join, that

L74: $(a \cap b)' = a' \cup b'$ and $(a \cup b)' = a' \cap b'$

and another corollary that the second half of L72 is redundant. [*Proof:* by L71 and the first half of L74, $(a \cup a') = (a'' \cup a') = (a' \cap a)' = \circledcirc'$, while under inversion of inclusion $\circledcirc$ evidently becomes $\square$.] This permits one to deduce L72 from the even weaker assumption that $a \subset a'$ implies $a = \circledcirc$. *Proof:* for any x, $(x \cap x')' = (x' \cup x'') = x' \cup x \supset x \cap x'$.

Hence if one admits as a postulate the assertion that *passage from an experimental proposition a to its complement a' is a dual automorphism of period two, and a implies a' is absurd,* one has in effect admitted L71–L74.

This postulate is independently suggested (and L71 proved) by the fact the "complement" of the proposition that the readings $x_1, \ldots, x_n$ from a series of compatible observations $\mu_1, \ldots, \mu_n$ lie in a subset S of $(x_1, \ldots, x_n)$-space, is by definition the proposition that the readings lie in the set-complement of S.

10. *The Distributive Identity*

Up to now, we have only discussed formal features of logical structure which seem to be common to classical dynamics and the quantum theory. We now turn to the central difference between them – the *distributive identity* of the propositional calculus:

L6: $a \cup (b \cap c) = (a \cup b) \cap (a \cup c)$ and $a \cap (b \cup c) = (a \cap b) \cup (a \cap c)$

which is a law in classical, but not in quantum mechanics.

From an axiomatic viewpoint, each half of L6 implies the other.[19] Further, either half of L6, taken with L72, implies L71 and L73, and to assume L6 and L72 amounts to assuming the usual definition of a Boolean algebra.[20]

From a deeper mathematical viewpoint, L6 is the characteristic property of set-combination. More precisely, every "field" of sets is isomorphic with a Boolean algebra, and conversely.[21] This throws new light on the well-known fact that the propositional calculi of classical mechanics are Boolean algebras.

It is interesting that L6 is also a logical consequence of the compatibility of the observables occurring in *a*, *b*, and *c*. That is, if observations are made by independent observers, and combined according to the usual rules of logic, one can *prove* L1–L4, L6, and L71–74.

These facts suggest that the distributive law *may* break down in quantum mechanics. That it *does* break down is shown by the fact that if *a* denotes the experimental observation of a wave-packet ψ on one side of a plane in ordinary space, a' correspondingly the observation of ψ on the other side, and *b* the observation of ψ in a state symmetric about the plane, then (as one can readily check):

$$b \cap (a \cup a') = b \cap \square = b > \circledcirc = (b \cap a) = (b \cap a')$$
$$= (b \cap a) \cup (b \cap a').$$

Remark. In connection with this, it is a salient fact that the *generalized* distributive law of logic:

$$\text{L6*:} \quad \prod_{i=1}^{m} \left(\sum_{j=1}^{n} a_{i,j} \right) = \sum_{j(i)} \left(\prod_{i=1}^{m} a_{i,j(i)} \right)$$

breaks down in the quotient algebra of the field of Lebesgue measurable sets by the ideal of sets of Lebesgue measure 0, which is so fundamental in statistics and the formulation of the ergodic principle.[22]

11. *The Modular Identity*

Although closed linear subspaces of Hilbert space and Cartesian *n*-space need not satisfy L6 relative to set-products and closed linear sums, the formal properties of these operations are not confined to L1–L4 and L71–L73.

In particular, set-products and straight linear sums are known[23] to satisfy the so-called "modular identity."

L5: If $a \subset c$, then $a \cup (b \cap c) = (a \cup b) \cap c$.

Therefore (since the linear sum of any two finite-dimensional linear subspaces of Hilbert space is itself finite-dimensional and consequently closed) set-products and *closed* linear sums of the *finite dimensional* subspaces of any topological linear space such as Cartesian n-space or Hilbert space satisfy L5, too.

One can interpret L5 directly in various ways. First, it is evidently a restricted associative law on mixed joins and meets. It can equally well be regarded as a weakened distributive law, since if $a \subset c$, then $a \cup (b \cap c) = (a \cap c) \cup (b \cap c)$ and $(a \cup b) \cap c = (a \cup b) \cap (a \cup c)$. And it is self-dual: replacing $\subset$, $\cap$, $\cup$ by $\supset$, $\cup$, $\cap$ merely replaces a, b, c, by c, b, a.

Also, speaking graphically, the assumption that a lattice L is "modular" (i.e., satisfies L5) is equivalent to[24] saying that L contains no sublattice isomorphic with the lattice graphed in Figure 1:

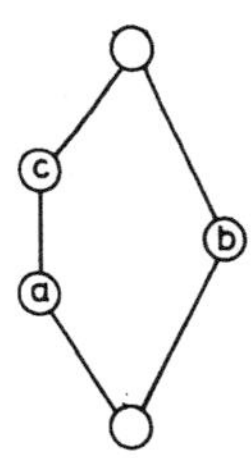

Fig. 1.

Thus in Hilbert space, one can find a counterexample to L5 of this type. Denote by $\xi_1, \xi_2, \xi_3, \ldots$ a basis of orthonormal vectors of the space, and by a, b, and c respectively the closed linear subspaces generated by the vectors $(\xi_{2n} + 10^{-n}\xi_1 + 10^{-2n}\xi_{2n+1})$, by the vectors ξ_{2n}, and by a and the vector ξ_1. Then a, b, and c generate the lattice of Figure 1.

Finally, the modular identity can be proved to be a consequence of the assumption that there exists a numerical dimension-function $d(a)$, with the properties

D1: If $a>b$, then $d(a)>d(b)$.

D2: $d(a)+d(b)=d(a\cap b)+d(a\cup b)$.

This theorem has a converse under the restriction to lattices in which there is a finite upper bound to the length n of chains[25] $\circledcirc<a_1<a_2<\cdots<a_n<\square$ of elements.

Since conditions D1–D2 partially describe the formal properties of probability, the presence of condition L5 is closely related to the existence of an "a priori thermo-dynamic weight of states." But it would be desirable to interpret L5 by simpler phenomenological properties of quantum physics.

12. *Relation to Abstract Projective Geometries*

We shall next investigate how the assumption of postulates asserting that the physical qualities attributable to any quantum-mechanical system $\mathfrak{S}$ are a lattice satisfying L5 and L71–L73 characterizes the resulting propositional calculus. This question is evidently purely algebraic.

We believe that the best way to find this out is to introduce an assumption limiting the length of chains of elements (assumption of finite dimensions) of the lattice, admitting frankly that the assumption is purely heuristic.

It is known[26] that any lattice of finite dimensions satisfying L5 and L72 is the direct product of a finite number of abstract projective geometries (in the sense of Veblen and Young), and a finite Boolean algebra, and conversely.

Remark. It is a corollary that a lattice satisfying L5 and L71–L73 possesses independent basic elements of which any element is a union, if and only if it is a Boolean algebra.

Again, such a lattice is a single projective geometry if and only if it is *irreducible* – that is, if and only if it contains no "neutral" elements.[27] $x\neq\circledcirc,\square$ such that $a=(a\cap x)\cup(a\cap x')$ for all a. In actual quantum mechanics such an element would have a projection-operator, which commutes with all projection-operators of observables, and so with all operators of observables in general. This would violate the requirement of "irreducibility" in quantum mechanics.[28] Hence we conclude that the *propositional calculus of quantum mechanics has the same structure as an abstract projective geometry.*

Moreover, this conclusion has been obtained purely by analyzing internal properties of the calculus, in a way which involves Hilbert space only indirectly.

13. *Abstract Projective Geometries and Skew-Fields*

We shall now try to get a fresh picture of the propositional calculus of quantum mechanics, by recalling the well-known two-way correspondence between abstract projective geometries and (not necessarily commutative) fields.

Namely, let F be any such field, and consider the following definitions and constructions: n elements $x_1, \ldots, x_n$ of F, not all $=0$, form a right-ratio $[x_1 : \ldots : x_n]_r$, two right-ratios $[x_1 : \ldots : x_n]_r$, and $[\xi_1 : \ldots : \xi_n]_r$ being called "equal," if and only if a $z \in F$ with $\xi_i = x_i z$, $i = 1, \ldots, n$, exists. Similarly, n elements $y_1, \ldots, y_n$ of F, not all $=0$, form a left-ratio $[y_1 : \ldots : y_n]_l$, two left-ratios $[y_1 : \ldots : y_n]_l$ and $[\eta_1 : \ldots : \eta_n]_l$ being called "equal," if and only if a z in F with $\eta_i = z y_i$, $i = 1, \ldots, n$, exists.

Now define an $n-1$-dimensional projective geometry $P_{n-1}(F)$ as follows: The "points" of $P_{n-1}(F)$ are all right-ratios $[x_1 : \ldots : x_n]_r$. The "linear subspaces" of $P_{m-1}(F)$ are those sets of points, which are defined by systems of equations

$$\alpha_{k1} x_1 + \cdots + \alpha_{kn} x_n = 0, \qquad k = 1, \ldots, m.$$

($m = 1, 2, \ldots$, the α_{ki} are fixed, but arbitrary elements of F). The proof, that this *is* an abstract projective geometry, amounts simply to restating the basic properties of linear dependence.[29]

The same considerations show, that the ($n-2$-dimensional) hyperplanes in $P_{m-1}(F)$ correspond to $m=1$, not all $\alpha_i = 0$. Put $\alpha_{1i} = y_i$, then we have

$$(*) \qquad y_1 x_1 + \cdots + y_n x_n = 0, \qquad \text{not all } y_i = 0.$$

This proves, that the ($n-2$-dimensional) hyperplanes in $P_{m-1}(F)$ are in a one-to-one correspondence with the left-ratios $[y_r : \ldots : y_n]_l$.

So we can identify them with the left-ratios, as points are already identical with the right-ratios, and (*) becomes the definition of "incidence" (point $\subset$ hyperplane).

Reciprocally, any abstract $n-1$-dimensional projective geometry Q_{n-1}

with $n=4, 5,\ldots$ belongs in this way to some (not necessarily commutative field $F(Q_{n-1})$, and Q_{n-1} is isomorphic with $P_{n-1}(F(Q_{n-1}))$.[30]

14. *Relation of Abstract Complementarity to Involutory Anti-Isomorphisms in Skew-Fields*

We have seen that the family of irreducible lattices satisfying L5 and L72 is precisely the family of projective geometries, provided we exclude the two-dimensional case. But what about L71 and L73? In other words, for which $P_{n-1}(F)$ can one define complements possessing all the known formal properties of orthogonal complements? The present section will be spent in answering this question.[30a]

First, we shall show that it is *sufficient* that F admit an *involutory antisomorphism* $W: \bar{x}=W(x)$, that is:

Q1. $w(w(u))=u$,
Q2. $w(u+v)=w(u)+w(v)$,
Q3. $w(uv)=w(v)\,w(u)$,

with a *definite diagonal Hermitian form* $w(x_1)\,\gamma_1\xi_1+\cdots+w(x_n)\,\gamma_n\xi_n$, where

Q4. $w(x_1)\,\gamma_1 x_1+\cdots+w(x_n)\,\gamma_n x_n=0$ implies $x_1=\cdots=x_n=0$,

the γ_i being fixed elements of F, satisfying $w(\gamma_i)=\gamma_i$.

Proof: Consider ennuples (not right- or left-ratios!) $x:(x_1,\ldots, x_n)$, $\xi:(\xi_1,\ldots, \xi_n)$ of elements of F. Define for them the vector-operations

$$xz:(x_1 z,\ldots, x_n z) \qquad (z \text{ in } F),$$
$$x+\xi:(x_1+\xi_1,\ldots, x_n+\xi_n),$$

and an "inner product"

$$(\xi_1 x)=w(\xi_1)\,\gamma_1 x_1+\cdots+w(\xi_n)\,\gamma_n x_n.$$

Then the following formulas are corollaries of Q1–Q4.

IP1 $(x, \xi)=w((\xi, x))$,
IP2 $(\xi, xu)=(\xi, x)\,u$, $(\xi u, x)=w(u)\,(\xi, x)$,
IP3 $(\xi, x'+x'')=(\xi, x')+(\xi, x'')$, $(\xi'+\xi'', x)=(\xi', x)+(\xi'', x)$,
IP4 $(x, x)=w((x, x))=[x]$ is $\neq 0$ if $x\neq 0$ (that is, if any $x_i\neq 0$).

We can define $x\perp\xi$ (in words: "x is orthogonal to ξ") to mean that $(\xi, x)=0$. This is evidently symmetric in x, ξ, and depends on the right-ratios $[x_1:\ldots:x_n]_r$, $[\xi_1:\ldots:\xi_n]_r$ only so it establishes the relation of "polarity," $a \perp b$, between the points

$$a:[x_1:\ldots:x_n]_r, \qquad b:[\xi_1:\ldots:\xi_n]_r \text{ of } P_{n-1}(F).$$

The polars to any point $b:[\xi_1:\ldots:\xi_n]_r$ of $P_{n-1}(F)$ constitute a linear subspace of points of $P_{n-1}(F)$, which by Q4 does not contain b itself, and yet with b generates whole projective space $P_{n-1}(F)$, since for any ennuple $x:(x_1,\ldots,x_n)$.

$$x = x' + \xi \cdot [\xi]^{-1}(\xi, x)$$

where by Q4, $[\xi] \neq 0$, and by IP $(\xi, x') = 0$. This linear subspace is, therefore, an $n-2$-dimensional hyperplane.

Hence if c is any k-dimensional element of $P_{n-1}(F)_1$ one can set up inductively k mutually polar points $b^{(1)},\ldots,b^{(k)}$ in c. Then it is easy to show that the set c' of points polar to every $b^{(1)},\ldots,b^{(k)}$ – or equivalently to every point in c – constitute an $n-k-1$-dimensional element, satisfying $c \cap c' = ◎$ and $c \cup c' = \square$. Moreover, by symmetry $(c')' \supset c$, whence by dimensional considerations $c'' = c$. Finally, $c \supset d$ implies $c' \subset d'$, and so the correspondence $c \to c'$ defines an involutory dual automorphism of $P_{n-1}(F)$ completing the proof.

In the Appendix it will be shown that this condition is also necessary. Thus the above class of systems is exactly the class of irreducible lattices of finite dimensions >3 satisfying L5 and L71–L73.

III. CONCLUSIONS

15. *Mathematical Models for Propositional Calculi*

One conclusion which can be drawn from the preceding algebraic considerations, is that one can construct many different models for a propositional calculus in quantum mechanics, which cannot be differentiated by known criteria. More precisely, one can take any field F having an involutory anti-isomorphism satisfying Q4 (such fields include the real, complex, and quaternion number systems [31]), introduce suitable notions of linear dependence and complementarity, and then construct for every dimension-number n a model $P_n(F)$, having all of the properties of the propositional calculus suggested by quantum-mechanics.

One can also construct infinite-dimensional models $P_\infty(F)$ whose elements consist of all closed linear subspaces of normed infinite-dimensional spaces. But philosophically, Hankel's principle of the "perseverance of formal laws" (which leads one to try to preserve L5)[32] and mathematically, technical analysis of spectral theory in Hilbert space, lead one to prefer a continuous-dimensional model $P_c(F)$, which will be described by one of us in another paper.[33]

$P_c(F)$ is very analogous with the model furnished by the measurable subsets of phase-space in classical dynamics.[34]

16. *The Logical Coherence of Quantum Mechanics*

The above heuristic considerations suggest in particular that the physically significant statements in quantum mechanics actually constitute a sort of projective geometry, while the physically significant statements concerning a given system in classical dynamics constitute a Boolean algebra.

They suggest even more strongly that whereas in classical mechanics any propositional calculus involving more than two propositions can be decomposed into independent constituents (direct sums in the sense of modern algebra), quantum theory involves irreducible propositional calculi of unbounded complexity. This indicates that quantum mechanics has a *greater logical coherence* than classical mechanics – a conclusion corroborated by the impossibility in general of measuring different quantities independently.

17. *Relation to Pure Logic*

The models for propositional calculi which have been considered in the preceding sections are also interesting from the standpoint of pure logic. Their nature is determined by quasi-physical and technical reasoning, different from the introspective and philosophical considerations which have had to guide logicians hitherto. Hence it is interesting to compare the modifications which they introduce into Boolean algebra, with those which logicians on "intuitionist" and related grounds have tried introducing.

The main difference seems to be that whereas logicians have usually assumed that properties L71–L73 of negation were the ones least able to withstand a critical analysis, the study of mechanics points to the *distributive identities* L6 as the weakest link in the algebra of logic. Cf. the last two paragraphs of Section 10.

Our conclusion agrees perhaps more with those critiques of logic, which find most objectionable the assumption that $a' \cup b = \square$ implies $a \subset b$ (or dually, the assumption that $a \cap b' = \circledcirc$ implies $b \supset a$ – the assumption that to deduce an absurdity from the conjunction of a and not b, justifies one in inferring that a implies b).[35]

18. *Suggested Questions*

The same heuristic reasoning suggests the following as fruitful questions.

What experimental meaning can one attach to the meet and join of two given experimental propositions?

What simple and plausible physical motivation is there for condition L5?

APPENDIX

1. Consider a projective geometry Q_{n-1} as described in Section 13. F is a (not necessarily commutative, but associative) field, $n = 4, 5, \ldots, Q_{n-1} = P_{n-1}(F)$ the projective geometry of all right-ratios $[x_1 : \ldots : x_n]_r$, which are the *points* of Q_{n-1}. The ($n-2$-dimensional) *hyperplanes* are represented by the left-ratios $[y_1 : \ldots : y_n]_l$, incidence of a point $[x_1 : \ldots : x_n]_r$ and of a hyperplane $[y_1 : \ldots : y_n]_l$ being defined by

$$(1) \qquad \sum_{i=1}^{n} y_i x_i = 0$$

All linear subspaces of Q_{n-1} form the lattice L, with the elements $a, b, c, \ldots$. Assume now that an operation a' with the properties L71–L73 in Section 9 exists:

L71 $\quad (a')' = a$

L72 $\quad a \cap a' = \circledcirc$ and $a \cup a' = \square$,

L73 $\quad a \subset b$ implies $a' \supset b'$.

They imply (cf. Section 9)

L74 $\quad (a \cap b)' = a' \cup b'$ and $(a \cup b)' = a' \cap b'$.

Observe, that the relation $a \subset b'$ is symmetric in a, b, owing to L73 and L71.

2. If $a:[x_1:\ldots:x_n]_r$ is a point, then a' is an $[y_1:\ldots:y_n]_l$. So we may write:

$$(2) \qquad [x_1:\ldots:x_n]'_r = [y_1:\ldots:\eta_n]_l,$$

and define an operation which connects right- and left-ratios. We know from Section 14, that a general characterization of a' (a any element of L) is obtained, as soon as we derive an algebraic characterization of the above $[x_1:\ldots:x_n]'_r$. We will now find such a characterization of $[x_1:\ldots:x_n]'_r$, and show, that it justifies the description given in Section 14.

In order to do this, we will have to make a rather free use of *collineations* in Q_{n-1}. A collineation is, by definition, a coordinate-transformation, which replaces $[x_1:\ldots:x_n]_r$ by $[\bar{x}_1:\ldots:\bar{x}_n]_r$,

$$(3) \qquad \bar{x}_j = \sum_{i=2}^{n} w_{ij} x_i \qquad \text{for } j=1,\ldots,n.$$

Here the ω_{ij} are fixed elements of F, and such, that (3) has an inverse.

$$(4) \qquad x_i = \sum_{j=1}^{n} \theta_{ij} \bar{x}_j, \qquad \text{for } i=1,\ldots,n,$$

the θ_{ij} being fixed elements of F, too. (3), (4) clearly mean

$$\delta_{kl} = \left\{ \begin{matrix} 1 & \text{if} & k=1 \\ 0 & \text{if} & k \neq 1 \end{matrix} \right\}:$$

$$(5) \qquad \sum_{j=1}^{n} \theta_{ij}\omega_{kj} = \delta_{ik}, \qquad \sum_{i=1}^{n} \omega_{ij}\theta_{ik} = \delta_{jk}.$$

Considering (1) and (5) they imply the contravariant coordinate-transformation for hyperplanes: $[y_1:\ldots:y_n]_l$ becomes $[\bar{y}_1:\ldots:\bar{y}_n]_l$, where

$$(6) \qquad \bar{y}_i = \sum_{i=1}^{n} y_i \theta_{ij}, \qquad \text{for } j=1,\ldots,n,$$

$$(7) \qquad y_i = \sum_{j=1}^{n} \bar{y}_j \omega_{ij}, \qquad \text{for } i=1,\ldots,n.$$

(Observe, that the position of the coefficients on the left side of the variables in (4), (5), and on their right side in (6), (7), is essential!)

3. We will bring about

$$(8)\qquad [\delta_{i1}:\ldots:\delta_{in}]_r' = [\delta_{i1}:\ldots:\delta_{in}]_l \qquad \text{for } i=1,\ldots,n,$$

by choosing a suitable system of coordinates, that is, by applying suitable collineations. We proceed by induction: Assume that (8) holds for $i=1,\ldots,m-1\,(m=1,\ldots,n)$, then we shall find a collineation which makes (8) true for $i=1,\ldots,m$.

Denote the point $[\delta_{i1}:\ldots:\delta_{in}]_r$ by p_i^*, and the hyperplane $[\delta_{i1}:\ldots:\delta_{in}]_l'$ by h_i^* our assumption on (8) is: $p_i^{*\prime}=h_i^*$ for $i=1,\ldots,m-1$. Consider now a point $a:[x_1:\ldots:x_n]_r$, and the hyperplane $a':[y_1:\ldots:y_n]_l$. Now $a \leqq p_i^{*\prime} = h_i^*$ means (use (1)) $x_i=0$, and $p_i^* \leqq a'$ means (use (8)) $y_i=0$. But these two statements are equivalent. So we see: If $i=1,\ldots,m-1$, then $x_i=0$ and $y_i=0$ are equivalent.

Consider now $p_m^*:[\delta_{m1}:\ldots:\delta_{mn}]_r$. Put $p_m':[y_1^*:\ldots:y_n^*]_l$. As $\delta_{mi}=0$ for $i=1,\ldots,m-1$, so we have $y_i^*=0$ for $i=1,\ldots,m-1$. Furthermore, $p_m^* \cap p_m^{*\prime}=0$, $p_m^* \neq 0$, so p_m^* not $\leqq p_m^{*\prime}$. By (1) this means $y_m^* \neq 0$.

Form the collineation (3), (4), (6), (7), with

$$\theta_{ii}=\omega_{ii}=1, \qquad \theta_{mi}=\omega_{im}=y_m^{*-1}y_i^* \qquad \text{for } i=m+1,\ldots,n,$$

all other θ_{ij}, $\omega_{ij}=0$.

One verifies immediately, that this collineation leaves the coordinates of the $p_1^*:[\delta_{i1}:\ldots:\delta_{in}]_r$, $i=1,\ldots,n$, invariant, and similarly those of the $p_i^{*\prime}:[\delta_{i1}:\ldots:\delta_{im}]_l$, $i=1,\ldots,m-1$, while it transforms those of

$$p_m^{*\prime}:[y_1^*:\ldots:y_n^*]_l$$

into $[\delta_{m1}:\ldots:\delta_{mn}]_l$.

So after this collineation (8) holds for $i=1,\ldots,m$.

Thus we may assume, by induction over $m=1,\ldots,n$, that (8) holds for all $i=1,\ldots,n$. This we will do.

The above argument now shows, that for $a:[x_1:\ldots:x_n]_r$, $a':[y_1:\ldots:y_n]_l$,

$$(9)\qquad x_i=0 \text{ is equivalent to } y_i=0, \qquad \text{for } i=1,\ldots,n.$$

4. Put $a:[x_1:\ldots:x_n]_r$, $a':[y_1:\ldots:y_n]_l$, and $b:[\xi_1:\ldots:\xi_n]_r$, $b':[\eta_1:\ldots:\eta_n]_l$.

Assume first $\eta_1=1$, $\eta_2=\eta$, $\eta_3=\cdots=\eta_n=0$. Then (9) gives $\xi_1\neq 0$, so we can normalize $\xi_1=1$, and $\xi_3=\cdots=\xi_n=0$. ξ_2 can depend on $\eta_2=\eta$ only, so $\xi_2=f_2(\eta)$.

Assume further $x_1=1$. Then (9) gives $y_1\neq 0$, so we can normalize $y_1=1$. Now $a\leqq b'$ means by (*i*) $1+\eta x_2=0$, and $b\leqq a'$ means $1+y_2 f_2(\eta)=0$. These two statements must, therefore, be equivalent. So if $x_2\neq 0$, we may put $\eta=-x_2^{-1}$, and obtain $y_2=-(f_2(\eta))^{-1}=-(f_2(-x_2^{-1}))^{-1}$. If $x_2=0$, then $y_2=0$ by (9). Thus, x_2 determines at any rate y_2 (independently of $x_3,\ldots,x_n$): $y_2=\varphi_2(x_2)$. Permuting the $i=2,\ldots,n$ gives, therefore:

There exists for each $i=2,\ldots,n$ a function $\varphi_i(x)$, such that $y_i=\varphi_i(x_i)$. Or:

(10) If $a:[1:x_2:\ldots:x_n]_r$, then $a':[1:\varphi_2(x_2):\ldots:\varphi_n(x_n)]_l$.

Applying this to $a:[1:x_2:\ldots:x_n]_r$ and $c:[1:u_1:\ldots:u_n]_r$ shows: As $a\leqq c'$ and $c\leqq a'$ are equivalent, so

(11) $\sum_{i=2}^{n}\varphi_i(u_i)\,x_i=-1$ is equivalent to $\sum_{i=2}^{n}\varphi(x_i)\,u_i=-1$.

Observe, that (9) becomes:

(12) $\varphi_i(x)=0$ if and only if $x=0$.

5. (11) with $x_3=\cdots=x_n=u_3=\cdots=u_n=0$ shows: $\varphi_2(u_2)\,x_2=-1$ is equivalent to $\varphi_2(x_2)\,u_2=-1$. If $x_2\neq 0$, $u_2=(-\varphi_2(x_2))^{-1}$, then the second equation holds, and so both do.

Choose x_2, u_2 in this way, but leave $x_3,\ldots,x_n$, $u_3,\ldots,u_n$ arbitrary. Then (11) becomes:

(13) $\sum_{i=3}^{n}\varphi_i(u_i)\,x_i=0$ is equivalent to $\sum_{i=3}^{n}\varphi_i(x_i)\,u_i=0$.

Now put $x_5=\cdots=x_n=u_5=\cdots=u_n=0$. Then (13) becomes:

$\varphi_3(u_3)\,x_3+\varphi_4(u_4)\,x_4=0$ is equivalent to

$$\varphi_3(x_3)\,u_3+\varphi_4(x_4)\,u_4=0,$$

that is (for x_4, $u_4\neq 0$):

(a) $x_3x_4^{-1}=\varphi_4(u_4)^{-1}\varphi_3(u_3)$

(14) is equivalent to

(b) $u_3u_4^{-1}=\varphi_4(x_4)^{-1}\varphi_3(x_3)$.

Let x_4, x_3 be given. Choose u_3, u_4 so as to satisfy (b). Then (a) is true, too. Now (a) remains true, if we leave u_3, u_4 unchanged, but change x_3, x_4 without changing $x_3x_4^{-1}$. So (b) remains too true under these conditions, that is, the value of $\varphi_4(x_4)^{-1}\varphi_3(x_3)$ does not change. In other words: $\varphi_4(x_4)^{-1}\varphi_3(x_3)$ depends on $x_3x_4^{-1}$ only. That is: $\varphi_4(x_4)^{-1}\varphi_3(x_3)= \varphi_{34}(x_3x_4^{-1})$. Put $x_3=xz$, $x_4=x$, then we obtain:

$$(15) \qquad \varphi_3(xz)=\varphi_4(x)\,\psi_{34}(z).$$

This was derived for x, $z\neq 0$, but it will hold for x or $z=0$, too, if we define $\psi_{34}(0)=0$. (Use (12).)

(15), with $z=1$ gives $\varphi_3(x)=\varphi_4(x)\,\alpha_{34}$, where $\alpha_{34}=\psi_{34}(1)\neq 0$, owing to (12) for $x\neq 0$. Permuting the $i=2,\ldots,n$ gives, therefore:

$$(16) \qquad \varphi_i(x)=\varphi_j(x)\,\alpha_{ij}, \quad \text{where } \alpha_{ij}\neq 0.$$

(For $i=j$ put $\alpha_{ii}=1$.)

Now (15) becomes

$$(17) \qquad \begin{aligned} &\varphi_2(zx)=\varphi_2(x)\,w(z)\\ &\quad w(z)=\alpha_{42}\psi_{34}(z)\,\alpha_{23}. \end{aligned}$$

Put $x=1$ in (17), write x for z, and use (16) with $j=2$:

$$(18) \qquad \begin{aligned} &\varphi_i(x)=\beta w(z)\,\gamma_i, \quad \text{where } \beta,\ \gamma_i\neq 0.\\ &\quad (\beta=\varphi_2(1),\ \gamma_i=\alpha_{i2}). \end{aligned}$$

6. Compare (17) for $x=1$, $z=u$; $x=u$, $z=v$; and $x=1$, $z=vu$. Then

$$(19) \qquad w(vu)=w(u)\,w(v)$$

results (12) and (18) give

$$(20) \qquad w(u)=0 \text{ if and only if } u=0.$$

Now write $w(z)$, γ_i for $\beta w(z)\,\beta^{-1}$, $\beta\gamma_i$. Then (18), (19), (20) remain true, (18) is simplified in so far, as we have $\beta=1$ there. So (11) becomes

$$(21) \qquad \sum_{i=2}^{n} w(u_i)\,\gamma_i x=-1$$

(21) is equivalent to

$$\sum_{i=2}^{n} w(x_i)\,\gamma_i u_i=-1$$

$x_2=x$, $u_2=u$ and all other $x_i=u_i=0$ give: $w(u)\gamma_2 x=-1$ is equivalent to $w(x)\gamma_2 u=-1$. If $x\neq 0$, $u=-\gamma_2^{-1}w(x)^{-1}$, then the second equation holds, and so the first one gives: $x=-\gamma_2^{-1}w(u)^{-1}=-\gamma_2^{-1}(w(-\gamma_2^{-1}w(x)^{-1}))^{-1}$. But (19), (20) imply $w(1)=1$, $w(w^{-1})=w(w)^{-1}$, so the above relation becomes:

$$x=-\gamma_2^{-1}(w(-\gamma_2^{-1}w(x_1^{-1}))^{-1}=-\gamma_2^{-1}w((-\gamma_2^{-1}w(x)^{-1})^{-1})$$
$$=-\gamma_2^{-1}w(w(x)(-\gamma_2))=-\gamma_2^{-1}w(-\gamma_2)\,w(w(x)).$$

Put herein $x=1$, as $w(w(1))=w(1)=1$, so $-\gamma_2^{-1}w(-\gamma_2)=1$, $w(-\gamma_2)=-\gamma_2$ results. Thus the above equation becomes

$$(22)\qquad w(w(x))=x,$$

and $w(-\gamma_2)=-\gamma_2$ gives, if we permute the $i=2,\ldots,n$,

$$(23)\qquad w(-\gamma_i)=-\gamma_i.$$

Put $u_i=-\gamma_i^{-1}$ in (21). Then considering (22) and (19)

$$(24)\qquad \sum_{i=2}^{n} x_i=1 \text{ is equivalent to } \sum_{i=2}^{n} w(x_i)=1$$

obtains. Put $x_2=x$, $x_3=y$, $x_4=1-x-y$, $x_5=\cdots=x_n=0$. Then (24) gives $w(x)+w(y)=1-w(1-x-y)$. So $w(x)+w(y)$ depends on $x+y$ only. Replacing x, y by $x+y$, 0 shows, that it is equal to $w(x+y)+w(0)=w(x+y)$ (use 20). So we have:

$$(25)\qquad w(x)+w(y)=w(x+y)$$

(25), (19) and (22) give together:

$w(x)$ is an involutory antisomorphism of F.

Observe, that (25) implies $w(-1)=-w(1)=-1$, and so (23) becomes

$$(26)\qquad w(\gamma_i)=\gamma_i.$$

7. Consider $a:[x_1:\ldots:x_n]_r$, $a':[y_1:\ldots:y_n]_l$. If $x_1\neq 0$, we may write $a:[1:x_2x_1^{-1}:\ldots:x_nx_1^{-1}]_r$, and so $a':[1:w(x_2x_1^{-1})\gamma_2:\ldots:w(x_nx_1^{-1})\gamma_n]_l$. But

$$w(x_ix_1^{-1})\gamma_i=w(x_1^{-1})\,w(x_i)\gamma_i=w(x_1)^{-1}w(x_i)\gamma_i,$$

and so we can write

$$a':[w(x_1):w(x_2)\gamma_2:\ldots:w(x_n)\gamma_n]_l$$

too. So we have

$$(27)\qquad y_i=w(x_i)\gamma_i \qquad \text{for } i=1,\ldots,n,$$

where the γ_i for $i=2,\ldots,n$ are those from 6., and $\gamma_1=1$. And $w(1)=1$, so (26) holds for all $i=1,\ldots,n$. So we have the representation (27) with γ_i obeying (26), if $x_i\neq 0$.

Permutation of the $i=1,\ldots,n$ shows, that a similar relation holds if $x_2\neq 0$:

$$(27^+)\qquad y_i=w^+(x_i)\gamma_i^+,$$

$$(26^+)\qquad w^+(\gamma_i^+)=\gamma_i^+,$$

$w^+(x)$ being an involutory antisomorphism of F. ($w^+(x)$, γ_i^+ may differ from $w(x)$, γ_i!) Instead of $\gamma_1=1$ we have now $\gamma_2^+=1$, but we will not use this.

Put all $x_i=1$. Then $a':[y_1:\ldots:y_n]_l$ can be expressed by both formulae (27) and (27^+). As $w(x)$, $w^+(x)$ are both antisomorphism, so $w(1)=w^+(1)=1$, and therefore $[y_1:\ldots:y_n]_l=[\gamma_1:\ldots:\gamma_n]_l=[\gamma_1^+:\ldots:\gamma_n^+]_l$ obtains. Thus $(\gamma_1^+)^{-1}\gamma_i^+=(\gamma_1)^{-1}\gamma_i=\gamma_i$, $\gamma_i^+=\gamma_1^+\gamma_i$ for $i=1,\ldots,n$.

Assume now $x_2\neq 0$ only. Then (27^+) gives $y_i=w^+(x_i)\gamma_i^+$, but as we are dealing with left ratios, we may as well put

$$y_i=(\gamma_1^+)^{-1}w^+(x_i)\gamma_i^+=(\gamma_1^+)^{-1}w^+(x)\gamma_1^+\gamma_i.$$

Put $\beta^+=\gamma_1^+\neq 0$, then we have:

$$(27^{++})\qquad y_i=\beta^{+-1}w^+(x_i)\beta^+\gamma_i.$$

Put now $x_1=x_2=1$, $x_3=x$, all other $x_i=0$. Again $a':[y_1:\ldots:y_n]_l$ can be expressed by both formulae (27) and (27^{++}), again $w(1)=w^+(1)$. Therefore

$$\begin{aligned}[y_1:y_2:y_3:y_4:\ldots:y_n]_l&=[\gamma_1:\gamma_2:w(x)\gamma_3:0:\ldots:0]_l\\&=[\gamma_1:\gamma_2:\beta^{+-1}w^+(x)\beta^+\gamma_3:0:\ldots:0]_l\end{aligned}$$

obtains. This implies $w(x)=\beta^{+-1}w(x)\beta^+$ for all x, and so (27^{++}) coincides with (27).

In other words: (27) holds for $x_2\neq 0$ too.

Permuting $i=2,\ldots,n$ (only $i=1$ has an exceptional rôle in (27)), we see: (27) holds if $x_i \neq 0$ for $i=2,\ldots,n$. For $x_1 \neq 0$ (27) held anyhow, and for some $i=1,\ldots,n$ we must have $x_i \neq 0$. Therefore:

(27) *holds for all points* $a:[x_1:\ldots:x_n]_r$.

8. Consider now two points $a:[x_1:\ldots:x_n]_r$ and $b:[\xi_1:\ldots:\xi_n]_r$. Put $a':[y_1:\ldots:y_n]_l$, then $b \leqq a'$ means, considering (1) and (27) (cf. the end of 7.):

$$(28) \qquad \sum_{i=1}^{n} w(x_i)\,\gamma_i \xi_i = 0.$$

$a \leqq a'$ can never hold ($a \cap a' = 0$, $a \neq 0$), so (28) can only hold for $x_i = \xi_i$, if all $x_i = 0$. Thus,

$$(29) \qquad \sum_{i=1}^{n} w(x_i)\,\gamma_i x_i = 0 \text{ implies } x_1 = \cdots = x_n = 0.$$

Summing up the last result of 6., and formulae (26), (29) and (28), we obtain:

There exists an involutory antisomorphism $w(x)$ *of* F *(cf.* (22), (25), (19)) *and a definite diagonal Hermitian form* $\sum_{i=1}^{n} w(x_i)\,\gamma_i \xi_i$ *in* F *(cf.* (26), (29)), *such that for* $a:[x_1:\ldots:x_n]_r$, $b:[\xi_1:\ldots:\xi_n]_r$ $b \leqq a'$ *is defined by polarity with respect to it:*

$$(28) \qquad \sum_{i=1}^{n} w(x_i)\,\gamma_i \xi_i = 0.$$

This is exactly the result of Section 14, which is thus justified.

The Society of Fellows, Harvard University,
The Institute for Advanced Study

NOTES

[1] If one prefers, one may regard a set of compatible measurements as a single composite "measurement" – and also admit non-numerical readings – without interfering with subsequent arguments.

Among conspicuous observables in quantum theory are position, momentum, energy, and (non-numerical) symmetry.

[2] L. Pauling and E. B. Wilson, *An Introduction to Quantum Mechanics*, McGraw-Hill, 1935, p. 422. Dirac, *Quantum Mechanics*, Oxford, 1930, §4.

[3] For the existence of mathematical causation, cf. also p. 65 of Heisenberg's *The Physical Principles of the Quantum Theory*, Chicago, 1929.
[4] Cf. J. von Neumann, *Mathematische Grundlagen der Quanten-mechanik*, Berlin, 1931, p. 18.
[5] Like systems idealizing the solar system or projectile motion.
[6] A similar situation arises when one tries to correlate polarizations in different planes of electromagnetic waves.
[7] Cf. J. von Neumann, 'Operatorenmethoden in der klassischen Mechanik,' *Annals of Math.* **33** (1932), 595–8. The difference of two sets S_1, S_2 is the set $(S_1+S_2) - S_1 \cdot S_2$ of those points, which belong to one of them, but not to both.
[8] F. Hausdorff, *Mengenlehre*, Berlin, 1927, p. 78.
[9] M. H. Stone, 'Boolean Algebras and Their Application to Topology', *Proc. Nat. Acad.* **20** (1934), 197.
[10] Cf. von Neumann, *op. cit.*, pp. 121, 90, or Dirac, *op. cit.*, **17**. We disregard complications due to the possibility of a continuous spectrum. They are inessential in the present case.
[11] By the "negative" of an experimental proposition (or subset S of an observation-space) is meant the experimental proposition corresponding to the set-complement of S in the same observation-space.
[12] I.e., that given such an operator α, one "could" find an observable for which the proper states were the proper functions of α.
[13] F. J. Murray and J. v. Neumann, 'On Rings of Operators', *Annals of Math.*, **37** (1936), 120. It is shown on p. 141, *loc. cit.* (Definition 4.2.1 and Lemma 4.2.1), that the closed linear sets of a ring M – that is those, the "projection operators" of which belong to M – coincide with the closed linear sets which are invariant under a certain group of rotations of Hilbert space. And the latter property is obviously conserved when a set-theoretical intersection is formed.
[14] Thus in Section 6, closed linear subspaces of Hilbert space correspond one-many to experimental propositions, but one-one to physical qualities in this sense.
[15] F. Hausdorff, *Grundzüge der Mengenlehre*, Leipzig, 1914, Chap. VI, §1.
[16] The final result was found independently by O. Öre, 'The Foundations of Abstract Algebra. I, *Annals of Math.* **36** (1935), 406–37, and by H. MacNeille in his Harvard Doctoral Thesis, 1935.
[17] The first reason is that this implies no restriction on the abstract nature of a lattice – any lattice can be realized as a system of its own subsets, in such a way that $a \cap b$ is the set-product of a and b. The second reason is that if one regards a subset S of the phase-space of a system $\mathfrak{S}$ as corresponding to the *certainty* of observing $\mathfrak{S}$ in S, then it is natural to assume that the combined certainty of observing $\mathfrak{S}$ in S and T is the certainty of observing $\mathfrak{S}$ in $S \cdot T = S \cap T$, – and assumes quantum theory.
[18] The following point should be mentioned in order to avoid misunderstanding: If a, b are two physical qualities, then $a \cup b$, $a \cap b$ and a' (cf. below) are physical qualities too (and so are $\circledcirc$ and $\square +$). But $a \subset b$ is not a physical quality; it is a relation between physical qualities.
[19] R. Dedekind, *Werke*, Braunschweig, 1931, vol. 2, p. 110.
[20] G. Birkhoff, 'On the Combination of Subalgebras', *Proc. Camb. Phil. Soc.* **29** (1933), 441–64, §§23–4. Also, in any lattice satisfying L6, isomorphism with respect to inclusion implies isomorphism with respect to complementation; this need not be true if L6 is not assumed, as the lattice of linear subspaces through the origin of Cartesian n-space shows.
[21] M. H. Stone, 'Boolean Algebras and Their Application to Topology', *Proc. Nat. Acad.* **20** (1934), 197–202.

[22] A detailed explanation will be omitted, for brevity; one could refer to work of G. D. Birkhoff, J. von Neumann, and A. Tarski.

[23] G. Birkhoff, *op. cit.*, §28. The proof is easy. One first notes that since $a \subset (a \cup b) \cap c$ if $a \subset c$, and $b \cap c \subset (a \cup b) \cap c$ in any case, $a \cup (b \cap c) \subset (a \cup b) \cap c$. Then one notes that any vector in $(a \cup b) \cap c$ can be written $\xi = \alpha + \beta [\alpha \in a, \beta \in b, \xi \in c]$. But $\beta = \xi - \alpha$ is in c (since $\xi \in c$ and $\alpha \in a \subset c$); hence $\xi = \alpha + \beta \in a \cup (b \cap c)$, and $a \cup (b \cap c) \supset (a \cup b) \cap c$, completing the proof.

[24] R. Dedekind, *Werke*, vol. 2, p. 255.

[25] The statements of this paragraph are corollaries of Theorem 10.2 of G. Birkhoff, op. cit.

[26] G. Birkhoff 'Combinatorial Relations in Projective Geometries', *Annals of Math.* **30** (1935), 743–8.

[27] O. Öre, *op. cit.*, p. 419.

[28] Using the terminology of footnote,[13] and of *loc. cit.* there: The ring *MM*' should contain no other projection-operators than 0, 1, or: the ring *M* must be a "factor." Cf. *loc. cit.*[13], p. 120.

[29] Cf. §§103–105 of B. L. Van der Waerden's *Moderne Algebra*, Berlin, 1931, Vol. 2.

[30] $n=4, 5, \ldots$ means of course $n-1 \geqq 3$, that is, that Q_{n-1} is necessarily a "Desarguesian" geometry. (Cf. O. Veblen and J. W. Young, *Projective Geometry*, New York, 1910, Vol. 1, page 41). Then $F = F(Q_{n-1})$ can be constructed in the classical way. (Cf. Veblen and Young, Vol. 1, pages 141–150). The proof of the isomorphism between Q_{n-1} and the $P_{n-1}(F)$ as constructed above, amounts to this: Introducing (not necessarily commutative) homogeneous coördinates $x_1, \ldots, x_n$ from F in Q_{n-1}, and expressing the equations of hyperplanes with their help. This can be done in the manner which is familiar in projective geometry, although most books consider the commutative ("Pascalian") case only. D. Hilbert, *Grundlagen der Geometrie*, 7th edition, 1930, pages 96–103, considers the non-commutative case, but for affine geometry, and $n-1=2, 3$ only.

Considering the lengthy although elementary character of the complete proof, we propose to publish it elsewhere.

[30a] R. Brauer, 'A Characterization of Null Systems in Projective Space', *Bull. Am. Math. Soc.* **42** (1936), 247–54, treats the analogous question in the opposite case that $X \cap X' \neq \circledcirc$ is postulated.

[31] In the real case, $w(x)=x$; in the complex case, $w(x+iy)=x-iy$; in the quaternionic case, $w(u+ix+jy+kz)=u-ix-jy-kz$; in all cases, the λ_i are 1. Conversely, A. Kolmogoroff, 'Zur Begründung der projektiven Geometrie', *Annals of Math.* **33** (1932), 175–6 has shown that any projective geometry whose k-dimensional elements have a *locally compact topology* relative to which the lattice operations are continuous, must be over the real, the complex, or the quaternion field.

[32] L5 can also be preserved by the artifice of considering in $P_\infty(F)$ only elements which either are or have complements which are of finite dimensions.

[33] J. von Neumann, 'Continuous Geometries', *Proc. Nat. Acad.* **22** (1936), 92–100 and 101–109. These may be a more suitable frame for quantum theory, than Hilbert space.

[34] In quantum mechanics, dimensions but not complements are uniquely determined by the inclusion relation; in classical mechanics, the reverse is true!

[35] It is not difficult to show, that assuming our axioms L1–5 and 7, the distributive law L6 is equivalent to this postulate: $a' \cup b = \square$ implies $a \subset b$.

MARTIN STRAUSS

THE LOGIC OF COMPLEMENTARITY AND THE FOUNDATION OF QUANTUM THEORY*

INTRODUCTION

Several attempts have been made to provide an axiomatic basis for the statistical transformation theory in quantum physics in the form of simple general principles. Thus, in his well-known book on *Quantum Mechanics* Dirac uses the *superposition principle* as a fundamental principle. This principle permits indeed to determine many characteristic features of the mathematical formalism. It does not, however, suffice to determine even the algebra of the state calculus, as Dirac has noticed himself. From the present point of view the superposition principle may be looked upon as an ingenious but rather artificial formulation of complementarity.

A complete axiomatic foundation of the statistical transformation theory is due to VON NEUMANN [1]. The present work is closely related to it; its critical discussion is a natural starting-point.

Von Neumann's postulates demand essentially two things: (A) a one-one correlation between physical quantities and hypermaximal Hermitean operators in Hilbert space, and (B) linearity of the mean value operator for these quantities.

This deduction of the statistical formulae of quantum mechanics appears to be remarkable and satisfactory in so far as it makes no use of hypotheses concerning equal or numerical probabilities – in contrast to other statistical theories, in particular classical statistical mechanics. From the point of view of the *Correspondence Principle* this had to be expected since numerical probabilities would have no analogues, in the sense of that Principle, in classical mechanics.

However, from a *physical* point of view it can hardly be called satisfactory to base the theory on a postulate whose connection with experimental facts is as little intelligible as is the case with postulate (A). Instead, one would like to see a principle directly suggested by experience,

C. A. Hooker (ed.), The Logico-Algebraic Approach to Quantum Mechanics, 27–44

as in Thermodynamics or Relativity Theory, such as the principle of indeterminacy or complementarity.

Even when adopting the *formal* point of view one is struck by the circumstance that the Principle of Complementarity, so closely connected[2] with that of Correspondence, is merely implicitly contained in postulate (A) but is not at all involved in the postulates concerning quantum mechanical probabilities. This would not be surprising if complementarity had no bearing on the mathematical theory of probability – a condition that is not satisfied. In fact, *complementarity restricts the validity or applicability of the ordinary theory of probability in a quite definite manner*. This may be seen even without the use of the formalism from the following examples.

Consider a statistical ensemble of hydrogen atoms all in the same energy state E_n. Then there exist the probabilities $\text{prob}(E_n; I_{\Delta q})$ and $\text{prob}(E_n; I_{\Delta p})$ to find the value of Q within the interval $I_{\Delta q}=(q, q+\Delta q)$ or the value of P within $I_{\Delta p}=(p, p+\Delta p)$, respectively. According to the ordinary theory of probability there would then also exist the probability $\text{prob}(E_n; I_{\Delta q} \text{ and } I_{\Delta p})$ for finding both the value of Q within $I_{\Delta q}$ and the value of P within $I_{\Delta p}$. In view of the complete[3] complementarity between P and Q this probability could not be tested. The formalism yields for it a two-valued complex expression – it gives a nonsensical answer to a senseless question. This violation of the ordinary calculus of probability (or rather: its rules of existence) does not destroy the internal consistency of the calculus. This consistency would be destroyed only if either the probability $\text{prob}(E_n \text{ and } I_{\Delta q}; I_{\Delta p})$ or $\text{prob}(E_n \text{ and } I_{\Delta p}; I_{\Delta q})$ would exist (which is not the case, due to the complementarity between H and Q or H and P, respectively) because then $\text{prob}\,(E_n; I_{\Delta q} \text{ and } I_{\Delta p})$ would be numerically determined by the general multiplication theorem:

$$\begin{aligned}\text{prob}(I_{\Delta a}; I_{\Delta b} \text{ and } I_{\Delta c}) &= \text{prob}(I_{\Delta a}; I_{\Delta b})\,\text{prob}(I_{\Delta a} \text{ and } I_{\Delta b}; I_{\Delta c})\\ &= \text{prob}(I_{\Delta a}; I_{\Delta c})\,\text{prob}(I_{\Delta a} \text{ and } I_{\Delta c}; I_{\Delta b})\end{aligned}$$

Cases where two (or all three) prob expressions with a logical conjunction exist do occur in quantum mechanics, e.g. in the case $A=M_x$, $B=Q_x$, $C=P_x$ (x-components of angular momentum, position and linear momentum vector, respectively). In this case, where A commutes with both B and C, the probabilities on the two right-hand sides exist while that on the left-hand side does not. [Hence the formalism should give a real one-valued expression for these probabilities, as indeed it does, but it can be interpreted only in the sense of the two right-hand sides.] (Of course, one may measure B in one half of an ensemble and C in the other half and multiply the relative frequencies corresponding to $\text{prob}(I_{\Delta a}; I_{\Delta b})$ and $\text{prob}(I_{\Delta a}; I_{\Delta c})$; but this is something quite different from a proper application of the general multiplication theorem.)

These simple examples demonstrating the limited applicability of the ordinary calculus of probability should make it clear that the *mean value postulate* [used by von Neumann] *is not an equivalent substitute for the rules of the calculus of probability*, and, hence, that it does not suffice to clarify the relation between that calculus and quantum mechanics which has been the original aim of von Neumann's work. (To be sure, this inequivalence has nothing to do with the question whether mean values are an equivalent substitute for a probability distribution: the latter is indeed determined by the mean values of all 'momenta' of the quantity in question.) The point is that the *logical operations* of the calculus of probability cannot be immitated by the averaging operation. The relation between the calculus of probability and the calculus of mean values is not one-one: only the former determines the latter.

Now it could happen that a physical theory permits only mean values to be measured. In that case a trunkated theory of probability, characterized by the postulate of the linearity of the mean value operation, would suffice. Contrary to a previous stage in the physical interpretation, quantum mechanics is not such a theory. True, the ordinary calculus of probability demands too much, but the mean value postulate [of von Neumann] demands too little; the ordinary calculus of probability is still needed if all quantities concerned commute.

It may seem paradoxical in view of all this that the mean value postulate suffices for deducing the correct formulae. In this connection it must be pointed out that in von Neumann's deduction essential use is made of an *extension* of the mean value postulate to 'quantities' that are represented by projection operators; as shown by von Neumann, these projection operators represent *statements* on the measured values of these quantities rather than the quantities themselves – (to the eigenvalues 1 and 0 of the projection operators correspond the truth values 'true' and 'false', respectively); thus, the calculus of the projection operators represents a kind of sentential calculus. Now it emerges from recent investigations concerning the axiomatics of probability theory that the algebra of the ordinary sentential calculus can be used as a substitute for certain axioms in the theory of probability; thus, given that all probabilities concerned exist, addition and multiplication theorem may be deduced from one another by employing the distributive laws of the sentential calculus. [[...]]. Although the state of affairs is somewhat different when we turn to quantum mechanical measurement statements and the projection operators representing them – the *logical meaning of complementarity* resides in just this difference – the difference, when properly formulated, does *not* concern the *algebraic formulae* as such but merely their *range of applicability*, viz., questions of *existence*. In this way it becomes intelligible that the mean value postulate, extended to projection operators, does suffice for deducing the statistical formulae and that, on the other hand, the anomalies mentioned above do remain. These anomalies present a violation of the axioms of probability theory only if the existential axioms corresponding to the ordinary sentential calculus are included (REICHENBACH[4]) or – what amounts to the same – if the region of definition of the probability function is supposed to be a set system [viz., the set of all subsets of a given set] (KOLMOGOROFF[5]). In other words: *the restricted applicability of the ordinary theory of probability is due solely to the invalidity of the ordinary sentential calculus for quantum mechanical measurement statements.* [[...]]

Since for clarifying the relation between probability theory and quantum mechanics it suffices to heed complementarity, the combination of complementarity and probability theory may be expected to be sufficient for building up the general formalism of the statistical transformation theory so that von Neumann's postulate (A) may be replaced by the Principle of Complementarity.

How far this expectation is justified will emerge from the following.

I. THE LOGICAL FORMULATION OF COMPLEMENTARITY

(1) If complementarity is to be used for an axiomatic reconstruction of quantum theory it has to be formulated in a way suitable for formal operation. As long as complementarity is conceived primarily as a relation between physical quantities it is difficult to see how this should be done; there is no obvious reason why quantities that cannot be measured

simultaneously should be represented by operators in Hilbert space.

The statement that two quantities cannot be measured simultaneously may be expressed thus: two statements concerning the results of measurement of the two quantities cannot be decided both, or: deciding the one makes it impossible to decide the other, or: decidability of one implies undecidability of the other. In this way complementarity becomes primarily a [semantic] *relation between statements*. This makes it possible to formulate complementarity in a formal way [viz., to formulate a non-classical sentential or predicate calculus to be called *complementarity logic*].

(2) What we need is not an axiomatic system for the sentential calculus but its algebra which can easily be recognized as rules of ordinary language; the semantic definition of complementarity as given under (i) then leads to a modified sentential calculus, this modification being the formal [syntactic] expression of complementarity.

As sentential variables we use the letters

$$R, S, T, \ldots$$

and for the negation and the sentential connectives we use Russell's symbols:

(a) $\sim$ for 'not' (negation)
(b) $\cdot$ for 'and' (conjunction, logical product)
(c) $\vee$ for 'or' (disjunction, logical sum)
(d) $\equiv$ for 'if and only if' (equivalence)

but, following Hilbert, we shall put the negation sign above the sentential symbol.

We then have the following equivalences [characteristic of Boolean algebra]:

n1. $\bar{\bar{R}} \equiv R$

$$\left.\begin{array}{ll} (1) & R \cdot R \equiv R \\ (2) & R \vee R \equiv R \\ (\text{k}1) & R \cdot S \equiv S \cdot R \\ (\text{k}2) & R \vee S \equiv S \vee R \end{array}\right\} \text{L}$$

$$\left.\begin{array}{ll} \text{(a1)} & R\cdot(S\cdot T) \equiv (R\cdot S)\cdot T \\ \text{(a2)} & R \vee (S \vee T) \equiv (R \vee S) \vee T \\ \text{(d1)} & R \vee (S\cdot T) \equiv (R \vee S)\cdot(R \vee T) \\ \text{(d2)} & R\cdot(S \vee T) \equiv (R\cdot S) \vee (R\cdot T). \end{array}\right\} \mathbf{L}$$

The algebraic significance of the negation consists in the fact that it permits to solve the equivalences

$$X \vee (R\cdot S) \equiv R \qquad\qquad Y\cdot(R \vee S) \equiv R$$

for X and Y:

$$X \equiv R\cdot\tilde{S} \qquad\qquad Y \equiv R \vee \tilde{S}.$$

Thereby the expressions

$$O =_{\text{df}} R\cdot\tilde{R} \qquad\qquad E =_{\text{df}} R \vee \tilde{R}$$

play the role of zero and unity:

$$\begin{array}{llll} \text{(n1.1)} & S \vee O \equiv S & \text{(n2.1)} & S\cdot E \equiv S \\ \text{(n1.2)} & S\cdot O \equiv O & \text{(n2.2)} & S \vee E \equiv E. \end{array}$$

O is called *contradition* and E *tautology*. [[. . . . .]]

The equivalences **L** can be handled in the same way as algebraic equations, which implies the following rule of substitution for the variables R, S, T, ...:

L Subst. 'R' may be replaced by (a) any other sentential symbol such 'S', (b) '$\tilde{R}$', (c) '$S\cdot T$', (d) '$S \vee T$', (e) '$S \equiv T$', [any such substitution for 'R' to take place everywhere where 'R' occurs within a given formula].

The calculus defined by the equivalences **L** and the rule **L Subst.** will be called **L**-*calculus*.

(3) Now the **L**-calculus is just that part of the ordinary sentential calculus that can be maintained if complementarity is taken into account, with the following proviso. According to the semantic definition of complementarity as given under (1) the sentential connection of two complementary sentences gives an undecidable statement and hence a meaningless sentence, contrary to what is implied in the ordinary sentential calculus. Hence the equivalences **L** must be interpreted thus: if one (and hence also the other) side of an equivalence is meaningful the equivalence is logically true; if this condition is not fulfilled the equivalence is not false but meaningless.

We now [take the *decisive step* and] *forbid* the formation of meaningless expressions. The resulting calculus will be called **L**′. Then the situation is as follows:

Though the equivalences **L** can be maintained [they are not violated semantically], the domain of definition of the sentential connectives is no longer a 'field'. Through this *change in algebraic structure* the **L**′-calculus is *not isomorphic with the set calculus*: while section and junction of two given sets always exist the corresponding conjunction and disjunction of two sentences may not exist in the **L**′-calculus. [[. . . . .]]

(4) The semantic justification for ruling out the formation of a compound sentence as given above breaks down if the two measurement statements refer to different instances of time: since measurements of complementary quantities can be performed at different instances of time, the corresponding statements can be decided both. Let us call such statements *simply complementary* to each other in contradistinction to two complementary statements referring to the same instant of time which will be called *strictly complementary*. Is there any reason for ruling out the sentential connection of simply complementary statements? The answer to this question is bound up with the following consideration.

The statement of a probability relation between strictly complementary statements has no obvious or direct meaning; it cannot be decided because complementary quantities cannot be measured simultaneously. An [experimental] meaning can be attached to it only by a limiting process $t_2 \to t_1$ when one of the two quantities is measured at t_1 and the other at t_2. Hence, if compounds of simply complementary statements were admitted [we would have a logical discontinuity for $t_2 \to t_1$, and] extending the non-admittance of compounds from strictly to simply complementary sentences could be justified only [by the wish to remove this discontinuity or] by reference to the calculus of probability, namely by the obvious demand that the prob expressions be continuous functions of time. No independent logistic meaning would then attach to simple complementarity.

Now, although the compound 'position of S at t_1 is within $I_{\Delta q}$ and momentum of S at t_2 is within $I_{\Delta p}$' is meaningful in sofar as it can be decided experimentally, it will not occur in a rationally constructed language [of quantum mechanics], because *the consequences* [predictions] *to be drawn from either part separately contradict each other even when the two parts are true.* Hence, only one *or* the other part can occur in any correct deduction. This is the formal expression of what is called *Nichtobjektivierbarkeit* of measurement results.

(The paradox resulting from handling complementary statements according to the rules of ordinary logic have lately been exposed by EINSTEIN[6] and SCHROEDINGER[7].) [[. . . .]]

(5) The sentence 'The momentum of the particle lies within $I_{\Delta p}$' does not characterize an individual state of affairs but a class [of particles]. Hence our $R, S, T, \ldots$ are to be regarded as *class* [or *predicate*] variables. This does not interfere with the equivalences and the substitution rule: the two calculi are isomorphic. The difference between the **L**- and the **L**′-calculus is of course transfered to the class [or predicate] calculus so that we have to distinguish between the ordinary class [or predicate] calculus corresponding to **L** and the one corresponding to **L**′; the latter may be called complementary class [or predicate] calculus. [[...]]

II. CALCULUS OF PROBABILITY

(6) We now turn to the calculus of probability. We show first why the (essentially equivalent) axiomatic systems of Reichenbach and Kolmogoroff cannot be used when complementarity is taken into account. According to Reichenbach (l.c.) a probability statement is a general implication between sentences stating class membership of elements and hence written in the form

$$\mathbf{R} \qquad (i)\,(x_i \in 0 \underset{p}{\Rrightarrow} y_i \in P)\,; \qquad \text{(R1)}$$

here, O and P are class variables, x and y are individual variables, $\Rrightarrow$ is the sign for the prob relation ('probability implication'), and p is the numerical value of the probability. A short-hand designation for (R1) is

$$\mathbf{R} \qquad (0 \underset{p}{\Rrightarrow} P) \quad \text{or} \quad w\,(0, P) = p\,. \qquad \text{(R2)}$$

The multiplication theorem (axiom IV) may then be written in the form

$$\mathbf{R} \qquad (0 \underset{p}{\Rrightarrow} P)\cdot(0\cdot P \underset{u}{\Rrightarrow} Q) \supset (0 \underset{\underline{w}}{\Rrightarrow} P\cdot Q)\cdot(w = pu)\,. \qquad \text{(R IV)}$$

Here, a compound of P and Q occurs only on the right-hand side of the implication, and this (together with the rule of inference referred to above) makes it possible to deduce from meaningful expressions an expression that may be meaningless; in other words: **R** IV has the property of transfering existence.

Similarly, the first axiom of Kolmogoroff's system demands that the domain of definition of the prob function be a field of sets, i.e., that with any two sets also their junction (set sum), their difference and their section belong to it, thus, the algebra of the ordinary sentential [or predicate] calculus is presupposed here, too. Hence this axiomatic system cannot be used either if complementarity is taken into account.

(7) For complementarity logic we have used a system of equivalences from ordinary logic; similarly we must use a system of equations from ordinary probability theory as basis for complementary probability theory. Such a system has already been given by Reichenbach (l.c.); it reads:

$$\mathbf{W}\begin{cases}\text{I.1} & w(R, R \vee S) = 1\\ \text{I.2} & w(R, S\cdot\tilde{S}) = 0\\ \text{I.3} & 0 \leq w(R, S)\\ \text{II.} & w(R, S \vee T) = w(R, S) + w(R, T) - w(R, S\cdot T)\\ \text{III.} & w(R, S\cdot T) = w(R, S)\, w(R\cdot S, T)\end{cases}$$

Though here (III), too, one side of the equation may be meaningless while the other one is not, this does no harm since the equation sign – contrary to the implication – does not transfer existence; it merely implies that certain probabilities are equal if they exist. Hence meaningless prob statements cannot be deduced from meaningful ones.

In order to compensate for the loss of deductive power in the transition from the **R**-system to the **W**-system one has to postulate in the ordinary prob calculus that a probability exists if its numerical value is determined by the equations of the calculus [and other probabilities known or assumed to exist] (Reichenbach's rule of existence, Kolmogoroff's 1. axiom). Similarly, we need the following *Existential Postulate*: if the numerical value of a prob function $w(R; S)$ is determined according to the **W**-system by other probabilities known to exist, then $w(R; S)$ exists *provided that both R and S exist.* [[...]]

Let us look for a moment at the [classical] system **LW** involving the ordinary sentential [or rather predicate] calculus. This system, as is well known, admits a set theoretical interpretation as follows:

To the tautology E corresponds a basic set $\underline{E}$, to the contradiction O the empty set $\underline{O}$, to the sentences [or rather predicates] $R, S, \ldots$ correspond subsets $\underline{R}, \underline{S}, \ldots$ of $\underline{E}$, to the negation $\tilde{R}$ corresponds the complementary set $\underline{E}-\underline{R}$, to the conjunction $R \cdot S$ corresponds the set section $\underline{R} \cdot \underline{S}$, and to the disjunction $R \vee S$ the union $\underline{R} \dot{+} \underline{S} = \underline{R} + \underline{S} - \underline{R} : \underline{S}$. Every additive set function

$$P(\underline{R} + \underline{S}) = P(\underline{R}) + P(\underline{S}) \tag{P1}$$

with

$$w(\underline{R}, \underline{S}) = P(\underline{R} \cdot \underline{S})/P(\underline{R}) \tag{P2}$$

then satisfies the axioms **W**, so that the system **W** may be replaced by (P1), (P2) together with suitable existential postulates (Kolmogoroff, loc. cit.).

III. QUANTUM THEORY

(8) We are now going to characterize the domain of definition of the [quantum mechanical] prob function. As can be seen directly from experimental experience, we are confronted with the following facts:

Q (a) To every measurement propostition R there exist an infinite number of other measurement propostitions, all noncomplementary to one another and to R (e.g., all those resulting from R by replacing the measurement interval refered to in R by a *larger* one).

Q (b) To every measurement proposition R there exist an infinite number of measurement propositions S_i all complementary to R (e.g., all those resulting from one such S by replacing the measurement interval referred to in S by a smaller one.

Q (c) The relation of [sentential or predicational] connectibility (σ) and the relation of inconnectibility (κ) are neither transitive nor intransitive; (i.e., all four possibilities of the scheme (see top of next page) are realized in nature, e.g., by the examples given in the last three columns; Q_x, P_x, M_x are components of position, momentum, and angular momentum, respectively, and for the pertaining intervals any finite intervals may be choosen.)

Taken together Q(a)–Q(c) imply that our $R, S, T, \ldots$, form an in-

RS	ST	RT	R	S	T
σ	σ	σ	Q_x	Q_y	Q_z
σ	σ	κ	Q_x	M_x	P_x
κ	κ	σ	M_x	M_y	Q_x
κ	κ	κ	M_x	M_y	M_z

finite (in fact: continuous) domain, with infinite 'islands' in which the ordinary sentential [or rather predicate] calculus and hence the unrestricted calculus of probability holds. [The algebraic structure of the domain is thus that of a *partial Boolean algebra*].

As the cardinality of the domain is that of a continuum, the following postulate appears adequate:

Q *con*. The prob function w is a continuous function of time [or rather the time interval(s) occurring] and of the measurement intervals.

(9) We are now going to show: *the calculus of projection operators over a linear vector space is isomorphic to the* L′*-calculus* under the following mapping:

		predicates	projection operators
Z	1.	R	$\boldsymbol{R}$
	2.	$\tilde{R}$	$\boldsymbol{I} - \boldsymbol{R}$
	3.	$R \cdot S$	$\boldsymbol{RS}$
	4.	$R \vee S$	$\boldsymbol{R} + \boldsymbol{S} - \boldsymbol{RS}$

($\boldsymbol{I}$ is the identity operator satisfying $\boldsymbol{IR} = \boldsymbol{R}$ for all $\boldsymbol{R}$).

(a) The logical equivalences **L** turn into mathematical identities if the predicates are replaced by projection operators according to **Z**; note that $\boldsymbol{RS}$ and $\boldsymbol{R} + \boldsymbol{S} - \boldsymbol{RS}$ are projectors only if $\boldsymbol{RS} = \boldsymbol{RS}$.

(b) If $\boldsymbol{RS} \neq \boldsymbol{SR}$, $\boldsymbol{RS}$ is not a projector, $(\boldsymbol{RS})(\boldsymbol{RS}) \neq \boldsymbol{RS}$, and hence the predicational compounds formed with $\boldsymbol{R}$ and $\boldsymbol{S}$ are not predicates either, i.e., $\boldsymbol{R}$ and $\boldsymbol{S}$ are complementary to one another. If this is taken into account the mapping **Z** is one-one.

[Confusion may arise from the fact that there is a one-one relation between projectors and closed linear subsets of the linear vector space concerned: this may suggest to take the calculus of closed linear subsets – instead of the calculus of projectors – as the mathematical model of quantum logic. The following paragraph shows why this is not feasible and thus refutes the much discussed 'Logic of Quantum Mechanics' of Birkhoff-von Neumann which likewise appeared in 1936.]

Since there is a one-one mapping between projectors and the [closed] linear subsets of the vector space $\mathfrak{R}$ concerned, **Z** implies a one-one relation between the predicates $R, S, \ldots$ of complementarity logic and the closed linear subsets of $\mathfrak{R}$. This corresponds to the isomorphism between the ordinary sentential calculus and the ordinary set calculus: instead of arbitrary sets we have now closed linear subsets of a vector space. However, this analogy is rather limited: the calculus of the closed linear subsets is *not isomorphic* to [the calculus of projectors and to] the calculus **L**′. True, the *junction* of two linear subsets is again a linear subset if and only if the pertaining projectors commute, but the *section* of two linear subsets is always a linear subset, even when the pertaining projectors do *not* commute, i.e., even when the predicates concerned are complementary so that the compound predicate does *not* exist.

Thus, it is *not* possible to satisfy the system **L**′ by linear subsets if isomorphism is to be maintained. This is decisive for the following treatment: the prob function w cannot be [represented by] a set function.

(10) By virtue of **Z** the domain of definition of the prob function w may be taken to be the set of projection operators [or rather the direct product of this set with itself]. This however does not imply that the numbers $w(R, S)$ can be determined otherwise than by explicite coordination [i.e., on a purely empirical basis]. To obtain a [physico-] mathematical theory we must demand that there exist a general function $W(\boldsymbol{R}, \boldsymbol{S})$ depending only on $\boldsymbol{R}$ and $\boldsymbol{S}$, which satisfies the system **W** with

$$w(R, S) = W(\boldsymbol{R}, \boldsymbol{S});$$

in other words: *the equations* **W** *are to be considered as functional equations for W over the set of projection operators.* As the values of W are to be real numbers, this implies that the vector space concerned is a metrical space. Though the metric is not uniquely determined by **W** alone a simple postulate to be given later will fix it.

We solve the functional equation **WIV** by

$$W(\boldsymbol{R}, \boldsymbol{S}) = \frac{f(\boldsymbol{RS})}{f(\boldsymbol{R})} \tag{W 1}$$

which corresponds to (P2).

Substituting this in **W** III gives, in consideration of **Z**,

$$f(\boldsymbol{RS} + \boldsymbol{RT} - \boldsymbol{RST}) = f(\boldsymbol{RS}) + f(\boldsymbol{RT}) - f(\boldsymbol{RST})$$

which yields the functional equation

$$f(\boldsymbol{R} + \boldsymbol{S}) = f(\boldsymbol{R}) + f(\boldsymbol{S}). \qquad \text{(W 2)}$$

Its [general] solution is

$$f(\boldsymbol{R}) = c\, Tr\, \boldsymbol{R} \quad (c = \text{constant independent of } \boldsymbol{R})$$

since the trace $Tr\, \boldsymbol{R}$ is the only linear invariant that depends only on $\boldsymbol{R}$. Hence

$$\boxed{W(\boldsymbol{R}, \boldsymbol{S}) = \frac{Tr\, \boldsymbol{RS}}{Tr\, \boldsymbol{R}}} \qquad \text{(W 4)}$$

(W4) is the general expression for the quantum mechanical probabilities. It merely remains to fix metric and number of dimensions of the underlying vector space $\mathfrak{R}$.

(11) It is obvious that the metric of $\mathfrak{R}$ must be either Euclidean [i.e., real][9] or unitary; otherwise the trace $Tr\, \boldsymbol{R}$, defined by $\boldsymbol{R}_i^i = \boldsymbol{R}_{ik} g^{ik}$, would depend on the metrical tensor g^{ik} for which there would be no physical interpretation.

For deciding between Euclidean [real] and unitary metric we consider the expression $Tr\, \boldsymbol{RST}$ which occurs in the general multiplication theorem. In the case of *Euclidean* [real] metric this expression is always real-valued, even if none of the projectors $\boldsymbol{R}$, $\boldsymbol{S}$, $\boldsymbol{T}$ commutes with any of the other two, i.e., if none of the three expressions of the multiplication theorem have any physical meaning. In the case of *unitary* metric the said expression is complex-valued (and the complex-conjugate of $Tr\, \boldsymbol{RTS}$) iff none of the three projectors commutes with any of the other two. Thus, *only the choice of unitary metric is in accoord with complementarity logic*.

[This result is of fundamental importance in two respects. For one, it answers the question, first put to the author by Reichenbach, whether the use of complex-valued state functions in quantum mechanics is a mathematical trick that could be avoided in principle (as often in classical physics), and if not, why not. Second, it shows why all attempts at interpreting the quantum mechanical formalism in terms of classical probability or statistics are doomed to failure.] [[...]]

(12) Finally, the number of dimensions [of the linear vector space $\mathfrak{R}$] can be determined by the well-known commutation rules for canonical quantities or else by the postulate that there exist continuous regions of measurable values; either postulate leads to an infinite number of dimensions. To the latter postulate correspond our axioms **Q**(a, b). It is easy to show that they demand an infinite number of dimensions. (Incidentally, it would be difficult to attach any physical meaning to a finite number of dimensions). Thus, besides (W4) we have also established the Hilbert space. [[...]]

(13) In conclusion it should be pointed out that nothing has been said about the connection between the projectors [which were introduced as a mathematical model of complementarity logic] on the one hand and the [hypermaximal Hermitean operators representing] physical quantities on the other hand: the question which projector corresponds to a given measurement statement [or predicate] has been left open. Quite naturally, this question can only be decided by considerations of correspondence. [It should be noted, however, that from our point of view the projectors are more fundamental than the hypermaximal Hermitean operators. This is in line with the fact that the later can be defined in terms of the former – a fact that would be merely a mathematical curiosity if the projectors had no fundamental significance.]

The considerations given above confirm and substantiate the often stressed analogy between the theory of special relativity and quantum mechanics: just as the world geometry of Einstein-Minkowski merely expresses the existence of a finite upper limit c of signal velocities, thus the general formalism of quantum mechanics merely reflects the unavoidability of complementarity resulting from the existence of the finite quantum of action h. This formalism thus appears as the appropriate mathematical language for expressing all special quantum mechanical experience. From this point of view it is rather obvious that the formalism has stood the test of the many-body problem and of relativistic generalization, and the same point of view may help to decide the question whether this general formalism is wide enough to encompass a future theory of elementary particles.

POSTSCRIPT 1971

The introduction of a non-classical logic in physics raises a number of philosophical questions, and the introduction of two competing logics for the same physical theory raises some additional questions of a more technical nature, but not without philosophical import. In the following notes I shall try to answer some of these questions.

(1) In the first place let me point out that there is no such thing as '*the* logic *of* quantum mechanics'. A physical theory is not given in the form of a formalized language but as the union of a mathematical formalism and its physical interpretation; the formalism does not contain any descriptive predicates (sentential functions) and hence no predicate connectives either. The connectives become part of a formal system only if the language of the theory is formalized. It follows that the logical syntax of a physical theory depends on the way the language of the theory is beeing formalized. Vice versa, advocating a particular 'logic' (viz. logical syntax) *for* a physical theory means advocating a particular way of formalizing its language. If different 'logics' are advocated for the same physical theory, it is only by comparing all consequences of the implied formalizations that a proper judgement on their relative merits can be given. True, even when we have a complete list of all relevant differences we may not agree on their relative merits but at least we are then compelled to state our reasons for any preferential decision we care to make.

(2) With this in view, I have carried out the two formalizations corresponding to complementarity logic (partial Boolean algebra) and 'quantum logic' (Birkhoff-von Neumann's nondistributive lattice algebra), respectively, in 1937 (doctor thesis, Prague 1939). Though all copies of this have been lost, one of its main results is easily established: *the Birkhoff-von Neumann logic leads to a language containing 'metaphysical' sentences*, namely the conjunction of sentences that are inconnectible in complementarity logic.[10]

(3) There are other – and perhaps more important – reasons for preferring complementarity logic to nondistributive lattice logic. Here are some of them: –

(a) *Giving up the distributive law for the sentential connectives implies giving up the (semantical) two-valuedness*: in any two-valued logic the two sides of the distributive law have the same truth-value. Now I have no philosophical objections against multi-valued logics, but none of the advocates of the Birkhoff-von Neumann logic seems to have noticed this implication.

(b) If the physical predicates are to be represented by subspaces (closed linear subsets) rather than by the projectors on these subspaces, one should expect that the quantum mechanical probabilities are functions of these subsets. However, they are functions of the projectors (and hence merely functionals of the subspaces).

(c) Most important of all, the Birkhoff-von Neumann logic *does not lead to the unitary metric*, even when combined with the prob calculus: it is equally well compatible with real (Euclidean) metric. On the other hand, complementarity logic demands unitary metric (complex valued state vectors), as shown above. As the unitary metric is one of the most important characteristics of the quantum mechanical state space, the Birkhoff-von Neumann logic, whatever it may be, is certainly not characteristic of quantum mechanics.

(4) In view of all this the question arises *why* the Birkhoff-von Neumann logic has attracted far more attention than complementarity logic. My answer: this is not just a case of authority against non-authority – after all, complementarity logic goes also back to von Neumann, if only by implication – but rather a case of fashion against unfashion. Indeed, lattice theory became quite fashionable in the nineteenthirties, thanks mainly to the work of Birkhoff, while partial Boolean algebra, of which the algebra of projectors and complementarity logic are examples, had to wait for another 30 years to become respectable among mathematicians. The quantum physicists, of course, have used complementarity logic all the time, even when not knowing it, and have paid no attention to 'the' 'quantum logic' of Birkhoff-von Neumann.

(5) Let me just add that complementarity logic has been rediscovered in recent years by several authors, among them P. SUPPES[11] and F. W. KAMBER[12].

(6) Does the use of a nonclassical logic in physics imply that logic is empirical, at least in the sense in which physical geometry is empirical, as argued, e.g., by H. PUTNAM[13]? My answer is 'no', as follows from what I have said above. The analogy with physical geometry breaks down because geometry belongs to the mathematical substructure of a

physical theory while logic does not; the question of 'logic' (viz., logical syntax) only arises in connection with an (implied or intended) formalization of physical language. Thus, logic is neither empirical, nor *a priori*, nor a matter of convention. Rather, it is a matter of optimal choice among a limited number of possibilities. It is only the whole set of possibilities that has some empirical content or significance. Of course, if we prescribe form and meaning of the (atomic) sentences the logic of the sentential connectives may only depend on the (semantic) meaning of the latter. But this is really a question that would require a separate paper, if not a monograph, for full treatment.

NOTES

*Translated from 'Zur Begründung der Statistischen Transformationstheorie der Quantenphysik', *Sitz. Ber. Berl. Akad. Wiss., Phys.-Math. Kl.* **27** (1936), 90–113.

[1] J. von Neumann, 'Wahrscheinlichkeitstheoretischer Aufbau der Quantenmechanik', *Goett. Nachr.* (1928), 245; *Mathematische Grundlagen der Quantenmechanik*, Berlin 1932, Kap. IV. See also M. Born und P. Jordan, *Elementare Quantenmechanik*, Berlin 1930, 6. Kap.

[2] [The true nature of this connection has only emerged much later in the study of the intertheory relations between quantum mechanics and classical Hamiltonian mechanics. The upshot is this: the Principle of Correspondence allows us to consider as meaningful statements of the form 'The value of a quantity Q of a physical system S lies within the interval (q_1, q_2)'; statements of this form referring to complementary quantities are then to be treated as complementary in the sense of complementarity logic, i.e., inconnectible, as shown in this paper. On the other hand, we may use instead statements of the form 'The physical system S is in a state where the quantity Q has a value between q_1 and q_2'; in this case statements referring to complementary quantities can be treated as contradictory (their conjunction would be allowed as meaningful but untrue) and the need for complementarity logic – or any other 'logic of quantum theory' – does not arise. The essential difference between the two statements emerges when their negations are considered: the negation of the first form would read 'The value of quantity Q of S lies outside the interval (q_1, q_2)' – this is indeed the proper negation for classical and the (improper) negation for quantum mechanics – but this is *not* equivalent to the (proper) negation of the second form which includes all states where the quantity Q has no value in any finite interval. The notion 'proper negation' as used here is a semantical one. A semantically adequate syntactic characterization of 'proper negation' for arbitrary languages has been attempted (e.g. by Carnap, *Logical Syntax of Language*, London 1937), but most attempts can be shown to be inadequate. In the present paper measurement statements are supposed to have the first form mentioned above.]

[3] [[...]] [Two quantities A and B are called *totally complementary* iff there is no state for which A and B have values within any finite intervals.]

[4] H. Reichenbach, 'Axiomatik der Wahrscheinlichkeitsrechnung', *Math. Z.* **34** (1932), 568; *Wahrscheinlichkeitslehre*, Leiden 1935.

[5] A. Kolmogoroff, 'Grundbegriffe der Wahrscheinlichkeitsrechnung', *Erg. Math.* II/3 (1933).

[6] A. Einstein, B. Podolsky, and N. Rosen, *Phys. Rev.* **47** (1935), 777.
[7] E. Schroedinger, *Naturwiss.* **23** (1935), 807, 823, 844.
[8] [If we take the *span* instead of the (set theoretical) sum we have the same situation as with the section: the span of two linear subsets is always again a linear subset (subspace). The set of subspaces then forms an orthocomplemented lattice in which – contrary to the Boolean lattice – the distributive laws do not hold. This is the 'Quantum Logic' advocated by Birkhoff and von Neumann. The negation in this logic has the same meaning as in our complementarity logic, i.e., it too is a nonproper negation, referring as it does to the orthogonal complement.]
[9] [To be sure, a Euclidean space is a point space, not a vector space. But we can define in it a vector space by distinguishing a fixed point as origin, i.e., by abandoning homogeneity (group of translations or displacements) while maintaining isotropy (group of rotations).]
[10] To be sure, the metaphysical nature of these compound sentences arises solely from the logical interpretation of the nondistributive lattice connectives as sentential (or predicate) connectives, and not by the admission of predicates correlated to the lattice elements (subspaces). Indeed, if we consider two subspaces X and Y with the (non-commuting) projectors P_X and P_Y, we have of course the further projector $P_{X \odot Y}$; but while $X \odot Y$ is interpreted by Birkhoff-von Neumann as conjunction of two predicates, $P_{X \odot Y}$ *cannot* be so interpreted because $P_{X \odot Y} \neq P_X \cdot P_Y$. Thus, complementarity logic does not omit any physically meaningful predicates (as is sometimes suggested) but prevents their misinterpretation.
[11] P. Suppes, 'Probability Concepts in Quantum Meohanics', *Phil. of Sc.* **28** (1961), 378–389; 'Logics Appropriate to Empirical Theories', in *The Theory of Models* (ed. by J. W. Addison, L. Henkin and A. Tarski), Amsterdam 1965, p. 364–375; 'Une logique nonclassique de la méchanique quantique', *Synthese* **10** (1966), 74–85.
[12] F. Kamber, 'Die Struktur des Aussagenkalkuels in einer physikalischen Theorie', *Nachr. Akad. Wiss. Goettingen* **10** (1964), 103–124.
[13] H. Putnam, 'Is Logic Empirical?' in *Boston Studies in the Philosophy of Science*, vol. **V** (ed. by R. S. Cohen and M. W. Wartofsky), Dordrecht 1969, pp. 216–241.

MARTIN STRAUSS

MATHEMATICS AS LOGICAL SYNTAX – A METHOD TO FORMALIZE THE LANGUAGE OF A PHYSICAL THEORY

I intend to explain a method how to formalize the language of a given physical theory. The essential part of this method consists in using the mathematical formalism of the theory in question in order to get the logical syntax of the coordinated physical language.

First it must be remarked that the mathematical formalism of a physical theory is not itself the formalized language of this theory. In order to prove this it is sufficient to remember that mathematics contains only logical sentences whereas a physical language must also contain synthetic sentences describing the results of experiments.

On the other hand the mathematical formalism together with its physical interpretation contains the whole theory just as a formalized language of this theory.

In this way it is plausible that the mathematical formalism together with its physical interpretation determines all essential features of a formalized language concerning the content of the physical theory in question.

Instead of describing our method in abstract terms we shall illustrate it by two examples: classical and quantum mechanics. It shall be emphasized that by formalizing the language of quantum mechanics we get a syntactical definition of the quantum mechanical complementarity concept suggested already in some previous communications (cf. [2], [3]).

I. CLASSICAL MECHANICS

Using the Hamiltonian form of classical mechanics, every subset M of the whole phase space E can be interpreted as follows: the phase point of a certain mechanical system s_0 lies at a certain moment t_0 in M. This fact allows us to introduce sentential functions

$$p_M(s, t)$$

having the indicated meaning, and to establish a one-to-one-correspondence between the subsets M and the predicates 'p_M'.

C. A. Hooker (ed.), The Logico-Algebraic Approach to Quantum Mechanics, 45–52.

Now we can define connective symbols called *predicational connections* as follows:

$$
\text{(C I)}\qquad
\begin{aligned}
\overline{p_M} &=_{df} p_{\bar{M}}\\
p_M \cdot p_{M'} &=_{df} p_{M\,M'}\\
p_M \vee p_{M'} &=_{df} p_{M+M'}\\
p_M \supset p_{M'} &=_{df} p_{\bar{M}+M'}
\end{aligned}
$$

using on the right hand the usual set-theoretical connective symbols.

By these definitions the following *displacing rules* are valid:

$$
\begin{aligned}
`\overline{\overline{p_M}}\text{'} &= `p_M\text{'}\\
`\overline{p_M \cdot p_N}\text{'} &= `\overline{p_M} \vee \overline{p_N}\text{'}\\
`\overline{p_M \vee p_N}\text{'} &= `\overline{p_M} \cdot \overline{p_N}\text{'}\\
`p_M \supset p_N\text{'} &= `\overline{p_M} \vee p_N\text{'}
\end{aligned}
$$

just as in the classical calculus of sentences or predicates.

A term consisting of two or more predicates connected with another by predicational connections is by definition (C I) a predicate too called *connected predicate*.

Now we can give the following FORMATIVE RULE:

(C FR) *A term of the C-language shall be a sentence, if it consists of a predicate of the C-language followed by two individual constants like 's_0' and 't_0' separated from each other by a comma and enclosed with another by brackets.*

In order to formulate the transformative rules we introduce first the mediate concept 'valable predicate' as follows:

(C df 1) *A predicate* 'p_{M_0}' is to be said *valable*, if M_0 is equal to E by means of mathematics.

Then we can formulate the TRANSFORMATIVE RULES as follows:

(C TR I) *A sentence of the C-language shall be valable if the corresponding predicate is valable.*

(C TR II) *A sentence* '$p_{M_1}(s_0, t_0)$' *shall be a consequence of the sentence* '$p_{M_0}(s_0, t_0)$' *if the predicate* '$p_{M_0} \supset p_{M_1}$' *is valable.*

These transformative rules obviously are in accordance with the physical interpretation of the formalism.

The language so obtained is an L-language the logical syntax of which forms a part of the classical calculus of predicates.

The physical laws of classical mechanics, i.e. the equations of motion.

$$q_i = \frac{dH}{dp_i} \qquad p_i = -\frac{dH}{dq_i}$$

H being the Hamiltonian, with its physical interpretation, can be formulated in this language as indetermined premisses or as an additional transformative rule obtaining a P-language instead of an L-language. Here we shall do the latter.

We use the fact that the equations of motion have an unique solution coordinating every point of the phase space and every time interval Dt another point of the phase space. Let $V(Dt)$ be the operator of this transformation and consequently

$$M^{Dt} =_{df} V(Dt)\, M$$

the subset coordinated to M and to Dt by the equations of motion, and let 'p_M^{Dt}' be new predicates defined by

(C df 2) $p_M^{Dt}(s_0, t_0) =_{df} p_M(s_0, t_0 + Dt)$.

Then the physical interpretation of the equations of motion can be formulated as the following TRANSFORMATIVE RULE:

(C TR III) *The sentence '$p_{M_1}^{Dt}(s_0, t_0)$' shall be a consequence of the sentence '$p_{M_0}(s_0, t_0)$' if the predicate '$p_{M_0}Dt \supset p_{M_1}$' is valable.*

This rule contains obviously the rule (C TR II) as a special case; thus the P- and the L-rules are here not essentially different. The reason for this is the use of mathematics, taking into account that M_{Dt} is defined by a mathematical operation. Thus, *the reduction of the P-rule to L-concepts* given by (C TR III) can be considered as the main result of our method to formalize the language of a physical theory by means of its mathematical formalism.

II. QUANTUM MECHANICS

In quantum mechanics we are not dealing with the phase space but – by means of the Schrödinger equation – with the set of all quadratic integrable functions $f(q_1, \ldots, q_f)$ – (q_i being the coordinates) – forming a so-called Hilbert space $\mathfrak{H}$. On account of the linearity of the Schrödinger equation a physical interpretation can not be given to *any* subset of $\mathfrak{H}$ *but* only *linear* subsets of $\mathfrak{H}$, and actually not to all linear subsets but only to those being coordinated to quantum mechanical quantities in a certain way.

Thus it seems that in order to formalize the quantum mechanical language by the method used above in the case of classical mechanics we have to use linear subsets instead of subsets at all. But it is not right to think so because there is an *ambiguity in the choice of mathematical entities which are to be coordinated to physical predicates*, each linear subset $\mathfrak{M}$ of $\mathfrak{H}$ being in a one-to-one-correspondence with a projection operator $P_{\mathfrak{M}}$ defined as coordinating to every element (function) of $\mathfrak{H}$ its projection upon $\mathfrak{M}$. This ambiguity would have no consequences for our problem only if the calculus of the linear subsets would be isomorphic to the calculus of the projection operators – a condition not fulfilled.

Thus, we can construct two languages – say the *M*- and the *P*-language – each containing the content of quantum mechanics. But as pointed out in another paper (4) only the *P*-language forming a sub-language of the *M*-language, is in full accordance with the physical interpretation of the quantum mechanical formalism, the sentences of the *M*-language which are not sentences of the *P*-language having a metaphysical or at least a partially metaphysical content. Therefore we shall consider here only the *P*-language, called in the following *Q-language*, being in agreement with Bohr's complementarity conception. (The *M*-language proposed indeed by Birkhoff and von Neumann (1) shall be remarked to disagree too with the classical logic by the lack of the distributive laws).

In order to construct the *Q*-language we introduce primitive sentential functions

$$p_{E^A_{Da}}(s, t)$$

having the following meaning: the value of the physical quantity A of a system s lies at a certain moment t in the interval Da, or more precisely: this quantity has been found or will be found in the interval Da by a measurement at the moment t. E^A_{Da} is a projection operator construction of which is not to be given here. Thus, we coordinate a primitive predicate '$p_{E^A_{Da}}$' to every projection operator E^A_{Da}, A being a physical quantity (an 'observable' in the terminology of Dirac) and Da a measurement interval.

For the sake of brevity we shall use

$$p_{E_0}, p_{E_1}, p_{E_2}, \ldots$$

as constant predicates, E_1 being projection operators of the mentioned kind, and

$$p_E, p_{E'}, p_{E''}, \ldots$$

as variable predicates.

Now we can define *predicational connections* as follows:

$$(Q\ I) \qquad \begin{array}{l} \overline{p_E} =_{df} p_{\bar{E}} \\ \left.\begin{array}{l} p_E \cdot p_{E'} =_{df} p_{EE'} \\ p_E \vee p_{E'} =_{df} p_{E+E'} \\ p_E \supset p_{E'} =_{df} p_{\bar{E}+E'} \end{array}\right\} \quad \textit{if} \quad EE' = E'E \end{array}$$

using on the right hand the abbreviations

$$E+E' =_{df} E+E'-EE' \quad \textit{if} \quad EE'=E'E$$
$$\bar{E} =_{df} I-E \qquad \text{(I being the identity: } I=P_{\mathfrak{H}}\text{).}$$

By these definitions the rules (CI) are valid also for the Q-language. But it must be emphasized that the predicational connections of the Q-language defined by $(Q\ I)$ are restricted to those argument predicates for which the coordinated projection operators are commutable with another: only in this case the connected projection operators form projection operators again. *The use of restricted predicational connections in the Q-language is the main difference between the classical logic and the logic of the Q-language.*

We call two predicates, connections of which do not form a predicate again, *inconnectable* or – following N. Bohr – *complementary* to each other.

Now the FORMATIVE RULE defining the concept 'sentence' by means of the concept 'predicate' can be given for the Q-language in the same manner as for the C-language (cf. [$C\ FR$]).

The TRANSFORMATIVE RULES for the Q-language are to be given analogous to these of the C-language:

(*Q df* 1) A *predicate* of the Q-language 'p_E' is to be said *valable* if E is equal to I by means of mathematics.

(*Q TR* I) *A sentence of the Q-language shall be valable if the corresponding predicate is valable.*

(*Q TR* II) *The sentence* '$p_{E_1}(s_0, t_0)$' *shall be a consequence of the sentence* '$p_{E_0}(s_0, t_0)$' *if* '$p_{E_0} \supset p_{E_1}$' *is a valable predicate.*

These transformative rules agree with the physical interpretation of the formalism as is easily seen.

In order to formulate the physical content of the Schrödinger equation as a transformative rule we use the fact that the Schrödinger equation

$$\left(H + ih\frac{d}{dt}\right) f = 0$$

has the unique solution

$$f(t) = U(t) f(0), \; U(t) =_{df} \exp(iHt/h)$$

H being the Hamiltonian operator and h being Planck's constant divided by 2π. Thus, $U(t)$ being an unitarian operator the Schrödinger equation coordinates to every linear subset $\mathfrak{M}$ and every time interval Dt another linear subset $\mathfrak{M}^{Dt}$ defined by

$$f \in \mathfrak{M}^{Dt} \text{ equivalent } U(Dt)^{-1} f \in \mathfrak{M},$$

and therefore to every projection operator $P_{\mathfrak{M}}$ and every time interval Dt the projection operator

$$P_{\mathfrak{M}^{Dt}} = U(Dt) P_{\mathfrak{M}} U(Dt)^{-1}.$$

Thus, using the abbreviation

$$E^{Dt} =_{df} U(Dt) E U(Dt)^{-1}$$

and introducing new predicates

(*Q df* 2) $p_E^{Dt}(s_0, t_0) =_{df} p_E(s_0, t_0 + Dt)$

we can formulate the physical content of the Schrödinger equation as the following TRANSFORMATIVE RULE

(*Q TR* III) *A sentence* '$p_{E_1}^{Dt}(s_0, t_0)$' *shall be a consequence of the sentence* '$p_{E_0}(s_0, t_0)$' *if* '$p_{E_0^{Dt}} \supset p_{E_1}$' *is a valable predicate.*

We can also formulate the probability laws of quantum mechanics as a transformative rule. For this purpose we have to introduce in the Q-language a *probability functor*

$$W(p_E, p_{E'})$$

as suggested by Reichenbach and consequently also the arithmetics; then we accept the probability axioms given by Reichenbach (1), p. 118, as additional transformative rules determining equations of which form between probability terms shall be valid (or more precisely: *L*-valid).

Then we can formulate the probability laws of quantum mechanics as follows:

All sentences of the form
(*Q TR* IV) $W(p_{E_0} p_{E_1}^{Dt}) = Sp\, E_0^{Dt} E_1 / Sp\, E_0$
shall be valid.

In this '*Sp*' means a certain mathematical operation definition of which cannot be given here.

The meaning of 0 probability sentence (i.e. an equation between a probability functor with constant arguments and a number laying between 0 and 1) in the sense of semantics is a problem being not to be treated here because the formalization of a physical language has nothing to do with it directly. Only it may be remarked that the frequency interpretation of the probability functor can be given in agreement with the physical interpretation in a similar manner as given by Reichenbach (1) notwithstanding the fact that here the argument predicates of the probability functor are predicates with *two* arguments; this is pointed out in another paper (cf. [4]).

Prague

BIBLIOGRAPHY

Birkhoff, G. and Neumann, J. von, [1] 'The Logic of Quantum Mechanics', *Ann. of Math.* 1936.
Carnap, R., [1] *Logical Syntax of Language*, London, 1936.
Reichenbach, H., [1] *Wahrscheinlichkeitslehre*, Leiden, 1935.

Strauss, M., [1] 'Ungenauigkeit, Wahrscheinlichkeit und Unbestimmtheit', *Erkenntnis* **6** (1936).
Strauss, M., [2] 'Zur Begründung der statistischen Transformationstheorie der Quantenphysik', *Ber. d. Berl. Akad., phys.-math. Kl.* **27** (1936).
Strauss, M., [3] 'Komplementarität und Kausalität im Lichte der logischen Syntax', *Erkenntnis* **6** (1936).
Strauss, M., [4] *Mathematische und logische Beiträge zur quantenmechanischen Komplementaritätstheorie*, Dissertation, Prague, 1938.

H. REICHENBACH

THREE-VALUED LOGIC AND THE INTERPRETATION OF QUANTUM MECHANICS

The following is an excerpt from Hans Reichenbach's important *Philosophic Foundations of Quantum Mechanics* and comprising the bulk of §§29–37 in which Reichenbach sets forth his own interpretation of quantum theory. The preceding sections of the book concern themselves with general considerations concerning the analysis of physical theory and with a mathematical analysis of the theory. This excerpt is almost completely self-contained; for the reader's benefit I reproduce here four references to the earlier text:

(i) Heisenberg's Uncertainty Relations (Equation (2) of §3).

$$q \cdot p \geqq \frac{h}{4\pi}$$

(ii) DEFINITION 1, §25: *If the value u_i of an entity u has been observed in a measurement of u, this value u_i means the value of u immediately before and immediately after the measurement.*

(iii) DEFINITION 2, §25: *The probability of a combination $v_k w_m$ relative to a physical situation s is given by the product of the individual probabilities of v_k and w_m relative to s.*

(iv) DEFINITION 3, §27: *The value of an entity u measured in a situation s means the value u existing after the measurement; before the measurement, and thus in the situation s, the entity u has all its possible values simultaneously.*

§29. INTERPRETATION BY A RESTRICTED MEANING

The interpretation by a restricted meaning which we shall present here formulates, on the whole, ideas which have been developed by Bohr and Heisenberg. We therefore shall call this conception the *Bohr-Heisenberg*

C. A. Hooker (ed.), *The Logico-Algebraic Approach to Quantum Mechanics*, 53–97.

interpretation, without intending to say that every detail of the given interpretation would be endorsed by Bohr and Heisenberg.

This interpretation does not use our Definitions 1 and 2, §25. For the values of measured entities it uses the following definition:

DEFINITION 4. *The result of a measurement represents the value of the measured entity immediately after the measurement.*

This definition contains only the common part of our Definitions 1, §25, and 3, §27; a statement concerning the value of the entity *before* the measurement is omitted. In this interpretation we therefore can no longer say that the observed entity remains undisturbed; both the observed and the unobserved entity may be disturbed. On the other hand, such a disturbance of the measured entity is *not asserted*; this question is deliberately left unanswered by Definition 4....

It is an immediate consequence of the restriction to Definition 4 that simultaneous values cannot be measured. The considerations which we attached to Definition 1 are no longer applicable, and if we measure first q and then p, the obtained values of q and p do not represent simultaneous values; only q represents a value existing between these two measurements, whereas p represents a value existing after the second measurement, when the value q is no longer valid....

We said that if q has been measured, we do not know the value of p. This lack of knowledge is considered in the Bohr-Heisenberg interpretation as making a statement about *p meaningless*. It is here, therefore, that this interpretation introduces a rule restricting quantum mechanical language. This is expressed in the following definition.

DEFINITION 5. *In a physical state not preceded by a measurement of an entity u, any statement about a value of the entity u is meaningless.*

In this definition we are using the term "statement" in a sense somewhat wider than usual, since a statement is usually defined as having meaning. Let us use the term "proposition" in this narrower sense as including meaning. Then, Definition 5 states that not every statement of the form "the value of the entity is u", is a proposition, i.e., has meaning. Because it uses Definition 5, the Bohr-Heisenberg interpretation can be called an *interpretation by a restricted meaning*.

We must add a remark concerning the logical form of Definition 5. Modern logic distinguishes between *object language* and *metalanguage*; the first speaks about physical objects, the second about statements, which in turn are referred to objects.[1] The first part of the metalanguage, *syntax*, concerns only statements, without dealing with physical objects; this part formulates the structure of statements. The second part of the metalanguage, *semantics*, refers to both statements and physical objects. This part formulates, in particular, the rules concerning truth and meaning of statements, since these rules include a reference to physical objects. The third part of the metalanguage, *pragmatics*, includes a reference to persons who use the object language.[2]

Applying this terminology to the discussion of Definition 5, we arrive at the following result: Whereas Definition 4, and likewise Definitions 1–2, §25, and Definition 3, §27, determine terms of the object language, namely, terms of the form "value of the entity u", Definition 5 determines a term of the metalanguage, namely, the term "meaningless". It is therefore a semantical rule. We can express it in Table 1. We denote

TABLE I

Observational language		Quantum mechanical language
m_u	u	U
true	true	true
true	false	false
false	true	meaningless
false	false	meaningless

the statement "a measurement of u is made" by m_u, and the statement "the indication of the measuring instrument shows the value u" by u. These two statements belong to the observational language. Quantum mechanical statements are written with capital letters; and U expresses the statement "the value of the entity immediately after the measurement is u".[3] Table I shows the coordination of the two languages.

A meaningless statement is not subject to propositional operations; thus, the negation of a meaningless statement is equally meaningless. Similarly, a combination of a meaningful and a meaningless statement is not meaningful. If the statement *a* is meaningful, and *b* meaningless, then *a and b* is meaningless; so is *a or b*. Not even the assertion of the *tertium non datur, b or non-b*, is meaningful. The given restriction of meaning therefore cuts off a large section from the domain of quantum mechanical language....

The only justification of Definition 5 is that it eliminates the causal anomalies. This should be clearly kept in mind. It would be wrong to argue that statements about the value of an entity before a measurement are meaningless because they are not verifiable. Statements about the value after the measurement are not verifiable either. If, in the interpretation under consideration, the one sort of statement are forbidden and the other admitted, this must be considered as a rule which, logically speaking, is arbitrary, and which can be judged only from the standpoint of expediency. From this standpoint its advantage consists in the fact that it eliminates causal anomalies, but that is all we can say in its favor.

It is often forgotten that the Bohr-Heisenberg interpretation uses Definition 4. This definition seems so natural that its character as a definition is overlooked. But without it, the interpretation could not be given. Applying the language used in Part I we can say that Definition 4 is necessary for the transition from observational data to phenomena; it defines the phenomena. It is therefore incorrect to say that the Bohr-Heisenberg interpretation uses only verifiable statements. We must say, instead, that it is an interpretation using a weaker definition concerning unobserved entities than other interpretations, and using a restricted meaning, with the advantage of thus excluding statements about causal anomalies.

We now must consider the relations between noncommutative entities. We know that if a measurement of q is made, a measurement of p cannot be made at the same time, and vice versa. Statements about simultaneous values of noncommutative entities are called *complementary statements*. With Definition 5 we have therefore the following theorem:

THEOREM 1. *If two statements are complementary, at most one of them is meaningful; the other is meaningless.*

We say "at most" because it is not necessary that one of the two statements be meaningful; in a general situation s, determined by a

ψ-function which is not an eigen-function of one of the entities considered, both statements will be meaningless.

Theorem 1 represents a physical law; it is but another version of the commutation rule, or of the principle of indeterminacy, which excludes a simultaneous measurement of noncommutative entities. We see that with theorem 1 a physical law has been expressed in a semantical form; it is stated as a rule for the meaning of statements. This is unsatisfactory, since, usually, physical laws are expressed in the object language, not in the metalanguage. Moreover, the law formulated in theorem 1 concerns linguistic expressions which are not always meaningful; the law states the conditions under which these expressions are meaningful. Whereas such rules appear natural when they are introduced as conventions determining the language to be used, it seems unnatural that such a rule should assume the function of a law of physics. The law can be stated only by reference to a class of linguistic expressions which includes both meaningful and meaningless expressions; with this law, therefore, meaningless expressions are included, in a certain sense, in the language of physics.

The latter fact is also illustrated by the following consideration. Let $U(t)$ be the propositional function "the entity has the value u at the time t". Whether $U(t)$ is meaningful at a given time t depends on whether a measurement m_u is made at that time. We therefore have in this interpretation propositional functions which are meaningful for some values of the variable t, and meaningless for other values of t.

The question arises whether it is possible to construct an interpretation which avoids these disadvantages. An interesting attempt to construct such an interpretation has been made by M. Strauss.[4] Although the rules underlying this interpretation are not expressly stated, it seems that they can be construed in the following way.

Definitions 1–2, §25, and Definition 3, §27, are rejected. Definition 4 is maintained. Instead of Definition 5, the following definition is introduced.

DEFINITION 6. *A quantum mechanical statement U is meaningful if it is possible to make a measurement m_u.*

It follows that all quantum mechanical statements concerning individual entities are meaningful, since it is always possible to measure such an entity. Only when U stands for the combination of two complementary statements P and Q is it not possible to make the corresponding measurement; therefore a statement like *P and Q* is meaningless. Similarly, other combinations are considered meaningless, such as *P or Q*. The logic of quantum mechanics is constructed, according to Strauss, so that not all statements are *connectable*; there are *nonconnectable* statements.

It may be regarded as an advantage that this interpretation constructs the language of physics in such a way that it contains only meaningful elements. A determination concerning meaningless expressions is formulated only in terms of the rules for the connection of statements. On the other hand, the physical law of complementarity is once more expressed as a semantic rule, not as a statement of the object language.

Leaving it open whether or not the latter fact is to be considered as a disadvantage, we now must point out a serious difficulty with this interpretation resulting from Definition 6. If we consider U as a function of t, it is indeed always possible to measure the entity u, and therefore U is always meaningful. It is different when we consider U as a function of the general physical situation s characterized by a general function ψ. Is it possible to measure the entity u in a general situation s? Since we know that the measurement of u destroys the situation s, this is not possible. Thus, it follows that in a general situation s even the individual statement U is meaningless. The given interpretation therefore is reduced to the interpretation based on Definition 5. On the other hand, if the meaning of Definition 6 is so construed that it is possible to measure u even in a general situation s, the obtained value u must mean the value of the entity in the situation s and therefore before the measurement. The interpretation is thereby shown to use Definition 1, §25. But if the latter definition is used, it is possible to make statements about simultaneous values between two measurements, such as explained in §25; this means that the rules concerning nonconnectable statements break down.

If we are right, therefore, in considering Strauss's interpretation as given through Definition 4 and Definition 6, we come to the result that this interpretation leads back to the interpretation of Definition 4 and Definition 5.[5]

§30. INTERPRETATION THROUGH A THREE-VALUED LOGIC

The considerations of the preceding section have shown that, if we regard statements about values of unobserved entities as meaningless, we must include meaningless statements of this kind in the language of physics. If we wish to avoid this consequence, we must use an interpretation which excludes such statements, not from the domain of *meaning*, but from the domain of *assertability*. We thus are led to a three-valued logic, which has a special category for this kind of statements.

Ordinary logic is two-valued; it is constructed in terms of the truth values *truth* and *falsehood*. It is possible to introduce an intermediate truth value which may be called *indeterminacy*, and to coordinate this truth value to the group of statements which in the Bohr-Heisenberg interpretation are called *meaningless*. Several reasons can be adduced for such an interpretation. If an entity which can be measured under certain conditions cannot be measured under other conditions, it appears natural to consider its value under the latter conditions as indeterminate. It is not necessary to cross out statements about this entity from the domain of meaningful statements; all we need is a direction that such statements can be dealt with neither as true nor as false statements. This is achieved with the introduction of a third truth value of indeterminacy. The meaning of the term "indeterminate" must be carefully distinguished from the meaning of the term "unknown". The latter term applies even to two-valued statements, since the truth value of a statement of ordinary logic can be unknown; we then know, however, that the statement is either true or false. The principle of the *tertium non datur*, or of the *excluded middle*, expressed in this assertion, is one of the pillars of traditional logic. If, on the other hand, we have a third truth value of indeterminacy, the *tertium non datur* is no longer a valid formula; there is a tertium, a middle value, represented by the logical status *indeterminate*.

The quantum mechanical significance of the truth-value *indeterminacy* is made clear by the following consideration. Imagine a general physical situation s, in which we make a measurement of the entity q; in doing so we have once and forever renounced knowing what would have resulted if we had made a measurement of the entity p. It is useless to make a measurement of p in the new situation, since we know that the measurement of q has changed the situation. It is equally useless to construct another system with the same situation s as before, and to make a measurement of p in this system. Since the result of a measurement of p is determined only with a certain probability, this repetition of the measurement may produce a value different from that which we would have obtained in the first case. The probability character of quantum mechanical predictions entails an absolutism of the individual case; it makes the individual occurrence unrepeatable, irretrievable. We express this fact by regarding the unobserved value as indeterminate, this word being taken in the sense of a third truth value....

TABLE II

Observational language		Quantum mechanical language
m_u	u	U
T	T	T
T	F	F
F	T	I
F	F	I

The introduction of the truth-value *indeterminate* in quantum mechanical language can be formally represented by Table II, which determines the truth values of quantum mechanical statements as a function of the truth values of statements of the observational language. We denote truth by T, falsehood by F, indeterminacy by I. The meaning of the symbols m_u, u, and U is the same as explained on p. 55.[1]

Let us add some remarks concerning the logical position of the quantum mechanical language so constructed. When we divide the exhaustive interpretations into a *corpuscle language* and a *wave language*, the language introduced by Table II may be considered as a *neutral language*, since it does not determine one of these interpretations. It is true that we speak of the measured entity sometimes as the path of a particle, sometimes as the path of needle radiation; or sometimes as the energy of a particle, sometimes as the frequency of a wave. This terminology, however, is only a remainder deriving from the corpuscle or wave language. Since the values of unobserved entities are not determined, the language of Table II leaves it open whether the measured entities belong to waves or corpuscles; we shall therefore use a neutral term and say that the measured entities represent parameters of *quantum mechanical objects*. The difference between calling such a parameter an energy or a frequency then is only a difference with respect to a factor h in the numerical value of the parameter. This ambiguity in the interpretation of unobserved entities is made possible through the use of the category *indeterminate*. Since it is indeterminate whether the unmeasured entity has the value u_1, *or* u_2, *or* etc., it is also indeterminate whether it has the values u_1, *and* u_2, *and* etc., at the same time; i.e., it is indeterminate whether the quantum mechanical object is a particle or a wave.

The name *neutral language*, however, cannot be applied to the language of Table I, p. 55. This language does not include statements about unobserved entities, since it calls them meaningless; it is therefore not equivalent to the exhaustive languages, but only to a part of them. The language of Table II, on the contrary, is equivalent to these languages to their full extent; to statements about unmeasured entities of these languages it coordinates indeterminate statements.

Constructions of multivalued logics were first given, independently, by E. L. Post[2] and by J. Lucasiewicz and A. Tarski.[3] Since that time, such logics have been much discussed, and fields of applications have been sought for; the original publications left the question of application open and the writers restricted themselves to the formal construction of a calculus. The construction of a logic of probability, in which a continuous scale of truth values is introduced, has been given by the author.[4] This logic corresponds more to classical physics than to quantum mechanics. Since, in it, every proposition has a determinate probability, it has no room for a truth value of indeterminacy; a probability of $\frac{1}{2}$ is not what is meant by the category *indeterminate* of quantum mechanical statements. Probability logic is a generalization of two-valued logic for the case of a kind of truth possessing a continuous gradation.

Quantum mechanics is interested in such a logic only so far as a generalization of its categories *true* and *false* is intended, which is necessary in this domain in the same sense as in classical physics; the use of the "sharp" categories *true* and *false* must be considered in both cases as an idealization applicable only in the sense of an approximation. The quantum mechanical truth-value *indeterminate*, however, represents a topologically different category. The application of a three-valued logic to quantum mechanics has been frequently envisaged; thus, Paulette Février[5] has published the outlines of such a logic. The construction which we shall present here is different, and is determined by the epistemological considerations presented in the preceding sections.

[*Omitted here is a brief discussion of two-valued logic – labelled* §31, *Ed.*]

§32. The Rules of Three-Valued Logic

The method of constructing a three-valued logic is determined by the idea that the metalanguage of the language considered can be conceived as belonging to a two-valued logic. We thus consider statements of the form "*A* has the truth value *T*" as two-valued statements. The truth tables of three-valued logic then can be constructed in a way analgous to the construction of the tables of two-valued logic. The only difference is that in the vertical columns to the left of the double line we must assume all possible combinations of the three values *T*, *I*, *F*.

The number of definable operations is much greater in three-valued tables than in two-valued ones. The operations defined can be considered as generalizations of the operations of two-valued logic; we then however, shall have various generalizations of each operation of two-valued logic. We thus shall obtain various forms of negations, implications, etc. We confine ourselves to the definition of the operations presented in truth Tables IVA and IVB.[1] As before, three-valued propositions will be written with capital letters.

The negation is an operation which applies to one proposition; therefore only one negation exists in two-valued logic. In three-valued logic several operations applying to one proposition can be constructed. We call all of them negations because they change the truth value of a proposition. It is expedient to consider the truth values, in the order *T*, *I*,

F, as running from the *highest* value *T* to the *lowest* value *F*. Using this terminology, we may say that the cyclical negation shifts a truth value to the next lower one, except for the case of the lowest, which is shifted to the highest value. We therefore read the expression $\sim A$ in the form *next-A*. The diametrical negation reverses *T* and *F*, but leaves *I* unchanged. This corresponds to the function of the arithmetical minus sign when the value *I* is interpreted as the number 0; and we therefore call the expression $-A$ *the negative of A*, reading it as *minus-A*. The complete negation shifts a truth value to the higher one of the other two. We read $\bar{A}$ as *non-A*. The use of this negation will become clear presently.

TABLE IVA

A	Cyclical negation $\sim A$	Diametrical negation $-A$	Complete negation $\bar{A}$
T	I	F	I
I	F	I	T
F	T	T	T

Disjunction and conjunction correspond to the homonymous operations of two-valued logic. The truth value of the disjunction is given by the higher one of the truth values of the elementary propositions; that of the conjunction, by the lower one.

There are many ways of constructing implications. We shall use only the three implications defined in Table IVB. Our first implication is a three-three operation, i.e., it leads from three truth values of the elementary propositions to three truth values of the operation. We call it *standard implication*. Our second implication is a three-two operation, since it has only the values *T* and *F* in its column; we therefore call it *alternative implication*. Our third implication is called *quasi implication* because it does not satisfy all the requirements which are usually made for implications.

What we demand in the first place of an implication is that it makes possible the procedure of *inference*, which is represented by the rule: If

TABLE IVB

A	B	Disjunction $A \vee B$	Conjunction $A . B$	Standard implication $A \supset B$	Alternative implication $A \rightarrow B$	Quasi implication $A \ni\!\!- B$	Standard equivalence $A \equiv B$	Alternative equivalence $A \equiv B$
T	T	T	T	T	T	T	T	T
T	I	T	I	I	F	I	I	F
T	F	T	F	F	F	F	F	F
I	T	T	I	T	T	I	I	F
I	I	I	I	T	T	I	T	T
I	F	I	F	I	T	I	I	F
F	T	T	F	T	T	I	F	F
F	I	I	F	T	T	I	I	F
F	F	F	F	T	T	I	T	T

A is true, and *A implies B* is true, then *B* is true. In symbols:

$$\frac{\begin{array}{c}A\\ A\supset B\end{array}}{B} \qquad (1)$$

All our three implications satisfy this rule; so will every operation which has a *T* in the first line and no *T* in the second and third line of its truth table. In the second place, we shall demand that if *A* is true and *B* is false, the implication is falsified; this requires an *F* in the third line – a condition also satisfied by our implications. These two conditions are equally satisfied by the "and", and we can indeed replace the implication in (1) by the conjunction. If we do not consider the "and" as an implication, this is owing to the fact that the "and" says too much. If the second line of (1) is $A\,.\,B$, the first line can be dropped, and the inference remains valid. We thus demand that the implication be so defined that without the first line in (1) the inference does not hold; this requires that there are some *T*'s in the lines below the third line. This requirement is satisfied by the first and second implication, though not by the quasi implication. A further condition for an implication is that *a implies a* is always true. Whereas the first and second implication satisfy this condition, the quasi implication does not. The reason for considering this operation, in spite of these discrepancies, as some kind of implication will appear later (cf. §34).

It is usually required that *A implies B* does not necessarily entail *B implies A*, i.e., that the implication is nonsymmetrical. Our three implications fulfill this requirement. The latter condition distinguishes an implication from an equivalence (and is also a further distinction from the "and"). The equivalence is an operation which states equality of truth values of *A* and *B*; it therefore must have a *T* in the first, the middle, and the last line. Furthermore, it is required to be symmetrical in *A* and *B*, such that with *A equivalent B* we also have *B equivalent A*. These conditions are satisfied by our two equivalences. Since these conditions leave the definition of equivalence open within a certain frame, further equivalences could be defined; we need, however, only the two given in the tables.

To simplify our notation we use the following *rule of binding force* for our symbols:

		strongest binding force
complete negation	equal force	$\overline{}$
cyclical negation	equal force	$\sim$
diametrical negation	equal force	-
conjunction		.
disjunction		$\vee$
quasi implication		$-\!\!\!\supset\!\!\!-$
standard implication		$\supset$
alternative implication		$\rightarrow$
standard equivalence		$\equiv$
alternative equivalence		$\equiv\!\!\!\equiv$
		weakest binding force

If several negations of the diametrical or cyclical form precede a letter A, we convene that the one immediately preceding A has the strongest connection with A, and so on in the same order. The line of the complete negation extended over compound expressions will be used like parentheses.

Our truth values are so defined that only a statement having the truth value T can be asserted. When we wish to state that a statement has a truth value other than T, this can be done by means of the negations. Thus the assertion

$$\sim\sim A \qquad (2)$$

states that A is indeterminate. Similarly, either one of the assertions

$$\sim A \qquad\qquad -A \qquad (3)$$

states that A is false.

This use of the negations enables us to eliminate statements in the metalanguage about truth values. Thus the statement of the object language *next-next-A* takes the place of the semantical statement "A is indeterminate". Similarly, the statement of the metalanguage "A is false" is translated into one of the statements (3) of the object language, and then is pronounced, respectively, "next-A", or "minus-A". We thus can carry through the principle that *what we wish to say is said in a true statement of the object language*.

As in two-valued logic, a formula is called *tautological* if it has only

T's in its column; *contradictory*, if it has only F's; and *synthetic*, if it has at least one T in its column, but also at least one other truth value. Whereas the statements of two-valued logic divide into these three classes, we have a more complicated division in three-valued logic. The three classes mentioned exist also in the three-valued logic, but between synthetic and contradictory statements we have a class of statements which are never true, but not contradictory; they have only I's and F's in their column, or even only I's, and may be called *asynthetic* statements. The class of synthetic statements subdivides into three categories. The first consists of statements which can have all three truth values; we shall call them *fully synthetic statements*. The second contains statements which can be only true or false; they may be called *true-false* statements, or *plain-synthetic* statements. They are synthetic in the simple sense of two-valued logic. The use of these statements in quantum mechanics will be indicated on p. 73. The third category contains statements which can be only true or indeterminate. Of the two properties of the synthetic statements of two-valued logic, the properties of being sometimes true and sometimes false, these statements possess only the first; they will therefore be called *semisynthetic* statements.

The cyclical or the diametrical negation of a contradiction is a tautology; similarly, the complete negation of an asynthetic statement is a tautology. A synthetic statement cannot be made a tautology simply by the addition of a negation.

All quantum mechanical statements are synthetic in the sense defined. They assert something about the physical world. Conversely, if a statement is to be asserted, it must have at least one value T in its column determined by the truth tables. Asserting a statement means stating that one of its T-cases holds. Contradictory and asynthetic statements are therefore *unassertable*. On the other hand, tautologies and semisynthetic statements are *indisprovable*; they cannot be false. But whereas tautologies must be true, the same does not follow for semisynthetic statements. When a semisynthetic statement is asserted, this assertion has therefore a *content*, i.e., is not *empty* as in the case of a tautology. For this reason we include semisynthetic statements in the synthetic statements; all synthetic statements, and only these, have a content.

The unique position of the truth value T confers to tautologies of the three-valued logic the same rank which is held by these formulae in two-valued logic. Such formulae are always true, since they have the value T for every combination of the truth values of the elementary propositions. As before, the proof of tautological character can be given by case analysis on the base of the truth tables; this analysis will include combinations in which the elementary propositions have the truth value I. We now shall present some of the more important tautologies of three-valued logic, following the order used in the presentation of the two-valued tautologies (1)–(12), § 31.

The *rule of identity* holds, of course:

$$A \equiv A \tag{4}$$

The *rule of double negation* holds for the diametrical negation:

$$A \equiv --A \tag{5}$$

For the cyclical negation we have a *rule of triple negation*:

$$A \equiv \sim\sim\sim A \tag{6}$$

For the complete negation the rule of double negation holds in the form

$$\bar{A} \equiv \bar{\bar{\bar{A}}} \tag{7}$$

It should be noticed that from (7) the formula $A \equiv \bar{\bar{A}}$ cannot be deduced, since it is not permissible to substitute A for $\bar{A}$; and this formula, in fact, is not a tautology. We shall therefore say that the rule of double negation does not hold *directly*. A permissible substitution is given by substituting $\bar{A}$ for A; in this way we can increase the number of negation signs in (7) correspondingly on both sides. This peculiarity of the complete negation is explained by the fact that a statement above which the line of this negation is drawn is thus reduced to a semisynthetic statement; further addition of such lines will make the truth value alternate only between truth and indeterminacy.

Between the cyclical and the complete negation the following relation holds:

$$\bar{A} \equiv \sim A \vee \sim\sim A \tag{8}$$

The tertium non datur does not hold for the diametrical negation, since $A \vee -A$ is synthetic. For the cyclical negation we have a *quantum non datur*:

$$A \vee \sim A \vee \sim \sim A \tag{9}$$

The last two terms of this formula can be replaced by $\bar{A}$, according to (8); we therefore have for the complete negation a formula which we call a *pseudo tertium non datur*:

$$A \vee \bar{A} \tag{10}$$

This formula justifies the name "complete negation" and, at the same time, reveals the reason why we introduce this kind of negation; the relation (8), which makes (10) possible, may be considered as the definition of the complete negation. The name which we give to this formula is chosen in order to indicate that the formula (10) does not have the properties of the tertium non datur of two-valued logic. The reason is that the complete negation does not have the properties of an ordinary negation: It does not enable us to infer the truth value of A if we know that $\bar{A}$ is true. This is clear from (8); if we know $\bar{A}$, we know only that A is either false or indeterminate. This ambiguity finds a further expression in the fact that for the complete negation no converse operation can be defined, i.e., no operation leading from $\bar{A}$ to A. Such an operation is impossible, because its truth tables would coordinate to the value T of $\bar{A}$, sometimes the value I of A, and sometimes the value F of A.

The *rule of contradiction* holds in the following forms:

$$\overline{A . \bar{A}} \tag{11}$$

$$\overline{A . \sim A} \tag{12}$$

$$\overline{A . - A} \tag{13}$$

The *rules of De Morgan* hold only for the diametrical negation:

$$-(A . B) \equiv -A \vee -B \tag{14}$$

$$-(A \vee B) \equiv -A . -B \tag{15}$$

The two *distributive rules* hold in the same form as in two-valued logic:

$$A . (B \vee C) \equiv A . B \vee A . C \tag{16}$$

$$A \vee B . C \equiv (A \vee B) . (A \vee C) \tag{17}$$

The *rule of contraposition* holds in two forms

$$-A \supset B \equiv -B \supset A \tag{18}$$
$$\bar{A} \rightarrow B \equiv \bar{B} \rightarrow A \tag{19}$$

Since for the diametrical negation the rule of double negation (5) holds, (18) can also be written in the form

$$A \supset B \equiv -B \supset -A \tag{20}$$

This follows by substituting $-A$ for A in (18). For (19), however, a similar form does not exist, since for the complete negation the rule of double negation does not hold directly.

The *dissolution of equivalence* holds in its usual form only for the standard implication in combination with the standard equivalence:

$$(A \equiv B) \equiv (A \supset B).(B \supset A) \tag{21}$$

The corresponding relation between alternative implication and alternative equivalence is of a more complicated kind:

$$(A \equiv B) \equiv (A \rightleftarrows B).(-A \rightleftarrows -B) \tag{22}$$

By the double arrow implication we mean implications in both directions. This double implication does not have the character of an equivalence, since it has values T in its column aside from the first, the middle, and the last line. By the addition of the second term these T's are eliminated, and the column of the alternative equivalence is reached. For a double standard implication, and similarly for two-valued implication, a second term of the form occurring on the right hand side of (22) is dispensable, because such a term follows from the first by means of the rule of contraposition (20). For the double alternative implication this is not the case. This relation states only that B is true if A is true, and that A is true if B is true; but it states nothing about what happens when A and B have one of the other truth values. A corresponding addition is given by the second term on the right hand side of (22).

The *dissolution of implication* holds for the alternative implication in the form

$$A \rightarrow B \equiv \sim -(\bar{A} \vee B) \tag{23}$$

The *reductio ad absurdum* holds in the two forms:

$$(A \supset \bar{A}) \supset \bar{A} \tag{24}$$

$$(A \rightarrow \bar{A}) \rightarrow \bar{A} \tag{25}$$

Next to the tautologies, those formulae offer a special interest which can only have two truth values. Among these the *true-false* statements, or plain-synthetic statements, are of particular importance. An example is given by the formula

$$\sim \sim (\sim A \vee \sim \sim A) \tag{26}$$

which assumes only the truth values T and F when A runs through all three truth values. The existence of such statements shows that the statements of three-valued logic contain a subclass of statements which have the two-valued character of ordinary logic. For the formulae of this subclass the tertium non datur holds with the diametrical negation. Thus, if D is a true-false formula, for instance the formula (26), the formula

$$D \vee -D \tag{27}$$

is a tautology.

The other two-valued formulae can easily be transformed into true-false formulae by the following device. An asynthetic formula A, which has the two values I and F in its truth table, can be transformed into the true-false formula $\sim A$. A semisynthetic formula A, which has the two values I and T, can be transformed into the true-false formula $\sim \sim A$.

We now turn to the formulation of complementarity. We call two statements *complementary* if they satisfy the relation

$$A \vee \sim A \rightarrow \sim \sim B \tag{28}$$

The left hand side is true when A is true and when A is false; in both these cases, therefore, the right hand side must be true. This is the case only if B is indeterminate. When A is indeterminate the left hand side is indeterminate; then we have no restriction for the right hand side, according to the definition of the alternative implication. Therefore (28) can be read: If A is true or false, B is indeterminate.

Substituting in (8) $\sim \sim A$ for A, and using (6), we derive

$$\overline{\sim \sim A} \equiv A \vee \sim A \tag{29}$$

We therefore can write (28) also in the form

$$\overline{\sim\sim A} \rightarrow \sim\sim B \tag{30}$$

Applying (19) we see that (30) is tautologically equivalent to

$$\overline{\sim\sim B} \rightarrow \sim\sim A \tag{31}$$

Substituting B for A in (29) we can transform (31) into

$$B \vee \sim B \rightarrow \sim\sim A \tag{32}$$

It follows that (32) is tautologically equivalent to (28).[2] The condition of complementarity is therefore symmetrical in A and B; *if A is complementary to B, then B is complementary to A.*

The relation of complementarity, opposing the truth value of indeterminacy to the two values of truth and falsehood, is a unique feature of three-valued logic, which has no analogue in two-valued logic. Since this relation determines a column in the truth tables of A and B, it can be considered as establishing a logical operation of complementarity between A and B, for which we could introduce a special sign. It appears, however, to be more convenient to dispense with such a special sign and to express the operation in terms of other operations, in accordance with the corresponding procedure used for certain operations of two-valued logic.

The *rule of complementarity of quantum mechanics* can now be stated as follows: If u and v are noncommutative entities, then

$$U \vee \sim U \rightarrow \sim\sim V \tag{33}$$

Here U is an abbreviation for the statement, "The first entity has the value u"; and V for, "The second entity has the value v". Because of the symmetry of the complementarity relation, (33) can also be written

$$V \vee \sim V \rightarrow \sim\sim U \tag{34}$$

Furthermore, the two forms (30) and (31) can be used.

With (33) and (34) we have succeeded in formulating the rule of complementarity in the object language. This rule is therefore stated as a physical law having the same form as all other physical laws. To show this let us consider as an example the law: If a physical system is closed (statement a), its energy does not change (statement $\bar{b}$). This law, which belongs to two-valued logic, is written symbolically:[3]

$$a \supset \bar{b} \tag{35}$$

This is a statement of the same type as (33) or (34). It is therefore not necessary to read (33) in the *semantical form*: "If U is true or false, V is indeterminate." Instead, we can read (33) in the object language: "U or next-U implies next-next-V".

The law (33) of complementarity can be extended to propositional functions. Our statement U can be written in the functional form

$$Vl(e_1, t) = u \tag{36}$$

meaning: "The value of the entity e_1 at the time t is u". The symbol "$Vl(\ )$" used here is a *functor*, meaning "the value of...",[4] and will be similarly applied to other entities. Then the law of complementarity can be expressed in the form:

$$(u)\,(v)\,(t)\,\{[Vl(e_1, t) = u] \vee \sim [Vl(e_1, t) = u] \rightarrow \sim\sim [Vl(e_2, t) = v]\} \tag{37}$$

The symbols (u), (v), (t), represent all-operators and are read, as in two-valued logic: "for all u" ... "for all t".

The relation of complementarity is not restricted to two entities; it may hold between three or more entities. Thus the three components of the angular momentum are noncommutative, i.e., each is complementary to each of the two others. In order to express this relation for the three entities u, v, w, we add to (33) the two relations:

$$V \vee \sim V \rightarrow \sim\sim W \qquad W \vee \sim W \rightarrow \sim\sim U \tag{38}$$

Each of these relations can be reversed, as has been shown above. The three relations (33) and (38) then state that, if one of the three statements is true or false, the other two are indeterminate.

Since the alternative implication is the *major operation* in (33) and (34),[5] it follows that these formulae can only be true or false, but not indeterminate. The rule of complementarity, although it concerns all three truth values, is therefore, in itself, a true-false formula. Since the rule is maintained by quantum mechanics as true, it has the truth of a two-valued synthetic statement. We see that this interpretation of the rule of complementarity, which is implicitly contained in the usual conception of quantum mechanics, appears as a logical consequence of our three-valued interpretation.[6]

This result shows that the introduction of a third truth value does not make all statements of quantum mechanics three-valued. As pointed out above, the frame of three-valued logic is wide enough to include a class of true-false formulae. When we wish to incorporate all quantum mechanical statements into three-valued logic, it will be the leading idea to put into the true-false class those statements which we call quantum mechanical *laws*. Furthermore, statements about the form of the ψ-function, and therefore about the *probabilities* of observable numerical values, will appear in this class. Only statements about these numerical values themselves have a three-valued character, determined by Table II.

§33. Suppression of Causal Anomalies through a Three-Valued Logic

In the given formulae we have outlined the interpretation of quantum mechanics through a three-valued logic. We see that this interpretation satisfies the desires that can be justifiably expressed with respect to the logical form of a scientific theory, and at the same time remains within the limitations of knowledge drawn by the Bohr-Heisenberg interpretation. The term "meaningless statement" of the latter interpretation is replaced, in our interpretation, by the term "indeterminate statement". This has the advantage that such statements can be incorporated into the object language of physics, and that they can be combined with other statements by logical operations. Such combinations are "without danger", because they cannot be used for the derivation of undesired consequences.

Thus, the and-combination of two complementary statements can never be true. This follows in our interpretation, because the formula

$$[A \vee {\sim} A \rightarrow {\sim}{\sim} B] \rightarrow \overline{A \cdot B} \qquad (1)$$

is a tautology. This is not an equivalence; therefore the condition of complementarity cannot be replaced by the condition $\overline{A \cdot B}$. But the implication (1) guarantees that two complementary statements cannot be both true. Such a combination can be false; but only if the statement about the measured entity is false. Now if a measurement of q has resulted in the value q_1, it will certainly be permissible to say that the statement "the value of q is q_2, and the value of p is p_1", is false. Similarly, it is without danger when we consider the or-combination, "the value of q is q_1 or the value of p is p_1", as true *after* a measurement of q has furnished the value q_1. With such a statement nothing is said about the value of p.

Furthermore, the reductio ad absurdum (24), § 32, or (25), § 32, cannot be used for the construction of indirect proofs. If we have proved by means of the reductio ad absurdum that $\bar{A}$ is true, we cannot infer that A is false; A can also be indeterminate. Similarly, we cannot construct a disjunctive derivation of a statement C by showing that C is true both when B is true and when B is false; we then have proved only the relation

$$B \vee -B \supset C \tag{2}$$

Since the implicans need not be true, we cannot generally infer that C must be true.

This shows clearly the difference between two-valued and three-valued logic. In two-valued logic a statement c is proved when the relation

$$b \vee \bar{b} \supset c \tag{3}$$

has been demonstrated, since here the implicans is a tautology. The analogue of (3) in three-valued logic is the relation

$$B \vee \bar{B} \supset C \tag{4}$$

which according to (8), § 32, is the same as

$$B \vee {\sim} B \vee {\sim}{\sim} B \supset C \tag{5}$$

If (5) is demonstrated, C is proved, since here the implicans is a tautology. But this means that in order to prove C we must prove that C is true in the three cases that B is true, false, or indeterminate. An analysis of quantum mechanics shows that such a proof cannot be given if C formulates a causal anomaly; we then can prove, not (5), but only the relation (2), or a generalization of the latter relation which we shall study presently.

For this purpose we must inquire into some properties of *disjunctions*.

Let us introduce the following notation, which applies both to the two-valued and the three-valued case.

A disjunction of n terms is called *closed* if, in case $n-1$ terms are false, the n-th term must be true.

A disjunction is called *exclusive* if, in case one term is true, all the others must be false.

A disjunction is called *complete* if one of its terms must be true; or what is the same, if the disjunction is true.

For the two-valued case the first two properties are expressed by the following relations:

$$\begin{aligned} b_1 &\equiv \bar{b}_2 \cdot \bar{b}_3 \ldots \bar{b}_n \\ b_2 &\equiv \bar{b}_1 \cdot \bar{b}_3 \ldots \bar{b}_n \\ &\cdot \cdot \cdot \cdot \cdot \cdot \cdot \cdot \cdot \cdot \\ b_n &\equiv \bar{b}_1 \cdot \bar{b}_2 \ldots \bar{b}_{n-1} \end{aligned} \tag{6}$$

That a disjunction for which these relations hold, is closed, follows when we read the equivalence in (6) as an implication from right to left; that it is exclusive follows when we read the equivalence as an implication from left to right. Now it can easily be shown that if relations (6) hold, the disjunction

$$b_1 \vee b_2 \vee \cdots \vee b_n \tag{7}$$

must be true. This result can even be derived if we consider in (6) only the implications running from right to left. In other words: A two-valued disjunction which is closed is also complete, and vice versa. This is the reason that in two-valued logic the terms "closed" and "complete" need not be distinguished. Furthermore, it can be shown that one of the relations (6) can be dispensed with, since it is a consequence of the others.

For the three-valued case a closed and exclusive disjunction is given by the following relations:

$$\begin{aligned} B_1 &\rightleftarrows -B_2 . -B_3 \ldots -B_n \\ B_2 &\rightleftarrows -B_1 . -B_3 \ldots -B_n \\ &\cdot \cdot \cdot \cdot \cdot \cdot \cdot \cdot \cdot \cdot \cdot \cdot \cdot \\ B_n &\rightleftarrows -B_1 . -B_2 \ldots -B_{n-1} \end{aligned} \tag{8}$$

As before, the closed character of the disjunction follows when we use the implications from right to left; and the exclusive character follows when we use the implications from left to right.

We now meet with an important difference from the two-valued case. From the relations (8) we cannot derive the consequence that the disjunction

$$B_1 \vee B_2 \vee \ldots \vee B_n \tag{9}$$

must be true. This disjunction can be indeterminate. This will be the case if some of the B_i are indeterminate and the others are false. All that follows from (8) is that not all B_i can be false simultaneously; the disjunction (9) therefore cannot be false. But since it can be indeterminate, we cannot derive that the disjunction is complete. In three-valued logic we must therefore distinguish between the two properties *closed* and *complete*; a closed disjunction need not be complete. A further difference from the two-valued case is given by the fact that the conditions (8) are independent of each other, i.e., that none is dispensable. This is clear, because, if the last line in (8) is omitted, the remaining conditions would be satisfied if the $B_1 \ldots B_{n-1}$ are false and B_n is indeterminate, a solution excluded by the last line of (8).

The disjunction $B \vee -B$ is a special case of a closed and exclusive disjunction. Corresponding to (2), the proof of the relation

$$B_1 \vee B_2 \vee \ldots \vee B_n \supset C \tag{10}$$

does not represent a proof of C if the $B_1 \ldots B_n$ constitute a closed and exclusive disjunction, since the implicans can be indeterminate. Only a complete disjunction in the implicans would lead to a proof of C.

We now shall illustrate, by an example, that the distinction of closed and complete disjunctions enables us to eliminate certain causal anomalies in quantum mechanics.

Let us consider the interference experiment of § 7 in a generalized form in which n slits $B_1 \ldots B_n$ are used. Let B_i be the statement: "The particle passes through slit B_i". After a particle has been observed on the screen we know that the disjunction $B_1 \vee B_2 \ldots \vee B_n$ is closed and exclusive; namely, we know that if the particle did not go through $n-1$ of the slits, it went through the n-th slit, and that if it went through one of the slits, it did not go through the others. In other words, the observation of a particle on the screen implies that relations (8) hold. But since from these relations the disjunction (9) is not derivable, we cannot maintain that this disjunction is *complete*, or *true*: all we can say is that it is *not false*. It can be indeterminate. This will be the case if no observation of the particle at one of the slits has been made.

The disjunction will also be indeterminate if an observation is made at the n-th slit with the result that the particle did not go through this slit. Such observations will, of course, disturb the interference pattern on the screen. But we are concerned so far only with the question of the truth character of the disjunction. The fact that, if observations with negative result are made in less than $n-1$ slits, the remaining slits will still produce a common interference pattern finds its expression in the logical fact that in such a case the remaining disjunction is indeterminate. Only when an observation with negative result is made at $n-1$ slits, do we know that the particle went through the n-th slit; then the disjunction will be true. On the other hand, if the particle has been observed at one slit, we know that the particle did not go through the others; the disjunction then is also true.

Although the disjunction (9) may be indeterminate, our knowledge about the relations holding between the statements $B_1 \ldots B_n$ will not be indeterminate, but true or false. This follows because relations (8), whose major operation is the alternative implication, represent true-false formulae.[1]

For the case $n=2$, relations (8) are simplified. We then have

$$\left.\begin{array}{l} B_1 \rightleftarrows -B_2 \\ B_2 \rightleftarrows -B_1 \end{array}\right\} \tag{11}$$

Using relations (22), § 32, and (5), § 32, we can write this in the form:

$$B_1 \equiv -B_2 \tag{12}$$

This means that B_1 is equivalent to the negative, or the diametrical negation of B_2. This relation, which we shall call a *diametrical disjunction*, can be considered as a three-valued generalization of the exclusive "or" of two-valued logic. In the case of a diametrical disjunction we know that B_1 is true if B_2 is false, that B_1 is false if B_2 is true, and that B_1 is indeterminate if B_2 is indeterminate. This expresses precisely the physical situation of such a case. If an observation at one slit is made, the statement about the passage of the particle at the other slit is no longer indeterminate, whether the result of the observation at the first slit is positive or negative. We see that this particular form of the disjunction results automatically from the general conditions (8) if we put $n=2$.[2]

Now let us regard the bearing of this result on probability relations. Using the notation introduced in §7, we can determine the probability that a particle leaving the source A of radiation and passing through slit B_1 or slit B_2 or ... or slit B_n will arrive at C by the formula[3]

$$P(A.[B_1 \vee B_2 \vee \ldots \vee B_n], C) = \frac{\sum_{i=1}^{n} P(A, B_i) \cdot P(A.B_i, C)}{\sum_{i=1}^{n} P(A, B_i)} \tag{13}$$

This formula is the mathematical expression of the principle of corpuscular superposition; it states that the statistical pattern occurring on the screen, when all slits are open simultaneously, is a superposition of the individual patterns resulting when only one slit is open. Formula (13), however, can only be applied when the two statements A and $B_1 \vee B_2 \vee \ldots \vee B_n$ are true. Now A is true, since it states that the particle came from the source A of radiation. But we saw that $B_1 \vee B_2 \vee \ldots \vee B_n$ cannot be proved as true. Therefore (13) is not applicable for the case that all slits are open. The probability holding for this case must be calculated otherwise, and is not determined by the principle of corpuscular superposition. We see that no causal anomaly is derivable.

We can interpret this elimination of the anomaly as given by the impossibility of an inference based on the implication (10) when we regard the C of this relation as meaning: The probability holding for the particle has the value (13). The inference then breaks down because our knowledge, formulated by (8), does not prove the implicans of (10) to be true.

In the example of the grating we applied the relations (8) to a case where the localization of the particle is given in terms of discrete positions. The same relations can also be applied to the case of a continuous sequence of possible positions, such as will result when we make an unprecise determination of position. We then usually say: The particle is localized inside the interval Δq, but it is unknown at which point of this interval it is. The latter addition represents the use of an exhaustive interpretation. Within a restrictive interpretation we shall also say that the particle is localized inside the interval Δq. But we shall not use the additional statement because we cannot say that the particle is at a specific point of the interval. Rather, we shall define: The phrase "inside the interval Δq" means that when we divide this interval into n small intervals $\delta q_1 \ldots \delta q_n$ adjacent to each other, the relations (8) will hold for the statements B_i made in the form "the particle is situated in δq_i". We see that the statement "the position of the particle is measured to the exactness Δq" is thus given an interpretation in terms of three-valued logic. The statement itself is true or false, since the relations (8) are so. But since we can infer from (8) only that the disjunction is closed, although it need not be complete, the statement is not translatable into the assertion "the particle is at one and only one point of the interval Δq". The fact that the latter consequence cannot be derived makes it impossible to assert causal anomalies.

In a similar way other anomalies are ruled out. Let us consider, as a further example, the anomaly connected with potential barriers. A potential barrier is a potential field so oriented that particles running in a given direction are slowed down, as the electrons emitted from the filament of a radio tube are slowed down by a negative potential of the grid. In classical physics a particle cannot pass a potential barrier unless its kinetic energy is at least equal to the additional potential energy H_0 which the particle would have acquired in running up to the maximum of the barrier. In quantum mechanics it can be shown that particles which, if measured inside the barrier, possess a kinetic energy smaller than H_0 can later be found with a certain probability outside the barrier. This result is not only a consequence of the mathematics of quantum mechanics, derivable even in so simple a case as a linear oscillator, but its validity is, according to Gamow, proved by the rules of radioactive disintegration. It is important to realize that the paradox cannot be eliminated by a suitable assumption about a disturbance through the measurement. Let us consider a swarm of particles having the same energy H, with $H < H_0$. That each of these particles has this energy can be shown by an energy measurement applied to each, or by taking fair samples out of a swarm originating in sufficiently homogeneous conditions. According to the considerations given on p. 54, we must consider the measured value H as the value of the energy *after* the measurement. After passing through the zone of measurement the particles enter the field of the potential barrier. Beyond the barrier, even at a great distance from it, measurements of position are made which localize particles at that place. Because of the distance, the latter measurements cannot have introduced an additional energy into the particle before it reached the barrier; this means we cannot assume that the measurement of position pushes the particle across the barrier, since such an assumption would itself represent a causal anomaly, an action at a distance. We must rather say that the paradox constitutes an intrinsic difficulty of the corpuscle interpretation; it is one of the cases in which the corpuscle interpretation cannot be carried through without anomalies. The anomaly in this case represents a violation of the principle of the conservation of energy, since we cannot say that in passing the barrier the particle possesses a negative kinetic energy. In view of the fact that the kinetic energy is determined by the square of the velocity, such an assumption would lead to an imaginary velocity of the particle, a consequence incompatible with the spatio-temporal nature of particles.

If, however, we use the restrictive interpretation by a three-valued logic, this causal anomaly cannot be stated. The principle requiring that the sum of kinetic and potential energy be constant connects simultaneous values of momentum and position. If one of the two is measured, a statement about the other entity must be indeterminate, and therefore a statement about the sum of the two values will also be indeterminate. It follows that the principle of conservation of energy is eliminated, by the restrictive interpretation, from the domain of true statements, without being transformed into a false statement; it is an indeterminate statement.

What makes the paradox appear strange is this: It seems that we need not make a measurement of velocity in order to know that in passing the barrier the particle violates the principle of energy. If only we know that the velocity at this point is any real number, zero included, it follows that the principle of energy is violated. The mistake in this inference originates from the assumption, discarded by the restrictive interpretation, that an unmeasured velocity must at least have one determinate real number as its value. It is true that we know the velocity cannot be an imaginary number; but from this we can only infer that the statement, "the velocity has one real number as its value", is not false. The statement, however, will be indeterminate if the velocity is not measured. This is clear when we consider the statement as given by the closed and exclusive disjunction, "the value of the velocity is v_1 or v_2 or ...",[4] which is indeterminate for the reasons explained in the preceding example.

We see that a three-valued logic is the adequate form of a system of quantum mechanics in which no causal anomalies can be derived.

§34. Indeterminacy in the Object Language

We said above that the observational language of quantum mechanics is two-valued. Although this is valid on the whole, it needs some correction. We shall see this when we consider a question concerning a test of predictions based on probabilities, such as raised in §30. For such questions the relation of complementarity introduces an indeterminacy even into the observational language.

Let us consider the two statements of observational language: "If a measurement m_q is made, the indicator will show the value q_1", and "if a measurement m_p is made, the indicator will show the value p_1". We know that it is not possible to verify both these statements for simultaneous values. The case is different from the example given in §30 concerning a throw of the die by Peter or by John; we said that in the latter case the statement about a throw of Peter's can be verified in principle even if Peter does not throw the die, by means of physical observations of another kind. For the combination of the two observational statements about measurements, however, a verification is not even possible in principle. We must therefore admit that we have in the observational language complementary statements.

Now the complementary statements of the observational language are *not* given by the two statements, "the indicator will show the value q_1", and "the indicator will show the value p_1". These statements are both verifiable, since the indicator will, or will not, show the said value even if the measurement is not made. It is rather the implications m_q *implies* q_1 and m_p *implies* p_1 which are complementary. We therefore have here in the observational language an implication which is three-valued and can have the truth-value indeterminate.

What is the nature of this implication? It is certainly not the material implication of the two-valued truth table, since this implication is true when the implicans is false. Thus, the statement

$$m_p \supset p_1 \tag{1}$$

conceived as written in terms of the material implication, will be true if a measurement m_q is made, since, then, the statement m_p is false. This difficulty cannot be eliminated by the attempt to interpret the implication (1) as a *tautological implication* or as a *nomological implication*, i.e., the implication of physical laws.[1] Although such an interpretation has turned out satisfactory for other cases in which the material implication appears unreasonable, it cannot be used with respect to (1), since the implication of this formula carries no necessity with it.

Now if we try to interpret the implication of (1) by the standard or the alternative implication of three-valued logic, the same difficulties as in the case of the two-valued material implication obtain. Since both m_p and p_1 are two-valued statements, we can use in the three-valued truth table (Table IVB, p. 64) only those lines which do not contain the value I in the first two columns; but for these lines both the first and second implication coincide with the material implication of the two-valued truth table (Table IVB). There remains therefore only the quasi implication, and we find that we must write instead of (1) the relation

$$m_p \twoheadrightarrow p_1 \qquad (2)$$

This implication has the desired properties, since by canceling all lines containing an I in the first two columns of the table, we obtain the implication noted in Table V. Therefore, (2) corresponds to what we want

TABLE V

a	b	Quasi implication $a \twoheadrightarrow$
T	T	T
T	F	F
F	T	I
F	F	I

to say, since we consider the statement m_p *implies* p_1 as verified or falsified only if m_p is true, whereas we consider it as indeterminate if m_p is false.

This shows that the observational language of quantum mechanics is not two-valued throughout. Although the elementary statements are two-valued, this language contains combinations of such statements which are three-valued, namely, the combinations established by the quasi implication. The truth-table of two-valued logic must therefore be complemented by the three-valued truth table (Table V) of quasi implication.[2]

We see that the three-valued logical structure of quantum mechanics penetrates to a small extent even into the observational language. Although the observational language of quantum mechanics is statistically complete, it is incomplete with respect to strict determinations. It contains a three-valued implication. If there were no relation of indeterminacy in the microcosm, this three-valued implication could be eliminated; the implication in "m_q implies q_1" then could be interpreted as a nomological implication which, in principle, could be verified or falsified. In observational relations of the kind considered, however, the uncertainty of the microcosm penetrates into the macrocosm. The same holds for all other arrangements in which an atomic occurrence releases macrocosmic processes. Such arrangements need not be measurements; they may consist as well in the lighting of lamps, or the throwing of bombs. The fact that no strict predictions can be made in microcosmic dimensions thus leads to a revision of the logical structure of the macrocosm.

§35. The Limitation of Measurability

Our considerations concerning exhaustive and restrictive interpretations lead to a revision of the formulation given to the principle of indeterminacy. We stated this principle as a limitation holding for the measurement of simultaneous values of parameters; this corresponds to the form in which the principle has been stated by Heisenberg. We now must discuss the question whether the principle, in this formulation, is based on an exhaustive or a restrictive interpretation.

If we start from a general situation *s*, and consider the two probability distributions $d(q)$ and $d(p)$ belonging to *s*, these distributions refer to the results of measurements made in systems of the type *s*. Therefore, if we apply Definition 4, §29, i.e., if we regard the measured values as holding only after the measurement, the obtained values do not mean values existing in the situation *s*, and thus the two distributions are not referred to numerical values belonging to the same situation. The values q and p, to which these distributions refer, rather pertain, respectively, to the two different situations m_q and m_p. We therefore cannot say that the inverse correlation holding for the distributions $d(q)$ and $d(p)$ states a limitation of simultaneous values; instead, the relation of uncertainty then must be formulated as a limitation holding for values obtainable in two different situations. It follows that the usual interpretation of Heisenberg's principle as a limitation holding for the measurement of simultaneous values presupposes, not Definition 4, §29, but Definition 1, §25, since only the use of this definition enables us to interpret the results of measurements m_q and m_p as holding before the operation of measurement and thus as holding for the situation s. Therefore, if Heisenberg's inequality (2), §3, is to be regarded as a cross-section law limiting the measurability of simultaneous values of parameters, this interpretation is based on the exhaustive interpretation expressed in Definition 1, §25.

Now we saw in §25 that if the latter definition is assumed, we can speak of an exact ascertainment of simultaneous values when we consider a situation between two measurements and add the restriction that the situation to which the obtained combination of values belongs no longer exists at the moment when the values are known. For such values, therefore, Heisenberg's principle does not hold. It follows that the principle of uncertainty must be formulated with a qualification: The principle states a limitation holding for the measurability of simultaneous values *existing at the time when we have knowledge of them.* There is no limitation for the measurement of *past* values; only *present* values cannot be measured exactly, but are bound to Heisenberg's inequality (2), §3. The limitation so qualified is, of course, sufficient to limit the predictability of future states; for the knowledge of past values cannot be used for predictions.

These considerations show that the conception of Heisenberg's principle as a limitation of measurability must be incorporated into an exhaustive interpretation. Within a restrictive interpretation we cannot speak of a limitation of exactness, since then the standard deviations Δq and Δp, used in the inequality (2), §3, are not referred to the same situation. Within such an interpretation we must say that, if we regard a situation for which q is known to a smaller or greater degree of exactness Δq, p is completely unknown for this situation and cannot even be said to be at least within the interval Δp coordinated to Δq by the Heisenberg inequality. The corresponding statement holds for the reversed case. By *completely unknown* we mean here, either the category *indeterminate* of our three-valued interpretation, or the category *meaningless* of the Bohr-Heisenberg interpretation. We see that for a restrictive interpretation Heisenberg's principle, in its usual meaning, must be abandoned.

§36. Correlated systems

In an interesting paper A. Einstein, B. Podolsky, and N. Rosen[1] have attempted to show that if some apparently plausible assumptions concerning the meaning of the term "physical reality" are made, complementary entities must have reality at the same time, although not both their values can be known. This paper has raised a stimulating controversy about the philosophical interpretation of quantum mechanics. N. Bohr[2] has presented his views on the subject on the basis of his principle of complementarity with the intension of showing that the argument of the paper is not conclusive. E. Schrödinger[3] has been induced to present his own rather skeptical views on the interpretation of the formalism of quantum mechanics. Other authors have made further contributions to the discussion.

In the present section we intend to show that the issues of this controversy can be clearly stated without any metaphysical assumptions, when we use the conceptions developed in this inquiry; it then is easy to give an answer to the questions raised.

In their paper Einstein, Podolsky, and Rosen construct a special kind of physical systems which may be called *correlated systems*. These are given by systems which for some time have been in physical interaction, but are separated later. They then remain correlated in such a way that the measurement of an entity u in one system determines the value of an entity v in the other system, although the latter system is not physically influenced by the act of measurement. The authors believe that this fact proves an independent reality of the entity u considered. This result appears even more plausible by a proof, given in the paper, stating that an equal correlation holds for entities other than u in the same systems, including noncommutative entities.

Translating these statements into our terminology, we can interpret the thesis of the paper as meaning that by means of correlated systems a proof can be given for the necessity of Definition 1, §25, which states that the measured value holds before and after the measurement. If we refuse to consider the measured value as holding before the measurement, such as does the Bohr-Heisenberg interpretation with Definition 4, §29, we are led to causal anomalies, since, then, a measurement in one system would physically produce the value of an entity in another system which is physically not affected by the measuring operations. This is what is claimed to be proved in the paper.

In order to analyze this argument, let us first consider the mathematical form in which it is presented. Let us assume two particles which for some time enter into an interaction; their ψ-function will then be a function $\psi(q_1 \ldots q_6)$ of six coordinates, which include the 3-position coordinates of each particle. When, after the interaction, the particles separate, the ψ-function will be given by a product of ψ-functions of the individual particles (cf. (12), §27). Let us expand the individual ψ-functions in eigen-functions φ_i of the same entity u; we then have

$$\psi(q_1 \ldots q_6) = \sum_i \sum_k \sigma_{ik} \varphi_i(q_1, q_2, q_3)\, \varphi_k(q_4, q_5, q_6) \tag{1}$$

Here the σ_{ik} determine the probability $d(u_i, u_k)$ that the value u_i is measured in the first system *and* the value u_k is measured in the second system:

$$d(u_i, u_k) = |\sigma_{ik}|^2 \tag{2}$$

We have, of course,

$$\sum_i \sum_k |\sigma_{ik}|^2 = 1 \tag{3}$$

Now let us assume that a measurement of u is made in the first system, and furnishes the value u_1. The subscript 1 is not meant here to denote the first or "lowest" eigen-value, but the value obtained in the measurement. We then shall have a new ψ-function such that

$$\sum_k |\sigma_{ik}|^2 = \begin{cases} 1 & \text{for} \quad i = 1 \\ 0 & \text{for} \quad i \neq 1 \end{cases} \tag{4}$$

Since the second system is not involved in the measurement, its part in (1) remains unchanged; therefore the new ψ-function results from (1) simply by canceling all terms possessing a σ_{ik} with $i \neq 1$. Thus, the new ψ-function will have the form

$$\psi(q_1 \dots q_6) = \sum_k \sigma_{1k} \varphi_1(q_1, q_2, q_3)\, \varphi_k(q_4, q_5, q_6)$$
$$= \varphi_1(q_1, q_2, q_3) \cdot \sum_k \sigma_{1k} \varphi_k(q_4, q_5, q_6) \tag{5}$$

with

$$\sum_k |\sigma_{1k}|^2 = \sum_k d(u_1, u_k) = d(u_1) = 1 \tag{6}$$

Let us put

$$\tau_{12} \chi_2(q_4, q_5, q_6) = \sum_k \sigma_{1k} \varphi_k(q_4, q_5, q_6) \tag{7}$$

Then (5) assumes the form

$$\psi(q_1 \dots q_6) = \tau_{12} \varphi_1(q_1, q_2, q_3)\, \chi_2(q_4, q_5, q_6) \tag{8}$$

This introduction of a new ψ-function is sometimes called the *reduction of the wave packet*. Now the physical conditions of the systems, and of the measurement of u in the first system, can be so chosen that $\chi_2(q_4, q_5, q_6)$ represents an eigen-function of an entity v. Then (8) represents a situation which is determined both in u and v. This means that the situation depicted by (8) corresponds to a situation which would result from measurements of both u and v, although only a measurement of u has been made. We therefore know: If we were to measure v in the second system, we would obtain the value v_2.

We can simplify the consideration by choosing the entity v identical with the entity u. It can be proved that this is physically possible. This means that it is possible to construct physical conditions such that, after a measurement of u in the first system, a ψ-function results for which all coefficients σ_{ik} of (1) vanish except for the value σ_{12}. We then have

$$\psi(q_1 \dots q_6) = \sigma_{12} \varphi_1(q_1, q_2, q_3)\, \varphi_2(q_4, q_5, q_6) \qquad |\sigma_{12}|^2 = 1 \tag{9}$$

Here the measurement of u in the first system, resulting in u_1, has made the second system definite in u, for the value u_2; i.e., if we were to measure u in the second system we would obtain the value u_2.

We now see the way the conclusion of Einstein, Podolsky, and Rosen is introduced: We must assume that the value u_2 exists in the second system *before* a measurement of u is made in this system; otherwise we are led to the consequence that a measurement of u in the first system *produces*, not only the value u_1 in the first system, but also the value u_2 in the second system. This would represent a causal anomaly, an action at a distance, since the measurement in the first system does not physically affect the second system.

With this inference the main thesis of the paper is derived. It then goes on to show that similar results can be obtained for an entity w which is complementary to u. Let ω be the eigen-functions of w; then it is possible to make, instead of a measurement of u, a measurement of w in the first system which results in the production of an eigen-function

$$\psi(q_1 \ldots q_6) = \rho_{12} \cdot \omega_1(q_1, q_2, q_3) \cdot \omega_2(q_4, q_5, q_6) \tag{10}$$

We therefore have a free choice, either to measure u in the first system and thus to make the second system definite in u, or to measure w in the first system and to make the second system definite in w. This is considered as further evidence for the assumption that the value of an entity must exist before the measurement.

In his reply to the Einstein-Podolsky-Rosen paper, Nils Bohr sets forth the opinion that an assumption of this kind is illegitimate. Within this exposition he gives a physical interpretation of the formalism developed for correlated systems, i.e., of the above formulae (1)–(10). Let us consider this illustration before we turn to a logical analysis of the problem.

Bohr assumes that the systems considered consist of two particles, each of which passes through a slit in the same diaphragm. If we include measurements concerning the diaphragm, he continues, it is possible to determine momentum or position of the second particle by measurements of the first particle *after* the particles have passed through the slits. For the determination of the momentum of the second particle, we would have to measure:

(1) the momentum of each particle *before* the particles hit the diaphragm

(2) the momentum of the diaphragm *before* the particles hit

(3) the momentum of the diaphragm *after* the particles hit

(4) the momentum of the first particle *after* the particle hit the diaphragm.

The momentum of the second particle is then determined by subtracting the change in the momentum of the first particle from the change in the momentum of the diaphragm, and adding this result to the initial momentum of the second particle.

In order to determine the position of the second particle, we would measure:

(1) the distance between the slits in the diaphragm

(2) the position of the first particle immediately after passing the slit.

From the second result we here would infer the position of the diaphragm, which is determined because the position of the particle tells us the position of the slit through which it went (we consider here the position of the plane of the diaphragm as known, and allow only a shifting of the diaphragm within its plane, caused by the impact of the particles). Since the distance between the slits is known, the position of the second slit and with this the position of the second particle in passing, or immediately after passing the slit, is determined.

It is interesting to see that Nils Bohr in these derivations uses the very definition which Einstein, Podolsky, and Rosen want to prove as necessary, namely, our Definition 1, §25. This definition, although not assumed for the measurements 1 and 2 of our first list, is assumed for 3 and 4, and likewise for the measurement 2 of our second list. Otherwise, for instance, the difference between the measurements 2 and 3 of the momentum of the diaphragm could not be interpreted as equal to the amount of momentum which the diaphragm has received through the impacts of both particles. If the measurement 3 changes the momentum of the diaphragm, the latter inference could not be made. Bohr does not mention his use of a definition which considers the measured value as holding before the measurement.[4] Fortunately, the use of this definition does not make Bohr's argument contradictory, as can be seen when we incorporate his answer into an analysis given in terms of our conception.

Using this conception, we shall answer the criticism of Einstein, Podolsky, and Rosen in a way different from Bohr's consideration. We shall not maintain that the use of Definition 1, §25, is *impermissible*; we shall say, instead, that this definition is *not necessary*. It may be used; thereby the corpuscle interpretation is introduced, and in the case of correlated systems of the kind considered it is this interpretation which is free from causal anomalies. Thus, even Bohr uses this exhaustive interpretation which makes his inferences plausible; he follows here the well-established habit of the physicist of switching over to the interpretation which is free from anomalies. What can be derived in such an interpretation must hold for all interpretations; this is the principle at the base of Bohr's inferences.

It would be wrong, however, to infer that because Definition 1, §25, furnishes here an interpretation free from anomalies, this definition *must* be chosen. We should be glad if we could identify this view of the problem, resulting from our inquiries, with Bohr's opinions. The latter do not seem to us to be stated sufficiently clearly to admit of an unambiguous interpretation; in particular, we should prefer to disregard Bohr's ideas about the arbitrariness of the separation into subject and object, which do not appear to us relevant for the logical problems of quantum mechanics. Let us therefore continue the analysis by the use of our own notation, applying the three-valued logic developed in §32.

What is proved by the existence of correlated systems is that it is not permissible to say that the value of an entity before the measurement is different from the value resulting in the measurement. Such a statement would lead to causal anomalies, since it would involve the consequence that a measurement on one system produces the value of an entity in a system which is in no physical interaction with the measuring operation. In the paper of Einstein, Podolsky, and Rosen, the inference is now made that we must say that the entity before the measurement has the same value as that found in the measurement. It is this inference which is invalid.

The inference under consideration would hold only in a two-valued logic; in a three-valued logic, however, it cannot be made. Let us denote by A the statement "the value of the entity before the measurement is different from the value resulting in the measurement"; then what the existence of correlated systems proves is that, if causal anomalies are to be avoided, the statement $\bar{A}$ must hold. This statement $\bar{A}$, which states that A is not true, does not mean, however, that A is false; A can also be indeterminate. The statement, "the value of the entity before the measurement is equal to the value resulting in the measurement", is to be denoted by $-A$, i.e., by means of the diametrical negation, since this statement is true when A is false. If we could infer from the existence of correlated systems the statement $-A$, the restrictive interpretation would indeed be shown to be contradictory. But this is not the case; all that can be inferred is $\bar{A}$, and such a statement is compatible with the restrictive interpretation, since it leaves open the possibility that A is indeterminate.

We see that the paper of Einstein, Podolsky, and Rosen leads to an important clarification of the nature of restrictive interpretations. It is not permissible to understand the restrictive Definition 4, § 29, as meaning that the value of the entity before the measurement is *different* from the results of the measurement; such a statement leads to the same difficulties as a statement about this value being *equal* to the result of the measurement. Every statement determining the value of the entity before the measurement will lead to causal anomalies, though these anomalies will appear in different places according as the determination of the value before the measurement is given. The anomalies appearing if equality of the values is stated are described in the interference experiment of § 7;[5] the anomalies resulting if difference of the values is stated are given in the case of correlated systems.

The given considerations constitute an instructive example for the nature of interpretations. They show the working of an exhaustive interpretation, and make it clear that restrictive interpretations are introduced for the purpose of avoiding causal anomalies; they prove, on the other hand, that the restrictive interpretation is consistent if all statements involved are dealt with by the rules of a three-valued logic.

Within the frame of the restrictive interpretation it is even possible to express the condition of correlation holding between the two systems after their interaction has been terminated. Using the functor $Vl()$ introduced on p. 73 and indicating the system I or II within the parentheses of this symbol we can write:

$$(u)\{[Vl(e_1, I)=u]\equiv[Vl(e_1, II)=f(u)]\} \qquad (11)$$

where the function f is known. Similarly we can write for a noncommutative entity v:

$$(v)\{[Vl(e_2, I)=v]\equiv[Vl(e_2, II)=g(v)]\} \qquad (12)$$

where the function g is known. Both relations hold so long as no measurement is made in one of the systems. After a measurement has been made in one of them only that relation continues to hold which concerns the measured entity. For instance, if u has been measured in system I, only (11) continues to hold.

It would be a mistake to infer from (11) or (12) that there exists a determinate value of u, or v, in one of the systems so long as no measurement is made. This would mean that the expressions in the brackets must be true or false; but the equivalence will hold also if these expressions are indeterminate. In (11) and (12) we thus have a means of expressing the correlations of the systems without stating that determinate values of the respective entities exist.[6] This represents an advantage of the interpretation by a three-valued logic over the Bohr-Heisenberg interpretation. For the latter interpretation the statements (11)–(12) would be meaningless. Only the three-valued logic gives us the means to state the correlation of the systems as a condition holding even before a measurement is made, and thus to eliminate all causal anomalies. We need not say that the measurement of u in system I produces the value of u in the distant system II. The predictability of the value u_2 in the system II, after u has been measured in the system I, appears as a consequence of condition (11), which in its turn is a consequence of the common history of the systems.

§37. Conclusion

We can summarize the results of our inquiry as follows. The relation of indeterminacy is a fundamental physical law; it holds for all possible physical situations and therefore involves a disturbance of the object by the measurement. Since the relation of indeterminacy makes it impossible to verify statements about the simultaneous values of complementary entities, such statements can be introduced only by means of definitions. The physical world therefore subdivides into the world of phenomena, which are inferable from observations in a rather simple way and therefore can be called observable in a wider sense; and the world of interphenomena, which can be introduced only by an interpolation based on definitions. It turns out that such a supplementation of the world of phenomena cannot be constructed free from anomalies. This result is not a consequence of the principle of indeterminacy; it must be considered as a second fundamental law of the physical world, which we call the principle of anomaly. Both these principles are derivable from the basic principles of quantum mechanics.

Instead of speaking of the structure of the physical world, we may consider the structure of the languages in which this world can be described; such analysis expresses the structure of the world indirectly, but in a more precise way. We then distinguish between observational language and quantum mechanical language. The first shows practically no anomalies with the exception of unverifiable implications occurring in some places. Quantum mechanical language can be formulated in different versions; we use in particular three versions: the corpuscle language, the wave language, and a neutral language. All three of these languages concern phenomena and interphenomena, but each of them shows a characteristic deficiency. Both the corpuscle language and the wave language show a deficiency so far as they include statements of causal anomalies, which occur in places not corresponding to each other and therefore can be transformed away, for every physical problem, by choosing the suitable one of the two languages. The neutral language is neither a corpuscle language nor a wave language, and thus does not include statements expressing causal anomalies. The deficiency reappears here, however, through the fact that the neutral language is three-valued; statements about interphenomena obtain the truth-value *indeterminate.*

The stated deficiencies are not due to an inappropriate choice of these languages; on the contrary, these three languages represent optima with respect to the class of all possible languages of quantum mechanics. The deficiencies must rather be regarded as the linguistic expression of the structure of the atomic world, which thus is recognized as intrinsically different from the macro-world, and likewise from the atomic world which classical physics had imagined.

NOTES

§29

[1] A means of indicating the transition from object language to metalanguage is in the use of quotes; similarly, italics can be used. We shall use, for our presentation of logic, italics instead of quotes for symbols denoting propositions, in combination with the rule that symbols of operations standing between symbols of propositions are understood to apply to the propositions, not to their names (i.e., autonomous usage of operations, in the terminology of R. Carnap). Thus, we shall write "*a* is true", not "'a' is true", and "*a.b* is true", not "'a.b' is true". All formulae given by us are therefore, strictly speaking, not formulae of the object language, but descriptions of such formulae. For most practical purposes, however, it is permissible to forget about this distinction.

[2] This distinction has been carried through by C. W. Morris, 'Foundations of the Theory of Signs', *International Encyclopedia of United Science*, Vol. I, No. 2 (Chicago, 1938).

[3] More precisely: "immediately after the time for which the truth value of the statement

m_u is considered". If m_u is true, this means the same as "immediately after the measurement"; and if m_u is false, we thus equally coordinate to U a time at which its stipulated truth value holds. (The expression "after the measurement" then would be inapplicable.)

[4] M. Strauss, 'Zur Begründung der statistischen Transformationstheorie der Quantenphysik', *Ber. d. Berliner Akad., Phys.-Math. Kl.* **27** (1936), and 'Formal Problems of Probability Theory in the Light of Quantum Mechanics', *Unity of Science Forum, Synthese* (The Hague, Holland, 1938), p. 35; (1939), pp. 49, 65. In these writings Strauss also develops a form of probability theory in which my rule of existence for probabilities is changed with respect to complementary statements. Such a change is necessary, however, only if an incomplete notation is used, such as is done by us in the beginning of §22. In a notation which uses the term m_u in the first place of probability expressions, such as introduced in (13), §22, a change of the rule of existence can be avoided.

[5] Mr. Strauss informs me that he is planning to publish a new and somewhat modified presentation of his conceptions.

§30

[1] As to the use of the functor, "the value of the entity," cf. fn. 4 of §32 (see below – ed.)

[2] E. L. Post, 'Introduction to a General Theory of Elementary Propositions', *Am. Journ. of Math.* **43** (1921), 163.

[3] J. Lucasiewicz, *Comptes rendus Soc. d. Sciences Varsovie*, **23** (1930), Cl. III, p. 51; J. Lucasiewicz and A. Tarski, *op. cit.*, p. 1. The first publication by Lucasiewicz of his ideas was made in the Polish journal *Ruch Filozoficzny*, **5** (Lwow, 1920), pp. 169–170.

[4] H. Reichenbach, 'Wahrscheinlichkeitslogik', *Ber. d. Preuss. Akad., Phys.-Math. Kl.* (Berlin, 1932).

§32

[1] Most of these operations have been defined by Post, with the exception of the complete negation, the alternative implication, the quasi implication, and the alternative equivalence, which we introduce here for quantum mechanical purposes. Post defines some further implications which we do not use. Our standard implication is Post's implication $\supset^{\mu}_{m}$ with $m=3$ and $\mu=1$, i.e., for a three-valued logic and $t_\mu = t_1 =$ truth.

[2] This result could not be derived if we were to use the standard implication, instead of the alternative implication, in (28) and (32).

[3] We simplify this example. A complete notation would require the use of propositional functions.

[4] The use of functors in three-valued logic differs from that in two-valued logic in that the existence of a determinate value designated by the functor can be asserted only when the statement (36) is true or false, whereas the indeterminacy of (36) includes the indeterminacy of a statement about the existence of a value. We shall not give here the formalization of this rule.

[5] I.e., the operation which divides these formulae into two major parts.

[6] It can be shown that the true-false character of (28) and (32) is not bound to the particular form which we gave to the arrow implication, but ensues if the following postulates are introduced: one, the relation of complementarity is symmetrical in A and B, i.e., (28) is equivalent to (32); two, if A in (28) is indeterminate, B can have any one of the three truth values; three, the implication used in (28) is verified if both implicans and implicate are true, and is falsified if the implicans is true and the implicate is false. We shall only indicate the proof here. Postulate two requires that the arrow implication have a T in the three cases where A is indeterminate; postulate three requires a T in the first value of the column of the arrow implication, and an F in the third. It turns out that with this

result, seven of the nine cases of (28) are determined, and contain only *T*'s and *F*'s. The missing case *T*, *F* of (28) then must be equal to the case *F*, *T*, according to the first postulate, and is thus determined as *F*. It then can be shown that in order to furnish this result, the arrow implication must have an *F* in the second value from the top. With this, the last case of (28), the *F*, *F*-case is determined as *F*. The arrow implication is not fully determined by the given postulates; its last three values can be arbitrarily chosen.

§33

[1] As before, the true-false character of these formulae is not introduced by us deliberately, but results from other reasons. If we were to use in (8) a double standard implication, this would represent a standard equivalence according to (21), §32; then the only case in which the disjunction (9) is indeterminate is the case that all B_i are indeterminate. But we need also cases in which some B_i are false and the others indeterminate, for the reasons explained above.

[2] Dr. A. Tarski, to whom I communicated these results, has drawn my attention to the fact that it is possible also to define a similar generalization of the inclusive "or". We then replace in the column of the disjunction in Table IVB the "I" of the middle row by a "T", leaving all other cases unchanged. This "almost or", as it may be called, thus means that at least one of two propositions is true or both are indeterminate. This operation can be shown to be commutative and associative, while it is not distributive or reflexive (the latter term meaning that "almost A or A" is not the same as "A"). Applied to more than two propositions, the "almost or" means: "At least one proposition is true or at least two are indeterminate". A disjunction in terms of the "almost or", therefore, represents what we called above a closed disjunction. It can be shown that if relations (8) hold, the $B_1 \ldots B_n$ will constitute a disjunction of this kind. The latter statement, of course, is not equivalent to the relations (8), but merely a consequence of the latter.

[3] Cf. the author's *Wahrscheinlichkeitslehre* (Leiden, 1935), (4), §22.

[4] It would be more correct to speak here of an existential statement instead of a disjunction of an infinite number of terms. It is clear, however, that the considerations given can be equally carried through for existential statements.

§34

[1] A complete definition of nomological implication will be given in a later publication by the author.

[2] The quasi implication of the latter table is identical with an operation which has been introduced by the author, by the use of the same symbol, in the frame of probability logic (cf. *Wahrscheinlichkeitslehre* (Leiden, 1935), p. 381, table IIc). It can be considered as the limiting case of a probability implication resulting when only the probabilities 1 and 0 can be assumed. It can also be considered as the individual operation coordinated to probability implication as a general operation; the probability then is determined by counting only the *T*-cases and *F*-cases of the quasi implication, the *I*-cases being omitted. In this sense the quasi implication has been used, under the name of comma-operation, or operation of selection, in my paper 'Ueber die semantische und die Objectauffassung von Wahrscheinlichkeitsausdrücken', *Journ. of Unified Science, Erkenntnis* **8** (1939), 61–62.

§36

[1] A. Einstein, N. Podolsky, N. Rosen, 'Can Quantum Mechanical Description of Physical Reality Be Considered Complete?', *Phys. Rev.* **47** (1935), 777.

[2] N. Bohr, 'Can Quantum Mechanical Description of Physical Reality Be Considered Complete?', *Phys. Rev.* **48** (1935), 696.

[3] E. Schrödinger, 'Die gegenwärtige Situation in der Quantenmechanik', *Naturwissenschaften* **23** (1935), 807, 823, 844. We shall use the term "correlated systems" as a translation of Schrödinger's "verschränkte Systeme".

[4] Once this definition is chosen, the correlated systems can even be used for a measurement of simultaneous values of noncommutative entities. We then measure u in the first system, and w in the second; then the obtained values u_i and w_k represent simultaneous values in the second system. But the possibility of measuring such simultaneous values has been pointed out already in §25 as being a consequence of Definition 1, §25. If only Definition 4, §29, is used, the two measurements on correlated systems would not furnish simultaneous values.

[5] This is to be understood as follows. When we put a Geiger counter at the place of each of the two slits B_1 and B_2 we shall always locate the particle either in one or in the other of these two counters. (This measurement, of course, disturbs the interference pattern on the screen.) The assumption that the particle was at that place before it hits the counter leads to the consequence that the particle would have been there also when no measurement was made. This result implies that, when no observation at the slits is made, the particle will go either through one or the other slit. We showed in §7 that this assumption leads to causal anomalies.

[6] Cf. fn. 4 of §32 (see above, ed.).

HILARY PUTNAM

THREE-VALUED LOGIC

Let us make up a logic in which there are three truth-values, T, F, and "M," instead of the two truth-values T and F. And, instead of the usual rules, let us adopt the following:

(a) If either component in a disjunction is true ("T"), the disjunction is true; if both components are false, the disjunction is false ("F"); and in all other cases (both components middle, or one component middle and one false) the disjunction is middle ("M").

(b) If either component in a conjunction is false ("F"), the conjunction is false; if both components are true, the conjunction is true ("T"); and in all other cases (both components middle, or one component middle and one true) the conjunction is middle ("M").

(c) A conditional with true antecedent has the same truth-value as its consequent; one with false consequent has the same truth-value as the denial of its antecedent; one with true consequent or false antecedent is true; and one with both components middle ("M") is true.

(d) The denial of a true statement is false; of a false one true; of a middle one middle.

These rules are consistent with all the usual rules with respect to the values T and F. But someone who accepts three truth values, and who accepts a notion of tautology based on a system of truth-rules like that just outlined, will end up with a different stock of tautologies than some one who reckons with just two truth values.

Many philosophers will, however, want to ask: *what could the interpretation of a third truth-value possibly be*? The aim of this paper will be to investigate this question. It will be argued that the words "true" and "false" have a certain "core" meaning which is *independent* of *tertium non datur*, and which is capable of precise delineation.

I

To begin with, let us suppose that the word 'true' retains at least this

C. A. Hooker (ed.), The Logico-Algebraic Approach to Quantum Mechanics, 99–107.

much of its usual force: if one ever says of a (tenseless) statement that it is true, then one is committed to saying that it was always true and will always be true in the future. For example, if I say that the statement 'Columbus crosses[1] the ocean blue in fourteen hundred and ninety-two' is true, then I am committed to the view that it was true, e.g., in 1300, and will be true in 5000 (A.D.). Thus 'true' cannot be identified with *verified*, for a statement may be verified at one time and not at another. But if a statement is *ever* accepted as verified, then at that time it must be said to have been true also at times when it was not verified.

Similarly with 'false' and 'middle'; we will suppose that if a statement is ever called 'false,' then it is also said never to have been true or middle; and if a statement is ever said to be middle, it will be asserted that it was middle even at times when it may have been incorrectly called 'true' or 'false.' In other words, we suppose that 'true' and 'false' have, as they ordinarily do have, a *tenseless* character; and that 'middle' shares this characteristic with the usual truth-values.

This still does not tell one the "cash value" of calling a statement 'middle.' But it does determine a portion of the syntax of 'middle,' as well as telling one that the words 'true' and 'false' retain a certain specified part of *their* usual syntax. To give these words more content, we may suppose also that, as is usually the case, statements that are accepted[2] as verified are called 'true,' and statements that are rejected, that is whose denials are accepted, are called 'false.' This does not determine that any particular statements must be called 'middle'; and, indeed, someone could maintain that there are some statements which have the truth-value middle, or some statements which could have the truth-value middle, without ever specifying that any particular statement has this truth-value. But certain limitations have now been imposed on the use of the word 'middle.'

In particular, statements I call 'middle' must be ones I do not accept or reject at the present time. However, it is not the case that 'middle' *means* "neither verified nor falsified at the present time." As we have seen, 'verified' and 'falsified' are *epistemic* predicates – that is to say, they are relative to the *evidence* at a particular time – whereas 'middle,' like 'true' and 'false' is not relative to the evidence. It makes sense to say that 'Columbus crosses the ocean blue in fourteen hundred and ninety-two' was verified in 1600 and not verified in 1300, but not that it was true in 1600 and false in 1300.

Thus 'middle' cannot be *defined* in terms of 'verified,' 'falsified,' etc. What difference does it make, then, if we say that some statements – in particular some statements not now known to be true or known to be false – may not be either true or false because they are, in fact, middle? The effect is simply this: that one will, as remarked above, end up with a different stock of tautologies than the usual.

Someone who accepts the three-valued logic we have just described will accept a disjunction when he accepts either component, and he will reject if when he rejects both components. Similarly, he will accept a conjunction when he accepts both components, and he will reject it when he rejects either component. This is to say that the behavior of the man who uses the particular three-valued logic we have outlined is not distinguishable from the behavior of the man who uses the classical two-valued logic in cases wherein they know the truth or falsity of all the components of the particular molecular sentences they are considering.

However, they will behave differently when they deal with molecular sentences some of whose components have an unknown truth-value.[3] If it is known that snow is white, then the sentence 'snow is white $\vee \sim$ snow is white' will be *accepted* whether one uses classical two-valued logic or the particular three-valued logic we have described. But if one does not know whether or not there are mountains on the other side of the moon, then one will *accept* the sentence 'there are mountains on the other side of the moon $\vee \sim$ there are mountains on the other side of the moon' if one uses the classical two-valued logic, but one will say 'I don't know whether that's true or not' if one uses three-valued logic, or certain other nonstandard logics, e.g., "Intuitionist" logic.[4]

II

At this point the objection may be raised: "But then does this notion of a 'middle' truth-value make sense? If having a middle truth-value does not mean having what is ordinarily called an *unknown* truth value; if, indeed, you can't tell us *what* it does mean, then does it make sense at all?"

Analytic philosophers today normally reject the demand that concepts be translatable into some kind of "basic" vocabulary in order to be meaningful. Yet philosophers often reject the possibility of a three-valued logic (except, of course, as a mere formal scheme, devoid of interesting interpretations), just on the ground that no satisfactory *translation* can be offered for the notion of having a "middle" truth-value. Indeed, if the notion of being a statement with a middle truth-value is defined explicitly in terms of a two-valued logic or metalogic, then one usually obtains a *trivial* interpretation of three-valued logic.

Does a middle truth-value, within the context of a system of three-valued logic of the kind we have described, have a use? The answer is that it does, or rather that it belongs to a *system* of uses. In other words, to use three-valued logic makes sense in the following way: to use a three-valued logic means to adopt a different way of using logical words. More exactly, it corresponds to the *ordinary* way in the case of molecular sentences in which the truth-value of all the components is known (i.e., we "two-valued" speakers say it is known); but a man reveals that he is using three-valued logic and not the ordinary two-valued logic (or partially reveals this) by the way he handles sentences which contain components whose truth-value is not known.

There is one way of using logical words which constitutes the ordinary two-valued logic. If we are using three-valued logic,[5] we will behave in exactly the same way except that we will employ the three-value rules and the three-valued definition of 'tautology.' Thus 'using three-valued logic' means adopting a systematic way of using the logical words which agrees in certain respects with the usual way of using them, but which also disagrees in certain cases, in particular the cases in which truth-values are unknown.

III

Of course one might say: "Granted that there is a consistent and complete way of using logical words that might be described as 'employing a three-valued logic.' But this alternative way of using logical words – alternative to the usual way – doesn't have any *point*."

And perhaps this is what is meant when it is said that three-valued logic does not constitute a real alternative to the standard variety: it exists as a calculus, and perhaps as a nonstandard way of using logical words, but there is no *point* to this use. This objection, however, cannot impress anyone who recalls the manner in which non-Euclidean geometries were first regarded as absurd; later as mere mathematical games;

and are today accepted as portions of fully interpreted physical hypotheses. In exactly the same way, three-valued logic and other nonstandard logics had first to be shown to exist as consistent formal structures; secondly, uses have been found for some of them – it is clear that the Intuitionist school in mathematics, for example, is, in fact, systematically using logical words in a nonstandard way, and it has just been pointed out here that one might use logical words in still another nonstandard way, corresponding to three-valued logic (that is, that this would be a form of linguistic behavior reasonably represented by the formal structure called 'three-valued logic'). The only remaining question is whether one can describe a physical situation in which this use of logical words would have a point.

Such a physical situation (in the microcosm) has indeed been described by Reichenbach.[6] And we can imagine worlds such that even in *macrocosmic* experience it would be physically impossible to either verify or falsify certain empirical statements. For example, if we have verified (by using a speedometer) that the velocity of a motor car is such and such, it might be impossible in such a world to verify or falsify certain statements concerning its position at that moment. If we *know* by reference to a physical law together with certain observational data that a statement as to the position of a motor car can *never* be falsified or verified, then there may be some point to not regarding the statement as true or false, but regarding it as "middle." It is only because, in macrocosmic experience, everything that we regard as an empirically meaningful statement seems to be at least potentially verifiable or falsifiable that we prefer the convention according to which we say that every such statement is either true or false, but in many cases we don't know which.

Moreover, as Reichenbach shows, adopting a three-valued logic permits one to preserve both the laws of quantum mechanics and the principle that no causal signal travels with infinite speed – "no action at a distance." On the other hand, the laws of quantum mechanics are logically incompatible with this principle if ordinary two-valued logic is used.[7] This inconsistency is not usually noticed, because in quantum mechanics no causal signal is ever *detected* traveling faster than light. Nevertheless it can be shown – as Einstein and others have also remarked[8] – that the mathematics of quantum mechanics entails that in certain situations a causal signal *must have* traveled faster than light.

A working physicist can dismiss this as "just an anomaly" – and go on to accept *both* quantum mechanics and the "no action" principle. But a logician cannot have so cheerful an attitude toward logical inconsistency. And the suggestion advanced by Bohr, that one should classify the troublemaking sentences as "meaningless" (complementarity) involves its own complications. Thus the suggestion of using a three-valued logic makes sense in this case, as a move in the direction of simplifying a whole system of laws.

To return to the macrocosmic case (i.e., the "speedometer" example), Bohr's proposal amounts to saying that a syntactically well-formed sentence (e.g., 'my car is between 30 and 31 miles from New York') is in certain cases *meaningless* (depending on whether or not one uses a speedometer). Reichenbach's suggestion amounts to saying that it is meaningful, but neither true nor false (hence, "middle"). There seems little doubt that it would be simpler in practice to adopt Reichenbach's suggestion. And I suspect that beings living in a world of the kind we have been describing would, in fact, regard such statements as *neither true nor false*, even if no consideration of preserving simple physical laws ("no action at a distance") happened to be involved. This "suspicion" is based on two considerations: (a) The sentences admittedly have a very clear cognitive use; hence it is unnatural to regard them as "meaningless." (b) There is no reason why, in such a world, one should even consider adopting the rule that every statement is either true or false.

On the other hand, in our world (or in any world in which Planck's constant h has a small value) it would be very unnatural to adopt three-valued logic for describing ordinary macrocosmic situations. Suppose we did. Then there would be two possibilities: (a) We maintain that certain sentences are "middle," but we never say which ones. This seems disturbingly "metaphysical." (b) We say that some particular sentence S is middle.

This last course is, however, fraught with danger. For, although "S is middle" does not mean "S will never be either verified or falsified," it *entails* "S will never be either verified or falsified." And the prediction that a particular sentence will never be either verified or falsified is a strong empirical prediction (attention is confined to synthetic sentences for the sake of simplicity); and one that is itself always potentially falsifiable in a world where no physical law prohibits the verification of the sentence S, regardless of what measurements may have antecedently been made.

Thus, the reason that it is safe to use three-valued logic in the Reichenbachian world (the microcosm) but not in the "actual" world (the macrocosm) is simply that in the Reichenbachian world one can, and in the "actual" world one cannot, *know in advance* that a particular sentence will never be verified or falsified. It is not that in a "Reichenbachian" world one *must* call sentences that will never be verified or falsified "middle," but, rather, that in *any* world only (but not necessarily *all*) such sentences must be classified as "middle." This follows from the fact that sentences that are said to be verified are also said to be true; sentences that are said to be falsified are also said to be false; and the truth values are "tenseless." Thus it would be a contradiction to say that a sentence is middle, but may someday be verified.

These features of the use of "true" and "false" seem indeed to be constitutive of the meaning of these words. *Tertium non datur* might also be said to be "true from the meaning of the words 'true' and 'false'" – but it would then have to be added that these words have a certain core meaning that can be preserved even if *tertium non datur* is given up. One can abandon two-valued logic without changing the meaning of 'true' and 'false' *in a silly way*.

IV

Analytic philosophers – both in the "constructivist" camp and in the camp that studies "the ordinary use of words" – are disturbingly unanimous in regarding two-valued logic as having a privileged position: privileged, not just in the sense of corresponding to the way we *do* speak, but in the sense of having no serious rival for *logical* reasons. If the foregoing analysis is correct, this is a prejudice of the same kind as the famous prejudice in favor of a privileged status for Euclidean geometry (a prejudice that survives in the tendency to cite "space has three dimensions" as some kind of "necessary" truth). One can go over from a two-valued to a three-valued logic without *totally* changing the meaning of 'true' and 'false'; and not just in *silly* ways, like the ones usually cited (e.g., equating truth with high probability, falsity with low probability, and middlehood with "in between" probability).

Indeed, so many strange things have been said about two- and three-valued logic by philosophic analysts who are otherwise of the first rank that it would be hopeless to attempt to discuss them all in one short paper. But two of these deserve special mention:

(a) It has often been said that "even if one uses a three-valued object language, one must use two-valued logic in the metalanguage." In the light of the foregoing, this can hardly be regarded as a *necessary* state of affairs. Three-valued logic corresponds to a certain way of speaking; there is no difficulty in speaking in that way about any particular subject matter. In particular, one may assign truth-values to molecular sentences in the way we have discussed, whether one is talking about rabbits or languages or metalanguages.

(Of course, if one is *explaining* three-valued logic to someone who only uses two-valued logic one will employ a two-valued language as a medium of communication. This is like remarking that one uses French to teach Latin to French schoolboys.)

(b) It has been argued[9] that the meaning of 'true' has been made clear by Tarski for the usual two-valued system, but that no analogous clarification is available for 'true' in three-valued systems. The obvious reply is that the famous biconditional *'snow is white' is true if and only if snow is white* is perfectly acceptable even if one uses three-valued logic. Tarski's criterion has as a consequence that one must accept *'snow is white'* is true if one accepts *snow is white* and reject *'snow is white'* is true if one rejects *snow is white*. But these (along with the "tenseless" character of the truth-values) are just the features of the use of 'true' and 'false' that we have preserved in our three-valued logic. It is, for instance, just because *tertium non datur* is independent of these features that it is possible for Intuitionist logicians to abandon it without feeling that they are changing the "meaning" of 'true' and 'false.'

Princeton University

NOTES

[1] 'Crosses' is used here "tenselessly" – i.e., in the sense of "crossed, is crossing, or will cross."

[2] More precisely, S is accepted if and only if 'S is true' is accepted.

[3] The distinction between sentences and statements will be ignored, because we have passed over to consideration of a formalized language in which it is supposed that a given sentence can be used to make only one statement.

[4] Cf. Alonzo Church's *Introduction to Mathematical Logic*, Princeton University Press, Princeton, N.J., 1956, p. 141. Intuitionist logic is not a truth-functional logic (with any finite number of truth-values). However, the rules given above hold (except when both components are "middle" in the case of rules (b) and (c)) provided truth is identified with intuitionist "truth," falsity with "absurdity," and middlehood with being neither "true" nor "absurd."

[5] In this paper, 'three-valued logic' means the system presented at the beginning. Of course, there are other systems, some of which represent a more radical change in our way of speaking.

[6] *Philosophic Foundations of Quantum Mechanics*, University of California, 1944.

[7] *Ibid.*, pp. 29–34.

[8] A. Einstein, B. Podolsky, and N. Rosen, 'Can Quantum Mechanical Description of Reality Be Considered Complete?', *Physical Review* **47** (1935), 777.

[9] For example, Hempel writes in his review of the previously cited work by Reichenbach (*Journal of Symbolic Logic* **10**, 99): "But the truth-table provides a (semantical) interpretation only because the concept of *truth* and *falsity*, in terms of which it is formulated, are already *understood*: they *have the customary meaning* which can be stated in complete precision by means of the *semantical definition of truth*." (Italics mine.)

PAUL FEYERABEND

REICHENBACH'S INTERPRETATION OF QUANTUM-MECHANICS

In Section III of his paper H. Putnam [1] deals with Reichenbach's attempt to interpret quantum mechanics on the basis of a three-valued logic, and he uses some arguments of his own in order to show that this attempt is "a move in the direction of simplifying the whole system of laws" (104). I believe that the Reichenbach-Putnam procedure cannot be defended and that it leads to undesirable consequences. These are the reasons for my belief:

I. THREE-VALUED LOGIC AND CONTACT-ACTION

Putnam asserts (a) that "the laws of quantum mechanics... are logically incompatible with" the principle of contact-action "if ordinary two-valued logic is used"; and (b) that "adopting a three-valued logic permits one to preserve both the laws of quantum mechanics and the principle that no causal signal travels with infinite speed" (104). Assuming for a moment that (a) and (b) give a correct statement of Reichenbach's position (which they do not – see Section III) and that (a) is true we can at once say that adopting the procedure suggested in (b) would violate one of the most fundamental principles of scientific methodology, namely, the principle to take refutations seriously. The statement that there is no velocity greater than the velocity of light is a well-corroborated statement of physics. If, as is asserted in (a), quantum mechanics implies the negation of that statement, we should consider it as refuted and look for a better theory. This is what has in fact happened. Ever since the invention of elementary quantum mechanics (which is not relativistically invariant) physicists have tried to design a *two-valued* relativistic theory. These attempts, although by no means completely successful, have yet led to some promising results such as, for example, Dirac's theory of the electron and the prediction of the existence of the positron.

Now consider the alternative suggested in (b). This alternative removes the need to modify either quantum mechanics or the principle of contact-

C. A. Hooker (ed.), The Logico-Algebraic Approach to Quantum Mechanics, 109–121.

action as it devises a language in which the statement that both are incompatible cannot be asserted. It thereby presents a defective theory (quantum mechanics) in such a way that its defects (it is not Lorentz-invariant) do not become apparent and that no need is felt to look for a better theory.

It is evident that this sly procedure is only one (the most "modern" one) of the many devices which have been invented for the purpose of saving an incorrect theory in the face of refuting evidence and that, consistently applied, it must lead to the arrest of scientific progress and to stagnation. In a private communication H. Putnam has evoked the example of non-Euclidian geometry as a case where it was suggested to change the formal structure of physical theories and he has said that it is analogous to the present case. But this comparison is altogether misleading. The application of non-Euclidean geometry to physics led to fruitful new theories; it suggested new experiments and enabled physicists to explain phenomena which so far had defied any attempt at explanation (the advance of the perihelion of Mercury is one of them). Nothing of that kind results from the application of a three-valued logic to quantum mechanics. On the contrary, important problems (how to relativize elementary quantum mechanics?) are covered up, objectionable theories (elementary quantum mechanics) are preserved, fruitful lines of research (attempts to find a general relativistic theory of micro-objects) are blocked. Hence, no physicist in his senses would adopt procedure (b).

II. EXHAUSTIVE INTERPRETATIONS AND THEIR ANOMALIES

Reichenbach's main problem is the interpretation of the unobservables of quantum mechanics. In this connection he considers what he calls 'exhaustive interpretations.' At least two nonsynonymous explanations are given for interpretations of that kind. According to the first explanation an exhaustive interpretation is an interpretation which "includes a complete description of interphenomena," i.e., of quantum-mechanical entities (PF 33).[2] An exhaustive interpretation in this sense does not employ any special assumption about the nature of the things to be interpreted. The only conditions to be satisfied are that the interpretation be consistent as well as compatible with the theory used. According to another explanation, an exhaustive interpretation is an interpretation which "attributes definite values to the unobservables" (PA 342; cf. PF 139). An exhaustive interpretation in this more specific sense (silently) employs an assumption (we shall call it assumption C) which may be expressed as follows: (a) Divide the class of all the properties which the entities in question may possess *at some time* into subclasses comprising only those properties which exclude each other. These subclasses will be called the *categories* belonging to the entities in question. Then each entity possesses *always* one property out of each category. (b) The categories to be used are the classical categories.[3] Applied to the case of an electron C asserts that the electron always possesses a well-defined position and a well-defined momentum.

It is evident that an exhaustive interpretation of the first kind (an E_1) i.e., an attempt to state what the nature of quantum-mechanical entities is, need not be an exhaustive interpretation of the second kind (an E_2), i.e., it need not be an attempt to represent quantum-mechanical systems as things which always possess some property out of each classical category relevant to them. An E_1 need not even comply with assumption Ca: It is not the case that water has always a well-defined surface tension (it possesses a surface tension only if it is in its fluid state); nor is it the case that it has always a well-defined value on the Mohs-scale (it possesses such a value only if it is in its solid state). Yet one can explain what kind of entity water is.

Reichenbach shows that all E_2 lead to causal anomalies.[4] Those anomalies are not unusual *physical processes* although Reichenbach's wording sometimes suggests that they are.[5] Assume, for example, that we try to interpret the behavior of electrons by waves. As soon as an electron is localized (at P) the wave collapses (into a narrow bundle around P). This sudden collapse cannot be understood on the basis of the wave equation which means that electrons are not (classical) waves.

On the other hand, consider the particle picture. If we want to explain interference (in the two-slit experiment) on the basis of the particle picture we must assume that the particle can 'know' what happens at distant points (cf. the discussion in 7 of PF). This 'knowledge' cannot be provided by any physical means (e.g., by a signal traveling with infinite speed) since (a) there is no independent evidence of the existence of such signals (hence the hypothesis that they exist would be an *ad hoc* hypothesis) and since (b) in the case of the existence of such signals the wave picture (which does not assume their existence) would lead to incorrect results even in those situations where it has been found to be correct. One may, of course, say that the wave picture provides us with a *description* of the dependences, existing between the state of a particle and some distant event (such as the event 'opening of the second slit') – but this amounts to saying that the particle picture is incorrect. Result: the so-called 'anomalies' are nothing but facts which show that quantum mechanics, interpreted in accordance with an E_2, leads to incorrect predictions. And the 'principle of anomaly' (in its second interpretation; see note 5) must be read as saying *that for any theory which consists of the mathematical formalism of quantum mechanics together with some E_2 there exist refuting instances.*

Reichenbach discusses four methods to solve this difficulty: *Method* 1 suggests that we should "become accustomed" to the anomalies (PF 37), i.e., it expects us not to be worried by the fact that the interpretation used turns quantum mechanics into a false theory. *Method* 2 advises us to use a certain interpretation only for describing those parts of the world where it works and to switch over to another interpretation as soon as a difficulty arises. The only difference between this method and method 1 is that the former uses alternatively two or more anomalous interpretations where the latter uses only one. It also leads to a renunciation of the idea that nature is uniform in the sense that the same laws apply to both observables and unobservables (PF 39, bottom paragraph). *Method* 3 suggests that we stop interpreting altogether and that we regard the statements of the theory as cognitively meaningless instruments of prediction (PF 40). *Method* 4., which is the one adopted by Reichenbach, suggests that we change the laws of logic in such a way that the statements which show the inadequacy of one of the chosen interpretations "can never be asserted as true" (PF 42). Reichenbach seems to assume that the principle of anomaly forces us to adopt one of those four methods.

This wild conclusion is completely unjustified. The principle of anomaly applies to E_2 only and shows that they are untenable. There is no reason to assume that they should be tenable being based as they are upon the (classical) principle C. Classical theory is incorrect. Therefore, it is to be expected that also the more general notions of classical thought such as are incorporated in C (which is not even correct in all classical cases – see above) will turn out to be true only in a restricted number of cases.

It is only when one does not realize that this assumption is part and parcel of classical thinking (rather than an a priori principle to be satisfied by any interpretation, whether E_1 or E_2) that one will be inclined to sense a breakdown of realism, of logic, or of the simple idea that theories are not only instruments of prediction but also descriptions of the world. Reichenbach is one of those thinkers who are prepared to give up realism and even classical logic because they cannot adjust themselves to the fact that a familiar and well-understood theory has turned out to be false (that not all interpretations are E_2).[6]

But the methods suggested by Reichenbach can be criticized also independently of these more general remarks. The criticism of 1 and 2 is evident. Method 4 was criticized already in Section II. Some further criticism will be developed in Section IV.

III. ANOMALIES AND THE PRINCIPLE OF CONTACT-ACTION

The above discussion (especially note 6) shows that the difficulty of the E_2 cannot be described by saying that "*the laws of quantum mechanics* ... are logically incompatible with" contact-action (cf. the beginning of the first paragraph of Section I). The reason is first that those difficulties arise only if we use the laws of quantum mechanics *together with* assumption C (which is not a law of quantum mechanics). Second, it would even be incorrect to assume that the conjunction of the laws of quantum mechanics with C is incompatible with contact-action. The principle of contact-action applies to fields which can be used for the transmission of signals. Neither the collapse of a wave (in the wave interpretation) nor the telepathic information conveyed to particles (in the particle interpretation) can be used in this way. Neither of these phenomena contradicts contact-action (as should be seen from the discussion in Section II and especially from note 5). It follows that the above description of Reichenbach's point of departure (due to H. Putnam) is incorrect.

IV. THE POSITION OF LAWS IN THE SUGGESTED INTERPRETATION

A criterion of adequacy of the interpretation by means of a three-valued logic is this: (a) every statement expressing an 'anomaly' should have the truth-value 'indeterminate' (PF 42); (b) every law of quantum mechanics should have either the truth-value 'true' or the truth-value 'false,' but never the truth-value 'indeterminate' (PF 160; FL 105). It turns out (PF 158f) that in the special system of three-valued logic used, formulas can be constructed which satisfy (b). The question arises whether every law of quantum mechanics satisfies (b).

Consider for that purpose the law of conservation of energy and assume that it is formulated as saying that the sum of the potential energy and the kinetic energy (both taken in their classical sense) is a constant. Now according to (a)[7] either the statement that the first part of the sum has a definite value is indeterminate, or the statement that the second part has a definite value is indeterminate, or both statements are indeterminate from whence it follows that of the conversation of energy will itself be a statement which has always the value 'indeterminate.' The same results if we use the statement in the form in which it appears in quantum mechanics. In this form the statement asserts that the sum of various operators, not all of them commuting, will disappear. According to (a) the statement that an operator has a certain value is indeterminate unless the operator is diagonal. As the only statements admitted to E_2 are statements to the effect that an operator *has* a certain value it follows again that the law of conservation of energy can only possess the truth-value 'indeterminate.'

The last argument admits of generalization: Every quantum-mechanical statement containing noncommuting operators can only possess the truth-value 'indeterminate.' This implies that *the commutation rules* which range among the basic laws of quantum mechanics *as well as the equations of motion* (consider them in their Heisenberg form) *will be indeterminate* and hence "neither verifiable nor falsifiable" (PF 42). We have to conclude that Reichenbach's interpretation does not satisfy his own criterion of adequacy since it violates (b).[8]

V. THE COPENHAGEN INTERPRETATION

Reichenbach considers his interpretation as superior to what he thinks is the Copenhagen interpretation. According to him this interpretation admits statements about (classically describable) phenomena only and calls meaningless all statements about unmeasured entities (PF 40).

Although this view is supported by many more or less vague pronouncements made by members of the Copenhagen circle, it is yet somewhat misleading. The correct account of the matter seems to be somewhat like this: In his earlier writings Bohr ascribed the change of the state of a system due to measurement to the *interaction* between the measuring device and the system measured. However, later on[9] he made a distinction between *physical changes* ("mechanical disturbances") of the state of a system which are caused by physical fields of force, and changes of "the very conditions which define the possible types of predictions regarding the future behavior of the system."[10] The fact that as an example of conditions of this latter kind Bohr mentions the reference systems introduced by the theory of relativity seems to indicate that what is meant here is a *logical property* of the state function and hence of any statement which ascribes a certain value to some variable of a quantum-mechanical system. On measurement the state S of a system s changes, not only because forces are acting upon the system, but also because it is a *relation between the system and a certain kind of physical preparation.* The analogues of classical properties are defined for a restricted class of states only, i.e., properties of a (classically explained) category are applicable to a system only if the system has been prepared in a certain way. If this interpretation is carried through consistently, that is, if it is separated from the instrumentalistic philosophy which is an altogether independent (though never clearly separated) element of the Copenhagen view, it may be used as an E_1, though not as an E_2 (since it contradicts C). In this interpretation statements about 'interphenomena' are meaningful (hence it is an E_1) but statements such as 'the position of the electron at time t is x' ('$P(s, t)=x$') are occasionally meaningless (hence it is not an E_2). It is mainly for the latter reason that Reichenbach finds the Copenhagen interpretation unsatisfactory. In the next section we shall consider some of his (and Putnam's) arguments which are intended to show that their interpretation is better than the Copenhagen interpretation.

VI. ARGUMENTS AGAINST IT CONSIDERED

Three arguments are used in FL to show that "three-valued logic, and it alone, provides an adequate interpretation of quantum physics." They seem to be the only arguments which Reichenbach has at his disposal. The first argument may be formulated thus: The statement that '$P(s, t) = x$' is meaningful only if certain conditions are realized amounts to saying that a statement is meaningless at the time t if no observer is testing it at time t. "The whole domain of unknown truth would thus be eliminated from physical language" (PA 347). H. Putnam uses a similar argument: the Copenhagen interpretation is unsatisfactory since it allows us to call a statement meaningful only if we actually *look* at some measuring apparatus used for the testing of the statement.

Two points must be distinguished in this argument, namely (a) the (correct) point that predications belonging to classical categories can be applied to a physical system only if certain conditions are fulfilled; and (b) the (incorrect) assertion that these conditions include an observation by a conscious observer. To deal with (b) first it must be pointed out that actual observation is by no means necessary in order to enable one to say that a certain predicate (like 'position') applies to a physical system. If the system is in such (physically definable) conditions that its state may be represented as a superposition of spatially well-defined bundles with negligible interference between them, then we may say that it possesses some position. If on the other hand the physical conditions are such that narrow trains with approximately the same frequency are fairly well isolated, then we may say that the system possesses some (perhaps unknown) momentum. Hence, whether or not a system possesses momentum or position depends on the existence of physical conditions (which may have been realized by a measuring instrument: every measuring instrument is devised to provide a separation of wave trains such that predicates of a given classical category become applicable to the system) and not on the presence of an observer. But is the fact that *such* a dependence exists in itself unsatisfactory (point a)? Reichenbach's and Putnam's answer to this question is that sentences like 's has the position x at time t' "admittedly have a very clear cognitive use; hence it is unnatural to regard them as 'meaningless'" (Putnam 78; cf. also Reichenbach FL 105). This answer smells dangerously of apriorism. It also overlooks that in our search for better theories we frequently discover that situations we thought would obtain universally do in fact exist only under special conditions, which implies that the properties of these situations are applicable in those conditions only. Within the theory this new dependence is then expressed by introducing a relation for something which was so far described by a property. The theory of relativity is a familiar example of this procedure which in the above argument is described as 'unnatural.'[11] Is it perhaps more 'natural' to stick to the notions of an overthrown theory at the expense of epistemology (transition to idealism) and even of classical logic?

Reichenbach's second argument runs as follows: If it is meaningless to say, in the case of the two-slit experiment, that an electron has passed through, say slit one, then it is also meaningless to say that it passed through one of the two slits. Yet we would like to assert the latter statement. In Reichenbach's interpretation such an assertion is possible since the disjunctive statement 'the electron passed through slit one or it passed through slit two' may be true (PF 41, 163f; FL 104). Hence this interpretation is preferable to the usual interpretation. Here it must be pointed out that the statement that every electron which arrives at the photographic plate has passed either through slit one or through slit two is of course excluded by quantum mechanics since it would imply that there is no interrelation between the situation at slit one and the situation at slit two. But it does not follow, as Reichenbach seems to assume (FL 104) that on that account we are unable to say that what has arrived at the photographic plate has passed through the slits and not through the wall in between; for it is not true that only particles can pass through slits and be intercepted by walls. The correct description consists in saying that within a certain interval of time (to be determined by the latitude of knowledge of energy) interfering parts of the electron passed through the (not simply connected) opening 'slit-one-plus-slit-two.'

Reichenbach's third argument consists in pointing out (FL 105) that in this interpretation all the laws of quantum mechanics are statements which are either true or false, but never indeterminate (whereas within the Copenhagen interpretation they may be meaningless). We have shown above that this statement is incorrect.

We may summarize this section by saying that none of Reichenbach's arguments in favor of his own position and against the Copenhagen interpretation are tenable.

VII. FORMALIZATION

The arguments of the above sections have mainly been formulated in the 'material mode of speech' to use an apt expression of Carnap's, i.e., they have been formulated as arguments about the properties of quantum-mechanical systems. These arguments may also be expressed in the formal mode of speech, i.e., as arguments "concerning the structure of the language in which this world can be described" (PF 177) and, more especially, as arguments concerning the structure of the language of quantum mechanics. Both kinds of arguments may be found in Reichenbach's book. It is H. Putnam's merit to have separated them more clearly by describing the problem at issue as *the problem to find an adequate formalization of quantum mechanics*.

Now one must realize that by using the formal mode of speech the problem at issue has not been removed from the domain of physical argument. Assume that we use a formalization in which (a) the logic is two-valued and (b) for every s some atomic sentences are of the form '$P(s, t) = x$.' Any theory which has been formalized in this way is committed to the assertion that a system with a single degree of freedom has always a well-defined position, i.e., it is committed to the particle interpretation. Similar remarks apply to other types of E_2. Hence, all our arguments against E_2 can be repeated against the corresponding formalizations and they show that a theory formulated in accordance with some such rule as (a) and (b) above will lead to 'anomalies.' Quantum mechanics prior to any formalization does not lead into anomalies. Hence, the formalizations considered are inadequate (here I have used the principle T' is an adequate formalization of T only if there is no empirical statement which follows from T and does not follow from T' and vice versa). And Reichenbach's interpretation may now be described as an interpretation which shows that even an inadequately formalized quantum mechanics can be made compatible with some very distressing facts, if a three-valued logic is used. However, it would obviously be more "natural" to use an adequate formalization.

As against this argument Putnam has asserted (in a private communication) that we cannot say what 'follows' from a theory unless we are already using some formalization. And that what I have done above amounts to nothing else but to comparing the results of one formalization with the results of another formalization. Now even if this were the case I would still maintain that my arguments are good arguments and show that my (alleged) formalization is preferable to Reichenbach's and

Putnam's. But a brief consideration of the arguments used in Section II will, I think, convince everybody that the point at issue here does not presuppose any specific formalization (although it may be possible to present it formally in a more satisfactory way).

I must conclude, then, that in spite of H. Putnam's arguments, in his paper and also in private discussion, I still feel that Reichenbach's suggestions cannot be regarded as a step toward a better understanding of the 'logic' (in a not strictly formal sense) of quantum mechanics.

University of Bristol

NOTES

[1] 'Three-Valued Logic', this volume, p. 99–107.

[2] The following abbreviations will be used: PF for *Philosophic Foundations of Quantum Mechanics*; PA for 'The Principle of Anomaly in Quantum Mechanics', *Dialectica* **7/8** (1948) 337; FL for 'Les fondements logiques de la théorie des quanta', in *Applications Scientifiques de la Logique Mathématique*, Paris, 1954, pp. 103ff.

[3] This omits spin.

[4] This is expressed by Reichenbach's *Principle of Anomaly* (which he assumes to be independent of the uncertainty principle PF 44). It is worthwhile considering the transformations this principle undergoes in Reichenbach's book. It is introduced as saying that "the class of descriptions of interphenomena contains no normal system" (PF 33) which means, when decoded, that the laws for quantum-mechanical objects cannot be *formulated* in such a way that they coincide with the laws governing the behavior of observable objects, viz. the classical laws (cf. PF 19). As it stands the principle is obviously refuted by the fact that formulations of quantum mechanics and of classical physics exist which are identical. We may, however, interpret the principle as saying that the *laws* (not their formulations) for quantum-mechanical objects are not the same as the laws governing the behavior of observed objects or, to use Reichenbach's terminology (which is supposed to express "the quantum mechanical analogue of the distinction between observed and unobserved objects" (PF 21), that the laws of interphenomena are not the same as the laws of phenomena. In this case the truth of the principle follows from the definitions of 'phenomenon' and 'interphenomenon' provided by Reichenbach which say that the "phenomena are determinate in the same sense as the unobserved objects of classical physics" (PF 21) whereas the introduction "of interphenomena can only be given within the frame of quantum mechanical laws" (PF 21). For according to these definitions the principle of anomaly asserts that the laws of classical physics are different from the laws of quantum mechanics, which is of course true but does not justify the introduction of the principle as an independent assumption (see the beginning of this note). However, this is not the sense in which the principle is used at other places of the book where it is meant to say that "every exhaustive interpretation" (in the second sense) "leads to causal anomalies" (136 PF). Having introduced this latter sense of the principle and having announced that it will be proved later on the basis of the principles of quantum mechanics, Reichenbach swiftly returns to the first interpretation (in which, as we have seen, the principle follows trivially from the definitions given for its two main terms together with the fact that

classical physics is not quantum mechanics) and derives from it that the idea of the uniformity of nature (same laws for observables and unobservables) must be given up (PF 39). These are only some of the confusions found in a book which demands that "the philosophy of physics should be as neat and clear as physics itself" (PF vii).

5 Reichenbach realizes that his 'anomalies' are not simply physical phenomena which exist in addition to the phenomena implied by some E_2; he calls them "pseudoanomalies" which are of a "ghostlike character... They can always be banished from the part of the world in which we happen to be interested although they cannot be banished from the world as a whole" (PF 40). This means that they are not physical phenomena (physical phenomena cannot be 'banished' from the part of the world in which they happen to occur) but are due to a deficiency of the picture chosen (which, of course, can be explained away). More especially the existence of anomalies is not the same as the existence of signals with over-light velocity as may be seen from the fact that Reichenbach counts among the anomalies of the theory of the pilot wave that "this wave field possesses no energy" (PF 32) and that he refers to the "anomaly connected with potential barriers" (PF 165) which simply consists in the violation of the principle of conservation of energy.

6 It appears that the principle of complementarity owes its existence to the same reluctance to part with classical ideas.

7 In applications to concrete cases, such as the one under review, Reichenbach uses the stronger condition that statements about unobserved values should be indeterminate (PF 145). The following arguments will be based upon this stronger condition.

8 Reichenbach admits that in his interpretation "the principle of the conservation of energy is eliminated... from the domain of true statements" (PF 166). Six pages earlier he asserts that "it will be the leading idea" (of the interpretation used) "to put into the true-false class those statements which we call quantum mechanical *laws*" (his italics). From this I can only conclude either that on page 166 Reichenbach has already forgotten what he said on page 160, or that for him the principle of conservation of energy (and, we shall have to add, the quantum conditions as well as the equations of motion) are not quantum mechanical *laws*. It seems that the latter is the case; for Reichenbach mentions as a case where (b) is satisfied the "*law* of complementarity" (FL 105; PF 159; my italics), which has certainly never been listed as a physical law. In his article H. Putnam asserts that in the case of quantum mechanics "the suggestion of using a three-valued logic makes sense... as a move in the direction of *simplifying*" (my italics) "the whole system of laws" (104). This statement is of course true, but it is true in an unexpected sense. Quantum mechanics is 'simplified' indeed as all important laws are "eliminated... from the domain of true statements."

9 Cf. especially *Phys. Rev.* **48** (1935), 696, and here especially the last paragraph, as well as *Albert Einstein, Philosopher-Scientist*, Evanston, 1949, especially pp. 231ff.

10 *Albert Einstein*, p. 234.

11 Reichenbach and Putnam express in the 'formal mode of speech' a type of argument which has frequently been used by 'traditional' philosophers against the conceptions introduced through new theories: it was regarded as 'unnatural' to let simultaneity depend on the coordinate system chosen (and to assume that 'Sim(xy)' is not well formed and hence, meaningless); yet it had to be admitted that special relativity was more successful than prerelativistic physics. In order to solve this difficulty traditional philosophers usually adopted what we have called method 3 (Section II above; cf. Philipp Frank, *Relativity, a Richer Truth*). That is, they regarded relativity proper as a set of cognitively meaningless sentences which nevertheless could be used as parts (cogwheels, so to speak) of a good prediction machine. However objectionable this method may be, traditional philosophers

took contradictions seriously and tried to remove them. It was left to Reichenbach (who argued against "speculative philosophy which must appear outmoded in the age of empiricism," PF vii) to provide the above approach with two further methods, viz. the 'method' to call contradictions 'anomalies' and "to become accustomed to them" (his methods 1 and 2) and the 'method' to drop two-valued logic.

ANDREW M. GLEASON*

MEASURES ON THE CLOSED SUBSPACES OF A HILBERT SPACE

1. Introduction

In his investigations of the mathematical foundations of quantum mechanics, Mackey[1] has proposed the following problem: Determine all measures on the closed subspaces of a Hilbert space. A measure on the closed subspaces means a function μ which assigns to every closed subspace a non-negative real number such that if $\{A_i\}$ is a countable collection of mutually orthogonal subspaces having closed linear span B, then

$$\mu(B)=\sum \mu(A_i).$$

It is easy to see that such a measure can be obtained by selecting a vector v and, for each closed subspace A, taking $\mu(A)$ as the square of the norm of the projection of v on A. Positive linear combinations of such measures lead to more examples and, passing to the limit, one finds that, for every positive semi-definite self-adjoint operator T of the trace class,

$$\mu(A)=\operatorname{trace}(TP_A),$$

where P_A denotes the orthogonal projection on A, defines a measure on the closed subspaces. It is the purpose of this paper to show that, in a separable Hilbert space of dimension at least three, whether real or complex, every measure on the closed subspaces is derived in this fashion.

If we regard the measure as being defined, not on the closed subspaces, but on the orthogonal projections corresponding, then the problem can be significantly generalized as follows: Determine all measures on the projections in a factor. We solve this problem for factors of type I, but our methods are not applicable to factors of types II and III.

For factors of type I the problem is simplified by the existence of minimal subspaces. In view of the complete additivity demanded in the definition, it is quite obvious that a measure on the closed subspaces of a separable Hilbert space is determined by its values on the one-dimen-

C. A. Hooker (ed.), The Logico-Algebraic Approach to Quantum Mechanics, 123–133.

sional subspaces. This leads us to the study of what we shall call frame functions.

DEFINITION. *A frame function of weight W for a separable Hilbert space $\mathscr{H}$ is a real-valued function f defined on the (surface of the) unit sphere of $\mathscr{H}$ such that if $\{x_i\}$ is an orthonormal basis of $\mathscr{H}$ then*

$$\sum f(x_i) = W.$$

While we are ultimately interested in non-negative frame functions, it is convenient to consider those with negative values, particularly in the finite-dimensional case. Here no convergence questions can arise since the sum in the definition is finite.

If S is a closed subspace of $\mathscr{H}$, then any frame function for $\mathscr{H}$ becomes one for S by restriction, the weight being probably changed. A one-dimensional S leads us immediately to the following observation: If f is any frame function and $|\lambda| = 1$, then $f(\lambda x) = f(x)$.

DEFINITION. *A frame function f is regular if and only if there exists a self-adjoint operator T defined on $\mathscr{H}$ such that*

$$f(x) = (Tx, x) \qquad \textit{for all unit vectors } x.$$

Our objective is to prove that all frame functions are regular, at least with suitable additional hypotheses. In dimension one it is obvious that every frame function is regular. In dimension two a frame function can be defined arbitrarily on a closed quadrant of the unit circle in the real case, and similarly in the complex case. In higher dimensions the orthonormal sets are intertwined and there is more to be said. However, if f is a finite-dimensional frame function and g is a discontinuous endomorphism of the real numbers as an additive group, $g \circ f$ is also a frame function. This construction produces a great class of wildly discontinuous frame functions, all of which are unbounded. This suggests the additional hypothesis bounded. Slightly stronger, and closer to our goals, is the hypothesis non-negative. In the finite-dimensional case these are essentially equivalent since a constant function is a frame function and the frame functions form a linear set. We shall show that every non-negative frame function in three or more dimensions is regular.

2. FRAME FUNCTIONS IN THREE-DIMENSIONAL REAL HILBERT SPACES

2.1. LEMMA. *In a finite-dimensional real Hilbert space a frame function is regular if and only if it is the restriction to the unit sphere of a quadratic form.*

Proof. Obvious.

2.2. LEMMA. *Consider the functions on the unit circle in R^2 given in the usual angular coordinate by* $\cos n\theta$, *n an integer. Such a function is a frame function if and only if $n=0$ or $n\equiv 2$* (mod 4).

Proof. If the weight is W we must have

$$\cos n\theta + \cos n\left(\theta + \frac{\pi}{2}\right) = \left(1 + \cos n\,\frac{\pi}{2}\right)\cos n\theta - \sin n\,\frac{\pi}{2}\sin n\theta = W$$

for all θ. This is true if and only if $n=0$ or $1+\cos\frac{1}{2}n\pi=0$.

2.3. THEOREM. *Every continuous frame function on the unit sphere in R^3 is regular.*

Proof. Let $\mathscr{C}$ denote the space of continuous functions on the unit sphere S of R^3 endowed with the usual norm. The rotation group G of R^3 is represented as a group of linear operators on $\mathscr{C}$ if we define

$$U_\sigma h = h \circ \sigma^{-1} \quad \text{where} \quad \sigma \in G,\ h \in \mathscr{C}.$$

Let Q_n denote the space of surface harmonics of degree n; these functions may be characterized as the restrictions to S of the homogeneous polynomial functions of degree n which satisfy Laplace's equation in R^3. These spaces are irreducible G-invariant subspaces of $\mathscr{C}$ and furthermore they are the only irreducible G-invariant subspaces.

Let F be the space of continuous frame functions on S. It is readily seen that F is a closed G-invariant linear subspace of $\mathscr{C}$. From the general theory of representations of compact groups (or the older theory of continuous functions on S) it follows that F is the closed linear span of certain of the Q_n. Now Q_0 consists of constant functions, so $Q_0 \subset F$. Since Q_1 is made up of linear functions on R^3 restricted to S, $Q_1 \not\subset F$ (these functions change sign on passing to antipodes, frame functions do not). The space Q_2 contains the restrictions to S of quadratic forms of trace 0; these are all frame functions of weight 0, so $Q_2 \subset F$. Suppose $n > 2$; then, using the characterization given in the previous paragraph, we may check that Q_n contains the restrictions to S of the explicit functions on R^3 given in *cylindrical* coordinates by

$$\rho^n \cos n\theta \quad \text{and} \quad [\rho^2 - 2(n-1)\, z^2]\, \rho^{n-2} \cos(n-2)\, \theta.$$

If $Q_n \subset F$, then both of these functions would restrict to be frame functions not only on S but on the unit circle in the x-y plane. This contradicts lemma 2.2, so $Q_n \not\subset F$. This proves that F is the closed linear span of Q_0 and Q_2. Since these spaces have finite dimension $F = Q_0 + Q_2$. The functions of Q_0, although constant, are nonetheless restrictions of quadratic forms because $x^2 + y^2 + z^2 = 1$ on S. This proves that every continuous frame function is the restriction of a quadratic form, which is a paraphrase of the theorem.

2.4. It is convenient to describe constructions on the sphere in terms of the ordinary latitude-longitude coordinates. When such a system has been selected we designate by N the closed northern hemisphere. Through every point q of N other than the north pole there is a unique great circle tangent to the circle of latitude through q; we shall call it the EW great circle through q.

If f is a real-valued function defined on the set X we write osc (f, X) for $\sup\{f(x) \mid x \in X\} - \inf\{f(x) \mid x \in X\}$.

2.5. Lemma. *Suppose z is a point of N other than the north pole. Consider the set X of all points $x \in N$ such that for some y*

(a) *y is on the EW great circle through x,*

(b) *z is on the EW great circle through y.*

Then X has a non-empty interior.

Proof. This is very easily seen geometrically. Analytically we may take the point z to have orthonormal coordinates $\langle \cos\theta, 0, \sin\theta \rangle$ where $0 < \theta < \frac{1}{2}\pi$, and then we verify that the locus L of points $y = \langle \xi, \eta, \zeta \rangle$ satisfying (b) is given by

$$\psi = (\xi^2 + \eta^2) \sin\theta - \xi\zeta \cos\theta = 0.$$

Now if x is any point at which the form ψ is negative, then the EW circle through x must cross the locus L, since on the equator $\psi > 0$. Thus X contains at least the open set of points at which ψ is negative, which is not empty because $\langle \cos\phi, 0, \sin\phi \rangle$ is in it if $\theta < \phi < \frac{1}{2}\pi$.

2.6. LEMMA. *Suppose that f is a frame function on the unit sphere S in R^3 and that, for a certain neighborhood U of p,* $\operatorname{osc}(f, U) = \alpha$. *Then every point of the great circle with pole p has a neighborhood V for which* $\operatorname{osc}(f, V) \leqq 2\alpha$.

Proof. We take latitude-longitude coordinates with p as north pole. Suppose U contains all points in latitudes above $\frac{1}{2}\pi - \theta$. Let q_0 be any point on the equator and let r be the point in latitude $-\frac{1}{2}\theta$ due south of q_0. Let C_0 be the great circle connecting r and q_0 and let r_0' and q_0' be orthogonal to r and q_0 respectively in $C_0 \cap N$. Both of these points fall in U; furthermore, the same will be true if q_0 is replaced by any point q in a certain neighborhood V of q_0, keeping r fixed.

If now q_1 and q_2 are in V, let C_i be the great circle connecting r and q_i and take r_i' and q_i' on $C_i \cap N$ so that $r_i' \perp r$, $q_i' \perp q_i$ $(i = 1, 2)$. Then we will have

$$f(r) + f(r_i') = f(q_i) + f(q_i'), \qquad i = 1, 2.$$

Subtracting these equations

$$|f(q_1) - f(q_2)| = |f(r_1') - f(r_2') + f(q_2') - f(q_1')| \leqq 2\alpha$$

since $r_1', r_2', q_1', q_2' \in U$. This shows that $\operatorname{osc}(f, V) \leqq 2\alpha$.

2.7. LEMMA. *Suppose that f is a frame function on the unit sphere S in R^3 and that, for a certain non-empty open set U,* $\operatorname{osc}(f, U) = \alpha$. *Then every point of S has a neighborhood W such that* $\operatorname{osc}(f, W) \leqq 4\alpha$.

Proof. From any point p of U we can reach any point of S in two steps of arc length $\frac{1}{2}\pi$; hence this lemma follows from the preceding.

2.8. THEOREM. *Every non-negative frame function on the unit sphere S in R^3 is regular.*

Proof. Let f be a non-negative frame function of weight W on S. We may subtract a constant from f and it will remain a frame function; hence it is no loss of generality to suppose that $\inf f(x) = 0$. The proof would be considerably shortened if we knew that f achieved the value 0, but we consider the general case.

Let ε be a positive number and put $\eta=\varepsilon/88$. We can find a point p such that $f(p)\leqq\eta$. Take latitude-longitude coordinates with p at the north pole. Let σ be the polar rotation through angle $\frac{1}{2}\pi$, and set

$$g(x)=f(x)+f(\sigma x).$$

Evidently g is a non-negative frame function of weight $2W$. For any point q on the equator, p, q, and σq form an orthonormal set so $g(q)=f(q)+f(\sigma q)=W-f(p)$; thus g is constant on the equator.

Consider any point $r\in N-\{p\}$. Let C be the EW great circle through r; it meets the equator at a point q orthogonal to r; therefore $2W\geqq g(r)+g(q)=g(r)+W-f(p)$, whence

$$(1) \qquad g(x)\leqq W+f(p)\leqq W+\eta \quad \text{for all} \quad x\in N-\{p\}.$$

Containing, consider any point $s\in C\cap N$ and an orthogonal point $t\in C\cap N$. We have $g(r)+W-f(p)=g(s)+g(t)\leqq g(s)+W+\eta$ giving

$$(2) \qquad g(r)\leqq g(s)+2\eta$$

for any point $r\in N-\{p\}$ and any point s on the EW circle through r.

Let $\beta=\inf\{g(x)\mid x\in N-\{p\}\}$ and take a point $z\in N-\{p\}$ for which $g(z)\leqq\beta+\eta$. If $x\in N-\{p\}$ is a point such that for some y

(a) y is on the EW great circle through x,

(b) z is on the EW great circle through y,

then

$$g(x)\leqq g(y)+2\eta,$$
$$g(y)\leqq g(z)+2\eta$$

and therefore

$$\beta\leqq g(x)\leqq g(z)+4\eta\leqq\beta+5\eta.$$

The set of points x satisfying the condition has a non-void interior U by Lemma 2.5 and the last display shows $\operatorname{osc}(g, U)\leqq 5\eta$. By Lemma 2.7 there is a neighborhood V of p such that $\operatorname{osc}(g, V)\leqq 20\eta$. Since $g(p)=2f(p)\leqq 2\eta$, $\sup\{g(x)\mid x\in V\}\leqq 22\eta$. Since f is non-negative and $f\leqq g$ pointwise, $\operatorname{osc}(f, V)\leqq 22\eta$. Applying 2.7 once again, any point $u\in S$ has a neighborhood W such that $\operatorname{osc}(f, W)\leqq 88\eta=\varepsilon$. Since ε can be arbitrarily small this proves that f is continuous and the theorem now follows from Theorem 2.3.

3. HIGHER DIMENSIONS AND COMPLEX HILBERT SPACES

3.1. *We shall say that a real-linear subspace $\mathscr{K}$ of a Hilbert space $\mathscr{K}$ is completely real if the inner product takes only real values on $\mathscr{K} \times \mathscr{K}$.*

A closed completely real subspace is itself a real Hilbert space with respect to the restriction of the inner product of $\mathscr{H}$. It is clear that if every pair of vectors in a set X has real inner product, then the real-linear subspace spanned by X is completely real and so is its closure. In particular, an orthonormal set of vectors spans a completely real subspace. Conversely, an orthonormal subset of a completely real subspace is an orthonormal subset of $\mathscr{H}$. It follows from these remarks that a frame function for $\mathscr{H}$ becomes a frame function when restricted to a completely real subspace.

3.2. LEMMA.[2] *If f is a non-negative regular frame function of weight W on a real Hilbert space, then for any unit vectors x and y*

$$|f(x)-f(y)| \leqq 2W \|x-y\|.$$

Proof. Since f is regular there is a symmetric operator T such that $f(x)=(Tx, x)$. Because f is non-negative we have $0 \leqq (Tx, x) \leqq W$ for all unit vectors x and therefore $\|T\| \leqq W$.

Now, for any unit vectors x and y, $(Tx, y)=(Ty, x)$, so $f(x)-f(y)=(T(x+y), x-y)$ and therefore

$$|f(x)-f(y)| \leqq \|T\| \; \|x+y\| \; \|x-y\| \leqq 2W\|x-y\|.$$

3.3. LEMMA. *Suppose that f is a non-negative frame function on a two-dimensional complex Hilbert space which is regular on every completely real subspace. Then f is regular.*

Proof. Suppose W is the weight of f and M is its least upper bound. We can choose unit vectors x_n so that $f(x_n) \to M$ and we can arrange that $x_n \to y$, because the unit sphere is compact. Let $\lambda_n = (y, x_n)/|(y, x_n)|$; we have $\lambda_n \to 1$ and $\lambda_n x_n \to y$. Since $|\lambda_n| = 1$, $f(\lambda_n x_n) = f(x_n)$. Moreover, $(\lambda_n x_n, y)$ is real, so $\lambda_n x_n$ and y are in a completely real subspace. By Lemma 3.2 we have

$$\begin{aligned} |f(y)-M| &\leqq |f(y)-f(\lambda_n x_n)| + |f(x_n)-M| \\ &\leqq 2W \|y-\lambda_n x_n\| + |f(x_n)-M|, \end{aligned}$$

from which we see that $f(y)=M$.

Define F on H by

$$F(v) = \|v\|^2 f\left(\frac{v}{\|v\|}\right) \quad \text{if} \quad v \neq 0,$$
$$F(0) = 0.$$

The hypotheses concerning f imply that F becomes a quadratic form when restricted to any completely real subspace. Furthermore, since $f(\lambda v) = f(v)$ whenever $|\lambda| = 1$, $F(\lambda v) = |\lambda|^2 F(v)$ for all scalars λ and vectors v.

Let z be any unit vector orthogonal to y. Then $F(y) = f(y) = M$ and $F(z) = f(z) = W - f(y) = W - M$. On the completely real subspace determined by y and z, F is a quadratic form whose maximum value on the unit circle is attained at y; therefore the matrix for F relative to the basis y, z is diagonal. Hence

$$F(\alpha y + \beta z) = \alpha^2 F(y) + \beta^2 F(z) = M\alpha^2 + (W - M)\,\beta^2$$

if α and β are real.

If λ and μ are non-zero complex numbers and $z' = (\mu/|\mu|)\,(|\lambda|/\lambda)\,z$, then z' is also a unit vector orthogonal to y; therefore

$$\begin{aligned} F(\lambda y + \mu z) &= F((|\lambda|/\lambda)\,(\lambda y + \mu z)) \\ &= F(|\lambda|\, y + |\mu|\, z') = M|\lambda|^2 + (W - M)\,|\mu|^2. \end{aligned}$$

The exceptional cases, λ or μ zero, present no difficulty, so we see that

$$F(x) = (Tx, x)$$

for any vector x, where T is the self-adjoint operator whose matrix, relative to y, z, is

$$\begin{Vmatrix} M & 0 \\ 0 & W - M \end{Vmatrix}.$$

This shows that f is regular.

3.4. Lemma.[3] *Suppose that f is a non-negative frame function for a Hilbert space $\mathscr{H}$ (either real or complex) and that f is regular when restricted to any two-dimensional subspace of $\mathscr{H}$. Then f is regular.*

Proof. We give the proof in a form which covers both the real and complex cases simultaneously. Define F as in Lemma 3.3. On each two-dimensional subspace S of $\mathscr{H}$, there is a form A_s (either bilinear or Hermitian) such that $F(x)=A_s(x, x)$ for $x \in S$. We define A on all of $\mathscr{H} \times \mathscr{H}$ by

$$A(x, y)=A_s(x, y) \quad \text{if} \quad x \in S,\ y \in S.$$

(Usually there will only be one two-dimensional subspace S containing both x and y, but if say $x=\lambda y$, then $A_s(x, y)=\lambda A_s(y, y)=\lambda F(y)$ which is independent of the choice of S.) Because only two-dimensional subspaces of $\mathscr{H}$ are involved we derive the following relations from the forms A_s:

$$\begin{array}{ll}
(1) & A(\alpha x, y)=\alpha A(x, y) \\
(2) & A(x, y)=\overline{A(y, x)} \\
(3) & 4\,\mathrm{Re}\,A(x, y)=F(x+y)-F(x-y) \\
(4) & 2F(x)+2F(y)=F(x+y)+F(x-y)
\end{array}$$

for all vectors x, y and scalars α.

Now

$$\begin{aligned}
&8\,\mathrm{Re}\,A(x, z)+8\,\mathrm{Re}\,A(y, z) \\
&\quad=2F(x+z)-2F(x-z)+2F(y+z)-2F(y-z) \\
&\quad=F(x+y+2z)+F(x-y)-F(x+y-2z)-F(x-y) \\
&\quad=4\,\mathrm{Re}\,A(x+y, 2z) \\
&\quad=8\,\mathrm{Re}\,A(x+y, z).
\end{aligned}$$

Replacing x and y by ix and iy and using (1) we find also

$$\mathrm{Im}\,A(x, z)+\mathrm{Im}\,A(y, z)=\mathrm{Im}\,A(x+y, z)$$

giving

$$(5) \qquad A(x, z)+A(y, z)=A(x+y, z)$$

which, together with (1) and (2), shows that A is bilinear or Hermitian on all of $\mathscr{H} \times \mathscr{H}$.

Take vectors x and y with $\|x\| \leqq 1$, $\|y\| \leqq 1$; with a proper choice of ω, where $|\omega|=1$, we have

$$\begin{aligned}
4|A(x, y)|&=4A(\omega x, y)=4\,\mathrm{Re}\,A(\omega x, y) \\
&=F(\omega x+y)-F(\omega x-y) \leqq M(\|\omega x+y\|^2 \\
&\quad+\|\omega x-y\|^2)=2M(\|\omega x\|^2+\|y\|^2) \leqq 4M
\end{aligned}$$

where $M=\sup\{|f(u)| \mid \|u\|=1\}$. Thus A is bounded and there exists a bounded self-adjoint operator T such that

$$A(x, y)=(Tx, y) \quad \text{for all} \quad x, y \in \mathscr{H}.$$

Finally, $f(x)=F(x)=A(x, x)=(Tx, x)$ for all unit vectors x which concludes the proof.

3.5. THEOREM. *Every non-negative frame function on either a real or complex Hilbert space of dimension at least three is regular.*

Proof. A frame function for $\mathscr{H}$ becomes a frame function for any completely real subspace of $\mathscr{H}$ by restriction. Every completely real two-dimensional subspace of $\mathscr{H}$ can be embedded in a completely real three-dimensional subspace, since $\dim \mathscr{H} \geqq 3$. Therefore Theorem 2.8 shows that any non-negative frame function f is regular on every completely real two-dimensional subspace of $\mathscr{H}$. Lemma 3.3 shows that f is regular on every two-dimensional subspace; hence f is regular by the last lemma.

4. THE MAIN RESULT

4.1. THEOREM. *Let μ be a measure on the closed subspaces of a separable (real or complex) Hilbert space $\mathscr{H}$ of dimension at least three. There exists a positive semi-definite self-adjoint operator* 1 *of the trace class such that for all closed subspaces A of $\mathscr{H}$*

$$\mu(A)=\operatorname{trace}(TP_x)$$

where P_A is the orthogonal projection of $\mathscr{H}$ onto A.

Proof. If B_x is the one-dimensional subspace spanned by the unit vector x, then $f(x)=\mu(B_x)$ defines a non-negative frame function f. There is a self-adjoint operator T such that $\mu(B_x)=(Tx, x)$, for all unit vectors x. Since $(Tx, x)\geqq 0$ for all unit vectors x, T is positive semi-definite. If $\{x_i\}$ is an orthonormal basis for $\mathscr{H}$,

$$\mu(\mathscr{H})=\sum \mu(B_{x_i})=\sum (Tx_i, x_i).$$

Since the latter sum converges, T is in the trace class, indeed trace $T=\mu(\mathscr{H})$.

If A is an arbitrary closed subspace, we can choose an orthonormal basis $\{y_i\}$ for A and adjoin further vectors $\{z_j\}$ so that $\{y_i, z_j\}$ is an orthonormal basis for $\mathscr{H}$. Then $P_A y_i = y_i$ for all i and $P_A z_j = 0$ for all j so

$$\mu(A) = \sum \mu(B_{y_i}) = \sum_i (Ty_i, y_i)$$
$$= \sum_i (TP_A y_i, y_i) + \sum_j (TP_A z_j, z_j) = \operatorname{trace}(TP_A).$$

The theorem is proved.

Harvard University, Cambridge, Massachusetts

NOTES

* The author has been partially supported by the Office of Ordnance Research, Contract No. DA 19-020-ORD-3778. He is indebted also to R. V. Kadison for helpful remarks.

[1] See his forthcoming article, 'Quantum Mechanics and Hilbert Space', *Amer. Math. Monthly*.

[2] The present version of this lemma and its proof are due to R. S. Palais, who was kind enough to read the first draft of this paper.

[3] This lemma is due to Jordan and von Neumann, On inner products in linear metric space, *Annals of Math.* **36** (1935), pp. 719–723.

E. P. SPECKER
(Translated by Allen Stairs*)

THE LOGIC OF PROPOSITIONS WHICH ARE NOT SIMULTANEOUSLY DECIDABLE**

La logique est d'abord une science naturelle.
F. GONSETH

The prefacing motto of this essay is the subtitle of the chapter 'La physique de l'objet quelconque' from the work *Les Mathématiques et la Réalité*. This physics shows itself essentially to be a form of classical propositional logic, which on the one hand receives a typical interpretation, and which on the other hand divests itself in an almost obvious manner of the claim to absoluteness with which it was occasionally associated. The following paper associates itself with this point of view and should be understood in the same empirical sense.

To begin with, let us consider the problem of investigating the structure of a domain B of propositions. Such a structural description is possible if, among the elements of B, certain relations or operations are defined. The simplest relation seems to be the relation of implication $a \rightarrow b$ (a, b in B), and will be taken as the basis for what follows. We do not assume that the proposition $a \rightarrow b$ is itself a member of B, but this is not excluded either. Let us consider the following example: the domain B consists of the ten propositions, It is warm, It is cold, It is raining, It is snowing, The sun is shining, It is not warm, It is not cold, It is not raining, It is not snowing, The sun is not shining. For certain a, b the implication $a \rightarrow b$ holds, for certain pairs it does not hold, and for others it is uncertain whether or not it holds. Examples are: If it is warm then it is not cold, If it is cold then it snows, and If it rains, then it does not snow. We will ignore the third case, i.e. the case where it is uncertain whether or not the implication obtains: for any a, b in B it is certain whether $a \rightarrow b$ does or does not hold. Special reference should be made to the second example, If it is cold then it is snowing. This is not a valid implication. Of course, this does not mean that it could not be cold and snowing, but only that it is not always snowing when it is cold. It should also be understood that the propositions It is cold, etc. are not meant as abbreviations for It is cold at 11:50 A.M. May 1, 1960 at the garden gate of Liegenschaft, Gelderstrasse 60 in Zurich (with whatever else might be necessary for a completely precise specification), rather the propo-

C. A. Hooker (ed.), The Logico-Algebraic Approach to Quantum Mechanics, 135–140.

sitions are general, like propositions which enter into the formulation of natural laws (i.e. they are propositional forms).

On the basis of the relation of implication, it is possible to state when a proposition c may be treated as a conjunction of the propositions a, b in B. First it is required that the implications $c \rightarrow a$ and $c \rightarrow b$ hold (if a and b then a, and if a and b then b); and secondly that the following condition be met: for any c' such that $c' \rightarrow a$ and $c' \rightarrow b$ hold, $c' \rightarrow c$ also holds (if c' implies a and c' implies b then c' implies a and b). It is not obvious that every domain of propositions contains an element with these characteristics. For example, B does not contain conjunctions of distinct elements. But this does not exclude the possibility that there is a domain B' which includes B and satisfies the conditions just defined. This is basically all that is required by the statement that for any two propositions there exists a proposition which is their conjunction. Before pursuing this question, it is necessary to determine whether the conjunction of two propositions is unique. If c_1 and c_2 are both conjunctions of a and b, then, as we have seen, the implications $c_1 \rightarrow c_2$ and $c_2 \rightarrow c_1$ are valid (in this case we write $c_1 \leftrightarrow c_2$ and say that c_1 and c_2 are equivalent). Now, equivalent propositions need not be identical (e.g. There is lightning and thunder and There is thunder and lightning), so the conjunction of two propositions is not unique. For this reason we shall consider equivalence classes of propositions, instead of the propositions themselves. [One shows that the equivalence class of the conjunction of two propositions depends on the equivalence classes of the conjoined propositions]. In the case of classical logic one thus arrives at a Boolean lattice. However, an analogous procedure is possible in all other logical calculi (e.g. intuitionist, modal, and many-valued logic). The possibility of substituting equivalence classes of propositions for propositions requires that the relation $c \leftrightarrow d$ is an equivalence relation, i.e., it is reflexive ($c \leftrightarrow c$), symmetric (if $c \leftrightarrow d$, then $d \leftrightarrow c$), and transitive (if $c \leftrightarrow d$ and $d \leftrightarrow e$, then $c \leftrightarrow e$). Of these properties, that of symmetry is satisfied on the basis of the definition of $\leftrightarrow$ in terms of implication $\rightarrow$, reflexiveness follows from the holding of the implication $c \rightarrow c$. Since we have placed no restrictions on implication, it is obvious that $c \rightarrow c$ cannot be proved; but our subsequent analysis of implication provides no reason for rejecting $c \rightarrow c$. The transitivity of the relation $\leftrightarrow$ is usually inferred from the transitivity of implication: if $c \rightarrow d$ and $d \rightarrow e$, then also $c \rightarrow e$. It may very well seem that transitivity is so intimately bound up with the concept of implication that it would be senseless to call a non-transitive relation "implication". That this is not entirely so may be shown by the following story, which in any case took place long ago and in a far-off land.

At the Assyrian School of Prophets in Arba'ilu in the time of King Asarhaddon, there taught a seer from Nineva. He was a distinguished representative of his faculty (eclipses of the sun and moon) and aside from the heavenly bodies, his interest was almost exclusively in his daughter. His teaching success was limited, the subject proved to be dry, and required a previous knowledge of mathematics which was scarcely available. If he didn't find the student interest which he desired in class, he did find it elsewhere in overwhelming measure. His daughter had hardly reached a marriageable age when he was flooded with requests for her hand from students and young graduates. And though he didn't believe that he would always have her by his side, she was in any case still too young and her suitors in no way worthy. In order that they might convince themselves of their worthiness, he promised her to the one who could solve a "prediction problem" which he set. The suitor was taken before a table on which three little boxes stood in a row and was asked to say which boxes contained a gem and which didn't. But no matter how many tried the task seemed impossible. In accordance with his prediction, each of the suitors was requested by the father to open two boxes which he had marked as both empty or both full. But it always turned out that one contained a gem and the other one didn't, and furthermore the stone was sometimes in the first box and sometimes in the second. But how should it be possible, given three boxes, neither to mark two as empty nor two as full? The daughter would have remained single until her father's death had she not followed the advice of a prophet's son and quickly opened two boxes, one of which was marked full and the other empty. Following the weak protest of her father that he had wanted two *other* boxes opened, she tried to open the third. But this proved impossible whereupon the father grudgingly admitted that the prediction was correct.

To give a logical analysis of the prediction-problem consider the following six propositions: A_i, A_i^* $(i=1, 2, 3)$ where A_i means that the ith box is full, A_i^* that it is empty. From the attempts of the suitors it is clear that the following implications hold in the domain of these propositions:

$A_i \rightarrow A_j^*$ (for each pair of distinct i and j ($1 \leqslant i, j \leqslant 3$); also the implications $A_1 \rightarrow A_2^*$, $A_2^* \rightarrow A_3$ hold while $A_1 \rightarrow A_3$ does not hold, but rather $A_1 \rightarrow A_3^*$ holds. It is clear that of these three implications none can be refuted because it is impossible to open all three boxes. We have found a necessary condition for inferring the implication $a \rightarrow c$ from the implications $a \rightarrow b$ and $b \rightarrow c$: the three propositions must be co-testable. (The implication $a \rightarrow b$ should be understood as follows: a and b are co-testable and whenever a holds, so does b).

The difficulties that arise with propositions which are not co-decidable appear especially clearly with propositions concerning a quantum mechanical system. In accordance with the customary terminology, we will refer to such collections of propositions as not simultaneously decidable. The logic of quantum mechanics was first investigated by Birkhoff and von Neumann in [1]. (We will return to their results.) In a certain sense, however, these issues were anticipated by scholastic speculations concerning "infuturabilien", [future contingencies – Transl.], i.e. the question of whether God's omniscience includes events which would occur if something were to happen which in fact does not happen. (Cf. [3], vol. 3, p. 363).

If not all propositions of B are simultaneously decidable, then, besides specifying the structure of the implication relation the description of B requires that we specify the set Γ of subsets of propositions of B which are simultaneously decidable. If for two elements a, b of B, $a \rightarrow b$ holds, then $\{a, b\}$ is in Γ. In particular, if we assume that for each a in B $a \rightarrow a$ then $\{a\}$ is in Γ; i.e., B contains no undecidable propositions. We shall assume further that implication is transitive and that B is partitioned into classes of propositions which are equivalent with respect to $\leftrightarrow$. However, to go from B to the set B' of equivalence classes, we need the additional assumption that Γ is compatible with this partition; this means that if $\{a, b\}$ is in Γ and $a \leftrightarrow a'$, then $\{a', b\}$ is in Γ. Making this assumption, we obtain a domain B' which is partially ordered by a relation $\rightarrow$, together with a set Γ' of subsets of B', such that Γ' contains all unit sets, the subsets of each set it contains, and if $a \rightarrow b$, the set $\{a, b\}$. Birkhoff and von Neumann have shown that for a quantum mechanical system the set B', over which B is interpreted, is isomorphic to the set of all linear subspaces of a complex Hilbert space (which in certain cases can be a unitary space); where implication corresponds to the subspace relation. A maximal set C of subspaces corresponds exactly to a maximal subset of Γ' if it forms a unitary basis for the space and includes a basis for each of its subspaces. It can be shown that this is the case if such a basis exists for every two subspaces of C. This condition is satisfied when the subspaces are mutually orthogonal in the sense of elementary geometry; i.e., when the orthogonal complement of the intersection of the subspaces separates them into orthogonal spaces. A collection of propositions about a quantum mechanical system is simultaneously decidable exactly when every two propositions of the collection are simultaneously decidable. Further, it can easily be shown that every such collection of propositions is contained in a Boolean lattice, i.e., for these propositions, classical logic holds. With each proposition there is associated a negation $\sim a$; $\sim a$ and b are simultaneously decidable exactly when a and b are simultaneously decidable. With each pair of simultaneously decidable propositions, there is associated a proposition which is their conjunction and a proposition which is their disjunction; moreover, all of these propositions are simultaneously decidable. In this characterization, conjunctions, and analogously, disjunctions, are not defined for propositions which are not simultaneously decidable. In the set of all subspaces of Hilbert space, these operations correspond, respectively, to the intersection and span of the associated subspaces. In contradistinction to the work of Birkhoff and von Neumann, the following problem requires only that these operations be defined for simultaneously decidable propositions: To determine whether it is possible to imbed the set of (closed) subspaces of a Hilbert space into a Boolean lattice so that the operations of negation, conjunction and disjunction are preserved. Clearly, this may be formulated as follows: if, for simultaneously decidable propositions, negation, conjunction and disjunction retain their standard meaning, can the description of a quantum mechanical system be enlarged by means of additional propositions in such a way that classical propositional logic is valid in the enlarged domain?

The answer to this question is negative except in the case of Hilbert (i.e. unitary) spaces of dimensions 1 and 2. In the 1-dimensional case the lattice of subspaces is just the Boolean lattice of two elements. In the 2-dimensional case, the lattice of subspaces may be represented as follows: there are subspaces H (the whole space), 0 (the null space), $A\alpha$, and $B\alpha$, (where α ranges over a set of the power of the continuum). H and 0 are orthogonal to all subspaces; $A\alpha$ and $B\alpha$ exactly to H, 0, $A\alpha$ and $B\alpha$. The complement (the negation) of $A\alpha$ is $B\alpha$ and conversely; the negation of H is 0, and conversely. The conjunction of $A\alpha$ and $B\alpha$ is 0, their disjunction is H. H and 0 are unit and null-elements of the sub-lattices: $0 \vee C = C$, $0 \wedge C = 0$, $H \vee C = H$, $H \wedge C = C$ (C an arbitrary subspace). It is easy to see that this structure can be imbedded into a Boolean lattice. The impossibility of such an imbedding for spaces of more than three dimensions follows from the fact that there is no imbedding for the 3-dimensional case. For the sake of simplicity, we shall restrict ourselves to the real orthogonal space contained in the unitary space. For this case the imbedding requires the existence of a one-to-one mapping from the set of all linear subspaces of a 3-dimensional orthogonal vector space into a Boolean lattice such that for each pair a, b of orthogonal subspaces: $f(a \wedge b) = f(a) \wedge f(b)$ (b), $f(a \vee b) = f(a) \vee f(b)$, and the image of the null space is the zero, while the image of the whole space is the unit, of the Boolean lattice. Because the two-element Boolean lattice is the homomorphic image of every Boolean lattice, the solution to the following "prediction problem" follows from the existence of an imbedding. For a three dimensional orthogonal vector space, exactly one of the values t(rue) and f(alse) is assigned to each linear subspace so that the following conditions are satisfied: t is assigned to the whole space, and f to the null space; if a and b are orthogonal subspaces, then the value t is assigned to their intersection $a \wedge b$ exactly when the value t is assigned to both of them; and the value t is assigned to their span $a \vee b$, if the value t is assigned to at least one of the subspaces a, b.

An elementary geometrical argument shows that such an assignment is impossible, and that therefore (aside from the exceptions noted above) no consistent prediction concerning a quantum mechanical system is possible.

Zurich

NOTES

* I wish to thank Mr. William Demopoulos for suggesting this project to me.

** From *Dialectica* **14** (1960), 239–246.

BIBLIOGRAPHY

[1] Birkhoff, G. and Neumann, J. v., 'The Logic of Quantum Mechanics', *Annals of Math.* **37** (1936), 823–843.

[2] Gonseth, F., *Les mathématiques et la réalité*, Feliz Alcan, Paris, 1936.

[3] Solana, M., *Historia de la filosofía española*, Asociacion española para el progreso de las ciencias, Madrid, 1941.

DAVID J. FOULIS

BAER *-SEMIGROUPS†

I. INTRODUCTION

Modern mathematics is replete with instances of semigroups S which are equipped with involutory antiautomorphisms $*: S \to S$, two noteworthy examples being multiplicative groups on the one hand, and the multiplicative semigroups of Baer *-rings [1, Chapter III, Definition 2] on the other. In this paper we take the second example cited above as our point of departure, setting forth certain postulates which determine what we will call a Baer *-semigroup, and showing that such semigroups provide a more or less natural "coordinatization" of the orthocomplemented weakly modular lattices employed by Loomis [2] in his version of the dimension theory of operator algebras.

The author would like to express his indebtedness to Professor F. B. Wright for his encouragement during the writing of this paper.

II. BAER *-SEMIGROUPS

By an *involution semigroup* we mean a multiplicatively written semigroup S equipped with a mapping $*: S \to S$, (called the *involution*), such that for $x, y \in S$, $(xy)^* = y^* x^*$ and $(x^*)^* = x^{**} = x$. An element $e \in S$ with the property that $e = e^2 = e^*$ will be called a *projection.*

If K is a two sided ideal in the involution semigroup S, i.e., if $SK \subset K$ and $KS \subset K$, then we will call K a *focal ideal* in case it is so that for each element $x \in S$, the set $\{y \mid y \in S \text{ and } xy \in K\}$ is a principal right ideal generated by a projection. A *Baer *-semigroup* is a pair (S, K) consisting of an involution semigroup S and a focal ideal K in S. Whenever no confusion can result, we will refer to S itself as being the Baer *-semigroup, rather than using the more cumbersome expression (S, K).

Henceforth, we will regard the symbol S as representing a Baer *-semigroup with focal ideal K. We denote by $P = P(S)$ the set of all projections in S, and we partially order P by decreeing that for $e, f \in P$, $e \leqq f$ means that $ef = e$, (or, what is the same thing, that $fe = e$).

It is clear that if a principal right ideal I in S is generated by a projection e, then this projection e is uniquely determined by I. Consequently, each element $x \in S$ determines a unique projection x' such that $\{y \mid y \in S$ and $xy \in K\} = x'S$. We call the mapping $' : S \to P$ the *focal mapping* induced by the focal ideal K. One easily verifies that the focal mapping has the following properties: (i) For $e, f \in P$, $e \leqq f$ implies that $f' \leqq e'$, (ii) for $e \in P$, $e \leqq e''$, (iii) for $e \in P$, $e' = e'''$, and (iv) for $a, b \in S$, $ab = a$ implies that $a'' \leqq b''$. Moreover, we remark that for each element $a \in S$, $a = aa''$.

Say that a projection $e \in P$ is *K-closed* if $e = e''$, and denote the set of all K-closed projections in S by $P' = P'(S)$. Notice that P' is exactly the range of the focal mapping $' : S \to P$. Furthermore, for each projection $f \in P$, f'' is the smallest K-closed projection containing f.

One noteworthy feature of the focal ideal K is that if $a \in S$ and $aa^* \in K$, then $a \in K$. In fact, if $aa^* \in K$, then $a^* = a'a^*$, so $a = aa' \in K$. One consequence of the fact just proved is that $K = K^*$; for if $a \in K$, then certainly $a^*a = a^*(a^*)^* \in K$, hence $a^* \in K$.

A question which arises naturally from time to time in the development of our theory is whether a given projection does or does not commute with various elements of S. This question can frequently be settled by an appeal to the fact that if the projection $e \in P$ commutes with the element $a \in S$, then the projection e' will also commute with a. Indeed, if $ae = ea$, then $eae' = aee' \in K$, so $ae' = e'ae'$. Also, if $ae = ea$, then $a^*e = ea^*$, and the above argument gives $a^*e' = e'a^*e'$, whence, $e'a = e'ae'$, so $ae' = e'a$.

We remark that the necessary and sufficient condition that every element of P' commutes with every element of S is that K is a radical ideal, i.e., if y is any element of S some positive integral power of which belongs to K, then y belongs to K.

In the following theorem it will come to light that S must contain a multiplicative unit. In general, of course, it is possible for a semigroup to admit more than one right (or more than one left) unit, but this cannot occur in an involution semigroup. Actually, if u is a right (or left) unit in an involution semigroup, then u is a two-sided unit and $u = u^*$. An analogous assertion can be made for right (or left) zeros in an involution semigroup.

THEOREM 1. *If (S, K) is a Baer *-semigroup, then S has a unit* 1, *and* 1 *is the largest projection in $P'(S)$. Moreover,* $1'$ *is the smallest projection* *in P', $1'$ is a central projection in S, and $K=1'S=S1'$. Consequently, the focal ideal in a Baer *-semigroup is a principal ideal generated by a central projection.*

Proof. Let k be any element of K, so that for any element $e\in P'$, $ke\in K$, and hence, $e\leqq k'$. It follows that k' is a right unit, hence a unit for S, so we write $k'=1$. The remainder of the theorem is clear as soon as we observe that $1'=1\cdot 1'\in K$.

The following lemma is an easy generalization of an analogous result in the standard theory of Baer *-rings, so its proof is omitted:

LEMMA 2. *Let M be a nonempty subset of S. Then, the set $\{y \mid y\in S$ and $My\subset K\}$ is a principal right ideal generated by a projection if and only if $\{m' \mid m\in M\}$ has an infimum in P. Moreover, if $f=\inf_P\{m' \mid m\in M\}$, then $f\in P'$ and $\{y \mid y\in S$ and $My\subset K\}=fS$.*

Let us agree to call the focal ideal K *complete* in case for each nonempty subset M of S, the set $\{y \mid y\in S$ and $My\subset K\}$ is a principal right ideal generated by a projection. If K is a complete focal ideal, we will call the Baer *-semigroup (S, K) a *complete* Baer *-semigroup.

We are now ready to come to grips with the question of the existence of the infimum in P' of two elements $e, f\in P'$. In the special case in which e commutes with f, it is clear that $\inf_{P'}\{e, f\}$ exists and equals ef. The general case is easily settled as follows: Since $f'ee'\in K$, then $e'\leqq(f'e)'$, so e', hence also $e=e''$, commutes with $(f'e)'$. Consequently, $\inf_{P'}\{(f'e)', e\}$ exists and equals $(f'e)'e$. Since $f'(f'e)'e=f'e(f'e)'\in K$, then $(f'e)'e\leqq f$, and $(f'e)'e$ is a lower bound in P' for $\{e, f\}$. We assert that $(f'e)'e$ is, in fact, the infimum in P' of e and f. Indeed, if $q\in P'$ and if $q\leqq e, f$, then $f'eq=f'q=f'fq\in K$, i.e., $q\leqq(f'e)'$. Consequently, $q\leqq(f'e)'e$, and we have proved that $\inf_{P'}\{e, f\}$ exists and equals $(f'e)'e$.

Henceforth, we will use the notation $e\wedge f=(f'e)'e$ for the infimum in P' of the elements $e, f\in P'$. It is immediate that for $e, f\in P'$, the projection $(e'\wedge f')'=[(fe')'e']'$ is the supremum in P' of e and f, and we will accordingly write this supremum as $e\vee f=(e'\wedge f')'$. It follows from these considerations that P' is a lattice with greatest element 1 and smallest element $1'$, and that the mapping $e\rightarrow e'$ from P' onto P' provides the lattice P' with an orthocomplementation.

In [2], Loomis calls an orthocomplemented lattice L *weakly modular* in case $e, f \in L$ with $e \leqq f$ implies that $f = e \vee (f \wedge e')$. We observe that our lattice P' is automatically weakly modular. In fact, for $e, f \in P'$ with $e \leqq f$, we have $f = [(fe')'e']' = [(fe')' \wedge e']' = (fe')'' \vee e$. Since f commutes with e, it also commutes with e', hence, $fe' = f \wedge e' = (fe')''$, proving that $f = (f \wedge e') \vee e$.

We observe in passing that a subset N of P' has an infimum in P if and only if it has an infimum in P', and that the two infima, if they exist, must coincide. An analogous assertion cannot be made for suprema.

The following theorem constitutes a summary of the most important results obtained so far:

THEOREM 3. *Let (S, K) be a Baer *-semigroup with induced focal mapping $x \to x'$. Then, S has a unit* 1, 1′ *is a central projection, and the focal ideal K is a principal two-sided ideal generated by the projection* 1′. *Moreover, the set P' of K-closed projections in S forms an orthocomplemented weakly modular lattice with $e \to e'$ as orthocomplementation. The lattice P' is complete if and only if K is a complete focal ideal, i.e., if and only if (S, K) is a complete Baer *-semigroup.*

In the following section we will show that given any orthocomplemented weakly modular lattice L, we can always find a Baer *-semigroup (S, K) whose lattice of K-closed projections is isomorphic to L. Thus, it turns out that the orthocomplemented weakly modular lattices can be characterized as those lattices which arise as lattices of K-closed projections in Baer *-semigroups.

III. ORTHOCOMPLEMENTED WEAKLY MODULAR LATTICES

In the present section, the symbol L will always represent an orthocomplemented weakly modular lattice with orthocomplementarion $e \to e'$. A mapping $\phi: L \to L$ will be said to be *monotone* in case $e, f \in L$ with $e \leqq f$ implies that $e\phi \leqq f\phi$. We will denote by $M(L)$ the semigroup (under function composition) of all monotone maps on L. Borrowing some nomenclature from Halmos [3, p. 231], we will call a mapping $\phi: L \to L$ a *hemimorphism* of L in case $(e \vee f)\phi = e\phi \vee f\phi$ for $e, f \in L$ and $0\phi = 0$. We remark that a hemimorphism ϕ of L is automatically monotone and that it is also submultiplicative, i.e., $(e \wedge f)\phi \leqq e\phi \wedge f\phi$ for $e, f \in L$.

Given two elements ϕ, ϕ^* of $M(L)$, we will say that ϕ and ϕ^* are *mutually adjoint* in case the inequalities $(e'\phi)'\phi^* \leqq e$ and $(e'\phi^*)'\phi \leqq e$ hold for every element $e \in L$. We claim that if ϕ, ϕ^*, $\phi^+ \in M(L)$, and if both ϕ and ϕ^*, as well as ϕ and ϕ^+, are mutually adjoint, then $\phi^* = \phi^+$. In fact, let e be any element of L and put $f = e\phi^*$. Then, $f'\phi = (e\phi^*)'\phi \leqq e'$, i.e., $e \leqq (f'\phi)'$, hence $e\phi^+ \leqq (f'\phi)'\phi^+ \leqq f = e\phi^*$. Similarly, $e\phi^* \leqq e\phi^+$, so $e\phi^* = e\phi^+$. It follows that $\phi^* = \phi^+$.

Denote by $S(L)$ the subset of $M(L)$ consisting of all those monotone maps ϕ such that there exists at least one, hence exactly one, monotone map ϕ^* with the property that ϕ and ϕ^* are mutually adjoint. It is clear that if $\phi \in S(L)$, then $\phi^* \in S(L)$ and $\phi^{**} = \phi$.

THEOREM 4. *$S(L)$ is an involution semigroup (under function composition) with involution $\phi \to \phi^*$. $S(L)$ has a zero element and every element $\phi \in S(L)$ is a hemimorphism of the lattice L.*

Proof. Let ϕ, $\psi \in S(L)$, and let $e \in L$. Then, $(e\phi\psi)'\psi^*\phi^* \leqq (e\phi)'\phi^* \leqq e'$ and $(e\psi^*\phi^*)'\phi\psi \leqq (e\psi^*)'\psi \leqq e'$, proving that $(\phi\psi)^* = \psi^*\phi^*$. The constant mapping $e \to 0$, (henceforth denoted by the symbol 0), serves as a zero element for $S(L)$. Finally, let e, f be arbitrary elements in L and put $g = e \vee f$. If $\phi \in S(L)$, then, since ϕ is monotone, $e\phi \leqq g\phi$, $f\phi \leqq g\phi$. But, if $h \in L$ is such that $e\phi$, $f\phi \leqq h$, then, $h' \leqq (e\phi)'$, $(f\phi)'$ and $h'\phi^* \leqq (e\phi)'\phi^*$, $(f\phi)'\phi^*$. It follows that $h'\phi^* \leqq e', f'$, i.e., that $e, f \leqq (h'\phi^*)'$. Consequently, $g \leqq (h'\phi^*)'$, so $g\phi \leqq (h'\phi^*)'\phi \leqq h$. This proves that $(e \vee f)\,\phi = e\phi \vee f\phi$. Finally, let us prove that for $e \in L$, $e\phi = 0$ if and only if $e \leqq (1\phi^*)'$. Indeed, if $e\phi = 0$, then $1\phi^* = (e\phi)'\phi^* \leqq e'$, so $e \leqq (1\phi^*)'$. Conversely, $e \leqq (1\phi^*)'$ implies $e\phi \leqq (1\phi^*)'\phi \leqq 1' = 0$, hence $e\phi = 0$. In particular, then, $0\phi = 0$, so ϕ is a hemimorphism.

LEMMA 5. *Let ϕ, $\psi \in S(L)$. Then, $\phi\psi = 0$ if and only if $1\phi \leqq (1\psi^*)'$.*

Proof. If $1\phi \leqq (1\psi^*)'$, then $e \in L$ implies $e\phi\psi \leqq 1\phi\psi = 0$, so $\phi\psi = 0$. Conversely, if $\phi\psi = 0$, then $(1\phi)\,\psi = 0$, so $1\phi \leqq (1\psi^*)'$.

For each element $g \in L$, we now define a mapping $\phi_g \in M(L)$ in accordance with the prescription $e\phi_g = (e \vee g') \wedge g$ for $e \in L$. We will prove that ϕ_g is a projection in the involution semigroup $S(L)$. To this end, we first notice that for h, $g \in L$, $h \leqq g$ implies that $h = h\phi_g$, because $h \leqq g$ implies $g' \leqq h'$, and the weak modularity of L gives $h = (h' \wedge g) \vee g'$, i.e., $h = (h \vee g') \wedge g = h\phi_g$. It follows immediately that $\phi_g = \phi_g^2$. Moreover, for $e \in L$, $(e\phi_g)'\phi_g =$

$[(e \vee g') \wedge g]' \phi_g = [(e' \wedge g) \vee g'] \phi_g = [(e' \wedge g) \vee g'] \wedge g = (e' \wedge g) \phi_g = e' \wedge g \leqq e'$; hence, ϕ_g^* exists and equals ϕ_g.

We have now assembled all the information needed to prove the following theorem, which is the main theorem of the paper:

THEOREM 6. *If L is any orthocomplemented weakly modular lattice, then* $(S(L), \{0\})$, *is a Baer *-semigroup, and the correspondence* $g \leftrightarrow \phi_g$ *between L and* $P'(S(L))$ *is an orthocomplementation preserving lattice isomorphism.*

Proof. Let $\phi \in S(L)$, and put $g = (1\phi)'$. By Lemma 5, $\phi\phi_g = 0$. On the other hand, suppose that $\psi \in S(L)$ is such that $\phi\psi = 0$. Again by Lemma 5, we have $1\psi^* \leqq g$, hence for $e \in L$, $e\psi^* \leqq g$, so $e\psi^*\phi_g = e\psi^*$, and consequently, $\psi^*\phi_g = \psi^*$, i.e., $\psi = \phi_g^*\psi = \phi_g\psi$. It follows that $\{\psi \mid \phi\psi = 0\}$ is a principal right ideal in $S(L)$ generated by the projection ϕ_g, i.e., that $\{0\}$ is a focal ideal in $S(L)$. It is evident that for $e, f \in L$, $e \leqq f$ if and only if $\phi_e \phi_f = \phi_e$, so that the correspondence $g \leftrightarrow \phi_g$ between L and $P'(S(L))$ is a lattice isomorphism. Furthermore, $(\phi_g)' = \phi_{g'}$, so that this lattice isomorphism preserves orthocomplementation.

In [4], von Neumann has shown that if L is an orthocomplemented *modular* lattice with four or more independent perspective elements, then there exists a *-regular ring R (uniquely determined up to an isomorphism), called the *coordinate ring* for R, such that L is isomorphic to the lattice of all projections of R. Note that if S represents the multiplicative semigroup of R, then $(S, \{0\})$ is a Baer *-semigroup, and the lattice $P'(S)$ is the lattice of all projections of R; hence $P'(S)$ is isomorphic to the lattice L.

We are thus led to define a *coordinate Baer *-semigroup* for an orthocomplemented weakly modular lattice L to be a Baer *-semigroup $(S, \{0\})$ with the property that the lattice $P'(S)$ is isomorphic to L and the isomorphism in question preserves orthocomplementation. The content of Theorem 6, then, is that *any orthocomplemented weakly modular lattice L possesses at least one coordinate Baer *-semigroup, namely* $(S(L), \{0\})$. The coordinate Baer *-semigroup $S(L)$ is a weak substitute for the coordinate ring of von Neumann in the case in which L is not modular, but only weakly modular.

Incidentally, no use was made of the weak modularity of L up to and including Lemma 5 in the present section, hence the involution semigroup $S(L)$ is available whenever L is any orthocomplemented lattice. It is not difficult to prove that $\{0\}$ is a focal ideal in $S(L)$ if *and only if* L is a weakly modular lattice, but we will not give the "only if" proof in this paper.

IV. THE NATURAL HOMOMORPHISM FROM S INTO $S(P'(S))$

If L is an orthocomplemented modular lattice, its coordinate ring is determined up to an isomorphism, but this is manifestly not the case for the coordinate Baer *-semigroups of a weakly modular L. Thus, if S is a Baer *-semigroup and $L=P'(S)$, the coordinate Baer *-semigroup $S(L)$ for L need not be isomorphic to S; however, there does exist – as we will demonstrate in the present section – a natural involution preserving homomorphism ϕ from S into $S(L)$.

For each element $x\in S$, define a mapping $\phi_x : L\to L$ in accordance with the prescription $e\phi_x=(ex)''$ for $e\in L$. We will see that if g is a projection in $P'(S)$, the mapping ϕ_g as just defined coincides with the hemimorphism $\phi_g\in S(L)$ defined in the previous section, so there will be no notational conflict. It is easy to verify that $1'\phi_x=1'$ and that $e\phi_x\leqq x''$ for every $x\in S$, $e\in L$. The following lemma shows that ϕ_x is a hemimorphism of L:

LEMMA 7. *Let $\{e_\alpha\}\subset L$ and suppose that $e=\bigvee\{e_\alpha\}$ exists in L. Then, for any element $x\in S$, $\bigvee\{e_\alpha\phi_x\}$ exists in L and is equal to $e\phi_x$.*

Proof. Let K be the focal ideal for S. For any α, $e_\alpha x(ex)'=e_\alpha ex(ex)'\in K$, so $(ex)'\leqq(e_\alpha x)'$ and $e_\alpha\phi_x\leqq e\phi_x$. On the other hand, if $q\in L$ is such that $e_\alpha\phi_x\leqq q$ for every α, then for every α, $e_\alpha xq'\in K$, $q'x^*e\in K$, hence $e_\alpha\leqq(q'x^*)'$. It follows that $e\leqq(q'x^*)'$, $q'x^*e\in K$, $exq'\in K$, $q'\leqq(ex)'$, and consequently, $e\phi_x\leqq q$.

Now, let $x\in S$, $e\in L$. Note that since $(ex)'x^*e\in K$, then $e\leqq[(ex)'x^*]'$, i.e., $(e\phi_x)'\phi_{x^*}\leqq e'$. The latter inequality shows that $\phi_x\in S(L)$, in fact, that $(\phi_x)^*=\phi_{x^*}$.

THEOREM 8. *For $x, y\in S$, $\phi_{xy}=\phi_x\phi_y$, hence, the mapping $\phi : S\to S(L)$ defined by $\phi(x)=\phi_x$ for $x\in S$ is an involution preserving semigroup homomorphism from S into $S(L)$. Moreover, for $e, g\in L$, $e\phi_g=(eg)''=(e\vee g')\wedge g$.*

Proof. Let $x, y\in S$, $e\in L$, and put $ex=a$. Then, $ay(a''y)'=aa''y(a''y)'\in K$, so $(ay)''\leqq(a''y)''$. Also, $a[a''y(ay)']=ay(ay)'\in K$, hence $a''y(ay)'=a'a''y(ay)'\in K$, and so $(a''y)''\leqq(ay)''$, proving that $e\phi_{xy}=(e\phi_x)\phi_y$. Finally, the mappings $e\to(eg)''$ and $e\to(e\vee g')\wedge g$ are both now known to be projections in the Baer *-semigroup $S(L)$, and elementary calculation reveals that each majorizes the other.

Notice that if the homomorphism ϕ of the previous theorem is restricted to the subset L of S, one obtains an isomorphism of the lattice L onto the lattice of all $\{0\}$-closed projections in $S(L)$. One consequence of this fact is that if e, f are projections in $P'(S)$, then one can decide whether or not e and f commute in S by checking to see whether or not the hemimorphisms ϕ_e and ϕ_f commute in $S(L)$; hence, if L is any orthocomplemented weakly modular lattice, the commutativity of two elements of L in any coordinate Baer *-semigroup for L implies their commutativity in any other coordinate Baer *-semigroup for L. This suggests a natural way in which von Neumann's notion of the center of a complemented modular lattice [4] can be carried over to the case of an orthocomplemented weakly modular lattice. It turns out that the center of such an L is a Boolean algebra, complete if L is complete, and hence that L itself is a Boolean algebra if and only if it is a subsemigroup of every one of its coordinate Baer *-semigroups.

Lehigh University and
Wayne State University

NOTE

† This paper contains part of the author's doctoral dissertation (Tulane, 1958), written under the direction of Professor F. B. Wright.

BIBLIOGRAPHY

[1] Kaplansky, I., *Rings of Operators*, University of Chicago Mimeographed Notes, 1955.
[2] Loomis, L. H., 'The Lattice Theoretic Background of the Dimension Theory of Operator Algebras', *Memoirs Amer. Math. Soc.*, No. 18, 1955.
[3] Halmos, P. R., 'Algebraic Logic, I, Monadic Boolean Algebras, *Compositio Math.* **12** (1955), 217–249.
[4] von Neumann, J., *Continuous Geometry*. Parts I, II, III, Princeton University Planographed Notes, 1937.

NEAL ZIERLER

AXIOMS FOR NON-RELATIVISTIC QUANTUM MECHANICS*

INTRODUCTION

In the approach to the axiomatization of quantum mechanics of George W. Mackey [7], a series of plausible axioms is completed by a final axiom that is more or less *ad hoc*. This axiom states that a certain partially ordered set – the set P of all two-valued observables – is isomorphic to the lattice of all closed subspaces of Hilbert space. The question arises as to whether this axiom can be deduced from others of a more *a priori* nature, or, more generally, whether the lattice of closed subspaces of Hilbert space can be characterized in a physically meaningful way. Our central result is a characterization of this lattice which may serve as a step in the indicated direction, although there is not now a precise sense in which our axioms are more plausible than his. Its principal features may be described as follows.

Suppose that P is an atomic lattice, define an element to be *finite* if it is the join of a finite number of points, and suppose that the unit element is not finite, but is the join of a countable set of points. Suppose for the moment that

(F) The lattice under every finite element of P is a real (or complex) projective geometry.

Then one additional axiom, which appears to be particularly mild from an operational viewpoint, is sufficient and necessary for us to show that P is isomorphic to the lattice of closed subspaces of a separable, infinite dimensional real (or complex) Hilbert space.

Of course, (F) is not taken as an axiom, but is deduced from more primitive assumptions. This part of the development follows well-known lines, but the structure of P (and its set S of states) permits us to give it a rather simple form. For example, in order to conclude that the lattice under every finite element of P is a projective geometry, we need make,

C. A. Hooker (ed.), The Logico-Algebraic Approach to Quantum Mechanics, 149–170.

in addition to the atomicity of P, only the following three assumptions: P is not a Boolean algebra; the lattices under any pair of finite elements of the same dimension are isomorphic; a certain weak (and rather intuitive) form of the modular law holds under finite elements (Theorem 2.1).

In a preliminary chapter we examine the interrelation of a number of regularity properties which a pair P, S satisfying a slight refinement of Mackey's basic axioms might have, and show that a few of the more plausible properties imply all the others (Theorem 1.1).

This work is a modification of part of a thesis submitted to the Department of Mathematics of Harvard University in partial fulfillment of the requirements for the degree of Doctor of Philosophy.

1. Events and states: preliminaries

Let P be a partly ordered set with least and greatest elements 0 and 1 respectively. If the greatest lower bound or least upper bound of elements a and b of P exists in P it is denoted ab or $a \vee b$ respectively. Let $a \to a'$ be an orthocomplementation in P; that is, for each $a \in P$, $a' \in P$ and

(1) $(a')' = a$,

(2) $a < b$ if and only if $b' < a'$,

(3) a' is a complement of a; i.e., $a'a$ and $a \vee a'$ exist and equal 0 and 1 respectively.

Two elements a and b of P are said to be *orthogonal*, $a \perp b$, if and only if $a \leqq b'$. Clearly $a \perp b$ is equivalent to $b \perp a$. If Q is a set of pairwise orthogonal elements of P we shall say, for short, that Q is orthogonal. It is easy to see that De Morgan's law holds in P: $(ab)' = a' \vee b'$ in the sense that if either ab or $a' \vee b'$ exists, so does the other and the equality holds.

We assume that P satisfies

(L1) If $\{a_1, a_2, \ldots\}$ is orthogonal, then $\bigvee a_i$ exists in P.

It follows readily that a variety of sups and infs exists in P: e.g., $b'c'$, ba' and $ba' \vee a$ if $b \perp c$ and $a \leqq b$; if $b_1 \leqq b_2 \leqq \ldots$ then $\bigvee b_i = b_1 \vee b_2 b_1' \vee b_3 b_2' \vee \ldots$.

Consider the following three properties for P.

(W) $a \leqq b$ implies $b = ba' \vee a$,
(W1) $a \leqq b$ and $ba' = 0$ imply $a = b$,
(W2) $a \leqq c$ and $b \perp c$ imply $(a \vee b)\, c = a$.[1]

LEMMA 1.1. *If* (W) *holds then* $a \perp b$ *implies* $b = (a \vee b)\, a'$.

Proof. $a \leqq b'$ so $b' = b'a' \vee a$ by (W) and $b = (b'a' \vee a)' = (a \vee b)\, a'$

LEMMA 1.2. *If* (W) *holds and a, b and c are pairwise orthogonal, then* $(a \vee b)(a \vee c) = a$ *and* $(a \vee b)(a \vee c)' = b$.

Proof. $b \leqq a'$, $b \leqq c'$ imply $b \leqq a'c'$ so $a'c' = a'c'b' \vee b$. Then $a = a(a \vee c \vee b) = (a \vee b)\, b'(a \vee c \vee b)$ by Lemma 1.1

$$= (a \vee b)(a'c'b' \vee b)' = (a \vee b)(a'c')' = (a \vee b)(a \vee c).$$

Since $b = (a \vee b)\, a'$ by Lemma 1.1 and $b \leqq c'$, $b = bc' = (a \vee b)\, a'c' = (a \vee b)(a \vee c)'$.

LEMMA 1.3. (W), (W1) *and* (W2) *are equivalent.*

(W1) implies (W). Suppose $a \leqq b$. Then $a \vee ba' \leqq b$ holds trivially and $b(a \vee ba')' = b(a'(ba')') = ba'(ba')' = 0$ so $b = a \vee ba'$ by (W1).

(W) implies (W2). If $a \leqq c$ and $b \perp c$, then ca', a and b are orthogonal so $a = (a \vee b)(a \vee ca')$ Lemma 1.2 (since (W) holds) $= (a \vee b)\, c$ by (W).

(W2) implies (W1). Suppose $a \leqq b$ and $ba' = 0$. Then $b \perp b'$ so, by (W2), $a = (a \vee b')\, b = (ba')'b = 0'b = 1b = b$.

P is said to be *weakly modular* (relative to the given orthocomplementation) if it satisfies any and hence all of (W), (W1) and (W2). We assume now that P is weakly modular and, borrowing a traditional term from the theory of probability, we call its members *events*.

Two events a_1, a_2 are said to *commute* or to be *simultaneously measurable* if there exist pairwise orthogonal events b_1, b_2 and c such that $a_i = b_i \vee c$. The set of all events which commute with all other events is called the center $\mathscr{C}$ of P. If $\mathscr{C} = P$, P is said to be *commutative* or *deterministic*. It is an easy consequence of Lemma 1.2 that a and b commute if and only if ab, ab' and $a'b$ exist, $a = ab \vee ab'$ and $b = ab \vee a'b$, and hence that P is deterministic if and only if it is a Boolean algebra.

LEMMA 1.4. *Suppose* ab *and* ab' *exist and* $a = ab \vee ab'$. *Then a and b commute.*

Proof. $a' = (ab \vee ab')' = (ab)'(ab')' = (ab)'(b \vee a') \geqq (ab)'b$ while $b \geqq (ab)'b$ holds trivially. On the other hand, if $a' \geqq c$ and $b \geqq c$ then $(ab)' \geqq a' \geqq c$ so $(ab)'b \geqq c$. Hence $(ab)'b = a'b$ and so $b = (ab)'b \vee ab = a'b \vee ab$.

COROLLARY. *If a and b commute, so do a and b'.*

Proof. The statement of the lemma is symmetric in b and b'.

LEMMA 1.5. *Suppose P is a lattice. Then P is a Boolean algebra if and only if $ab = 0$ implies $a \perp b$.*

Proof. If P is a Boolean algebra and $ab = 0$, then $a1 = a = a(b \vee b') = ab \vee ab' = ab'$ so $a \leqq b'$. Conversely, for any a and b, $a(ab)'b = 0$ so $a(ab)' \leqq b'$ by hypotheses. Then $a = ab \vee a(ab)' = ab \vee a(ab)'b' = ab \vee ab'$ since $b' \leqq (ab)'$.

If we interpret the weakly modular lattice P as the logic of an abstract physical system,[2] $a \leqq b$ means "a implies b" and a' is the event "not a". If $a \perp b$, it is natural to say that a and b are "mutually exclusive" – a implies not b and b implies not a – and in this case the question of the simultaneous occurrence of a and b is completely settled. If, however, $ab = 0$ but a and b are not orthogonal, no experiment exists for the system whose outcome can indicate that a and b have both occurred even though a and b are not mutually exclusive. According to Lemma 1.5, the absence of this uncertainty is equivalent to the commutativity of P.

Digression. It may be shown that the notion of determinacy is further characterized in the following three ways (the statements depend on definitions which appear below). We suppose given a system of states and events $\mathscr{S}$, P.

(i) Let X denote the real linear space of signed measures on P generated by $\mathscr{S}$. P is deterministic if and only if X is a pre-L-space in a certain natural sense (see [4]).

(ii) Define an *observable*, as in [7], to be a function A from the Borel subsets of the real line R to P such that $A_\phi = 0$, $A_R = 1$, $A_E \perp A_E$ if $E \cap F = \phi$ and $A_{\cup E_i} = \sum A_{E_i}$ if $E_i \cap E_j = \phi$ for $i \neq j$; A is *bounded* if $A_E = 1$ for some bounded Borel set E. Given $x \in X$ (see (i) above) and a bounded observable A, let $\mu_{x,A}$ denote the Borel measure on the line: $\mu_{x,A}(E) = x(A_E)$ and let L_A denote the functional on X: $L_A(x) = \int_{-\infty}^{\infty} \lambda d\mu_{x,A}(\lambda)$. The set Y of all such L_A is partially ordered as a subset of the dual of the partially ordered linear space X. P is deterministic if and only if Y is a lattice.

(iii) Suppose P has a unit. Then P is deterministic if and only if every pair A, B of observables is simultaneously measurable in the following intuitive sense: there exist an observable C and Borel functions α and β from R to R (depending on A and B) such that $A=\alpha(C)$ and $B=\beta(C)$ (where, by definition, $\alpha(C)_E=C_{\alpha^{-1}(E)}$).

A function f from the weakly modular partially ordered set P to the closed real unit interval is said to be a *state* for P if $f(1)=1$ and f is countably additive in the sense that whenever $\{a_i\}$ is orthogonal, $f(\bigvee a_i)=$ $=\sum f(a_i)$. It is easy to see that if f is a state and $\{b_i\}$ is an increasing (decreasing) sequence of events with sup (inf) b, then $f(b_i)\to f(b)$.

Now suppose there exists a set $\mathscr{S}$ of states such that

(D) $\quad a\leqq b$ if and only if $f(a)\leqq f(b)$ for all f in $\mathscr{S}$.

Of course, if $a\leqq b$ and f is any state, $f(a)=f(b)-f(ba')\leqq f(b)$. We observe that

E1. If $f(a)=f(b)$ for all f in $\mathscr{S}$, $a=b$,

E2. For each $a\in P$ there exists $b\in P$ such that $f(b)=1-f(a)$ for all f in $\mathscr{S}$; there exists $c\in P$ such that $f(c)=0$ for all f in $\mathscr{S}$,

E3. Let $\{a_1, a_2,\ldots\}$ be a sequence of elements of P such that $i\neq j$ and $f\in\mathscr{S}$ imply $f(a_i)+f(a_j)\leqq 1$. Then there exists $a\in P$ such that $f(a)=\sum f(a_i)$ for all $f\in\mathscr{S}$.

Indeed, E1 is immediate from (D) and in E2 we need only set $b=a'$, $c=0$. In E3, $f(a_i)\leqq 1-f(a_j)=f(a_j')$ implies $a_i\leqq a_j'$ by (D) so the $\{a_i\}$ are mutually orthogonal and we may set $a=\bigvee a_i$.

Suppose, on the other hand, that we are given a set P (without any *a priori* structure) and a set $\mathscr{S}$ of functions from P to the closed unit interval satisfying E1–E3. The elements b and c of E2 are unique by E1 and are denoted a' and 0 respectively and we let $1=0'$; the element a of E3 is also unique by E1 and is denoted $\sum a_i$. Let a partial ordering be defined in P by (D); evidently 0 and 1 are the least and greatest elements of P and $a\to a'$ is an orthocomplementation. We shall show that the orthocomplemented partly ordered set P is weakly modular and $\mathscr{S}$ is a collection of states for P (which trivially satisfies (D)).

Let $\{a_i\}$ be orthogonal, let $I_1, I_2,\ldots$ be a partition of the positive integers and let $b_i=\sum_{j\in I_i} a_j$ where $\sum$ denotes the sum of the a_j in the sense of E3. It follows at once from the fact that the sum of a convergent series of nonnegative numbers is unaffected by a rearrangement of its terms that the b_i are pairwise orthogonal and $\sum b_i = \sum a_i$. As a particular case we have,

LEMMA 1.6. *If* $a_1, a_2, \ldots$ *are pairwise orthogonal and* $b \perp a_i$ *for every* i, *then* $b \perp \sum a_i$.

LEMMA 1.7. *If* $a_1, a_2, \ldots$ *are pairwise orthogonal, then* $\sum a_i = \bigvee a_i$.

Proof. Clearly $a = \sum a_i \geqq a_j$ for all j. If $b \geqq a_j$ for all j, $a_j \perp b'$ so $a \perp b'$ by Lemma 1.6; i.e., $b \geqq a$.

Now suppose $a \leqq b$. Then $a \perp b'$ so $a + b'$ exists by E3 and equals $a \vee b'$ by Lemma 1.7; hence $ba' = (a \vee b')'$ exists. Since $ba' \perp a$, $ba' \vee a$ exists by E3 and Lemma 1.7. Then if $f \in \mathscr{S}$,

$$f(ba' \vee a) = f(ba') + f(a) = 1 - f((ba')') + f(a) = 1 - f(b' \vee a) + f(a) = 1 - f(b') - f(a) + f(a) = 1 - f(b') = f(b)$$

and it follows from E1 that $b = ba' \vee a$, i.e., (W) holds and P is weakly modular. If $\{a_i\}$ is orthogonal, $f(\bigvee a_i) = f(\sum a_i)$ by Lemma 1.7 $= \sum f(a_i)$ and so f is a state for P.

Let P be a weakly modular partially ordered set and let $\mathscr{S}$ be a family of states for P which determines the order relation in P (as in (D)). The pair $\mathscr{S}$, P will be called a *system of state and events*, or simply a *system*, if it has the following five properties.

E4. (Axiom of separability) Every orthogonal subset of P contains at most countably many non-zero elements.

E5. P is a lattice.

S1. $\mathscr{S}$ is closed under countable convex combination; i.e., if $f_1, f_2, \ldots$ are in $\mathscr{S}$ and $\lambda_1, \lambda_2, \ldots$ are nonnegative real numbers with $\sum \lambda_i = 1$, then $\sum \lambda_i f_i \in \mathscr{S}$.

S2. If a is a non-zero event, there exists $f \in \mathscr{S}$ such that $f(a) = 1$.

S3. If $f(a) = 0$ and $f(b) = 0$, then $f(a \vee b) = 0$.

The following series of lemmas, culminating in Theorem 1.1, develop a number of regularity properties that systems enjoy; interrelations among the properties are exhibited in accompanying remarks.

LEMMA 1.8. *Suppose P is separable* (i.e., satisfies E4). *Then if Q is a nonempty chain in P, a sequence $\{a_1, a_2, \ldots\}$ of elements of Q may be found such that $\bigvee a_i = \sup Q$; in particular,* $\sup Q$ *exists in P.*

Proof. Let Q be a nonempty chain in P and let T be the set of all events of the form ab' where $a \in Q$ and $b < a$. Let $\{c_i\}$ be a maximal set of pairwise orthogonal non-zero elements of T which exists by Zorn's lemma and is countable by E4. Say $c_i = a_i b_i'$ where $a_i \in Q$, $a_1 < a_2 < \cdots$ and $b_i < a_i$ and let $a = \bigvee a_i$. Suppose there exists $b \in Q$ such that $b \nleqq a$. Then since $b \nless a_i$, $a_i < b$ holds for all i since Q is a chain. Hence $a < b$ and the non-zero event ba' belongs to T and is orthogonal to all the c_i contrary to the maximality of $\{c_i\}$.

A *cut* in P is a subset of P which contains all lower bounds of the set of its upper bounds. If $Q \subseteqq P$, we denote by $\bar{Q}$ the smallest cut containing Q. Thus, for $a \in P$, $\bar{a} = \{b \in P : b \leqq a\}$ and for $Q \subseteqq P$, $\bar{Q} = \bigcap \bar{a} : Q \subseteqq \bar{a}$. The mapping $Q \to \bar{Q}$ is evidently a closure operation in the power class $\mathscr{B}(P)$ of P (see [1]); hence the set $\bar{P}$ of all cuts in P is a complete lattice under inclusion.

LEMMA 1.9. *If P is a lattice and every chain in P has a sup in P, then P is a complete lattice.*

Remark. If P is a lattice and $\{b_i\} \subseteqq P$, $\bigvee b_i = \bigvee (b_1 \vee \cdots \vee b_i)$, i.e., P is σ-complete.

Proof. Suppose $Q \subseteqq P$, let Q_1 be a chain in $\bar{Q}$ and let $b = \sup Q_1$. If $a \in P$ such that $Q \subseteqq \bar{a}$, then $Q_1 \subseteqq \bar{a}$ so $b \leqq a$, i.e., $b \in \bar{Q}$. It follows now from Zorn's lemma that $\bar{Q}$ contains a maximal element b. The assumption that P is a lattice clearly implies that $\bar{Q}$ is a sublattice of P and so if $a \in Q$, $a \vee b \in \bar{Q}$. Then by the maximality of b in $\bar{Q}$, $a \vee b = b$ so $a \leqq b$; thus, $Q \subseteqq \bar{b}$ and $b = \sup Q$. Dually, $\inf Q = (\sup a' : a \in Q)'$.

For $Q \subseteqq P$ let $Q^\circ = \{f \in \mathscr{S} : f(a) = 0$ for all $a \in Q\}$ and if $T \subseteqq \mathscr{S}$ let $T^\circ = \{a \in \mathscr{S} : f(a) = 0$ for all $f \in T\}$. Clearly $Q = Q^{\circ\circ}$ and if $Q_1 \subseteqq Q_2$, then $Q_2^\circ \subseteqq Q_1^\circ$, and similarly for the subsets of $\mathscr{S}$. The first relation implies $Q^\circ \subseteqq Q^{\circ\circ\circ}$ and the second applied to the first yields $Q^{\circ\circ\circ} \subseteqq Q^\circ$; thus $Q^\circ = Q^{\circ\circ\circ}$ and similarly, $T^\circ = T^{\circ\circ\circ}$. A subset H of P or of $\mathscr{S}$ such that $H = H^{\circ\circ}$ is called an *annihilator*, and the mapping $H \to H^\circ$ is a one-to-one inclusion inverting correspondence between the annihilators in $\mathscr{B}(P)$ and those in $\mathscr{B}(\mathscr{S})$. In this notation, S3 is: $a^\circ \cap b^\circ = (a \vee b)^\circ$. It is easy to see that if $\mathscr{S}$, P has any one of the following three properties, it has the others.

(4) $a^{\circ\circ} \subseteqq b^{\circ\circ}$ implies $a \leqq b$,
(5) $b^{\circ} \subseteqq a^{\circ}$ implies $a \leqq b$,
(6) if $f(a)=1$ whenever $f(b)=1$, then $b \leqq a$.

LEMMA 1.10. *If* E5, S2 *and* S3 *hold for* $\mathscr{S}$, P, *so do* (4)–(6).

Proof. Suppose $b^{\circ} \subseteqq a^{\circ}$. Then $b^{\circ} = b^{\circ} \cap a^{\circ} = (a \vee b)^{\circ}$ by S3. If $a \vee b \neq b$, then $(a \vee b)\, b' \neq 0$ by (W1) so, by S2, there exists $f \in \mathscr{S}$ such that $f((a \vee b)\, b') = 1$. Then $f \in b^{\circ}$ so $f \in a^{\circ}$ and $f(a \vee b) = 0$ by S3. But $f(a \vee b) \geqq f((a \vee b)\, b') = 1$, so $a \vee b = b$, $a \leqq b$ must hold.

LEMMA 1.11. *Suppose P is a separable lattice. Then P is a complete lattice and $Q \subseteqq P$ implies there exists $Q_1 \subseteqq Q$ with at most countably many elements such that* $\sup Q_1 = \sup Q$.

Proof. P is a complete lattice by Lemmas 1.8 and 1.9. Let Q be a non-empty subset of P and let $a = \sup Q$. Let T denote the set of all joins of countable subsets of Q. If T_1 is a chain in T, its join is obtainable as the join of a countable subsets of T_1 by Lemma 1.8 and hence belongs to T. Hence, we may use Zorn's lemma to extract a maximal element a from T, and then, clearly, $a = \sup Q$.

Remark. The converse is also true. Indeed, suppose $\{\alpha_\alpha\}$ is orthogonal and a is its join; by hypothesis $a = \bigvee a_{\alpha_i}$ for appropriate α_i. If $\alpha \notin \{\alpha_i\}$, then $a_\alpha \perp a$ by Lemma 1.6. Since $a_\alpha \leqq a$ by definition of a, $a_\alpha = 0$.

We consider now the general form of S3:

(7) If a is the sup of the subset Q of P and $f(b)=0$ for all $b \in Q$, then $f(a)=0$ (equivalently: $Q^{\circ} = a^{\circ}$).

It is easy to see that if E5 and S3 hold, so does (7) whenever Q has countably many elements.

LEMMA 1.12. *If* E4, E5 *and* S3 *hold for* $\mathscr{S}$, P, *so does* (7).

Proof. Let $Q \subseteqq P$, $a = \sup Q$ and let $f \in Q^{\circ}$. By Lemma 1.11 we may choose a sequence $\{a_i\} \subseteqq Q$ such that $a = \bigvee a_i$; let $b_n = a_1 \vee \cdots \vee a_n$. Then $b_1 \leqq b_2 \leqq \cdots$, $f(b_i) = 0$ imply $f(\bigvee b_i) = f(a) = 0$.

An event a is said to be a *carrier* of the state f on P if $f(b)=0$ is equivalent to $b \perp a$; if a carrier exists for f it is clearly unique and is denoted a_f. Evidently, if f is a state with a carrier, $f(b)=1$ if and only if $a_f \leqq b$, and $f^\circ = a_f'^{\circ\circ} = \overline{a_f'}$.

LEMMA 1.13. *Suppose P is a complete lattice. Then if $\mathscr{S}$, P satisfies* (7), *it also satisfies*

(8) Every $f \in \mathscr{S}$ has a carrier in P.

Proof. $a_f = (\sup f^\circ)'$.

Remark 1. Conversely, if P is a complete lattice and $\mathscr{S}$, P satisfies (8), then (7) holds. Indeed, if $Q \subseteqq P$ and $a = \sup Q$, let $f \in Q^\circ$. Then $b \leqq a_f'$ for all $b \in Q$ so $a \leqq a_f'$ and hence $f \in a^\circ$, i.e., $Q^\circ \subseteqq a^\circ$.

Remark 2. (8) is equivalent to the following: $Q \subseteqq P$, $f \in Q^\circ$, $f(a)=1$ imply there exists $b \leqq a$ such that $f(b)=1$ and $b \perp Q$. For if (8) holds, we may take $b = a_f$ while, conversely, given f, observe that since $f(1)=1$, the hypothesis implies the existence of b such that $f(b)=1$ and $b \perp f^\circ$; clearly $b = a_f$.

LEMMA 1.14. *Suppose $\mathscr{S}$, P satisfies* (4)–(6) *and* (8). *Then it also satisfies*

(9) $\bar{Q} = Q^{\circ\circ}$ *for every subset Q of P.*

Proof. $Q^{\circ\circ} = \bigcap \overline{a_f'} : f \in Q^\circ$ by definition and (8). But $Q \subseteqq \overline{a_f'}$ for all $f \in Q^\circ$ so $\bar{Q} = \{\bigcap \bar{a} : Q \subseteqq \bar{a}) \subseteqq (\bigcap \overline{a_f'} : f \in Q^\circ\} = Q^{\circ\circ}$. On the other hand, $b \in Q^{\circ\circ}$ implies $Q^\circ = Q^{\circ\circ\circ} \subseteqq b^\circ$ while $Q \subseteqq \bar{a}$ implies $a^\circ \subseteqq Q^\circ$. Hence $a^\circ \subseteqq b^\circ$ so $b \leqq a$ by (5). Thus, $Q^{\circ\circ} \subseteqq \bar{Q}$ so $Q^{\circ\circ} = \bar{Q}$.

Remark. If $\mathscr{S}$, P satisfies (9), it also satisfies (4)–(6) and (7). Indeed, (4)–(6) are immediate. To prove (7), suppose $a \in P$ is the sup of the subset Q of P. Then $\bar{a} = \bar{Q} = Q^{\circ\circ}$ by (9) so $Q^{\circ\circ\circ} = Q^\circ = a^\circ$.

LEMMA 1.15. *Suppose $\mathscr{S}$, P satisfies* E4, E5, S1, S2 *and* (8). *Then* (10) *Every non-zero event is the carrier of some $f \in \mathscr{S}$.*

Proof. We may use the conclusion of Lemma 1.11. Assuming $a \neq 0$, it follows from S2 that $a'^\circ \neq \phi$; let $b = \bigvee a_f : f \in a'^\circ$. Since $a_f \leqq a$ for all $f \in a'^\circ$,

$b \leqq a$. If $ab' \neq 0$, choose $g \in \mathscr{S}$ with $g(ab')=1$; then $0=g((ab')')=g(a' \vee b) \geqq g(a')$ so $g \in a'^{\circ}$ and $a_g \leqq b$ by definition of b. On the other hand, $g(b') \geqq g(ab')=1$ implies $a_g \leqq b'$ so $a_g=0$. Since 0 cannot be the carrier of a state, $ab'=0$ must hold and so $a=b$ by (W1). Choose $\{f_1, f_2, \ldots\} \subseteqq a'^{\circ}$ such that $a=\bigvee a_{f_i}$; $f_0=f_1/2+f_2/2^2+\cdots$ belongs to $\mathscr{S}$ by S1. Then $f_0(a)=1$ so $a_{f_0} \leqq a$; but, clearly, $f_i(a_{f_0})=1$ so $a_{f_i} \leqq a_{f_0}$ for $i=1,2,\ldots$ and $a=a_{f_i} \leqq a_{f_0}$. Hence $a=a_{f_0}$ and the proof is complete.

Remark. If $\mathscr{S}$, P satisfies (10), it also satisfies (4)–(6), for suppose $a^{\circ\circ} \subseteqq b^{\circ\circ}$. Now $a \leqq b$ holds trivially if $b=1$ so suppose $b \neq 1$ and choose $f \in \mathscr{S}$ in accordance with (10) such that $a_f=b'$. Then $\bar{a} \subseteqq a^{\circ\circ} \subseteqq b^{\circ\circ}=a_f'^{\circ\circ}=\overline{a_f'}=\bar{b}$ so $a \leqq b$.

A state f on P such that $f(a)=0$ implies $a=0$ is said to be a *unit* for P. It is easy to see that if P has a unit, it is separable.

LEMMA 1.16. *If $\mathscr{S}$, P satisfies* (10), *$\mathscr{S}$ contains a unit.*

Proof. $f \in \mathscr{S}$ such that $a_f=1$ is a unit.

We have proved, in particular:

THEOREM 1.1. *Let $\mathscr{S}$, P be a system of states and events. Then P is a complete lattice and the sup of any infinite family of its elements is obtainable as the sup of a countable subfamily. Furthermore, $\mathscr{S}$ contains a unit for P, and the pair $\mathscr{S}$, P has the following properties.*

(6) *If $f(a)=1$ whenever $f(b)=1$, then $b \leqq a$.*
(7) *If $Q \subseteqq P$ and $f(b)=0$ for all $b \in Q$, then $f(\sup Q)=0$.*
(8) *Every $f \in \mathscr{S}$ has a carrier in P.*
(9) *$\bar{Q}=Q^{\circ\circ}$ for every $Q \subseteqq P$.*
(10) *Every non-zero event is the carrier of some $f \in \mathscr{S}$.*

2. THE MODEL FOR NON-RELATIVISTIC QUANTUM MECHANICS

We shall show that certain further constraints on a system $\mathscr{S}$, P imply that P is isomorphic to the lattice of closed subspaces of a separable infinite dimensional Hilbert space.

We recall that *a covers b* means that $a>b$ and $a \geqq c>b$ implies $a=c$. A *point* is an element which covers 0 and P is *atomic* if each of its elements is the join of points. We shall call an event *finite* if it is the join of a finite number of points and let P_f denote the set of all finite events. Suppose now that $\mathcal{S}$, P is a system satisfying

(A) P is atomic; $1 \notin P_f$.

Let (a) denote the lattice under a; clearly (a) is weakly modular relative to the orthocomplementation $b \to ab'$. We assume

(M) Let $a \in P_f$ and suppose b, c and d are elements of (a) with $d \leqq c$ and $bc = 0$. Then $(d \vee b)\,c = d$.[3]

LEMMA 2.1. *If a is finite, (a) is modular.*

Proof. Let d, b and c be elements of (a) with $d \leqq c$. Then $d \vee bc \leqq c$ and $b(bc)'c = 0$ so writing $b = bc \vee b(bc)'$ (by weak modularity) and letting $d \vee bc$, $b(bc)'$ and c play the roles of d, b and c of (M) respectively in the last of the following equalities, $(d \vee b)\,c = (d \vee (bc \vee b(bc)'))\,c = ((d \vee bc) \vee b(bc)')\,c = d \vee bc$.

Remark. This result is valid for an arbitrary orthocomplemented lattice L; that is, if L has the property attributed to (a) in (M), it obviously satisfies (W2) of Section 1, hence is weakly modular (see Lemma 1.3), so the proof applies, and L is modular.

LEMMA 2.2. *Suppose $a > b$. Then a covers b if and only if ab' is a point.*

Proof. Suppose ab' is a point and $a \geqq c > b$. Then $0 < cb'$ by (W1) so $cb' \leqq ab'$ implies $cb' = ab'$. Hence $c = cb' \vee b = ab' \vee b = a$, i.e., a covers b. If ab' is not a point, $ab' > c > 0$ for some $c \in P$ and then $a = b \vee ab' = b \vee ab'c' \vee c > b \vee c > b$ so a does not cover b.

COROLLARY. *Let $a \in P$. The chain $0 = a_0 < a_1 < a_2 < \cdots$ is maximal in (a) if and only if $a_i a'_{i-1}$ is a point for $i = 1, 2, \ldots$ and $\bigvee a_i = a$.*

LEMMA 2.3. ([1, pp. 66, 67]) *Let $a \in P_f$ and suppose every orthogonal set of points in (a) is finite. Then if $b \leqq a$ and $\{a_1, \ldots, a_n\}$ and $\{b_1, \ldots, b_m\}$ are two maximal orthogonal sets of points in (b), $m = n$.*

LEMMA 2.4. ([1, p. 66]) *Let $a \in P_f$ and suppose b, c and d are elements of*

(a) such that b covers d, b and c are not comparable and $d<c$. Then $b \vee c$ covers c.

For $a \in P_f$ let $\dim a = -1 + \min\{n : a \text{ is the join of } n \text{ points}\}$ and let $P_i = \{a \in P_f : \dim a = i\}$, $i = -1, 0, 1, \ldots$. Clearly, $P_{-1} = \{0\}$, P_0 is the set of points and $P_f = \bigcup P_i$.

Suppose there exists $a \in P_f$ such that (a) contains an infinite orthogonal set $\{b_i\}_{i=0}^{\infty}$ of points, and assume that $n = \dim a$ is a minimum for a with this property; clearly $n > 0$. Let $a_0, \ldots, a_n$ be points with join a. Since $\dim a_0 \vee \cdots \vee a_{n-1} = n-1$, Lemma 2.3 implies the existence of orthogonal points $c_0, \ldots, c_{n-1}$ such that $c_0 \vee \cdots \vee c_{n-1} = a_0 \vee \cdots \vee a_{n-1}$. Then a_n covers 0 and is not comparable with $a_0 \vee \cdots \vee a_{n-1}$ so $a = a_0 \vee \cdots \vee a_{n-1} \vee a_n$ covers $a_0 \vee \cdots \vee a_{n-1}$ by Lemma 2.4 and hence $c_n = a(a_0 \vee \cdots \vee a_{n-1})'$ is a point by Lemma 2.2; clearly $a = c_0 \vee \cdots \vee c_n$ and the c_i are orthogonal. Now $c_0 \neq b_0$, for otherwise $c_1 \vee \cdots \vee c_n = b_1 \vee b_2 \vee \cdots$ contrary to the choice of a with minimum dimension. Hence $c_0 \vee b_0$ covers c_0, so $d_0 = (c_0 \vee b_0) c_0'$ is a point. For $i = 1, 2, \ldots$ let $d_i = (c_0 \vee b_0 \vee \cdots \vee b_i)(c_0 \vee b_0 \vee \cdots \vee b_{i-1})'$. If $b_i \leqq c_0 \vee b_0 \vee \cdots \vee b_{i-1}$, $d_i = 0$ while if not, $c_0 \vee b_0 \vee \cdots \vee b_i$ covers $c_0 \vee b_0 \vee \cdots \vee b_{i-1}$ by Lemma 2.4 so d_i is a point. Since all the d_i are orthogonal and lie under ac_0', all but a finite number must be 0, since $\dim ac_0' = n-1 < \dim a$. Since $\bigvee d_i = ac_0'$, exactly n of the d_i are points by Lemma 2.3 and we assume without essential loss of generality that $d_0, \ldots, d_{n-1}$ are points. But then $a = ac_0' \vee c_0 = d_0 \vee \cdots \vee d_{n-1} \vee c_0 = b_0 \vee \cdots \vee b_{n-1} \vee c_0$. Since c_0 is a point not comparable with $b_0 \vee \cdots \vee b_{n-1}$, a covers $b_0 \vee \cdots \vee b_{n-1}$ and so $e = a(b_0 \ldots b_{n-1})'$ is a point. But $b_i \leqq e$ for $i \geqq n$ and so all but one of these b_i must be zero. This contradiction completes the proof of

LEMMA 2.5. *If a is finite, every orthogonal set of points in (a) is finite.*

COROLLARY. *If a is finite and $\{a_i\}_{i=0}^{n}$ is an orthogonal set of points in (a) with join a then $n = \dim a$.*

We call the elements of P_1 *lines*, of P_2, *planes*, and use the following notation: if $a \in P$, $(a)_i = \{b \leqq a : \dim b = i\}$, $i = -1, 0, 1, \ldots$.

We make the following assumption of homogeneity:

(H) If a and b are finite elements of the same dimension, then (a) and (b) are isomorphic.

LEMMA 2.6. *Suppose P contains a pair of distinct points a_0, b_0 such that the line $a_0 \vee b_0$ contains no third point. Then P is deterministic.*

Proof. $(a_0 \vee b_0)\, b_0'$ is a point distinct from b_0 so is equal to a_0 by hypothesis and hence $a_0 \perp b_0$. It follows now from (H) that if a_1 and b_1 are any two distinct points, then $a_1 \perp b_1$. Hence if a and b are events with $ab=0$, $a=\bigvee a_1 : a_1 \in (a)_0 \leqq \bigwedge b_1' : b_1 \in (b)_0 = (\bigvee b_1 : b_1 \in (b)_0)' = b'$ so $a \perp b$ and P is deterministic by Lemma 1.5.

We assume

(ND) P is not deterministic.

COROLLARY. *Every line contains at least three distinct points.*

LEMMA 2.7. $\mathscr{C}=\{0, 1\}$.

Proof. Suppose $a \in \mathscr{C}$ with $0<a<1$. Then there exist points b_1 and b_2 such that $b_1 \leqq a$ and $b_2 \leqq a'$. Let c be a point in $b_1 \vee b_2$ distinct from b_1 and b_2. Then $c = ca \vee ca'$ so either $c \leqq a$ or $c \leqq a'$ since c is a point. But the former implies that $b_2 < a$ since then $b_1 \vee b_2 = b_1 \vee c \leqq a$ and similarly the latter implies that $b_1 < a'$; hence the assumption $0<a<1$ is untenable.

We have shown that for $a \in P_f$, (a) is an orthocomplemented, modular lattice of finite dimension with trivial center and at least three points on each line. Thus, we have (see e.g., [1, Theorem 6, p. 120]):

THEOREM 2.1. *Let $\mathscr{S}$, P be a system satisfying* (A), (M), (H) *and* (ND). *Then the lattice under every finite element of P is a projective geometry.*

It follows from (H) that there exists a division ring D such that a coordinatizing division ring[4] for any finite (a) is isomorphic to D. We shall make use of the natural metric ρ for P: $\rho(a, b) = \sup |f(a) - f(b)| : f \in \mathscr{S}$.

LEMMA 2.8. *Orthocomplementation is continuous in (a) for any $a \in P$. That is, if $\{b_n\} \subset (a)$, $b \in (a)$ and $b_n \to b$, then $ab_n' \to ab'$.*

Proof. Given $\varepsilon>0$ choose N so that $n>N$ implies that $\rho(b_n, b)<\varepsilon$. Then if $f \in \mathscr{S}$ and $n>N$,

$$\begin{aligned} \varepsilon > |f(b_n) - f(b)| &= |(1-f(b)) - (1-f(b_n))| \\ &= |f(b') - f(b_n')| = |f(b'a \vee a') - f(b_n'a \vee a')| \\ &= |f(b'a) + f(a') - f(b_n'a) - f(a')| = |f(b'a) - f(b_n'a)|. \end{aligned}$$

Thus, $\rho(ab_n', ab') < \varepsilon$ and the result follows.

We assume now

(C′) If a is finite and $0 \leqq i \leqq \dim a$, $(a)_i$ is compact.

Remark. It seems reasonable to suppose that there exists $\varepsilon > 0$ so small that if the probabilities of occurrence of two events b and c differ in every state by less than ε, then $b = c$, i.e., b and c are operationally identical. The completeness of $(a)_i$ is clearly weaker than this operational assumption. The assumption that $(a)_i$, in addition to being complete, is totally bounded, may be paraphrased as follows: for each $\varepsilon > 0$ there exists a finite set $\{b_1, \ldots, b_m\}$ of elements of $(a)_i$ such that given any b in $(a)_i$ and $f \in \mathcal{S}$ the probability of occurrence of the event b in the state f differs from the probability of occurrence of one of the b_j in f by an amount less than ε.

Lemma 2.9. *Let a be a finite event of dimension at least two. Let $0 \leqq i$, $j < \dim a$, let $\{b_n\} \subset (a)$, $\{c_n\} \subset (a)$ with* $\dim b_n = i$ *and* $\dim c_n = j$ *for all n. Suppose that $b_n \to b$ and $c_n \to c$ where b and c are in "general position,"* i.e., $\dim b \vee c = \min(\dim a, i + j + 1)$. *Then $b_n \vee c_n \to b \vee c$ and, dually, $b_n c_n \to bc$.*

Proof. $\{b_n \vee c_n\}$ clusters at some $d \leqq a$ by (C′); assume for convenience that $b_n \vee c_n \to d$. Let $\varepsilon > 0$ and choose N so that $n > N$ implies $\rho(b_n, b) < \varepsilon/2$ and $\rho(b_n \vee c_n, d) < \varepsilon/2$. Then if $f(b) = 1$ and $n > N$, $f(d) + \varepsilon/2 > f(b_n \vee c_n) \geqq f(b_n) > 1 - \varepsilon/2$ so $f(d) > 1 - \varepsilon$. Hence $f(d) = 1$ and so $b \leqq d$ by (6) of Theorem 1.1. Similarly $c \leqq d$ so $b \vee c \leqq d$. Since $\dim d \leqq \max_n \dim b_n \vee c_n$ $\dim b \vee c$, $b \vee c = d$ must hold. The dual follows from Lemma 2.8.

Corollary. *Let $a \in P_f$. Then, in (a), the lattice operations are continuous in both variables simultaneously.*

We have therefore

Lemma 2.10.[5] *D is a locally compact division ring.*

We now assume

(Co) For some $b \in P_f$ and real interval I there exists a continuous non-constant function $t \to a_t$ from I to (b).

Remark. Postulate (Co) may be obtained from the following "intuitive" assumptions. There exist a one-parameter family L_t of mappings of $\mathcal{S}$

on $\mathscr{S}$ (describing how the states change with time (regarded as a real parameter) – corresponding to certain assumptions concerning the dynamics of the system (see [6, 7])) and a state f such that, letting a_t denote the carrier of $L_t(f)$, a_t is continuous, non-constant and remains in some finite (b) for all t in an interval I.

For convenience assume $I=[0, 1]$, let $n=\dim b$, $m=\dim a_0$. It follows at once from the continuity of a_t and the compactness of $(b)_m$ that $\dim a_t=m$ for all $t\in I$. Suppose $m>0$. Without essential loss of generality we assume that $a_t \neq a_0$ for $t>0$ and choose a point $c<a_0$ such that $c \nleqq a_t$ for (again, for convenience) $t>0$. Let $d=c \vee a_0'$. Choose $\delta>0$ such that $0\leqq t<\delta$ implies $\rho(a_0, a_t)<\frac{1}{2}$. Then for such t, $a_t a_0'=0$, for otherwise there exists $f\in\mathscr{S}$ such that $f(a_t a_0')=1$ so $f(a_t)=f(a_0')=1$. But then $|f(a_0)-f(a_t)|=|f(a_0)-1|<\frac{1}{2}$ implies $f(a_0)>\frac{1}{2}$, a contradiction. Hence, taking $\delta=1$ for convenience, $da_t=d_t$ is a point for all t (for $d_t\neq 0$ by a count of dimension while $a_0' a_t=0$ implies $\dim da_t\leqq 0$). Since $d_0=c$ and $d_t\neq c$ for $t>0$, d_t is not constant, while it follows from Lemma 2.9 that d_t is a continuous function of t; in case $m=0$ we set $d_t=a_t$. Again by continuity and without essential loss of generality, we can find a point $e^{(1)}$ disjoint from $\{d_t\}_{t\in I}$ and hyperplane $h^{(1)}$ such that $(e^{(1)}\vee d_t)\, h^{(1)}=d_t^{(1)}$, which is automatically continuous, is not constant. Similarly, if $\dim h^{(1)}=n-1>1$, we can find $e^{(2)}\varepsilon h^{(1)}$ disjoint from $\{d_t^{(1)}\}$ and $h^{(2)}=h^{(1)}$ with $\dim h^{(2)}=n-2$ such that $d_t^{(2)}=(e^{(2)}\vee d_t^{(1)})\, h^{(2)}$ is non-constant in some subinterval of I. Continuing in this way, we arrive finally at a continuous non-constant function $d_t^{(n)}$ from some subinterval of I to a line $h^{(n)}$ in (b). Then for a sub-interval J of I, $\{d_t^{(n)}\}_{t\in J}$ omits a point p of $h^{(n)}$. But D is homomorphic to $h^{(n)}$ with p removed and hence contains a connected set, the image of $\{d_t^{(n)}\}_{t\in J}$ under such a homomorphism. Since a locally compact division ring is readily seen to be either connected or totally disconnected we have

LEMMA 2.11. *D is connected.*

It follows now from Pontrjagin's theorem that D is the real, complex or quaternion division ring.[6] We assume henceforward that the real or complex case has been singled out, e.g., by the assumption of simple ordering on the one hand or algebraic closure on the other, the quaternions having been set aside by postulating commutativity for D, i.e., that Pappus's theorem holds under finite elements. Turning now to the representation of P itself, we shall need the final postulate

(C) For each $i=0, 1, \ldots, P_i$ is complete.[7]

LEMMA 2.12. *Let L and Λ be complete weakly modular lattices and let $L_0(\Lambda_0)$ be a subset of $L(\Lambda)$ such that every element of $L(\Lambda)$ is a join of elements of $L_0(\Lambda_0)$. Suppose further that φ is a mapping of L_0 onto Λ_0 such that*

(1) $a \perp b$ *if and only if* $\varphi(a) \perp \varphi(b)$.

Then φ can be extended to an isomorphism of L onto Λ.

Define $\theta: L \to \Lambda$ by $\theta(a) = \bigvee \varphi(c); c \in [a]$ where $[a] = \{b \leqq a : b \in L_0\}$. Clearly θ preserves order and $\theta \mid L_0 = \varphi$. The lemma is proved in the following steps:

(2) $\theta(a') \leqq \theta(a)'$.
(3) $a < b$ implies $\theta(a) < \theta(b)$.
(4) Let A be a subset of $[a]$ such that $a = \sup A$. Then $\theta(a) = \sup \varphi(b): b \in A$.
(5) $\theta(a \vee b) = \theta(a) \vee \theta(b)$.
(6) θ is one-to-one.
(7) θ^{-1} preserves order.
(8) θ is onto.

The proofs are as follows.

(2) If $b \in [a']$ and $c \in [a]$, $\varphi(b) \leqq \varphi(c)'$ by (1) so $\theta(a') = \bigvee \varphi(b): b \in [a'] \leqq \bigwedge \varphi(c)' : c \in [a] = (\bigvee \varphi(c))' : c \in [a] = \theta(a)'$.

(3) If $a < b$, there exists $c \neq 0 \in [ba']$. Then $\varphi(c) \perp \varphi(a_1)$ for all $a_1 \in [a]$ by (1) so $\varphi(c) \perp \theta(a)$. Clearly $\varphi(c) \leqq \theta(b)$ and $\theta(a) \leqq \theta(b)$ so $\theta(a) < \theta(b)$.

(4) Let $\alpha = \sup \varphi(b): b \in A$; clearly $\alpha \leqq \theta(a)$. If $c \in L_0$ with $\varphi(c) \in [\theta(a)\, \alpha']$ then $c \perp b$ for every $b \in A$ by (1) so $c \perp a$. Hence $\varphi(c) \leqq \theta(a') \leqq \theta(a)'$ by (2). Since $\varphi(c) \leqq \theta(a)$, $c = 0$ and hence $\theta(a) = \alpha$ by weak modularity.

(5) Let $A = [a] \cup [b]$. Then $\sup A = a \vee b$ so $\theta(a \vee b) = \sup \varphi(c): c \in A$ by (4). Now if $c \in A$, $\varphi(c) \leqq \theta(a)$ or $\varphi(c) \leqq \theta(b)$ so $\varphi(c) \leqq \theta(a) \vee \theta(b)$; the opposite inequality is immediate.

(6) and (7). If $a \nleqq b$ then $b < a \vee b$ so $\theta(b) < \theta(a \vee b)$ by (3) $= \theta(a) \vee \theta(b)$ by (5) and hence $\theta(a) \nleqq \theta(b)$.

(8) Let $\alpha \in \Lambda$ and let $A = \{\varphi^{-1}(\beta) : \beta \in [\alpha]\}$. Then, by (4), $\theta(\sup A) = \bigvee \beta : \beta \in [\alpha] = \alpha$.

For each $a \in P_f$ we choose a distinct Euclidean space H_a over D of dimension $1 + \dim a$ and an isomorphism φ_a of (a) onto L_a, the lattice of subspaces of H_a. Assuming $n = \dim a > 0$, we wish to choose a scalar product for H_a so that the orthocomplementation $\varphi_a b \to \varphi_a ab'$ which is induced in L_a by that in (a) coincides with the one induced by the scalar product. First of all, there exists an involution σ of D and non-zero numbers (i.e., elements of D) $\gamma_0, \ldots, \gamma_n$ such that $\gamma_i^\sigma = \gamma_i$, $\sum_{i=0}^n x_i \gamma_i x_i^\sigma = 0$ implies all $x_i = 0$, and if $b \in (a)_0$ and $\varphi_a b = [(x_0, \ldots, x_n)]$ then $\varphi_a ab' = \{(y_0, \ldots, y_n) : \sum y_i \gamma_i x_i^\sigma = 0\}$.[8] In the real case, $\sigma = 1$ is the only automorphism; we shall show that σ is continuous, hence is either 1 or conjugation in the complex case, and the value 1 is excluded, for otherwise $(\gamma_0^{-1/2}, (-\gamma_1^{-1/2}), 0, \ldots, 0)$ would be self-orthogonal. Then all the γ_i must be positive real numbers and the desired scalar product is $(y, z) = \sum y_i \gamma_i \bar{z}_i$.

Let b and c be orthogonal points in (a), and choose x, y in H_a such that $\varphi_a b = [x]$, $\varphi_a c = [y]$. Let λ_m be a sequence of numbers with $\lambda_m \to 0$ and let $b_m = \varphi_a^{-1}[x + \lambda_m y]$. Then $b_m \to b$ so $(b \vee c)\, b'_m \to (b \vee c)\, b' = c$ by Lemma 2.8 and we may assume that $(b \vee c)\, b'_m \neq b$ holds for all m. Then a sequence μ_m of numbers with $\mu_m \to 0$ is determined by: $\varphi_a (b \vee c)\, b'_m = [\mu_m x + y]$. Since $b \perp c$, $\sum (\mu_m x_i + y_i)\, \gamma_i (x_i + \lambda_m y_i)^\sigma = 0$ so $0 = \mu_m \sum x_i \gamma_i x_i^\tau + \lambda_m^\sigma \sum y_i \gamma_i y_i y_i^\sigma$ and it follows from the fact that $\mu_m \to 0$ and $\sum y_i \gamma_i y_i^\sigma \neq 0$ that $\lambda_m^\sigma \to 0$. Thus, σ is continuous at 0 and hence, by its additivity, is continuous everywhere, and the proof is complete.

We assume now, in accordance with the foregoing, that each H_a has been provided with a scalar product such that $\varphi_a b \perp \varphi_a c$ for b, c in (a) if and only if $b \perp c$. If $a \leqq b \in P_f$, $\varphi_{ba} = \varphi_b \varphi_a^{-1}$ is clearly an orthogonality preserving isomorphism of L_a in L_b. It is well known that there then exists an isometric transformation ψ_{ba} of H_a in H_b, unique up to multiplication by a number of absolute value one, such that if $v \in H_a$, $\varphi_{ba}[v] = [\psi_{ba} v]$. We shall show that the ψ's may be chosen consistently, i.e., so that

(15) $a \leqq b \leqq c$ implies $\psi_{ca} = \psi_{cb} \psi_{ba}$.

We establish a one-to-one correspondence $\alpha \leftrightarrow a_\alpha$ between the elements of P_f and the ordinal numbers less than an ordinal ζ such that $\alpha < \beta$ implies $\dim a_\alpha \leqq \dim a_\beta$. Thus, in particular, $a_0=0$; it is understood that all ordinals $\alpha, \beta, \ldots$ which occur lie under ζ and, where no confusion can result, we shall write "α" for "a_α". In particular, we let $\alpha\beta$ represent (the index of) $a_\alpha a_\beta$. Let $\gamma<\zeta$ and suppose that ψ has already been defined so that (15) holds for $c=a_\alpha$ with $\alpha<\gamma$. Now choose α such that $a_\alpha<a_\gamma$ and $\dim a_\gamma-\dim a_\alpha=1$; we call such an α "maximal". Fix $\psi_{\gamma,\alpha}$ arbitrarily; then if $a_\eta<a_\alpha$, $\psi_{\gamma,\eta}$ is defined as $\psi_{\gamma,\alpha}\psi_{\alpha,\eta}$. If β is a second maximal element, and assuming $\dim a_\gamma>1$ (i.e., $\dim H_\gamma>2$), for otherwise there is nothing to prove, $\alpha\beta\neq 0$ and we define $\psi_{\gamma,\beta}$ by $\psi_{\gamma,\alpha\beta}=\psi_{\gamma,\beta}\psi_{\beta,\alpha\beta}$. Now let η be any ordinal with $a_\eta<a_\gamma$ and let β, ε both be maximal such that $a_\eta\leqq a_\beta$ and $a_\eta\leqq a_\varepsilon$. Assuming that $\dim a_\gamma\geqq 3$, we shall show

$$(16) \qquad \psi_{\gamma,\beta}\psi_{\beta,\eta}=\psi_{\gamma,\varepsilon}\psi_{\varepsilon,\eta} \ (\beta,\ \varepsilon \text{ maximal},\ a_\eta\leqq a_{\beta\varepsilon},\ \dim a_\gamma\geqq 3).$$

But if (16) holds for $\eta=\beta\varepsilon$ then, by the inductive hypothesis, it will hold for arbitrary η, for then $\psi_{\gamma,\beta}\psi_{\beta,\eta}=\psi_{\gamma,\beta}\psi_{\beta,\beta\varepsilon}\psi_{\beta\varepsilon,\eta}=\psi_{\gamma,\varepsilon}\psi_{\varepsilon,\beta\varepsilon}\psi_{\beta\varepsilon,\eta}=\psi_{\gamma,\varepsilon}\psi_{\varepsilon,\eta}$. To prove (16) in the case $\eta=\beta\varepsilon$ observe that $\psi_{\gamma,\alpha}\psi_{\alpha,\alpha\beta\varepsilon}=\psi_{\gamma,\alpha}\psi_{\alpha,\alpha\beta}\psi_{\alpha\beta,\alpha\beta\varepsilon}=\psi_{\gamma,\alpha\beta}\psi_{\alpha\beta,\alpha\beta\varepsilon}=\psi_{\gamma,\beta}\psi_{\beta,\alpha\beta}\psi_{\alpha\beta,\alpha\beta\varepsilon}=\psi_{\gamma,\beta}\psi_{\beta,\alpha\beta\varepsilon}=\psi_{\gamma,\beta}\psi_{\beta,\beta\varepsilon}\psi_{\beta\varepsilon,\alpha\beta\varepsilon}$. Similarly – interchanging β and ε – $\psi_{\gamma,\alpha}\psi_{\alpha,\alpha\beta\varepsilon}=\psi_{\gamma,\varepsilon}\psi_{\varepsilon,\beta\varepsilon}\psi_{\beta\varepsilon,\alpha\beta\varepsilon}$. In other words, $\psi_{\gamma,\varepsilon}\psi_{\varepsilon,\beta\varepsilon}=\psi_{\gamma,\beta}\psi_{\beta,\beta\varepsilon}$ on $\psi_{\beta\varepsilon,\alpha\beta\varepsilon}H_{\alpha\beta\varepsilon}$ and since $\alpha\beta\varepsilon\neq 0$ (by our assumption that $\dim a_\gamma\geqq 3$), this equality holds on all of $H_{\beta\varepsilon}$ and (16) is proved. Thus, if β is maximal and $a_\eta<a_\beta$, $\psi_{\gamma,\eta}$ is unambiguously defined by: $\psi_{\gamma,\eta}=\psi_{\gamma,\beta}\psi_{\beta,\eta}$. If $a_\eta<a_\delta<a_\gamma$, choose β maximal with $a_\delta<a_\beta$ and then $\psi_{\gamma,\eta}=\psi_{\gamma,\beta}\psi_{\beta,\eta}=\psi_{\gamma,\beta}\psi_{\beta,\delta}\psi_{\delta,\eta}=\psi_{\gamma,\delta}\psi_{\delta,\eta}$, completing the proof that ψ as extended to all γ, η with $a_\eta<a_\gamma$ satisfies (15) providing that $\dim a_\gamma\geqq 3$. We begin the induction and complete the proof by "constructing" all $\psi_{c,b}$ with $\dim c\leqq 2$ in the following way. Let A_i denote the set of all $\alpha<\zeta$ for which $\dim a_\alpha=i$, $i=0,1,\ldots$. Let $\beta_1\varepsilon A_1$, let $B_1=\{\beta\in A_1:\beta<\beta_1\}$ and make the inductive assumption that $\psi_{\gamma,\beta}$ and $\psi_{\beta,\alpha}$ (and consequently $\psi_{\gamma,\alpha}$) have already been consistently defined whenever $\beta\in B_1$, $\gamma\in A_2$, $\alpha\in A_0$ and $a_\alpha<a_\beta<a_\gamma$. For all $\gamma\in A_2$ such that $a_{\beta_1}<a_\gamma$, define ψ_{γ,β_1} arbitrarily and then, choosing $\alpha\in A_0$ with $a_\alpha<a_{\beta_1}$, define $\psi_{\beta_1,\alpha}$ by $\psi_{\gamma,\alpha}=\psi_{\gamma,\beta_1}\psi_{\beta_1,\alpha}$ if $\psi_{\gamma,\alpha}$ has already been defined for some $\gamma\in A_2$ with $a_\alpha<a_\gamma$ – i.e., if $a_\alpha<a_\beta<a_\gamma$ for some $\beta\in B_1$; otherwise define $\psi_{\beta_1,\alpha}$ arbitrarily and set $\psi_{\gamma,\alpha}=\psi_{\gamma,\beta_1}\psi_{\beta_1,\alpha}$ for all $\gamma\in A_2$ with $a_{\beta_1}<a_\gamma$. This procedure evidently extends ψ consistently to all γ, β_1; β_1, α and γ, α such that $\gamma\in A_2$, $\alpha\in A_0$ and $a_\alpha<a_{\beta_1}<a_\gamma$. It then follows inductively – beginning with $B_1=\phi$ – that ψ may be consistently defined for all $\psi_{c,a}$ such that $a\leqq c$ and $\dim c\leqq 2$.

Now let H be a separable, infinite dimensional Hilbert space over D, let L be its lattice of closed subspaces and let $\{v_i\}$ be a complete orthonormal set in H. Let $\{a_i\}$ be a maximal orthogonal subset of P_0 which exists by Zorn's lemma and is countable by E4 and (A), and for each i let u_i be a fixed unit vector in H_{a_i}. Let $a \in P_0$, let $u \in H_a$ and define

$$\begin{aligned}\lambda_i(u) &= (\psi_{a \vee a_i, a} u, \psi_{a \vee a_i, a_i} u_i),\\ \xi u &= \sum \lambda_i(u)\, v_i,\\ \theta(a) &= \{\xi u : u \in H_a\}.\end{aligned}$$

Thus, the domain of λ_i and ξ is $\bigcup_{a \in P_0} H_a$, that of θ is P_0 and their ranges are in D, H, and the set L_0 of one dimensional subspaces of H respectively.[9] We shall show that θ is one-to-one, onto and that θ and θ^{-1} preserve orthogonality. Hence by Lemma 2.12, θ can be extended to an isomorphism $\bar{\theta}$ of P on L. Then $f\bar{\theta}^{-1}$ will be a state for L and the characterization of $\mathscr{S}$ is given by the[10]

Theorem of Gleason. ([5]) Let μ be a state on the lattice L of closed subspaces of the separable real or complex Hilbert space H of dimension at least three. Then there exists an orthonormal basis $\{x_i\}$ for H and nonnegative real numbers λ_i with $\sum \lambda_i = 1$ such that if Q is the projection on $M \in L$, $\mu(M) = \sum \lambda_i (Qx_i, x_i)$.

Each L_a for $a \in P_f$ becomes a metric space under the definition: distance $(M_1, M_2) = \sup\{|\omega(M_1) - \omega(M_2)| : \omega \text{ a state for } L_a\}$. An immediate consequence of Gleason's theorem is that φ_a is an isometry of (a) on L_a.

LEMMA 2.13. *Let $a \in P_0$, $u \in H_a$. Then $\|\xi u\| = \|u\|$.*

Proof. For $n = 1, 2, \ldots$ let $b_n = a \vee a_1 \vee \ldots \vee a_n$. Then if $1 \leqq i \leqq n$,

$$\begin{aligned}\lambda_i(u) &= (\psi_{a \vee a_i, a} u, \psi_{a \vee a_i, a_j} u_i)\\ &= (\psi_{b_n, a \vee a_i} \psi_{a \vee a_i, a} u, \psi_{b_n, a \vee a_i} \psi_{a \vee a_i, a_i} u_i)\\ &= (\psi_{b_n, a} u, \psi_{b_n, a_i} u_i) \quad \text{so} \quad \sum_{i=1}^{n} |\lambda_i(u)|^2 \leqq \|\psi_{b_n, a} u\|^2\\ &= \|u\|^2 \text{ since the } \psi_{b_n, a_i} u_i \text{ are orthonormal in } H_{b_n} \text{ and}\\ &\quad\ \psi_{b_n, a} \text{ is an isometry.}\end{aligned}$$

Since ξ is linear, we assume without essential loss of generality that $\|u\| = 1$ and suppose that, contrary to the assertion of the lemma, $\|\xi u\| =$

$(\sum_{i=1}^{\infty} |\lambda_i(u)|^2)^{1/2} = \delta < 1$. Then, in particular, $(\psi_{b_n, a_i} u_i)_{i=1}^{n}$ must fail to be a basis in all but a finite number of the H_{b_n}, so, for convenience, we assume $b_n > a_1 \vee \ldots \vee a_n$ for all n and let $c_n = b_n a_1' \ldots a_n'$; evidently, $c_n \in (b_n)_0$. Let $w_n = \sum_{i=1}^{n} \lambda_i(u) \psi_{b_n, a_i}$, let $\alpha_n = \|\psi_{b_n, a} u - w_n\|$ and let $y_n = (\psi_{b_n, a} u - w_n) \alpha_n^{-1}$. Clearly $\alpha_n \to \sqrt{1-\delta^2}$ and $y_n \in \varphi_{b_n c_n}$. Then if $n > m$,

$$
\begin{aligned}
(y_n, \psi_{b_n, b_m} y_m) &= \alpha_n^{-1} \alpha_m^{-1} (\psi_{b_n, a} u - w_n, \psi_{b_n, a} u - \psi_{b_n, b_m} w_m) = \\
&= \alpha_n^{-1} \alpha_m^{-1} (\|\psi_{b_n, a} u\|^2 - (w_n, \psi_{b_n, a} u) - (\psi_{b_n, a} u, \psi_{b_n, b_m} w_m) + \\
&\quad (w_n, \psi_{b_n, b_m} w_m)) = \alpha_n^{-1} \alpha_m^{-1} \left(1 - \sum_{i=1}^{n} |\lambda_i|^2\right) - \sum_{i=1}^{m} |\lambda_i|^2 + \\
&\quad \sum_{i=1}^{m} |\lambda_i|^2 \Bigg) = \alpha_n^{-1} \alpha_m^{-1} \left(1 - \sum_{i=1}^{n} |\lambda_i|^2\right) \to \frac{1-\delta^2}{1-\delta^2} = 1.
\end{aligned}
$$

Thus, given $\varepsilon > 0$, we may choose N so that $n > m > N$ implies $\varepsilon > 1 - (y_n, \psi_{b_n, b_m} y_m) = \text{distance } ([y_n], [\psi_{b_n, b_m} y_m]) = \rho(c_n, c_m)$. Then, in virtue of (C), there exists a point c in P such that $c_n \to c$. We shall complete the proof by showing that $c \perp a_i$ for all i contrary to the maximality of $\{a_i\}$. Indeed, if $f \in \mathscr{S}$ with $f(a_i) = 1$ and $n > i$, then $c_n \perp a_i$, and if n is chosen so large that $\rho(c_n, c)$ is less than a preassigned $\varepsilon > 0$, $f(c) < f(c_n) + \varepsilon = \varepsilon$, i.e., $f(c) = 0$ so $c \perp a_i$ by (6) of Theorem 1.1 and the proof of Lemma 2.13 is complete.

Corollary 1. *Let a and b be points. Then $\theta a \perp \theta b$ if and only if $a \perp b$.*

Proof. For $u \in H_{a \vee b}$ let $\eta u = \sum (\psi_{a \vee b \vee a_i, a \vee b} u, \psi_{a \vee b \vee a_i, a_i} u_i) v_i$. Clearly η is linear and if $c = \varphi_{a \vee b}^{-1} [u]$ and

$$
\begin{aligned}
w = \psi_{c \vee a_i, c}^{-1} u,\ \eta u &= \sum (\psi_{a \vee b \vee a_i, c \vee a_i} \psi_{c \vee a_i, c} w, \psi_{a \vee b \vee a_i, c \vee a_i} \times \\
&\times \psi_{c \vee a_i, a_i} u_i) v_i = \sum (\psi_{c \vee a_i, c} w, \psi_{c \vee a_i, a_i} u_i) v_i = \xi w;
\end{aligned}
$$

clearly $\theta c = [\eta u]$. Hence $\|\eta u\| = \|\xi w\| = \|w\| = \|u\|$ so η is an isometry and then letting $0 \neq u \in \varphi_{a \vee b} a_i$; $0 \neq v \in \varphi_{a \vee b} b$, $a \perp b$ if and only if $u \perp v$ if and only if $\eta u \perp \eta v$ if and only if $\theta a \perp \theta b$.

Corollary 2. *θ is one-to-one.*

Proof. If $\theta a = \theta b$ and $c \in (b')_0$, $c \perp b$ so $\theta c \perp \theta b$ by Corollary 1, $\theta c \perp \theta a$ by our assumption and then $a \leqq c'$ by Corollary 1. Hence $a \leqq \bigwedge c' : c \in (b')_0 = (\bigvee c : c \in (b')_0)' = b$ by postulate (A). Similarly $b \leqq a$, so $a = b$.

COROLLARY 3. *Let $b_1 < b_2 < \cdots$ be a chain of finite elements and suppose $y_n \in H_{b_n}$ with $\|y_n\| = 1$ such that given $\varepsilon > 0$ there exists N such that $n > m > N$ implies $\|y_n - \psi_{b_n, b_m} y_m\| < \varepsilon$. Let $c_n = \varphi_{b_n}^{-1}[y_n]$. Then the sequence of points $\{c_n\}$ converges to a point c in P.*

LEMMA 2.14. *θ is onto.*

Proof. Let $M \in L_0$, $v = \sum \mu_i v_i$ a unit vector in M. Let $b_n = a_1 \vee \ldots \vee a_n$, $w_n = \sum_{i=1}^{n} \mu_i \psi_{b_n, a_i} u_i$, $y_n = w_n / \|w_n\|$ when $w_n \neq 0$ and $c_n = \varphi_{b_n}^{-1}[y_n]$. It follows at once from Corollary 3 above that there exists $c \in P_0$ with $c_n \to c$. Let $d_n = c \vee b_n$ and let y be a unit vector in H_c. Now $(\psi_{d_n, b_n} y_n, \psi_{d_n, c} y)$ tends to a limit η with $|\eta| = 1$ and $\|\psi_{d_n, b_n} y_n - \eta \psi_{d_n, c} y\| \to 0$. Hence, by Lemma 2.13 $\xi y_n \to \xi \eta y$. Since $\xi y_n \to v$ is obvious, $\xi \eta y = v$, $\theta c = M$ and the proof is complete.

θ is one-to-one from P_0 onto L_0 by Corollary 2 of Lemma 2.13 and the preceding lemma. Furthermore, $\theta a \perp \theta b$ if and only if $a \perp b$ by Corollary 1 of Lemma 2.13 and so we may apply Lemma 2.12 to obtain

THEOREM 2.2. *Suppose the system $\mathscr{S}$, P satisfies the following eight postulates:*

- (A) *P is atomic; $1 \notin P_f$.*
- (M) *If a is finite and b, c and d are elements of (a) such $d \leqq c$ and $bc = 0$, then $(d \vee b)\, c = d$.*
- (H) *If a and b are finite elements of the same dimension, then (a) and (b) are isomorphic.*
- (ND) *P is not deterministic.*
- (C′) *If a is finite and $0 \leqq i \leqq \dim a$, $(a)_i$ is compact.*
- (Co) *There exists a continuous, non-constant function from an interval of the real line to the lattice under a finite event.*
- (P) *If a is finite, Pappus's theorem holds in (a).*
- (C) *For each $i = 0, 1, \ldots, P_i$ is complete.*

Then P is isomorphic to the lattice L of closed subspaces of a separable, infinite dimensional Hilbert space over either the real or the complex field in such a way that the orthocomplementations in P and L correspond.

NOTES

* The work reported in this paper was performed while the author was a member of the staff of Lincoln Laboratory, Massachusetts Institute of Technology. He is now with the Arcon Corporation, Lexington, Massachusetts.

[1] That is, $(a \vee b)\, c$ exists and is equal to a. In general, when x exists a priori but y may not, the assertion $y = x$ is understood to include the assertion that y exists.

[2] Cf. [2].

[3] If $d \leqq c$ and $b \perp c$, $(d \vee b)\, c = d$ by weak modularity (cf. Lemma 1.3); thus, (M) asserts that, under finite elements, $bc = 0$ bears a certain resemblance to $b \perp c$.

[4] [1, Theorem 15, p. 131].

[5] See Kolmogorov [5].

[6] See [8] for a unified derivation of the classification of locally compact division rings.

[7] Cf. the remark following the statement of postulate (C').

[8] [2, Appendix]. $[(x_0, \ldots, x_n)]$ denotes the 1-dimensional subspace of H_a generated by the element $(x_0, \ldots, x_n)$.

[9] For the convergence of ξu, see the proof of Lemma 2.13.

[10] It follows then from S1 and S2 that $\mathscr{S}$ contains *all* states for P.

BIBLIOGRAPHY

[1] Birkhoff, G., *Lattice Theory*, Amer. Math. Soc. Colloq. Publ., 25, revised, 1948.

[2] Birkhoff, G. and Neumann, J. von, 'The Logic of Quantum Mechanics', *Ann. of Math.* **37** (1936), 823–843.

[3] Gleason, A. M., 'Measures on the Closed Subspaces of a Hilbert Space', *J. of Rat. Mech. and Analysis* **6** (1957), 885–894.

[4] Kakutani, S., 'Concrete Representation of Abstract (L)-Spaces and the Mean Ergodic Theorem', *Ann. of Math.* **42** (1941), 523–537.

[5] Kolmogorov, A., 'Zur Begründung der projektiven Geometrie', *Ann. of Math.* **33** (1932), 175–176.

[6] Mackey, G. W., *Lecture Notes for Math. 265, The Mathematical Foundations of Quantum Mechanics*, Harvard, Spring 55–56 (unpbl.).

[7] Mackey, G. W., 'Quantum Mechanics and Hilbert Space', *Amer. Math. Monthly* **64**, (1957), 45–57.

[8] Weiss, E. and Zierler, N., 'Locally Compact Division Rings', *Pacific J. Math.* **8** (1958), 369–371.

V. S. VARADARAJAN

PROBABILITY IN PHYSICS AND A THEOREM ON SIMULTANEOUS OBSERVABILITY

I. INTRODUCTION

It is nearly thirty years since A. N. Kolmogorov explicitly wrote down the axioms of modern probability theory in his celebrated monograph [10]. During the intervening decades this theory has seen remarkable development, both in its theoretical and practical aspects. The diverse theories of mathematical statistics, the rapidly developing field of information theory, the applications to thermodynamics and statistical mechanics are only some of a long list of fields which are dominated to a substantial degree by probability theory. Moreover, many of the mathematical questions raised and answered by this theory have given deep and subtle insights into some difficult problems of analysis. One has only to mention the modern theory of Markov processes which has given new insights into such classical problems as boundary values and potential theory.

The Kolmogorov axioms combine generality with simplicity. Probability becomes a part of measure theory, with its own special emphasis. In his basic monograph Kolmogorov himself proved many theorems indicating clearly the scope of this new discipline: the extension theorem which proves the existence of random variables with preassigned joint distributions, conditional probabilities and expectations, sequences of independent random variables – one can multiply these examples almost indefinitely. The subsequent work of Kolmogorov, Doob, Feller and others in the modern theory of stochastic processes convinces one of the enormous richness of this subject.

During this vigorous development, the basic notions themselves have however barely changed. The Kolmogorov axioms are precise, concise and lead to an extensive theory; the pure mathematician asks for nothing more. However there remains the problem of understanding the phenomenological background to these axioms. Kolmogorov himself was aware of this problem and in any event one possesses now a sound ex-

C. A. Hooker (ed.), The Logico-Algebraic Approach to Quantum Mechanics, 171–203.

planation of the motivation behind these axioms. This explanation consists essentially in observing that the set of events associated with an experiment is an abstract Boolean σ-algebra and may be roughly thought of, via the Loomis representation theorem, as a Boolean σ-algebra of subsets of some space – indeed the space Ω which hovers in the background in all investigations in probability theory. This is a convincing demonstration that there is very little that is *ad hoc* in the Kolmogorov axioms. A sparkling account of this circle of ideas is given by Halmos [8].

In spite of this it is found that when thus formulated, the theory of probability does not include the situations that arise in quantum physics. For example there is nothing in this theory that corresponds to the Heisenberg Uncertainty Principle; and yet we are told that the Heisenberg principle simply displays pairs of observables the product of whose variances always remains above a positive constant no matter what original probability distribution is assigned to the basic events. Expositions in physics always stress the fact that the intervention of probability in quantum theory is fundamentally different from its role in statistical mechanics. An interesting account of some of the differences that arise when probability is applied to the domain of quantum physics has been given by Feynman [4]. Feynman however does not examine too closely the precise mathematical reasons that cause these profound differences and the overall impression, to an orthodox probabilist at least, is one of bewilderment.

It is obvious that since the Kolmogorov axioms are rooted in empirical experience, any change in the theory, if by such change one wants to extend its applications to the physical world, should spring directly from some phenomenological considerations. Anticipating our discussions in the subsequent sections one might say that the point of departure for the contemplated change in the model can be traced to the remarkable discovery that the physical systems arising in quantum physics are of such nature that one is no longer entitled to make the assumption that the associated experimental propositions constitute a Boolean σ-algebra. As a consequence the conventional i.e. the Kolmogorov formalism of probability theory is inadequate for a precise description of these systems. As a spectacular instance of such a failure we may mention the facts that the notion of disjoint events is at a somewhat deeper level and that the identity $P(A+B)=P(A)+P(B)-P(AB)$ is not always true (the examples of Feynman are concerned with this failure among other things). In his classic treatise [17] and even more incisively, in an interesting paper written jointly with Birkhoff [3], J. von Neumann has clearly emphasized the fundamental fact that the *experimental propositions* associated with a physical system do not form a Boolean σ-algebra but only "some sort of projective geometry" (this description is von Neumann's). It is the view of the present writer that there is a need for an exposition of some of these ideas, in a language close to that of the probabilist.

In recent years there has been some interest in the mathematical features which such a "generalized probability theory" is likely to exhibit. Especially significant is a paper of Segal [20]. Segal takes the view that the algebra of all bounded random variables on a probability space Ω is mathematically a more cogent object than the σ-algebra of subsets of Ω and is equivalent to it so far as the theory is concerned. This algebra comes equipped with a distinguished linear functional, namely the expectation. The transition to the probabilistic schemes of quantum physics is now made by replacing this commutative algebra by a noncommutative one, in fact by an algebra of linear operators on a Hilbert space. Segal however does not link up these noncommutative algebras with the "projective geometries" of Birkhoff and von Neumann. The point is that the relation between probability of events and expected values of random variables is at a somewhat deeper technical level in the noncommutative theory. In particular it is not at all an easy matter to prove, in a suitably general context, that the additivity of the expectation as a functional on the set of observables is a consequence of the additivity properties of the probability measure on the events. In a recent paper, A. M. Gleason [5] has proved this in a very important special case and the arguments used to attain the proof are by no means trivial. Inasmuch as Segal straightaway assumes the expectation to be additive he is not concerned with this question of relating probabilities to expectations.

In his paper [12] and subsequently in his Harvard notes [13], G. W. Mackey has taken the decisive step towards a systematic exposition of these ideas. However there are two features of the present discussion which make it somewhat more complete than Mackey's. The first is that the results dealing with simultaneous observability, which Mackey (as does von Neumann) establishes in the special Hilbert space context of quantum physics, are established here in their appropriately general context. The second is that the present discussion emphasizes more strongly the implications of these ideas toward a more general formulation of the basic concepts of probability theory.

It is not the claim of this writer that the detailed examination of a model more general than the orthodox one is either easy or certain to lead to significant results. What is pointed out rather is that there is a more general way of formulating the fundamental axioms which subsumes the probabilistic aspects of quantum physics under its fold and that there arise some interesting mathematical problems in this context whose counterparts in the orthodox model are either trivial or nonexistent.

II. MOTIVATION

The center of the stage of the present discussion is occupied by a physical system $\mathfrak{S}$ and the experimental propositions that are associated with $\mathfrak{S}$. We shall take for granted the idea of a physical system. The hydrogen atom, particles of matter executing Brownian motion in a medium and so on are examples of physical systems. Probability theory deals with such systems and makes statistical assertions concerning numerical variables which are associated with the system. These assertions are generally of the following type: the probability that the value of the variable x lies between two real numbers a and b is p.

The conventional analysis of this situation, in fact the one which leads rather naturally to the Kolmogorov formalism, begins by first noting that there is a class $\mathscr{E}$ of meaningful propositions associated with the system $\mathfrak{S}$, and that it is the aim of the theory to assign a numerical probability for each one of these propositions being true or false. $\mathscr{E}$ is equipped with an intrinsic algebraic structure. Of certain pairs a, b of propositions one can say that a implies b – we then write $a \leqq b$. The properties of the relation of implication are: (1) $a \leqq a$, (2) if $a \leqq b$ and $b \leqq a$, then $a = b$, (3) if $a \leqq b$ and $b \leqq c$, then $a \leqq c$. Mathematically, $\leqq$ is a partial ordering on $\mathscr{E}$. One singles out two propositions 0 and 1 such that 0 implies every proposition while 1 is implied by every proposition. Further to any proposition a there is the contrary proposition a' – the negation of a; the process which associates with any proposition a its negation a' has the properties: (1) $(a')' = a$ for any a, (2) if $a \leqq b$, then $b' \leqq a'$, (3) $0' = 1$, $1' = 0$. One may observe that the idea of mutually disjoint or mutually exclusive propositions now lies near at hand; a and b are exclusive if $a \leqq b'$ (or $b \leqq a'$). Given any two propositions a and b which are disjoint there is another proposition c, called the sum of a and b. This corresponds to the proposition "a or b" and is defined entirely by means of $\leqq$; a and b both imply c and c implies every proposition implied by both a and b. In symbols, $c = a \dot{+} b$. It is natural to assume $a \dot{+} a' = 1$ and that if a implies b, there exists a proposition c disjoint from a such that b is the sum of a and c.

The point of departure that leads to the Kolmogorov formalism comes now. One's own built-in notions on the general structure of the set of experimental propositions lead one to postulate the following (crucial) assumption:

(1) Given *any* two propositions a and b, one can "split" them into disjoint propositions, i.e. there exist propositions a_1, b_1 and c, any two of which are disjoint, such that $a = a_1 \dot{+} c$ and $b = b_1 \dot{+} c$.

It will turn out later (Section III) that this assumption is equivalent to assuming that $\mathscr{E}$ is a Boolean algebra with $\leqq$ as its inclusion and $'$ as its complementation operations. As a technical concession one also adds the following assumption:

(2) Given a sequence of mutually exclusive propositions $a_1, a_2, \ldots$, one can find their sum, i.e. there is a proposition a such that each a_j implies a and a implies any other proposition with this property. With this the class $\mathscr{E}$ of propositions becomes a Boolean σ-algebra.

By a probability we then mean an assignment which associates with each proposition its probability. Mathematically it is a measure on $\mathscr{E}$, i.e. a real-valued function $p: a \to p(a)$ on $\mathscr{E}$ such that (1) $0 \leqq p(a) \leqq 1$ for all a, $p(0) = 0$, $p(1) = 1$, (2) if $a_1, a_2, \ldots$ are mutually disjoint with sum a, $p(a) = p(a_1) + p(a_2) + \cdots$. In this fashion we see that a crucial part is played by assumption I and that one is led inevitably to the fact that the propositions form a Boolean σ-algebra. It turns out that this assumption cannot be given convincing operational significance in quantum physics. The critiques of Heisenberg [9] and von Neumann [17, 3] are conclusive on this point. Even though individual experimental propositions associated with $\mathfrak{S}$ can be given operational significance, when several arbitrary propositions are involved, it is in general not possible to give any experimental meaning to the operation of "splitting" these up. In other words, propositions associated with a physical system in quantum theory are related, but not as the elements in a Boolean σ-algebra.

From our point of view an appropriately general model for probability theory would start with a class $\mathscr{E}$ of propositions of which however much less is assumed than the structure of a Boolean σ-algebra. In Section III we shall give a precise description of what is assumed of $\mathscr{E}$. In the meantime we proceed to examine, at an intuitive level, several questions which naturally arise concerning the calculus of this more general theory. We shall discuss these in greater detail in Section III and Section IV but some remarks might be worthwhile here.

Any probability theory must deal with random variables or observables[1] and their probability distributions. In the conventional model where $\mathscr{E}$ is taken to be a σ-algebra of subsets of a space Ω, we define a random variable as a real valued function f on Ω such that $f^{-1}(E)\in\mathscr{E}$ for all Borel sets E on the line. Given any probability measure p on $\mathscr{E}$ we can then obtain the distribution α_f^p of f under p; $\alpha_f^p(E)=p(f^{-1}(E))$.

In our general context where $\mathscr{E}$ is not assumed to be a σ-algebra (let alone a σ-algebra of subsets of a space) we cannot use any such definition. It turns out that we can give a precise meaning to the notion of an observable which reduces to the customary one in the conventional setup. Notice that observables have numerical values and hence they give rise to propositions which state whether these values lie in preassigned sets. Indeed, if x is an observable, to each Borel set E of the line is associated the proposition which states that the value of x lies in E. Calling this proposition $x(E)$ we thus see that there is a fundamental mapping $E\to x(E)$ associated with any observable which sends Borel subsets of the line into certain propositions. Moreover the fact that $x(E)$ is the proposition that the value of x lies in E enables us to conclude that the assignment $E\to x(E)$ must send disjoint sets into disjoint propositions of $\mathscr{E}$ and their unions into the corresponding sums in $\mathscr{E}$. In view of this it is natural to *define* an observable to be *this* mapping of the Borel sets into propositions. These mappings are usually known as σ-homomorphisms.

Thus while general arguments lead to the conclusion that observables are σ-homomorphisms of the class of Borel sets of the line – $\mathscr{B}(R^1)$ – into $\mathscr{E}$, in the Kolmogorov model we define them to be real valued functions on Ω. It is clear that if f is a measurable function on Ω, then, for any Borel set E, $f^{-1}(E)$ is a measurable set and $E\to f^{-1}(E)$ is a σ-homomorphism. It is urgent to verify at this stage that the most general σ-homomorphism of $\mathscr{B}(R^1)$ into the class of measurable subsets of Ω can arise only in this manner. Actually it is necessary to be a little more general. According to the Loomis representation theorem, if $\mathscr{E}$ is a Boolean σ-algebra, there is a space Ω, a σ-algebra $\mathscr{A}$ of subsets of Ω and a σ-homomorphism h of $\mathscr{A}$ onto $\mathscr{E}$; in general h is not an isomorphism. We shall prove in Section III that even in this general setup, every σ-homomorphism of $\mathscr{B}(R^1)$ into $\mathscr{E}$ arises naturally from an essentially unique point function (of course measurable) on Ω.

There is no difficulty in seeing how one ought to define a probability measure p in $\mathscr{E}$. To each $a \in \mathscr{E}$, $p(a)$ must be a number between 0 and 1 with $p(0)=0$, $p(1)=1$. Moreover if $a_1, a_2, \ldots$ is a sequence of mutually disjoint propositions with sum a, we must require that $p(a)=p(a_1)+$ $+p(a_2)+\cdots$. If then $x(E \to x(E))$ is an arbitrary observable and p a probability measure on $\mathscr{E}$, $\alpha_x^p : E \to p(x(E))$ will be a probability measure on the line. It is the distribution of x under p. In the conventional theory this reduces to the well known concept of the distribution of a random variable.

We have thus described, along with $\mathscr{E}$, the set of all observables. To every observable x and to a given probability measure p on $\mathscr{E}$ we have a distribution on the line – that of x under p. The calculus of probability to which these definitions lead is in many ways very remarkable. One of the most remarkable features is that in general two given observables need not have a joint distribution and that very special circumstances are needed to ensure that joint distributions exist. A number of examples are presented in Section IV which indicate clearly the various new and subtle features that are characteristic of this generalization. Especially noteworthy from the present point of view is an example given there which is the counterpart of a probability space having exactly n points. It must be emphasized that unlike the conventional theory, where the Boolean σ-algebra $\mathscr{E}$ is always replaced by a σ-algebra of subsets of Ω, there is no such "unique" concrete representation of the more general algebraic construct $\mathscr{E}$. In modern quantum physics one assumes that the partially ordered set of propositions is isomorphic with the class of closed linear subspaces of a complex Hilbert space $\mathscr{H}$ with negation in $\mathscr{E}$ corresponding to orthogonal complementation in $\mathscr{H}$. Under this assumption more concrete descriptions are possible for the observables, the probability measures on $\mathscr{L}$ and the entire calculus in fact. In Section IV some other concrete examples $\mathscr{L}$ are presented and the problems which arise when a complete analysis of a logic is attempted are discussed.

III. MATHEMATICAL FORMULATION

We now proceed to obtain precise formulations. We begin with a summary of definitions and results in the theory of partially ordered sets. For details and references see Birkhoff [2].

Let $\mathscr{L}$ be an abstract set. Elements of $\mathscr{L}$ will be denoted by $a, b, c, \ldots$, etc. We say that $\mathscr{L}$ is *partially ordered* if there is a relation $\leqq$ between certain pairs of elements of $\mathscr{L}$ such that (P1) $a \leqq a$ for all $a \in \mathscr{L}$, (P2) if $a, b \in \mathscr{L}$, $a \leqq b$ and $b \leqq a$, then $a = b$, (P3) if $a, b, c \in \mathscr{L}$, $a \leqq b$ and $b \leqq c$, then $a \leqq c$. The pair $(\mathscr{L}, \leqq)$ is a *partially ordered set*, $\leqq$ is the *partial ordering*. If $a \leqq b$ we write $b \geqq a$. By customary abuse of language we shall speak of $\mathscr{L}$ itself as the partially ordered set. If $\mathscr{L}$ is a partially ordered set there is at most one element a such that $a \leqq b$ for all $b \in \mathscr{L}$, we denote it by 0 whenever it exists. Similarly 1 denotes the unique element $\geqq b$ for all $b \in \mathscr{L}$, whenever it exists. A partially ordered set $\mathscr{L}$ is said to be *complemented* if (C1) 0 and 1 exist, (C2) there exists a mapping $\gamma: a \to a'$ of $\mathscr{L}$ into itself which is involutory and order-inverting, i.e. $a \leqq b$ implies $b' \leqq a'$ and $(a')' = a$ for all a. It is easy to show that then γ must be one-one and onto and that $0' = 1$, $1' = 0$. γ is called a *complementation* of $\mathscr{L}$. If $\mathscr{L}$ is a complemented partially ordered set and $a, b \in \mathscr{L}$, a and b are said to be *disjoint* whenever $a \leqq b'$; notice that $a \leqq b'$ if and only if $b \leqq a'$ so that the definition of disjointness is symmetric in a and b. Since $a \leqq a$ for any a, a and a' are always disjoint. If a and b are disjoint we write $a \perp b$. $0 \perp a$ for any $a \in \mathscr{L}$. If $\mathscr{L}$ is any partially ordered set and $\{a_\lambda\}$ is any indexed collection of elements of $\mathscr{L}$, we denote by $\bigcup_\lambda a_\lambda$ any element c of $\mathscr{L}$ with the properties: (i) $a_\lambda \leqq c$ for all λ, (ii) if $a_\lambda \leqq d$ for all λ, then $c \leqq d$. c need not exist; but if it does, it is unique. If the collection is $\{a_1, \ldots, a_n\}$ we also denote c by $a_1 \cup a_2 \cup \ldots \cup a_n$. $\bigcup_\lambda a_\lambda$ is called the *sum* of the a_λ. Replacing $\leqq$ by $\geqq$ in the above we obtain the definition of $\bigcap_\lambda a_\lambda$, the *product* of the a_λ. If $a \cup b$ and $a \cap b$ exist for all pairs $a, b \in \mathscr{L}$, then $\mathscr{L}$ is called a *lattice*. A lattice $\mathscr{L}$ is *distributive* if, whenever $a, b, c \in \mathscr{L}$, $a \cap (b \cup c) = (a \cap b) \cup (a \cap c)$ and $a \cup (b \cap c) = (a \cup b) \cap (a \cup c)$. If $\mathscr{L}$ is a complemented partially ordered set and $\{a_\lambda\}$ is an indexed collection of elements of $\mathscr{L}$, then $(\bigcup a_\lambda)' = \bigcap a'_\lambda$

in the sense that if either side exists so does the other and the two are equal. A complemented lattice for which $a \cap (b \cup c) = (a \cap b) \cup (a \cap c)$ is easily seen to be distributive. A lattice $\mathscr{E}$ is called *modular* if $a \cap (b \cup c) = (a \cap b) \cup (a \cap c) = (a \cap b) \cup c$ holds for all a, b, $c \in \mathscr{E}$ with $a \geqq c$.

Among distributive lattices a very important place is occupied by Boolean algebras and Boolean σ-algebras. A complemented distributive lattice is a *Boolean algebra* if the complementation $a \to a'$ has the properties: (i) if $a \leqq b$, there is a c disjoint from a such that $a \dot{+} c = b$, (ii) $a \dot{+} a' = 1$ for all a. A fundamental theorem of Stone asserts that any Boolean algebra is isomorphic and finite over X. Moreover for any real number s,

$$D_s = \bar{f}^{-1}((-\infty, s)) = \{w : \bar{f}(w) < s\}$$
$$= \bigcup_{\{n : r_n < s\}} E_n,$$

and hence

$$h(D_s) = \bigcup_{\{n : r_n < s\}} h(E_n) = \bigcup_{\{n : r_n < s\}} b_n = \bigcup_{\{n : r_n < s\}} x((-\infty, r_n))$$
$$= x((-\infty, s)).$$

If we now notice that the class of all Borel sets E on the line for which $h(\bar{f}^{-1}(E)) = x(E)$ is a σ-algebra including all intervals $(-\infty, s)$, we can conclude that $h(\bar{f}^{-1}(E)) = x(E)$ for all Borel sets E on the line. Define now f to be 0 outside X and $\bar{f}$ inside it. It is a trivial computation to check that $h(f^{-1}(E)) = h(\bar{f}^{-1}(E))$ for all E and hence $h(f^{-1}(E)) = x(E)$ for all E.

Suppose f' is another $\mathscr{A}$-measurable function such that $h(f'^{-1}(E)) = x(E)$ for all E. Write $A_n = f^{-1}((-\infty, r_n))$ and $B_n = f'^{-1}((-\infty, r_n))$. Since for any n, $h(A_n) = h(B_n)$,

$$h(A_n - B_n \cup B_n - A_n) = 0.$$

Write

$$N_0 = \bigcup_n ((A_n - B_n) \cup (B_n - A_n)).$$

Clearly $h(N_0) = 0$. We claim that $f = f'$ on $\Omega - N_0$. If $w \in \Omega - N_0$ and $f(w) \neq f'(w)$, then there will be an integer n such that r_n lies strictly between $f(w)$ and $f'(w)$. This shows that $w \in (A_n - B_n) \cup (B_n - A_n)$, i.e. $w \in N_0$, a contradiction. The proposition is proved.

Given a general logic $\mathscr{L}$ it is not at once obvious how to construct all the observables associated with $\mathscr{L}$. At this stage we shall be content to describe simple examples. Indeed let U be a set, $\mathscr{U}$ a σ-algebra of subsets of U and $t_1, t_2, \dots$ distinct elements of U. If $a_1, a_2, \dots$ are pairwise disjoint elements of $\mathscr{L}$ such that $1 = a_1 \dot{+} a_2 \dot{+} \cdots$, then there exists a unique $\mathscr{L}$-valued measure x based on $\mathscr{U}$ such that $x(E) = \bigcup_{j:t_j \in E} a_j$ for all $E \in \mathscr{U}$. If $\{t_j\} \in \mathscr{U}$, $x(\{t_j\}) = a_j$. The straightforward proof is omitted. The reader should notice that the construction given above, when $U = R^1$ and $\mathscr{U} = \mathscr{B}(R^1)$, defines what one might call the discrete observables of the theory. In fact, let x be an observable obtained as above from $t_1, \dots, t_k, \dots$. If p is *any* probability measure on $\mathscr{L}$, then α_x^p has mass concentrated at $t_1, t_2, \dots$. It is in this sense that x is the analogue of a discrete observable of the conventional model. We might call $t_1, t_2, \dots$ the values of x. One can discuss these ideas at a somewhat more general level. Given any observable x, a set E (Borel) on the line is called *x-null* if, for every probability measure p on $\mathscr{L}$, $\alpha_x^p(E) = 0$. Routine arguments show that the union of all open subsets of R^1 which are x-null is once again x-null and is the largest x-null open set. The complement of this set is called the *spectrum of* x. x is called a *bounded observable* if the spectrum is a compact subset of R^1. It is easy to prove that if $\mathscr{L}$ is a σ-algebra of subsets of a space Ω and if all single point sets are in $\mathscr{L}$, the spectrum of an observable x is the closure of the set of points which are the values of the unique $\mathscr{L}$-measurable function corresponding to x. In general if x is a bounded observable and Λ its spectrum, we define $\|x\| = \sup\{|\lambda| : \lambda \in \Lambda\}$.

Given a logic $\mathscr{L}$ one can associate with it the set of all observables. Our chief aim in the present section is to examine the principal features of this set and to compare them with the set of all observables in the Kolmogorov model. We begin with the notion of a *functional calculus*. Intuitively, given any physical quantity x and a Borel function u ($\equiv$ a real valued Borel measurable function on the real line R^1), there is an operational definition of the quantity $u(x)$; in fact if x has value ξ, $u(x)$ has value $u(\xi)$. It is easy to translate this into a precise definition if we notice that $u(\xi) \in E$ if and only if $\xi \in u^{-1}(E)$. Thus given any observable $x(E \to x(E))$ and any Borel function u, we define the observable $u(x)$ by the assignment $u(x): E \to x(u^{-1}(E))$. It is easily verified that $u(x)$ is an $\mathscr{L}$-valued measure based on $\mathscr{B}(R^1)$ so that $u(x)$ is in fact an observable. Standard juggling with composition mappings yields.

PROPOSITION 3.4. Let x be an observable, u_1, u_2 Borel functions and $u=u_1 \circ u_2$. Then $u(x)=u_1(u_2(x))$. If x has distribution α under p, then $u(x)$ has distribution β under p, where $\beta(E)=\alpha(u^{-1}(E))$ for all E.

In other words the rules of calculations of functions and distributions of functions of a given observable are the same as in the conventional formalism. However functions of more than one observable can in general only be formed under special circumstances. A consequence of the theorems of this section is a precise characterization of these circumstances.

Given an observable x and sets E, $F \in \mathscr{B}(R^1)$, one can write

$$x(E)=x(E-F)\dot{+}x(E\cap F) \quad \text{and} \quad x(F)=x(F-E)\dot{+}x(E\cap F).$$

Clearly $x(E-F)$, $x(E\cap F)$ and $x(F-E)$ are pairwise disjoint. According to the motivating remarks made in Section II the propositions $x(E)$ and $x(F)$ can be "split". (This is also natural in an intuitive sense since after all $x(E)$ (and $x(F)$) are propositions which state that the value of the "quantity" x lies in E (and F) and to these propositions the conventional rules of reduction should apply.) What is more interesting is that the converse is true. Given $a, b \in \mathscr{E}$ we write $a \leftrightarrow b$ if there are pairwise disjoint elements a_1, b_1 and c such that $a=a_1 \dot{+} c$ and $b=b_1 \dot{+} c$.

PROPOSITION 3.5. Let $\mathscr{E}$ be a logic and $a, b \in \mathscr{E}$. Then a necessary and sufficient condition that $a \leftrightarrow b$ is that there should exist an observable x and Borel sets E and F such that $a=x(E)$ and $b=x(F)$.

Proof. We have already settled the sufficiency. For the necessity write $a=a_1 \dot{+} c$, $b=b_1 \dot{+} c$. Clearly there exists an observable x such that $x(\{0\})=a_1$, $x(\{1\})=b_1$, $x(\{2\})=c$ and $x(\{3\})=(a_1 \dot{+} b_1 \dot{+} c)'$. Then $x(\{0, 2\})=a$ and $x(\{1, 2\})=b$.

This proposition motivates the following definition. Elements $a, b \in \mathscr{E}$ are called *simultaneously verifiable* if $a \leftrightarrow b$. Proposition 3.5 shows also that if $a \leftrightarrow b$, then any two of a, b, a', b' are simultaneously verifiable.

PROPOSITION 3.6. Let $\mathscr{E}$ be a logic, $a, b \in \mathscr{E}$ and $a \leftrightarrow b$. Then $a \cup b$ and $a \cap b$ exist. If a_1, b_1 and c are pairwise disjoint such that $a=a_1 \dot{+} c$ and $b=b_1 \dot{+} c$, then $a_1=a-a\cap b$, $b_1=b-a\cap b$ and $c=a\cap b$. Moreover $a'=(b-a\cap b)\dot{+}(a\cup b)'$, $b'=(a-a\cap b)\dot{+}(a\cup b)'$.

Proof. Write $a=a_1 \dot{+} c$ and $b=b_1 \dot{+} c$, where a_1, b_1 and c are pairwise disjoint. Write $d=a_1\dot{+}b_1\dot{+}c$. Now, $d\geqq a$ and $d\geqq b$; moreover if $\bar{d}$ is any element of $\mathscr{L}$ with $\bar{d}\geqq a$ and $\bar{d}\geqq b$, then $\bar{d}\geqq a_1\dot{+}b_1\dot{+}c$ and hence $\bar{d}\geqq d$. This shows that $a\cup b$ exists and is $=d$. Apply this to a' and b' to deduce that $a'\cup b'$ and hence $a\cap b$ exists. Further $a'=b_1\dot{+}d'$ and $b'=a_1\dot{+}d'$ so that the above argument yields $a'\cup b'=a_1\dot{+}b_1\dot{+}d'=c'$ and hence that $c=a\cap b$. The rest of the proposition is now obvious.

A useful technique of verifying simultaneous verifiability is contained in

PROPOSITION 3.7. In order that $a\leftrightarrow b$ it is necessary and sufficient that there should exist some $c\in\mathscr{L}$ such that (i) $c\leqq a$ and $c\leqq b$, (ii) $(b-c)\perp a$. In this case $c=a\cap b$.

Proof. Necessity is obvious if we take $c=a\cap b$. Sufficiency follows at once from the definition if we write $a=(a-c)\dot{+}c$, $b=(b-c)\dot{+}c$.

PROPOSITION 3.8. (1) If $a_1, a_2,\ldots\in\mathscr{L}$, if $a\leftrightarrow a_j$ for each j and if $\bigcup_j a_j$ and $\bigcup_j(a\cap a_j)$ both exist, then $a\leftrightarrow\bigcup_j a_j$; moreover, one has $a\cap(\bigcup_j a_j)=$ $=\bigcup_j(a\cap a_j)$. In particular, if $a_1, a_2,\ldots$ are pairwise disjoint and $a\leftrightarrow a_j$ for all j, then $a\leftrightarrow(a_1\dot{+}a_2\dot{+}\cdots)$ and one has

$$a\cap(a_1\dot{+}a_2\dot{+}\cdots)=(a\cap a_1)\dot{+}(a\cap a_2)\dot{+}\cdots.\text{*}$$

(2) If $b_1, b_2\in\mathscr{L}$, $b_1\leqq b_2$ and $a\leftrightarrow b_1$ and $a\leftrightarrow b_2$, then $a\leftrightarrow(b_2-b_1)$; moreover, one has

$$a\cap(b_2-b_1)=(a\cap b_2)-(a\cap b_1).$$

Proof. Write $c=\bigcup_j(a\cap a_j)$. Clearly $c\leqq a$ and $c\leqq\bigcup_j a_j$. Thus, to prove (1) it is enough to prove that $(a-c)\perp\bigcup a_j$. It will then follow that $a\bigcup_j a_j$ and $a\cap(\bigcup_j a_j)=c$. Now, for any fixed j, $a-(a\cap a_j)\perp a_j$ since $a\leftrightarrow a_j$, and

* ERRATUM, *Comm. Pure Appl. Math.* **15** (1962), 189–217:

It has been pointed out by Dr. S. P. Gudder of the University of Illinois that Proposition 3.8 in this paper is incorrect, thus vitiating the conclusions of the main theorems of Section III.

We wish to remark that Propositions 3.8 and 3.9 of this article, and hence all subsequent theorems are true if we add to the assumptions that $\mathscr{L}$ is actually a lattice. In other words, if the fundamental logic is assumed to be a lattice, then the proofs go through without any further changes. The main theorems of Section III are therefore valid for all logics $\mathscr{L}$ which are lattices.

$a-c \leqq a-(a \cap a_j)$ since $c \geqq a \cap a_j$. Therefore $(a-c) \perp a_j$. Since j is arbitrary we deduce that $(a-c) \perp \bigcup_j a_j$.

To prove the second part note that $a \leftrightarrow b_1$ and $a \leftrightarrow b_2'$ and hence $a \leftrightarrow b_1 \dot{+} b_2'$ from which we can conclude that $a \leftrightarrow (b_1 \dot{+} b_2')' = b_2 - b_1$. Since $b_1 \dot{+} (b_2 - b_1) = b_2$, we deduce from (1) that $(a \cap b_1) \dot{+} a \cap (b_2 - b_1) = a \cap b_2$ from which the formula for $a \cap (b_2 - b_1)$ follows.

We have hinted at various stages of the present discussion that it is the features centering around the mutual relations subsisting between several observables that indicate sharply the real significance of the present generalization of the Kolmogorov formalism. If x and y are two observables, $x(E)$ and $y(F)$ are elements of $\mathscr{L}$. But we have seen that $x(E)$ and $y(F)$ need not be simultaneously verifiable. It is natural to call x and y *simultaneously observable* if for any pair of Borel sets E and F, $x(E)$ and $y(F)$ are simultaneously verifiable. More generally, if $\{x_\lambda; \lambda \in \Delta\}$ is an indexed set of observables, it is said to be *simultaneously observable* if x_λ and $x_{\lambda'}$ are simultaneously observable for all λ, $\lambda' \in \Delta$. Intuitively, if x and y are simultaneously observable physical quantities, one can assign operational significance to numerical statements that involve x and y simultaneously and hence joint probabilities such as the probability of x lying between 0 and 1 and y lying between 3 and 4 can be computed.

Suppose now x is an observable and $\{u_\lambda : \lambda \in \Delta\}$ is an indexed set of Borel functions. If we set $x_\lambda = u_\lambda(x)$ and notice that $x_\lambda(E) = x(u_\lambda^{-1}(E))$, it is clear that the x_λ are simultaneously observable. If p is any probability measure on $\mathscr{L}$, and $\lambda_1, \lambda_2, \ldots, \lambda_k \in \Delta$, the definition $P^{\lambda_1, \ldots, \lambda_k}(C) = p(x(F))$, where $F = \{t : t \in R^1, (u_{\lambda_1}(t), \ldots, u_{\lambda_k}(t)) \in C\}$, gives us a probability measure on the Borel sets C of R^k. It is natural to define $P^{\lambda_1, \ldots, \lambda_k}$ to be the joint distribution of $\{x_{\lambda_1}, \ldots, x_{\lambda_k}\}$.

It was proved by von Neumann (see [17], p. 173 and the references cited there) that, when $\mathscr{L}$ is the logic of closed linear subspaces of a complex separable Hilbert space, then simultaneous observability can be secured only in the above manner. More precisely, for such a logic, if $\{x_\lambda : \lambda \in \Delta\}$ is an indexed set of observables, and these are simultaneously observable, then there exists an observable x and Borel functions u_λ for each $\lambda \in \Delta$ such that $x_\lambda = u_\lambda(x)$ for all λ. The question naturally arises as to whether such a theorem can be proved in our general context. In his Harvard lecture notes Mackey raises this question also but limits himself to stating the von Neumann theorem. Clearly if such a theorem is proved, then the way is open for the elucidation of the circumstances under which joint distributions and functions of several observables can be introduced. The main theorem (Theorem 3.3) of this paper is in fact the generalization of the von Neumann theorem to our general context of an arbitrary logic. Since an observation on a physical quantity yields at once observations on all of its functions (by computation) this theorem gives a justification of the term "simultaneously observable".

To proceed with our aim of obtaining our abstract version of von Neumann's theorem we introduce a few definitions. A subset $B \subset \mathscr{L}$ is called a *Boolean subalgebra of* $\mathscr{L}$ if there exists a space U, a Boolean algebra $\mathscr{U}$ of subsets of U and a mapping $u(C \to u(C))$ of sets in $\mathscr{U}$ into $\mathscr{L}$ such that (i) $u(\phi)=0$, $u(U)=1$, (ii) if C, $D \in \mathscr{U}$ and $C \cap D = \phi$, then $u(C) \perp u(D)$ and $u(C \cup D) = u(C) \dot{+} u(D)$. If we demand in the above definition that $\mathscr{U}$ be a σ-algebra and that in addition to (i) and (ii) the equation $u(\bigcup_n C_n) = u(C_1) \dot{+} u(C_2) \dot{+} \cdots$ hold for all $C_1, C_2, \ldots \in \mathscr{U}$ such that $C_i \cap C_j = \phi$ for $i \neq j$, then we have the notion of a Boolean sub σ-algebra of $\mathscr{L}$. Concisely, a *Boolean sub* σ*-algebra* of $\mathscr{L}$ is the range of an $\mathscr{L}$-valued measure u based on a σ-algebra $\mathscr{U}$ of subsets of some space U.

PROPOSITION 3.9. A subset $B \subset \mathscr{L}$ is a Boolean subalgebra of $\mathscr{L}$ if and only if it has the following properties: (i) $0, 1 \in B$, (ii) if $a, b \in B$, then $a \leftrightarrow b$ and further $a \cup b$, $a \cap b$ both[2] belong to B, (iii) if $a \in B$, then $a' \in B$.

Proof. The only if part is trivial. Suppose now B has the properties (i)–(iii). $\leqq$, when restricted to B, converts B into a complemented lattice. We claim that it is even distributive, i.e. $a \cap (b \cup c) = (a \cap b) \cup (a \cap c)$, whenever $a, b, c \in B$. Notice that, since the lattice B is complemented, the equations $a \cap (b \cup c) = (a \cap b) \cup (a \cap c)$ imply the equations $a \cup (b \cap c) = (a \cup b) \cap (a \cup c)$ for all $a, b, c \in B$. Proposition 3.8 implies at once that $a \cap (b \cup c) = (a \cap b) \cup (a \cap c)$. This proves that B is distributive. Since, if $a, b \in B$ and $a \leqq b$, one can write $b = a \dot{+} (a \dot{+} b')'$, it follows that B is a Boolean algebra. By Stone's theorem there exists a space U, a Boolean algebra $\mathscr{A}$ of subsets of U and a map $u: \mathscr{A} \to B$ such that u has the appropriate properties and B is the range of u. This proves the proposition.

An analogous reasoning which uses the Loomis theorem instead of Stone's leads to

PROPOSITION 3.10. A subset $S \subset \mathscr{L}$ is a Boolean sub σ-algebra of $\mathscr{L}$ if and only if it has the following properties: (i) 0, $1 \in S$, (ii) if a, $b \in S$, then $a \leftrightarrow b$ and $a \cup b$, $a \cap b$ both $\in S$, (iii) if $a \in S$, then $a' \in S$, (iv) if $a_1, a_2, \ldots \in S$ and $a_i \perp a_j$ whenever $i \neq j$, then $(a_1 \dot{+} a_2 \dot{+} \cdots) \in S$.

Propositions 3.9 and 3.10 tell us that if $\{A_\lambda\}$ is an indexed family of Boolean subalgebras (sub σ-algebras) of $\mathscr{E}$, then $\bigcap_\lambda A_\lambda$ is a Boolean subalgebra (sub σ-algebra) of $\mathscr{E}$. From this it follows that if A is any subset of $\mathscr{E}$, then, as soon as there exists one Boolean subalgebra (sub σ-algebra) of $\mathscr{E}$ containing A, there will be a smallest such subalgebra. Given a family $\{A_\lambda : \lambda \in \Delta\}$ of subsets of $\mathscr{E}$, the smallest Boolean subalgebra of $\mathscr{E}$ containing all the A_λ, provided it exists, is denoted by $[A_\lambda : \lambda \in \Delta]$. If $\Delta = \{\lambda_1, \ldots, \lambda_k\}$, we denote this by $[A_{\lambda_1}, \ldots, A_{\lambda_k}]$. The following technical preliminary is needed first.

PROPOSITION 3.11. Let A be a Boolean subalgebra of $\mathscr{E}$ and $B \subset A$ a subset such that A is the smallest Boolean subalgebra of $\mathscr{E}$ which includes B. If $a_0 \in \mathscr{E}$ is such that $a_0 \leftrightarrow b$ for all $b \in B$, then $a_0 \leftrightarrow a$ for all $a \in A$.

Proof. Let $A_1 = \{a : a \in A,\ a_0 \leftrightarrow a\}$. Clearly 0, $1 \in A_1$ and if $a \in A_1$, then $a' \in A_1$. Moreover from Proposition 3.8 it follows, on noting that lattice sums exist for pairs of elements of A_1, that if a_1, $a_2 \in A_1$, then $a_1 \cup a_2 \in A_1$ and hence $a_1 \cap a_2 = (a_1' \cup a_2')' \in A_1$. This proves that A_1 is a Boolean subalgebra of $\mathscr{E}$. Since $B \subset A_1 \subset A$ it follows that $A_1 = A$.

To formulate the subsequent propositions concisely we need a notation. If A and B are two subsets of $\mathscr{E}$ we write $A \leftrightarrow B$ whenever the relation $a \leftrightarrow b$ holds for arbitrary $a \in A$ and $b \in B$.

PROPOSITION 3.12. Let $A_1, A_2, \ldots, A_k$ be any Boolean subalgebras of $\mathscr{E}$. In order that there should exist a Boolean subalgebra of $\mathscr{E}$ containing all the A_j it is necessary and sufficient that $A_i \leftrightarrow A_j$ for all i and j.

Proof. Since the necessity of the condition is trivial we confine our attention to the sufficiency. We first use induction to reduce the theorem to $k=2$. Indeed if we have proved the sufficiency of the condition for $k=2, 3, \ldots, m-1$, then we can form $[A_1, A_2, \ldots, A_{m-1}]$ and, observing that $[A_i, \ldots, A_{m-1}]$ is the smallest Boolean subalgebra of $\mathscr{E}$ containing $\bigcup_{i=1}^{m-1} A_i$, conclude from Proposition 3.11 and the assumption $A_m \leftrightarrow \bigcup_{i=1}^{m-1} A_i$ that $A_m \leftrightarrow [A_1, \ldots, A_{m-1}]$. The case $k=2$ now shows that there is a Boolean subalgebra of $\mathscr{E}$ containing A_m and $[A_1, \ldots, A_{m-1}]$. It thus remains only to settle the case $k=2$. In other words we shall show that if $A_1 \leftrightarrow A_2$, there exists a Boolean subalgebra of $\mathscr{E}$ which contains both A_1 and A_2.

Consider *finite* Boolean subalgebras B_1 and B_2 of $\mathscr{E}$, with $B_1 \subset A_1$ and $B_1 \subset A_2$. Since B_1 and B_2 are finite, there are finite sets $U = \{t_1, \ldots, t_p\}$ and $V = \{s_1, \ldots, s_q\}$ and $\mathscr{E}$-valued measures u and v based, respectively, on the subsets of U and V such that B_1 is the range of u and B_2 that of v. Write $a_i = u(\{t_i\})$ and $b_j = v(\{s_j\})$. Clearly the a_i (and the b_j) are pairwise disjoint and

$$a_1 \dotplus \cdots \dotplus a_p = b_1 \dotplus \cdots \dotplus b_q = 1.$$

Since $a_i \leftrightarrow b_j$ for all i and j, $c_{ij} = a_i \cap b_j$ exists. If $(i, j) \neq (i', j')$, then either $i \neq i'$ or $j \neq j'$. If $i \neq i'$, $c_{ij} \leqq a_i$ and $c_{i'j'} \leqq a_{i'}$, while if $j \neq j'$, $c_{ij} \leqq b_j$ and $c_{i'j'} \leqq b_{j'}$. This shows that the c_{ij} are pairwise disjoint. Moreover, we obtain from Proposition 3.8 the equations

$$\begin{aligned} c_{i1} \dotplus \cdots \dotplus c_{iq} &= (a_i \cap b_1) \dotplus \cdots \dotplus (a_i \cap b_q) \\ &= a_i \cap (b_1 \dotplus \cdots \dotplus b_q) \\ &= a_i \end{aligned}$$

for each fixed $i = 1, 2, \ldots, p$ and the equations

$$c_{1j} \dotplus \cdots \dotplus c_{pj} = b_j$$

for each fixed $j = 1, 2, \ldots, q$. Finally, adding all these up we have also

$$\bigcup_{i, j} c_{ij} = 1.$$

If we now write $W = U \times V$, then the last equation shows that there exists an $\mathscr{E}$-valued measure w based on subsets of w such that $w(\{t_i, s_j\}) = c_{ij}$ for all i, j. The preceding equations then tell us that $w(U \times \{s_j\}) = b_j$ for $j = 1, 2, \ldots, q$ and $w(\{t_i\} \times V) = a_i$ for $i = 1, 2, \ldots, p$. If the range of w is C, then it is clear that B_1 and $B_2 \subset C$ and that C is a finite Boolean subalgebra of $\mathscr{E}$. In fact $C = [B_1, B_2]$.

What we have proved so far is that, if B_1 and B_2 are arbitrary but finite Boolean subalgebras of $\mathscr{E}$ such that $B_1 \subset A_1$ and $B_2 \subset A_2$, then $[B_1, B_2]$ exists and is finite. Let us now form the set union $A = \bigcup [B_1, B_2]$, where B_1 and B_2 run over all finite Boolean subalgebras of $\mathscr{E}$ with $B_1 \subset A_1$ and $B_2 \subset A_2$. We complete the proof by showing that A is a Boolean subalgebra of $\mathscr{E}$ containing both A_1 and A_2. If $a_1 \in A_1$ and $a_2 \in A_2$, then $\bar{B}_1 = \{0, 1, a_1, a_1'\}$ and $\bar{B}_2 = \{0, 1, a_2, a_2'\}$ are Boolean subalgebras of $\mathscr{E}$ and clearly $[\bar{B}_1, \bar{B}_2] \subset A$ so that in particular $a_1, a_2 \in A$. This proves that A

contains both A_1 and A_2. To prove that A is a Boolean subalgebra of $\mathscr{L}$, we notice first that if $a \in A$, $a' \in A$ trivially. Further suppose $a, b \in A$. Then $a \in [B_1, B_2]$ and $b \in [B_1', B_2']$, where B_1, B_1', B_2, B_2' are finite Boolean subalgebras of $\mathscr{L}$ with $B_1, B_1' \subset A_1$ and $B_2, B_2' \subset A_2$. If we now write $B_1'' = [B_1, B_1']$ and $B_2'' = [B_2, B_2']$, then it is easy to verify that a and b both belong to $[B_1'', B_2'']$ and hence $a \cup b$ and $a \cap b \in [B_1'', B_2''] \subset A$. This proves, in view of Proposition 3.9 that A is a Boolean subalgebra of $\mathscr{L}$. The proof is thus completed.

THEOREM 3.1. *Let $\mathscr{L}$ be a logic and $\{A_\lambda : \lambda \in \Delta\}$ be an indexed set of Boolean subalgebras of $\mathscr{L}$. Then, in order that there be a Boolean subalgebra of $\mathscr{L}$ including all the A_λ, it is necessary and sufficient that $A_\lambda \leftrightarrow A_\lambda'$ for all $\lambda, \lambda' \in \Delta$.*

Proof. Necessity is trivial. For sufficiency note that for each finite subset $\Delta' \subset \Delta$, $[A_\lambda : \lambda \in \Delta']$ exists in view of Proposition 3.11. Denote this by $B(\Delta')$. If $a \in B(\Delta_1')$ and $b \in B(\Delta_2')$, then $a \cup b$ and $a \cap b$ exist and in fact both belong to $B(\Delta_1' \cup \Delta_2')$. This shows that $\bigcup_{\substack{\Delta' \subset \Delta \\ \Delta' \text{ finite}}} B(\Delta') = B$ is a Boolean subalgebra of $\mathscr{L}$. Obviously $A_\lambda \subset B$ for each λ.

Our aim is to extend Theorem 3.1 to σ-algebras. To do this we need to introduce one more concept. A Boolean subalgebra A of $\mathscr{L}$ is said to be *maximal* if A is properly contained in no other Boolean subalgebra of $\mathscr{L}$.

PROPOSITION 3.13. Any Boolean subalgebra of $\mathscr{L}$ is contained in a maximal one. A maximal Boolean subalgebra of $\mathscr{L}$ is necessarily a Boolean sub σ-algebra of $\mathscr{L}$.

Proof. The first statement follows easily from Proposition 3.9 and Zorn's lemma. To prove the second, let M be a maximal Boolean subalgebra of $\mathscr{L}$. In order to prove that M is a Boolean sub σ-algebra of $\mathscr{L}$ it is enough if we show that for any sequence $a_1, a_2, \ldots$ of pairwise disjoint elements of M, $a_1 \dotplus a_2 \dotplus \cdots$ is in M. Suppose $a_1, a_2, \ldots \in M$ with $a_i \perp a_j$ whenever $i \neq j$. Write $a = a_1 \dotplus a_2 \dotplus \cdots$. Clearly $a_j \leftrightarrow b$ for any $b \in M$ so that from Proposition 3.8 we conclude that $a \leftrightarrow b$ for all $b \in M$. This shows that $M \leftrightarrow B_0$, where B_0 is the Boolean subalgebra of $\mathscr{L}$ consisting of 0, 1, a and a'. Then $[M, B_0]$ exists and since M is maximal, $M = [M, B_0]$. In particular, $a \in M$.

THEOREM 3.2. *Let $\{A_\lambda : \lambda \in \Delta\}$ be an indexed set of Boolean sub σ-algebras of $\mathscr{L}$. In order that there should exist a Boolean sub σ-algebra of $\mathscr{L}$ including all the A_λ, it is necessary and sufficient that $A_\lambda \leftrightarrow A_{\lambda'}$ for all λ, $\lambda' \in \Delta$.*

Proof. The necessity is trivial. For sufficiency we can, in view of Theorem 3.1, find a Boolean subalgebra of $\mathscr{L}$ including all the A_λ. By Proposition 3.12 this in turn is contained in a Boolean sub σ-algebra of $\mathscr{L}$.

Recall that we launched on these detailed discussions with a view to examining the simultaneous observability of a given collection of observables. Intuitively, if x is any physical quantity and $\{u_\lambda : \lambda \in \Delta\}$ is a collection of Borel functions, the collection $\{x_\lambda\}$ of observables, where $x_\lambda = u_\lambda(x)$, is simultaneously observable. (Indeed, if x has a value ξ, x_λ has the value $u_\lambda(\xi)$, λ being an arbitrary element of Δ.) We shall now proceed to prove that this is the only way to secure simultaneous observability. We need certain definitions. A Boolean σ-algebra $\mathscr{A}$ is said to be *separable* if there exists a countable set $\mathscr{D} \subset \mathscr{A}$ such that the smallest sub σ-algebra of $\mathscr{A}$ containing $\mathscr{D}$ is $\mathscr{A}$ itself. $\mathscr{D}$ is said to *generate* $\mathscr{A}$. A logic $\mathscr{L}$ is said to be *separable* if every Boolean sub σ-algebra of $\mathscr{L}$ is separable.

PROPOSITION 3.14. Let U be a space and $\mathscr{U}$ a separable σ-algebra of subsets of U. Then there exists a real valued $\mathscr{U}$-measurable function f on U such that

$$\mathscr{U} = \{f^{-1}(E) : E \in \mathscr{B}(R^1)\}.$$

Proof. Let $D_1, D_2, \ldots$ generate $\mathscr{U}$ and let ψ_{D_n} be the characteristic function of D_n. Then $\gamma : u \to (\psi_{D_1}(u), \psi_{D_2}(u), \ldots)$ is a $\mathscr{U}$-measurable map of U into the cartesian product X of a countable number of copies of the unit interval. If $\mathscr{B}(X)$ is the class of Borel subsets of X, it is obvious that $\mathscr{U} = (\gamma^{-1}(E) : E \in \mathscr{B}(X)\}$. Since there is an isomorphism between X and R^1, say $\beta : X \to [0, 1]$ which preserves the Borel structures on the two spaces [14, Section III], it follows that $f = \beta \circ \gamma$ is a real valued $\mathscr{U}$-measurable function on U such that $\mathscr{U} = \{f^{-1}(E) : E \in \mathscr{B}(R^1)\}$.

PROPOSITION 3.15. A Boolean σ-algebra $\mathscr{A}$ is separable if and only if there exists a σ-homomorphism of $\mathscr{B}(R^1)$ *onto* $\mathscr{A}$.

Proof. Let $h : \mathscr{B}(R^1) \to \mathscr{A}$ be a σ-homomorphism of $\mathscr{B}(R^1)$ onto $\mathscr{A}$ and let $\mathscr{D} = \{h((-\infty, r)) : r \text{ rational}\}$. Suppose $\mathscr{A}_0$ is the sub σ-algebra of $\mathscr{A}$

generated by $\mathscr{D}$. $\mathscr{A}_0$ is separable. Moreover since $\{E:h(E)\in\mathscr{A}_0\}$ is a σ-algebra of Borel subsets of R^1 including all intervals $(-\infty, r)$, it includes all Borel sets and hence $\mathscr{A}_0$ is the range of h. This proves that $\mathscr{A}=\mathscr{A}_0$ and hence is separable.

Conversely let $\mathscr{A}$ be separable and $\mathscr{D}\in\mathscr{A}$ any countable set generating $\mathscr{A}$. By the theorem of Loomis there is a space U, a σ-algebra $\mathscr{U}$ of subsets of U and a σ-homomorphism u of $\mathscr{U}$ onto $\mathscr{A}$. Let $C_1, C_2,\ldots$ be subsets of U such that $\mathscr{D}=\{u(C_j):j=1,2,\ldots\}$ and let $\mathscr{U}_0$ be the σ-algebra of subsets of U generated by the C_j. $\mathscr{U}_0$ is separable and hence (Proposition 3.13) there exists a real valued $\mathscr{U}_0$-measurable function f such that $\mathscr{U}_0=\{f^{-1}(E):E\in\mathscr{D}(R^1)\}$. If now $h(E)=u(f^{-1}(E))$ for all $E\in\mathscr{B}(R^1)$, then h is obviously a σ-homomorphism of $\mathscr{B}(R^1)$ into $\mathscr{A}$. Since the range of h includes all the $u(C_j)$, h is onto $\mathscr{A}$.

PROPOSITION 3.16. Let $\mathscr{A}$, $\mathscr{A}'$ be separable σ-algebras with $\mathscr{A}'\subset\mathscr{A}$, and let τ, τ' be σ-homomorphisms of $\mathscr{B}(R')$ onto $\mathscr{A}$ and $\mathscr{A}'$, respectively. Then there exists a Borel function u such that $\tau'(E)=\tau(u^{-1}(E))$ for all Borel sets E on the line.

Proof. Let $r_1, r_2,\ldots$ be any distinct enumeration of the rationals of the line and let $b_n=\tau'((-\infty, r_n))$. By using an inductive argument similar to the one employed in Proposition 3.3, we construct Borel sets $E_1, E_2\ldots$ on the line with the properties: (1) $b_n=\tau(E_n)$, $n=1,2,\ldots$, (2) $E_i\subset E_j$ whenever $r_i<r_j$, (3) $\bigcap_n E_n=\phi$. Put $X_0=\bigcup_n E_n$ and define for any $t\in X_0$, $\bar{u}(t)=\inf\{r_n:t\in E_n\}$. $\bar{u}$ is finite and well defined over X_0. Moreover, for any s,

$$\{t:\bar{u}(t)<s\}=\bigcup_{n:r_n<s} E_n,$$

so that $\bar{u}$ is a Borel function on X_0. If we now notice that

$$\begin{aligned}\tau(\{t:\bar{u}(t)<s\}) &= \bigcup_{n:r_n<s}\tau(E_n)=\bigcup_{n:r_n<s} b_n\\ &= \bigcup_{n:r_n<s}\tau'((-\infty, r_n))=\tau'((-\infty, s))\end{aligned}$$

for any real number s, we can easily conclude that $\tau(\bar{u}^{-1}(E))=\tau'(E)$ for all Borel sets E. There remains a small difficulty in that $\bar{u}$ is defined only over X_0. This however is not serious. Define u to be 0 outside the set X_0 and $\bar{u}$ inside. From the fact that $\tau(X_0)=1$, it follows easily that $\tau(u^{-1}(E))=\tau'(E)$ for all Borel sets E.

THEOREM 3.3. *Suppose $\{x_\lambda : \lambda \in \Delta\}$ to be an indexed set of observables. Suppose further that either Δ is denumerable or that $\mathscr{L}$ is a separable logic. Then a necessary and sufficient condition that the x_λ be simultaneously observable is that there should exist an observable x and a set $\{u_\lambda : \lambda \in \Delta\}$ of Borel functions such that $x_\lambda = u_\lambda(x)$ for all λ.*

Proof. Suppose x is an observable, u_λ are Borel functions and $x_\lambda = u_\lambda(x)$, then for any two Borel sets E and F and any $\lambda, \lambda' \in \Delta$, $x_\lambda(E) = x(u_\lambda^{-1}(E))$ and $x_{\lambda'}(F) = x(u_{\lambda'}^{-1}(F))$ so that $x_\lambda(E)\, x_{\lambda'}(F)$. This proves that x_λ and $x_{\lambda'}$ are simultaneously observable. Conversely let us assume that x_λ and $x_{\lambda'}$ are simultaneously observable for all $\lambda, \lambda' \in \Delta$. If we write $A_\lambda = \{x_\lambda(E) : E$ a Borel set on the lines, then A_λ is a Boolean sub σ-algebra of $\mathscr{L}$ for each λ. Moreover by our assumption on the x_λ, $A_\lambda \leftrightarrow A_{\lambda'}$ for all λ, λ' and hence by Theorem 3.2 there exists a Boolean sub σ-algebra of $\mathscr{L}$ containing all the A_λ. Let B denote the smallest such σ-algebra. We claim that, under our hypotheses, B is separable. In fact if $\mathscr{L}$ is separable, B is automatically separable. On the other hand, if Δ is denumerable and for each $\lambda \in \Delta D_\lambda$ is a countable subset of A_λ generating A_λ, the countable set $D = \bigcup_{\lambda \in \Delta} D_\lambda$ clearly generates B, showing that B is separable. By Proposition 3.15 we can find an observable x such that $B = \{x(E) : E$ a Borel set$\}$. By Proposition 3.16 we can find for each λ a Borel function u_λ such that $x_\lambda = u_\lambda(x)$. This proves the theorem.

We next proceed to a discussion of the circumstances under which joint distributions exist and the definitions of functions of several observables. Suppose x and y are two observables. We may then say that x and y have a joint distribution if there exists a σ-homomorphism z of the Borel sets of the plane R^2 into $\mathscr{L}$ such that $z(E \times R^1) = x(E)$ and $z(R^1 \times E) = y(E)$ for all Borel sets E. That such a z exists when we operate in the conventional framework is obvious. More generally let $\{x_\lambda : \lambda \in \Delta\}$ be an indexed set of observables. Following Kolmogorov [10], we define the Borel sets in the space R^Δ of all real valued functions on Δ to be the sets of the smallest σ-algebra containing all the cylinder sets. We denote this σ-algebra by $\mathscr{B}(R^\Delta)$. For any $\lambda \in \Delta$ we define π_λ as the "projection" $f \to f(\lambda)$ of R^Δ into R^1. We then say that the x_λ have a *joint distribution* whenever there exists a σ-homomorphism z of $\mathscr{B}(R^\Delta)$ into $\mathscr{L}$ such that $z(\pi_\lambda^{-1}(E)) = x_\lambda(E)$ for all λ and all real line Borel sets E. If z exists, it follows easily that it is unique. If p is any probability measure on $\mathscr{L}$ and we define $\alpha_p^\Delta(C) = p(z(C))$ for all $C \in \mathscr{B}(R^\Delta)$, it is obvious that α_p^Δ is a probability measure on $\mathscr{B}(R^{\Delta})$. We shall call it the *joint distribution of the* x_λ *under p.*

THEOREM 3.4. *Let* $\mathscr{L}$ *be any logic and* $\{x_\lambda : \lambda \in \Delta\}$ *an indexed set of observables. Then the following statements are equivalent:*

(1) *the* x_λ *are simultaneously observable;*

(2) *the* x_λ *have a joint distribution;*

(3) *there exists a space* Ω*, a* σ*-algebra* $\mathscr{A}$ *of subsets of* Ω*, a* σ*-homomorphism* θ *of* $\mathscr{A}$ *into* $\mathscr{L}$ *and* $\mathscr{A}$*-measurable real valued functions* f_λ, $\lambda \in \Delta$, *such that* $x_\lambda(E) = \theta(f_\lambda^{-1}(E))$ *for all* $\lambda \in \Delta$ *and real line Borel sets E.*

Proof. We shall first prove that (1) implies (3). Suppose the x_λ to be simultaneously observable. The range of each x_λ is a Boolean sub σ-algebra S_λ of $\mathscr{L}$ and hence by Theorem 3.2 there exists a Boolean sub σ-algebra S of $\mathscr{L}$ such that $S_\lambda \subset S$ for all λ. By the theorem of Loomis there exists a space Ω, a σ-algebra $\mathscr{A}$ of subsets of Ω and a σ-homomorphism θ of $\mathscr{A}$ onto S. For any fixed λ, $x_\lambda(E \to x_\lambda(E))$ is a σ-homomorphism of $\mathscr{B}(R^1)$ into S and hence, by Proposition 3.3, there exists an $\mathscr{A}$-measurable real-valued function f_λ such that $\theta(f_\lambda^{-1}(E)) = x_\lambda(E)$ for all $E \in \mathscr{B}(R^1)$. This is just the assertion (3).

Next we show that (3) implies (2). If we define $\phi : w \to \phi(w)$ as the mapping of Ω into R^{Δ} which sends $w \in \Omega$ into the function $\phi(w)$ on Δ whose value at $\lambda \in \Delta$ is $f_\lambda(w)$, it is obvious that for any $C \in \mathscr{B}(R^{\Delta})$, $\phi^{-1}(C) \in \mathscr{A}$ and that $f_\lambda^{-1}(E) = \phi(\pi_\lambda^{-1}(E))$ for all λ and all real line Borel sets E. If we now define z by setting $z(C) = \theta(\phi^{-1}(C))$, then z is a σ-homomorphism of $\mathscr{B}(R^{\Delta})$ into $\mathscr{L}$ such that $z(\pi_\lambda^{-1}(E)) = x_\lambda(E)$ for all $\lambda \in \Delta$ and real line Borel sets E. This proves that the x_λ have a joint distribution.

The implication (2)→(1) follows trivially and completes the proof of the theorem.

Roughly speaking Theorem 3.4 tells us that joint distributions of several observables may be defined only when the observables in question are simultaneously observable. The reader may compare Theorem 3.4 with the remarks of von Neumann [17, pp. 211–230].

The precise characterization of simultaneous observability that we have obtained enables us to give definitions of functions of several observables. Let $x_1, x_2, \ldots, x_n$ be observables which are simultaneously observable. Then by Theorem 3.4 there is a (unique) σ-homomorphism z of $\mathscr{B}(R^k)$ into $\mathscr{L}$ such that $z(\pi_j^{-1}(E)) = x_j(E)$ for all E and $j = 1, 2, \ldots, n$. If

ϕ is any Borel function on R^k, we define $\phi(x_1, x_2, \ldots, x_n)$ as the observable $E \to z(\phi^{-1}(E))$. We leave it to the reader to check that this definition leads to the natural properties which any functional calculus may be reasonably expected to have.

IV. EXAMPLES AND REMARKS

The entire discussion which has preceded has been very abstract. In particular we have given no examples. The purpose of this section is to give a few examples and also to make a few remarks on a number of questions of significant interest which naturally arise. For example, are there analogues of the Loomis theorem which describe the most general type of a logic? What is the logic in the formulation of quantum physics? Are any two logics isomorphic? And so on. Complete answers to these questions are not known in many instances. We shall confine ourselves to brief remarks.

First we examine the question of concrete representations. In the conventional formalism the theorem of Loomis asserts that every σ-algebra arises as a σ-homomorphic image of a σ-algebra of subsets of some space. If a probability measure is also given over a σ-algebra converting it into a measure algebra, then there is the theorem of von Neumann and Halmos which asserts that a nonatomic separable measure algebra is isomorphic to the standard measure algebra of the unit interval [7, p. 173]. Moreover, there are theorems which assert [14, Section III] that under certain conditions a σ-algebra $\mathscr{A}$ of subsets of a space Ω is σ-isomorphic to the σ-algebra of Borel subsets of a complete separable metric space and that two such σ-algebras $\mathscr{A}_1$ and $\mathscr{A}_2$ (in Ω_1 and Ω_2) are σ-isomorphic if and only if Ω_1 and Ω_2 have the same cardinality.

No such *general* results are known concerning arbitrary logics. Unlike the σ-algebra situation there is an interesting reduction problem to be solved before the known concrete representations can be formulated. Notice that, given a logic $\mathscr{L}$, we may ask which ones of its elements are simultaneously verifiable with all the elements of $\mathscr{L}$. Let $\mathscr{C}$ be the set of all such elements; $\mathscr{C} = \{a : a \leftrightarrow b \text{ for all } b \in \mathscr{L}\}$. We shall call $\mathscr{C}$ the *center* of $\mathscr{L}$.

PROPOSITION 4.1. $\mathscr{C}$ is a Boolean sub σ-algebra of $\mathscr{L}$.

Proof. Clearly $0, 1 \in \mathscr{C}$ and $a' \in \mathscr{C}$ whenever $a \in \mathscr{C}$. From Proposition 3.8 we see that if $a_1, a_2, \ldots \in \mathscr{C}$ and $a_i \perp a_j$ whenever $i \neq j$, then $a_1 \dot{+} a_2 \dot{+} \cdots \in \mathscr{C}$. Moreover, if a, $b \in \mathscr{E}$, then $a \leftrightarrow b$ in particular, so that $a \cup b$ and $a \cap b$ exist. Since $a \leftrightarrow e$ and $b \leftrightarrow e$ for all $e \in \mathscr{E}$, Proposition 3.8 implies that $a \cup b \leftrightarrow e$ and $a' \cup b' \leftrightarrow e$ for all $e \in \mathscr{E}$. This proves that $a \cup b$, $a \cap b$ both belong to C. Hence, by Proposition 3.10, $\mathscr{C}$ is a Boolean sub σ-algebra of $\mathscr{E}$. If $\mathscr{E}$ is a σ-algebra, one has $\mathscr{C} = \mathscr{E}$. Observables $x(E \to x(E))$ such that $x(E) \in \mathscr{C}$ for all E are simultaneously observable with any other observable.

A logic $\mathscr{E}$ is called *simple* if its center consists only of 0 and 1. Clearly a logic is simple if and only if only the constants are simultaneously observable with *all* the observables. In this sense the simple logics are antithetical to the σ-algebras that arise in the conventional model. The problem of reduction is that of "decomposing", in some natural sense, any logic into simple ones. In its general context no solution to this problem is known. In special cases solutions exist. A more or less trivial case may be settled at once. Suppose $\mathscr{C}$ is separable and atomistic, i.e. there is a sequence of pairwise disjoint elements $a_1, a_2, \ldots$ of $\mathscr{C}$ such that $\mathscr{C}$ is precisely the set of all elements of the form $a_{i_1} \dot{+} a_{i_2} \dot{+} \cdots$, $i_1 < i_2 < \cdots$ being a sequence of integers. Then given any $b \in \mathscr{E}$ we can write $b = b_1 \dot{+} b_2 \dot{+} \cdots$, where $b_j \leqq a_j$ and the b_j are uniquely determined (in fact $b_j = b \cap a_j$). If we define $\mathscr{E}_j = \{b : b \in \mathscr{E},\ b \leqq a_j\}$, then the $\mathscr{E}_j$ are *simple logics* (a_j is the "unit" of $\mathscr{E}_j$). The logic $\mathscr{E}$ may thus be thought of as a direct sum of the simple logics $\mathscr{E}_j$. In the general case, when $\mathscr{C}$ is nonatomic, it is natural to look for "integral-like" decompositions.

Much more is known about simple logics. However all the known results assume that the simple logic $\mathscr{E}$ is in fact a lattice. The first such theorem which characterized an extensive class of logics by intrinsic algebraic properties was obtained by Birkhoff and von Neumann [3]. We now describe this result. Let D be a field (not necessarily commutative) and V an n-dimensional vector space over D ($n < \infty$). Let $L_n(V, D)$ denote the lattice of linear subspaces of V. By an involutory anti-automorphism of D is meant a map $\theta(s \to s^\theta)$ of D onto itself such that (i) θ is one-one, (ii) $(s_1 + s_2)^\theta = s_1^\theta + s_2^\theta$ and $(s_1 s_2)^\theta = s_2^\theta s_1^\theta$ for all s_1, $s_2 \in D$, (iii) $(s^\theta)^\theta = s$ for all $s \in D$. By a θ-semilinear form on V is meant a map $\gamma: V \times V \to D$ such that (i) $\gamma(\cdot, \cdot)$ is additive in each variable when the other is fixed, (ii) $\gamma(s_1 u, s_2 v) = s_1 \gamma(u, v) s_2^\theta$ for all u, $v \in V$ and s_1, $s_2 \in D$, (iii) $\gamma(v, u) = (\gamma(u, v))^\theta$. A θ-semilinear form γ is called *definite* if $\gamma(u, u)=0$ implies $u=0$. If γ is any definite θ-semilinear form we define, for any linear subspace S of V,

$$S'=\{v:v\in V \quad \text{and} \quad \gamma(v,u)=0 \quad \text{for all } u\in S\}.$$

It is then easy to prove that $S\to S'$ is a complementation in $L_n(V, D)$ and that $'$ converts $L_n(V, D)$ into a logic. We shall denote it by $L_n(V, D; \gamma, \theta)$. Suppose $\mathscr{E}$ is a simple logic. A *chain* in $\mathscr{E}$ is any sequence $a_1, a_2, \ldots, a_{N-1}$ of elements of $\mathscr{E}$ with $0<a_1<a_2<\cdots<a_{N-1}<1$. N is called the length of the chain and we denote by $\nu(\mathscr{E})$ the upper bound of the lengths of chains in $\mathscr{E}$; $\nu(\mathscr{E})$ can be infinite. Birkhoff and von Neumann prove that if the simple logic $\mathscr{E}$ is a modular lattice with $3\leqq\nu(\mathscr{E})<\infty$, then there exists an isomorphism of $\mathscr{E}$ with some $L_n(V, D, \gamma, \theta)$ (where the complementation in $\mathscr{E}$ goes over to the complementation in L_n); moreover, D and θ are unique up to isomorphism and γ is unique up to a multiplicative constant. A very elegant proof of this result is given in Baer's book on projective geometry [1, p. 102].

It may be noted that the anti-automorphism θ may be quite complicated. If $D=C$, the complex field θ can be wildly discontinuous. If we demand in this case that θ be continuous, then θ is simply the conjugation $(*): z\to z^*$ which sends z to its complex conjugate. The logics that arise are the logics of linear subspaces of complex vector spaces equipped with positive definite Hermitian forms, complementation being the usual orthogonal complementation. If D is the real or quaternionic field, θ is continuous and in the real case it is the identity. For generalizations when the chain conditions are dropped and for other relevant literature the reader may be referred to von Neumann's book on continuous geometry [18]. We now give a few examples.

Logic of Quantum Mechanics. In almost all of quantum physics the probabilistic formalism is in terms of a logic of a very special type. Let $\mathscr{H}$ be a complex, separable, infinite dimensional Hilbert space (hereafter simply Hilbert space). The set $\mathscr{L}(\mathscr{H})$ of all its closed linear subspaces is clearly partially ordered under $\subset$ and is even a lattice. To any closed linear subspace S of $\mathscr{H}$ one can associate its orthogonal complement $S^{\perp}$. It is an elementary fact of Hilbert space theory (see [17] and [6] for all details concerning Hilbert spaces) that $S\to S^{\perp}$ is a complementation in $\mathscr{L}(\mathscr{H})$ and that $\mathscr{L}(\mathscr{H})$ becomes a logic under this complementation.

Notice that by the fundamental theorem of projective geometry [1, p. 44] the field (in this case, C) over which $\mathscr{H}$ is taken is already determined up to an isomorphism by the structure of $\mathscr{L}(\mathscr{H})$ as a partially ordered set. The fundamental assumption in quantum physics is then that the experimental propositions associated with a physical system $\mathfrak{S}$ form a logic which is isomorphic with $\mathscr{L}(\mathscr{H})$ for some Hilbert space $\mathscr{H}$.

We shall now describe briefly the special case when the logic $\mathscr{E}=\mathscr{L}(\mathscr{H})$ for some Hilbert space $\mathscr{H}$. To start with there is a natural one-one correspondence $S\rightarrow P^S$ between closed linear subspaces of $\mathscr{H}$ and the projection operators that project orthogonally on these. If $x(E\rightarrow x(E))$ is any observable, it gives rise to the projection valued measure $P^x(E\rightarrow P^{x(E)})$ and conversely every projection valued measure based on the Borel sets of the line so arises from a unique observable. By the spectral theorem the projection valued measures are in one-one correspondence with the (not necessarily bounded) self-adjoint operators in $\mathscr{H}$; to the operator A corresponds the projection valued measure $E\rightarrow P_E^A$ which gives the "spectral resolution" of A. We may thus conclude that the observables are in one-one correspondence with the self-adjoint operators in $\mathscr{H}$. If $x(E\rightarrow x(E))$ is an observable, then the self-adjoint operator A_x that corresponds to x is the unique one having $E\rightarrow P^{x(E)}$ as its spectral resolution.

The description of the probability measures on $\mathscr{L}(\mathscr{H})$ is somewhat less trivial. First of all, if $\phi\in\mathscr{H}$ is a unit vector, i.e. $\|\phi\|=1$, the assignment $p_\phi: S\rightarrow(P^S\phi, \phi)=\|P^S\phi\|^2$ is a probability measure on $\mathscr{L}(\mathscr{H})$. If $\phi_1, \phi_2, \dots$ are unit vectors and $\gamma_1, \gamma_2, \dots$ constants $\geqq 0$ with $\gamma_1+\gamma_2+\cdots=1$, $\gamma_1\cdot p_{\phi_1}+\gamma_2\cdot p_{\phi_2}+\cdots$ is also a probability measure on $\mathscr{L}(\mathscr{H})$. Gleason [5] has proved that every probability measure on $\mathscr{L}(\mathscr{H})$ is of the form $\gamma_1\cdot p_{\phi_1}+\gamma_2\cdot p_{\phi_2}+\cdots$ for a suitable choice of $\phi_1, \phi_2, \dots$ and $\gamma_1, \gamma_2, \dots$.

We are in a position to describe the calculus of observables and their distributions. If $x(E\rightarrow x(E))$ is any observable and A_x the associated operator, the distribution of x under p_ϕ (ϕ being a unit vector in $\mathscr{H}$) is the measure $E\rightarrow p_\phi(x(E))$ on the line. Now $p_\phi(x(E))=(P^{x(E)}\phi, \phi)$, and hence we reach the first proposition in our calculus that, if A_x is the operator corresponding to x and $E\rightarrow P_E^{A_x}$ its spectral resolution, then the distribution of x under p_ϕ is given by $E\rightarrow(P_E^{A_x}\phi, \phi)$. As another proposition in this calculus we may mention the characterization of bounded observables. From our point of view an observable $x(E\rightarrow x(E))$ is bounded if and only if, for some compact set K, $p(x(K))=1$ for all probability measures p on $\mathscr{L}(\mathscr{H})$. It follows at once from Gleason's theorem that x is bounded if and only if there is a compact set K such that $x(K)=\mathscr{H}$; this can happen if and only if the operator A_x corresponding to x is bounded. Easy calculations then show that if x is bounded, the spectrum of x coincides with the spectrum of the operator A_x and $\|x\|$ coincides with the norm $\|A_x\|$ of the operator. The special case of an observable x for which A_x has a pure point spectrum is especially illuminating. If $\lambda_1, \lambda_2, \ldots$ are the eigenvalues of A_x, then the distribution of x under any probability measure p on $\mathscr{L}(\mathscr{H})$ is concentrated on the λ_j. If the λ's are distinct and have all multiplicity one, then the distribution of x under p_ϕ has masses $|(\phi, \psi_1)|^2$, $|(\phi, \psi_2)|^2, \ldots$, respectively, at $\lambda_1, \lambda_2, \ldots$, where $\psi_1, \psi_2, \ldots$ are unit vectors with $A_x\psi_j=\lambda_j\psi_j$ for $j=1, 2, \ldots$, i.e. unit eigenvectors of A_x corresponding to its eigenvalues.

The notions of functional calculus and simultaneous observability can also be neatly described. Standard arguments in spectral theory show that if x is an observable, A_x the corresponding operator and u any Borel function, the observable $u(x)$ has $u(A_x)$ as its corresponding operator. Secondly, if S and T are elements of $\mathscr{L}(\mathscr{H})$, it can be easily shown that S and T are simultaneously verifiable if and only if the corresponding projection operators P^S and P^T commute. If we now recall the well-known result that two bounded self-adjoint operators A and B with respective spectral resolutions $E \rightarrow P_E^A$ and $E \rightarrow P_E^B$ commute if and only if $P_E^A P_F^B = P_F^B P_E^A$ for all E and F, we may conclude that the bounded observables x and y are simultaneously observable if and only if $A_x A_y = A_y A_x$. Theorem 3.3 then yields a well-known theorem of Hilbert space theory [17, p. 173].

In the case when $\mathscr{E}=\mathscr{L}(\mathscr{H})$, or even when $\mathscr{E} \subseteq \mathscr{L}(\mathscr{H})$, it is possible to examine somewhat more closely the circumstances under which observables have joint distributions. Let $x_1, \ldots, x_k$ be bounded observables and $A_{x_1}, \ldots, A_{x_k}$ the corresponding bounded self-adjoint operators. For any ordered k-tuple $(u_1, \ldots, u_k)$ of real numbers let $\pi(u_1, \ldots, u_k)$ be the map $(z_1, \ldots, z_k) \rightarrow u_1 z_1 + \cdots + u_k z_k$ of R^k into R^1 and for any probability measure μ on $\mathscr{B}(R^k)$ and any $(u_1, \ldots, u_k)$ let $\mu^{\pi(u_1, \ldots, u_k)}$ be the probability measure on the Borel sets of the line defined by $\mu^{\pi(u_1, \ldots u_k)}(E)=\mu(\pi(u_1, \ldots, u_k)^{-1}(E))$. We shall now say that $x_1, \ldots, x_k$ have a *joint distribution in the weak sense* if for each probability measure p on $\mathscr{E}$ there exists a probability measure μ_p on $\mathscr{B}(R^1)$ such that for each $(u_1, \ldots, u_k)$, the distribution of the observable whose operator is $u_1 A_{x_1} + \cdots + u_k A_{x_k}$ is given by $\mu_p^{\pi(u_1, \ldots, u_k)}$. (We denote this observable by $u_1 x_1 + \cdots + u_k x_k$.) An indexed family $\{x_\lambda : \lambda \in \Delta\}$ of bounded observables will be said to have a joint distribution in the weak sense if for any k and $\lambda_1, \ldots, \lambda_k \in \Delta$, $x_{\lambda_1}, \ldots, x_{\lambda_k}$ have a joint distribution in the weak sense.

PROPOSITION 4.2.[3] If $\mathscr{E} \subseteq \mathscr{L}(\mathscr{H})$ and $\{x_\lambda : \lambda \in \Delta\}$ is an indexed family of bounded observables, then a necessary and sufficient condition that they have a joint distribution in the weak sense is that they have a joint distribution (in the sense described in Section III).

Proof. The sufficiency is an obvious consequence of the remarks made just before Theorem 3.4. We now prove the necessity. It is enough to prove that if $\lambda_1, \lambda_2 \in \Delta$, then x_{λ_1} and x_{λ_2} are simultaneously observable. Since any subfamily of the family $\{x_\lambda : \lambda \in \Delta\}$ also has a joint distribution in the weak sense, we may (and do) assume that $\Delta = \{1, 2\}$. Writing $A_1 = A_{x_1}$, $A_2 = A_{x_2}$ and denoting by $E \to P_E^{(u_1, u_2)}$ the projection-valued measure associated with $u_1 A_1 + u_2 A_2$, we shall prove that for any two Borel sets E_1 and E_2 on the line, $P_{E_1}^{(1,0)}$ and $P_{E_2}^{(0,1)}$ commute.

For any $\phi \in \mathscr{H}$ we define a measure μ_ϕ on $\mathscr{B}(R^2)$ as follows: If $\phi = 0$, we set $\mu_\phi = 0$ and if $\phi \neq 0$, then we set $\mu_\phi = \|\phi\|^2 \cdot \mu_{p_{\phi'}}$, where $\phi' = (1/\|\phi\|) \cdot \phi$ and $\mu_{p_{\phi'}}$ is the measure on the plane that corresponds to the measure $p_{\phi'}$ on $\mathscr{E}$. From the definition of the distribution of $u_1 x_1 + u_2 x_2$ under $p_{\phi'}$ it follows at once that

$$(*) \qquad (P_E^{(u_1, u_2)} \phi, \phi) = \mu_\phi(\pi(u_1, u_2)^{-2}(E))$$

for all $\phi \in \mathscr{H}$ and all Borel sets E on the line. Define now, for $\phi, \psi \in \mathscr{H}$,

$$\mu_{\phi, \psi} = \tfrac{1}{4}\{\mu_{\phi+\psi} - \mu_{\phi-\psi} + i\mu_{\phi+i\psi} - i\mu_{\phi-i\psi}\}.$$

Clearly $\mu_{\phi, \psi}$ is a complex measure on $\mathscr{B}(R^2)$ for fixed ϕ, ψ and it follows from $(*)$ that

$$(**) \qquad (P_E^{(u_1, u_2)} \phi, \psi) = \mu_{\phi, \psi}(\pi(u_1, u_2)^{-1}(E))$$

for all $\phi, \psi \in \mathscr{H}$ and Borel sets E on the line. We claim that the map $\phi, \psi \to \mu_{\phi, \psi}$ is linear in ϕ and conjugate-linear in ψ and that $|\mu_{\phi, \psi}(F)| \leqq 2 \cdot \|\phi\| \cdot \|\psi\|$ for all $\phi, \psi \in \mathscr{H}$ and $F \in \mathscr{B}(R^2)$. Indeed, if a_1, a_2, b_1, b_2 are complex constants, it follows from $(**)$ that the complex measures

$$\beta_1 = \mu_{a_1\phi_1, b_1\psi_1 + b_2\psi_2} - a_1 b_1^* \mu_{\phi_1, \psi_1} - a_1 b_2^* \mu_{\phi_1, \psi_2}$$

and $$\beta_2 = \mu_{a_1\phi_1 + a_2\phi_2, b_1\phi_1} - a_1 b_1^* \mu_{\phi_1, \psi_1} - a_2 b_1^* \mu_{\phi_2, \psi_1}$$

vanish for all sets of the form $\pi(u_1, u_2)^{-1}(E)$ and hence, by a well-known theorem in measure theory, for all $F \in \mathscr{B}(R^2)$. This proves that the map $\phi, \psi \to \mu_{\phi,\psi}$ is linear in ϕ and conjugate-linear in ψ. For the second assertion it is enough to consider the case where $\|\phi\| = \|\psi\| = 1$; then it follows easily since $0 \leqq \mu_\xi(F) \leqq \|\xi\|^2$ for all F and $\xi \in \mathscr{H}$.

It follows that there exists a bounded linear operator P_F with $\|P_F\| \leqq 2$ such that $\mu_{\phi,\psi}(F) = (P_F\phi, \psi)$ for all $\phi, \psi \in \mathscr{H}$. Clearly

$$(P_F\phi, \phi) = \mu_{\phi,\phi}(F) = \mu_\phi(F) \geqq 0$$

for all $\phi \in \mathscr{H}$ and hence P_F is a self-adjoint non-negative operator for each F. The additivity of $\mu_{\phi,\psi}$ implies that if F_1 and F_2 are disjoint Borel sets in the plane and $F = F_1 \cup F_2$, then $P_F = P_{F_1} + P_{F_2}$. Clearly $P_F = P_E^{(u_1, u_2)}$ if $F = \pi(u_1, u_2)^{-1}(E)$. We would be finished if we knew at this stage that P_F is a projection for all F. This we do not. But notice that if $F_1 \subseteqq F_2$, then $P_{F_1} \leqq P_{F_2}$ (in the sense that $P_{F_2} - P_{F_1}$ is a non-negative operator).

We shall now complete the proof by showing that $P_{E_1}^{(1,0)}$ and $P_{E_2}^{(0,1)}$ commute for arbitrary Borel sets E_1, E_2 of the line. Now, $E_1 \times E_2 \subseteqq E_1 \times R^1$ and also $\subseteqq R^1 \times E_2$ so that $P_{E_1 \times E_2} \leqq P_{E_1}^{(0,1)}$ and $\leqq P_{E_2}^{(0,1)}$. This implies that $P_{E_1 \times E_2} \leqq P_{E_1}^{(1,0)} \wedge P_{E_2}^{(0,1)} = Q_1$, say ($\wedge$ denotes lattice intersection). Similarly we deduce that $P_{E_1 \times R^1 - E_2} \leqq P_{E_1}^{(1,0)} \wedge P_{R^1 - E_2}^{(0,1)} = Q_2$. Now the non-negativity of the operators P_F for all $F \in \mathscr{B}(R^2)$, the equation

$$P_{E_1 \times R^1} = P_{E_1}^{(1,0)} = P_{E_1 \times E_2} + P_{E_1 \times R^1 - E_2},$$

and the facts that Q_1 and Q_2 are orthogonal and $\leqq P_{E_1}^{(1,0)}$, imply that $P_{E_1 \times E_2} = Q_1$ and $P_{E_1 \times R^1 - E_2} = Q_2$. In other words we have shown that

$$P_{E_1}^{(1,0)} = (P_{E_1}^{(1,0)} \wedge P_{E_2}^{(0,1)}) + P_{E_1}^{(1,0)} \wedge (1 - P_{E_1}^{(0,1)})$$

which enables us to conclude that $P_{E_1}^{(1,0)}$ and $P_{E_2}^{(0,1)}$ commute. This complete the proof.

Remark 1. It also follows now that $F \to P_F$ is the unique projection-valued measure on $\mathscr{B}(R^2)$ such that $P_{E_1 \times R^1} = P_{E_1}^{(1,0)}$ and $P_{R^1 \times E_2} = P_{E_2}^{(0,1)}$ for all Borel sets E_1, E_2 of the line.

Remark 2. It may be noted that our overall assumption that $\mathscr{H}$ is a complex, separable, infinite-dimensional Hilbert space has played essentially no role in the above proof. Indeed the dimension of $\mathscr{H}$ is irrelevant to the above discussion and only trivial modifications are needed to take care of the case when $\mathscr{H}$ is a real Hilbert space.

Finally if A_1 and A_2 are bounded self-adjoint operators so is A_1+A_2; if $A_1=A_{x_1}$ and $A_2=A_{x_2}$ and $A_1+A_2=A_z$, it is natural to define z to be the "sum" of x_1 and x_2. Notice that x_1 and x_2 *need not* be simultaneously observable. Now a typical property of the sum of two random variables in the conventional theory is that the expected value of the sum is the sum of the expected values of the summands. For any probability measure p on $\mathscr{L}(\mathscr{H})$ and any bounded observable x, one may define the expected value $e_p(x)$ of x under p (it definitely exists) as $\int_{-\infty}^{\infty} t\, d\alpha_x^p(t)$, where α_x^p is the distribution of x under p. We might now ask whether $e_p(z)=e_p(x_1)+e_p(x_2)$ for *all* p. This is obvious if $p=p_\phi$ for a unit vector ϕ and hence by Gleason's theorem it is valid for all p. In other words the passage from the probability measure on the closed linear subspaces to the expectation functional on the bounded operators of $\mathscr{H}$ leads us to a *linear* functional. It is this aspect which constitutes one of the major sources of difficulties of our present theory. In this example for instance, the natural functional (the integral) on the bounded self-adjoint operators of $\mathscr{H}$ which one obtains starting from a probability measure on $\mathscr{L}(\mathscr{H})$ is in fact linear; but the proof of this fact leans heavily on the theorem of Gleason and is nontrivial. The reader may contrast this situation with the conventional one where the linearity of the integral is a matter of routine verification.

A "Finite" Model. A simpler example than the $\mathscr{L}(\mathscr{H})$ of the preceding discussion is provided by what one might describe as the analogue of a finite probability space. The discussion is valid for both the real and complex fields; we give only the complex case. Briefly, we take $\mathscr{E}$ as the lattice of linear subspaces of C^n, the n-dimensional complex space with inner product

$$(a, b)=a_1b_1^*+\cdots+a_nb_n^*,$$

where $a=(a_1,\ldots,a_n)$ and $b=(b_1,\ldots,b_n)\in C^n$. Orthocomplementation in $\mathscr{E}$ is the customary orthogonal complementation in the lattice of linear subspaces. The observables are in one-one correspondence with the $n\times n$ Hermitian matrices. If ϕ is a unit vector in C^n, the assignment $p_\phi:S\to \|P^S\phi\|^2=(P^S\phi,\phi)$ is a probability measure on $\mathscr{E}$. Every probability measure on $\mathscr{L}$ can be expressed as a convex combination of a finite number of the p_ϕ. If x is an observable and A_x is the corresponding matrix with eigen values $\lambda_1, \lambda_2, \ldots, \lambda_k$, then $k \leq n$ and the distribution of x under every probability measure p on $\mathscr{L}$ is concentrated on the λ_j; x has thus $\{\lambda_1, \ldots, \lambda_k\}$ as its spectrum. The probability measure $\bar{p}$ which assigns to each linear subspace S the number $\bar{p}(S) = \text{dimension } (S)/n$ is especially noteworthy. If x is an observable such that A_x has n distinct eigenvalues $\lambda_1, \ldots, \lambda_n$, then the distribution of x under $\bar{p}$ has masses $1/n$ at each of the n points $\lambda_1, \ldots,$ and λ_n. Clearly this is the analogue of the conventional model, where Ω has exactly n points; $\bar{p}$ is the generalization of the "uniform distribution" on Ω.

Projections in a Factor. Familiarity with Hilbert space theory enables one to generalize the first example considerably. We still keep to a Hilbert space $\mathscr{H}$ but take a weakly closed self-adjoint operator algebra W and take $\mathscr{L}$ to be the lattice $\mathscr{P}(W)$ of all projections in W. It is easily proved that $\mathscr{L}$ is a logic. If we want a simple logic, then we must choose W to be a factor in the sense of Murray and von Neumann [15]. The algebra of *all* operators in $\mathscr{H}$ is a factor and it is this choice of W that leads to the example of quantum theory. Murray and von Neumann have classified factors into various types I, II, III. If W is a factor, it follows as in the first example that to every observable corresponds a self-adjoint operator in $\mathscr{H}$. However, not all self-adjoint operators correspond to observables, only those the projections in whose spectral resolutions lie in W. These are precisely those operators that are left invariant by every unitary operator which commutes with all members of W. (These are the self-adjoint operators *affiliated* to W; see [15].) When W is of type II or type III, the problem of describing all probability measures on $\mathscr{L} = \mathscr{P}(W)$ is still open. So is the closely related problem of proving that the integral is a linear functional. More precisely, for any bounded self-adjoint $A \in W$ and any probability measure p on $\mathscr{P}(W)$, define

$$e_p(A) = \int_{-\infty}^{\infty} t \, d\alpha^p(t),$$

where α^p is the measure $E \to p(P_E^A)$. It seems plausible that $A \to e_p(A)$ is a real linear functional on the self-adjoint elements of W.

It may be noted that the results of Section III imply easily that if W is a ring of operators in a separable Hilbert space and $\{A_\lambda : \lambda \in \Delta\}$ is a family of self-adjoint operators in W such that any two of them commute, then there exists a self-adjoint operator C *in the ring* W and Borel functions u_λ, $\lambda \in \Delta$, such that $A_\lambda = u_\lambda(C)$ for all $\lambda \in \Delta$. The characterization of abelian operator rings to which this leads was first obtained by von Neumann (see the remarks of R. V. Kadison on p. 63 of the *Bull. A.M.S.* **64**, No. 3, Pt. 2, May, 1958).

In generalizing conventional probability theory so that descriptions such as those of quantum physics would be handled by the generalization, Segal [20] takes roughly the view that the probability theory should be described by a pair (A, e), where A is a complex algebra with a unit and a distinguished involution $(*)$ and e a positive definite linear functional on A. An important special case arises if one takes A to be a C^*-algebra with e as a linear functional on A for which $e(aa^*) \geqq 0$ for any $a \in A$ and with $e(1) = 1$. The bounded observables would then be the self-adjoint elements of A and for any bounded observable x, $e(x)$ will be its expectation value. Several remarks are now in order concerning the relation of these ideas with ours. First, in view of a well-known theorem of Gelfand and Neumark, we may (and do) assume that A is a uniformly closed self-adjoint algebra of bounded operators on a not necessarily separable Hilbert space. Since A need not be weakly closed, it need not contain any nontrivial projection. Secondly, even if A is weakly closed, the expectations need not give rise to countably additive probability distributions for all bounded observables. More precisely, let x be a bounded observable. The map $g \to e(g(x))$ is then a positive linear functional on the algebra of bounded continuous functions on the spectrum of x and hence we can find a unique measure μ such that $e(g(x)) = \int g \, d\mu$ for all continuous g. If x were to have a countably additive distribution, then clearly μ must be it and $e(g(x)) = \int g \, d\mu$ for all bounded Borel measurable g. But this need not always happen. Segal gives in his paper [21] the relevant counter-examples. As a final remark we mention the fact that in the case, where A is an arbitrary weakly closed self-adjoint algebra, it is not known whether every (countably additive) probability measure on the projections of A leads to a positive *linear* functional on the self-adjoint elements of A.

We make two concluding remarks. In their paper [16] Murray and von Neumann have singled out a class of type H_1-factors (the "approximately finite" ones) which are all mutually isomorphic so that the logics of the projections in these factors are mutually isomorphic. Secondly the reduction problem (mentioned in this section) of decomposing an arbitrary logic $\mathscr{L}$ into simple ones was solved by von Neumann by means of his direct integral theory [19] when $\mathscr{L}$ is the logic of projections in a weakly closed self-adjoint algebra of operators in a Hilbert space.

The writer would like to express his warm appreciation to Professor Warren Hirsch, of the Courant Institute of Mathematical Sciences, for his suggestions which have been responsible for many improvements in presentation and for his careful reading of the manuscript. He would also like to express his indebtedness to Professor Allan Birnbaum, also of the Courant Institute of Mathematical Sciences, for many conversations on the general subjects of the structure of science and the logic of scientific explanation.

NOTES

* This paper represents results obtained as a Temporary Member of the Courant Institute of Mathematical Sciences, under the sponsorship of the National Science Foundation, Contract No. NSF-G-14520. Reproduction in whole or in part permitted for any purpose of the United States Government.

1 Throughout this paper the terms "observable" and "random variable" are used interchangeably.

2 since $a \leftrightarrow b$, $a \cup b$ and $a \cap b$ exist (cf. Proposition 3.6).

3 Added in proof. In the special case when the operators A_{x_λ} have purely discrete spectra, Proposition 4.2 has been recently proved by K. Urbanik, *Studia Mathematica*, **21** (1962), 117–133 (see in particular Theorem 2).

BIBLIOGRAPHY

[1] Baer, R., *Linear Algebra and Projective Geometry*, Academic Press, New York, 1952.

[2] Birkhoff, G. *Lattice Theory*, Amer. Math. Soc. Colloq. Publ., 1948.

[3] Birkhoff, G., and von Neuman, J., 'The Logic of Quantum Mechanics', *Ann. of Math.* **37** (1936), 823–843.

[4] Feynman, R. P., 'The Concept of Probability in Quantum Mechanics', *Proc. Second Berkeley Symposium in Mathematical Statistics and Probability*, Berkeley, Calif., 1951, pp. 533–541.

[5] Gleason, A. M., 'Measures on the Closed Subspaces of a Hilbert Space', *J. Rat. Mech. Analysis* **6** (1957), 885–894.

[6] Halmos, P. R., *Introduction to Hilbert Space and the Theory of Spectral Multiplicity*, Chelsea Press, New York, 1957.

[7] Halmos, P. R., *Measure Theory*, Von Nostrand, New York, 1950.

[8] Halmos, P. R., 'The Foundations of Probability', *Amer. Math. Monthly* **51** (1944), 493–510.
[9] Heisenberg, W., *The Physical Principles of the Quantum Theory*, Dover Publications, New York, 1930.
[10] Kolmogorov, A. N., *Grundbegriffe der Wahrscheinlichkeitsrechnung*, Berlin, 1933.
[11] Loomis, L. H., 'On the Representation of σ-Complete Boolean Algebras', *Bull. Amer. Math. Soc.* **53** (1947), 757–760.
[12] Mackey, G. W., 'Quantum Mechanics and Hilbert Space', *Amer. Math. Monthly* **64** (1957), 45–57.
[13] Mackey, G. W., *The Mathematical Foundations of Quantum Mechanics*, Harvard University, lecture notes, 1960.
[14] Mackey, G. W., 'Borel Structures in Groups and Their Duals', *Trans. Amer. Math. Soc.* **85** (1957), 134–165.
[15] Murray, F. J. and von Neumann, J., 'On Rings of Operators', *Ann. of Math.* **37** (1936), 116–229.
[16] Murray, F. J., and von Neumann, J., 'On Rings of Operators. IV', *Ann. of Math.* **44** (1943), 716–808.
[17] von Neumann, J., *Mathematical Foundations of Quantum Mechanics* (transl. from the the German edition by R. T. Beyer), Princeton University Press, Princeton, 1955.
[18] von Neumann, J., *Continuous Geometry*, Princeton University Press, Princeton, 1960.
[19] von Neumann, J., 'On Rings of Operators. Reduction Theory', *Ann. of Math.* **50** (1949), 401–485.
[20] Segal, I. E., 'Abstract Probability Spaces and a Theorem of Kolmogoroff', *Amer. J. Math.* **76** (1954), 721–732.
[21] Segal, I. E., 'Postulates for General Quantum Mechanics', *Ann. of Math.* **48** (1947), 930–948.

JERZY ŁOŚ

SEMANTIC REPRESENTATION OF THE PROBABILITY OF FORMULAS IN FORMALIZED THEORIES

In probability theory, or rather in its foundations, there has long been a trend in favour of identifying events, i.e., objects to which probability is ascribed, with formulas of certain theories. Without adducing arguments in favour of that idea I shall confine myself to mentioning its principal representatives, namely J. M. Keynes, J. Nicod, H. Jeffreys, H. Reichenbach, R. Carnap, and in Poland J. Łukasiewicz and K. Ajdukiewicz.

It is, of course, formally possible to ascribe probability to formulas, since formulas form a Boolean algebra (the term "sentences" is sometimes used instead of "formulas", but I shall not use it since I want to make a distinct difference between sentences and sentential functions). But it does not seem that the interpretation of formulas of a language as events is always the same. Moreover, it seems that at least two interpretations can be distinguished, the confusion of which occasionally leads to errors. One such error has been discovered when Nicod's works were studied in my seminar in the Polish Academy of Sciences' Institute of Philosophy and Sociology in the academic year 1958/9. Roughly speaking, it consists in that Nicod confuses the probability of appearance of a causal relationship between two phenomena with the probability of existence of such a relationship. It seems that we have to do here with two quite different probabilities. In the first case we are concerned with a kind of a sentential function in the form of the implication: $A(x) \rightarrow B(x)$, the question of what is the probability of that relationship being the question of what is the probability (or frequency) of drawing by lot (or obtaining) such an individual (object, moment, point in space – according to interpretation) which would satisfy that sentential function. In the second case we are concerned with the probability of the world we live in possessing a certain characteristic (namely that A is the cause of B). It would be difficult to say in what sense the term "probability" might be used here, since we in no case draw at random the world we live in, but that is not the point. The point is that we face the

C. A. Hooker (ed.), The Logico-Algebraic Approach to Quantum Mechanics, 205–219.

necessity of making a distinction between the probability of a sentential function, understood as the probability of its being satisfied by elements chosen in a certain way, and the probability of a sentence whose truth depends not on such and such elements, but on the whole of the relationships among those objects which form the universe of discourse, in a word, on the model in question.

In less general considerations than those which refer to the principle of causality the issue loses its metaphysical aspect of "drawing a world at random". Let us consider the following example. Suppose we investigate the theories of the ordering of a set by the relation $<$. Let us reflect, how to interpret the statement that the sentential function $x_1 < x_2$ has the probability $1/2 : P(x_1 < x_2) = 1/2$. Apparently this means (in the frequency interpretation) that by drawing, in a given way, the elements x_1 and x_2 from a given ordered set we obtain elements which in one half of the cases will satisfy the sentential function in question.

But it is obvious that although $P(x_1 < x_2) = 1/2$, nevertheless $P(\prod_{x_1} \prod_{x_2} x_1 < x_2) = 0$, and that because the formula with the probability of which we are now concerned is a sentence, and moreover a sentence that is false in any ordered set.

But what about the probability of the sentence $\sum_{x_1} \prod_{x_2} x_1 \leqslant x_2$?

If the ordered set in question has been determined then that sentence, which expresses the existence of the least element in that set, will be true or false in that set, and hence will have the probability 1 or 0.

It seems possible for that sentence to have a probability other than 1 or 0, e.g., $P(\sum_{x_1} \prod_{x_2} x_1 \leqslant x_2) = 1/2$. But in such a case we must imagine that there is a given class of ordered sets and a given way of drawing its elements. The probability of the sentence $\sum_{x_1} \prod_{x_2} x_1 \leqslant x_2$ is the probability of drawing from that class such an ordered set in which the least element exists.

In this way we have as it were two probabilities, one for sentential function, when – given a certain model – we draw elements and inquire whether they satisfy that function, and the other for sentences, when we draw a model (in the example above: an ordered set) and inquire whether the sentence is true in that model.

Of course, nothing prevents us from applying the first case to sentences or the second case to sentential functions. But then in the first case all the sentences will have probabilities equal to 1 or 0, and in the second case the sentential functions will have the same probabilities as their generalizations (covering of all variables by universal quantifiers). Thus, we shall have for instance $P(x_1 < x_2) = P(\prod_{x_1} \prod_{x_2} x_1 < x_2) = 0$.

What an interpretation is then to be given to formulas if these two extreme cases are to be avoided? It seems that there is a middle course. We may interpret the probability of a formula as the probability of its being satisfied by a certain sequence of elements obtained by double drawing: first we draw a model in accordance with a probability given for the class of models, and next from that model, also in accordance with a probability given in that model, we draw a sequence of elements.

For that procedure we must have a class of models $\{\mathcal{M}_t\}_{t \in T}$ and a probability, to be symbolized μ, in that class, or rather in the set T, for the models belonging to the class concerned are indexed by elements of that set. Finally, we must have a probability in the class of sequences of every model $\mathcal{M}_t$; let that probability be symbolized ν_t. Then every formula α, whether a sentence or a sentential function, has its probability $\nu_t(\alpha)$ in the model $\mathcal{M}_t$. By fixing α and changing t in the set T we obtain changing values of $\nu_t(\alpha)$ (if α is a sentence, these values will be only 0 and 1), and thus we have to do with a function in the set T in which the probability μ is defined.

This is a random variable (certain conditions of measurability must be satisfied here) for which we can compute the expected value $E_\mu(\nu_t(\alpha))$. That expected value is a number which depends only on α.

Let us put

$$(0) \qquad P(\alpha) = E_\mu(\nu_t(\alpha)).$$

In this way we define a certain probability in the set of all formulas α.

In order to explain certain details of this way of defining the probability of formulas, and at the same time to impart precision to the concepts involved, we shall refer to a simplified example.

Let the class of models consist exactly of four models: $\{\mathcal{M}_i\}$, $i = 1, 2, 3, 4$. Every model, as usual, consists of the set A_i and the relations $R_j^{(i)}$ which are interpretations for the primitive signs of the theory: $\mathcal{M}_i = \langle A_i, R_1^{(i)}, R_2^{(i)}, \ldots \rangle$. Let the symbol S stand for the set of the formulas belonging to our theory, and $A_i^{\omega o}$ for the set of sequences of the elements of the set A_i, i.e., an element x of $A_i^{\omega o}$ is an infinite sequence $x = \langle x_1, x_2, \ldots \rangle$, where all x_j are elements of A_i.

For every α belonging to S, let $\sigma_i(\alpha)$ stand for the set of those sequences from $A_i^{\omega o}$ which satisfy α; let it further be supposed that in every set $A_i^{\omega o}$ there is given the probability $\bar{v}_i$, which is anyhow defined for all the sets $\sigma_i(\alpha)$.

Let us put $v_i(\alpha)=\bar{v}_i(\sigma_i(\alpha))$. In this way the probabilities in S are defined. It can easily be verified by the formula

$$(1) \qquad \sigma_i(\alpha \vee \beta)=\sigma_i(\alpha)+\sigma_i(\beta),$$

where $\vee$ symbolizes disjunction, and $+$ the addition of sets. If the conjunction $\alpha \wedge \beta$ is contradictory then the sets $\sigma_i(\alpha)$ and $\sigma_i(\beta)$ are disjoint, and the assumption that $\bar{v}_i$ is a probability function leads to additivity:

$$(2) \qquad v_i(\alpha \vee \beta)=v_i(\alpha)+v_i(\beta).$$

Other conditions required of probability (that $v_i(\alpha)$ should range between 0 and 1, and that $\bar{v}_i(\alpha)$ should be 1 for tautologies) result directly from the assumption that v_i is a probability.

Let it now be supposed that the probability μ is defined in the set of the indices of models, i.e., in our case, in the set of the numbers 1, 2, 3, 4. In this case, since the set is finite (it consists of four elements), this means simply that the numbers from 1 to 4 are correlated with four non-negative numbers $\mu(1)$, $\mu(2)$, $\mu(3)$ and $\mu(4)$ whose sum is 1. Hence every $v_i(\alpha)$ becomes a random variable of the parameter i. The expected value of that random variable is computed by taking the average of its values, weighted by the values of probability:

$$(3) \qquad E_\mu(v_i(\alpha))=\mu(1)\, v_1(\alpha)+\mu(2)\, v_2(\alpha)+\mu(3)\, v_3(\alpha)+\mu(4)\, v_4(\alpha).$$

The expected value – not only in this case, but in general – is additive, which means that the expected value of a sum is the sum of the expected values. Hence if we take two expressions, α and β, belonging to S, and on the strength of (3) compute the expected value of the function $v_i(\alpha)+v_i(\beta)$, we obtain

$$(4) \qquad E_\mu(v_i(\alpha)+v_i(\beta))=E_\mu(v_i(\alpha))+E_\mu(v_i(\beta)).$$

From this and from (2) it follows that when the conjunction $\alpha \wedge \beta$ is contradictory we have

$$(5) \qquad E_\mu(v_i(\alpha \vee \beta))=E_\mu(v_i(\alpha)+v_i(\beta))=E_\mu(v_i(\alpha))+E_\mu(v_i(\beta)).$$

The expected value of (3) no longer depends on i, as can be seen from the right side of that formula. For fixed probabilities μ, ν_1, ν_2, ν_3, ν_4 it depends only on α, so that we may put

$$P(\alpha)=E_\mu(\nu_i(\alpha)).$$

The function of a formula, when so defined, is, as can easily be verified, a probability. In particular, if $\alpha \wedge \beta$ is contradictory, then

$$P(\alpha \vee \beta)=P(\alpha)+P(\beta)$$

results from (5).

In order to give more precision to the example in question let us assume that we consider theories of a densely ordered set, and let the models $\mathcal{M}_i$ be: (1) the open segment $A_1=[0, 1]$; (2) the closed segment $A_2=[0, 1]$; (3) the segment $A_3=(0, 1]$, open on the left and closed on the right; (4) the segment $A_4=[0, 1)$, closed on the left and open on the right. All those segments are ordered by the ordinary relation "lesser than". (Since we consider theories with one primitive concept, the models include only one relation which interprets them). Let it further be assumed that $\mu(1)=1/8$, $\mu(2)=3/8$, $\mu(3)=2/8$, $\mu(4)=2/8$, and finally that the probabilities $\bar{\nu}_i$, defined in the sets of sequences of numbers from the corresponding segment A_i are probabilities connected with such a choice: there is such a way of drawing numbers from a segment that the probability of drawing a point from every subsegment is equal to the length of that subsegment, and in order to draw a sequence we perform infinitely many independent draws with replacement. This way of defining probability may appear inexact, but in fact it is quite precise. In mathematics probability thus defined is called product probability of Lebesgue's measure in the product $A_i^{\omega_0}$. It can easily be demonstrated that for every probability $\bar{\nu}_i$ there is $\bar{\nu}_i(\sigma_i(x_1 \leqslant x_2))=\nu_i(x_1 \leqslant x_2)=1/2$, whence

$$P(x_1 \leqslant x_2)=E_\mu(\nu_i(x_1 \leqslant x_2))=1/8 \cdot 1/2+3/8 \cdot 1/2+ \\ +2/8 \cdot 1/2+2/8 \cdot 1/2=1/2.$$

But of course $\sigma_i(\prod_{x_1} \prod_{x_2} x_1 \leqslant x_2)$ is an empty set for every $i=1, 2, 3, 4$, for $\prod_{x_1} \prod_{x_2} x_1 \leqslant x_2$ is a false sentence in every model. Hence it follows that

$$\bar{\nu}_i\left(\sigma_i\left(\prod_{x_1} \prod_{x_2} x_1 \leqslant x_2\right)\right)=\nu_i\left(\prod_{x_1} \prod_{x_2} x_1 \leqslant x_2\right)=0.$$

On the other hand, the sentence $\sum_{x_1}\prod_{x_2} x_1 \leqslant x_2$, which states the existence of a least element, is true in the second and the fourth model, and false in the first and the third. Thus $\sigma_i(\sum_{x_1}\prod_{x_2} x_1 \leqslant x_2) = A_i^{\omega_0}$ for $i=2, 4$, and $=0$ for $i=1, 3$. Consequently, $v_i(\sum_{x_1}\prod_{x_2} x_1 \leqslant x_2)=1$ for $i=2, 4$, and $=0$ for $i=1, 3$. On computing the probability P we obtain

$$P\Big(\sum_{x_1}\prod_{x_2} x_1 \leqslant x_2\Big) = 0\cdot 1/8 + 1\cdot 3/8 + 0\cdot 28 + 1\cdot 2/8 = 5/8.$$

Thus we see that both sentential functions and sentences can have here probabilities other than 0 and 1.

In the case of an infinite set T, and hence of an infinite set of models $\{\mathcal{M}_t\}_{t\in T}$, the computation of the expected value $E_\mu(v_t(\alpha))$ is not so simple as in the formula (3). It is expressed by the abstract integral

$$(6) \qquad E_\mu(v_t(\alpha)) = \int_T v_t(\alpha)\,\mu(dt),$$

and for the existence of that integral it suffices that the integrated function $v_t(\alpha)$ (which is anyhow bounded) should be measurable, that is, that for every number l the set of those t in T for which $v_t(\alpha) < l$ should have a definite probability μ. The condition is not trivial, since probability in infinite sets T is usually defined not for all the subsets of T, but only for a certain field of such subsets.

An additional condition for the existence of the expected value (6) is that the probability μ should be defined over a denumerably additive field of subsets of T and should itself be denumerably additive. Hence it must be so that if μ is defined for each of the sets $X_1, X_2, \ldots$ (ad inf.), then it is also defined for their sum $\bigcup_{i=1}^{\infty} X_i$, and moreover if the sets $X_1, X_2, \ldots$ are pairwise disjoint, then

$$\mu\Big(\bigcup_{i=1}^{\infty} X_i\Big) = \sum_{i=1}^{\infty} \mu(X_i).$$

The probabilities $\bar{v}_i$ need not satisfy these conditions; it suffices if they satisfy the condition of additivity for two pairwise disjoint sets.

What has been stated above and explained by examples is a certain semantic method of defining probability of formulas. It is semantic for we start from probability in models and in sets of models and then define the probability of formulas by making use of the semantical concepts. In order to avoid misunderstandings we shall describe that method once more, this time in a purely formal fashion.

Let S be the set of all formulas built of given constant predicates and such that individual variables are the only free variables. Let further $\{\mathcal{M}_t\}_{t\in T}$ be the class of models interpreting formulas belonging to S, and for every t belonging to T let v_t be a probability defined in the set of infinite sequences of elements of the model $\mathcal{M}_t$ and such that for every α belonging to S the set $\sigma_t(\alpha)$ of those sequences which satisfy α has a definite probability. Let finally μ be a probability in T, denumerably additive and defined over a denumerably additive field, such that for a given α from S, $v_t(\alpha)=\bar{v}_t(\sigma_t(\alpha))$ as a function of t is measurable with respect to μ. With these assumptions, the function P of the formula α, defined as the expected value of the function $v_t(\alpha)$ with respect to the probability μ: $P(\alpha)=E_\mu(v_t(\alpha))$, is a probability in S.

Let us now pass to the probabilistic intuitions connected with the procedure described above. "Probabilistic" means not connected with the fact that the events under consideration are formulas. For that purpose let us imagine that we have the bag W with cubes marked with the numbers 1, 2, 3, 4, each cube with only one number, and further four boxes U_1, U_2, U_3, U_4, containing balls marked with the letters a, b, c, d, also each ball with only one letter. In this scheme, which is known as the "bag and boxes scheme", when we know the probabilities of drawing the cubes marked with numbers and the probabilities of drawing from the various boxes balls marked with letters, we can compute the probability of drawing a ball marked with a given letter, assuming that we first draw a cube from the bag W, and next a ball from that box which bears the number that marks the cube that has been drawn first.

If we draw "honestly" both from the bag and from the boxes, i.e., if the probability of drawing a given letter or a given number depends on what a given box or the bag contains, then from the data pertaining to the experiment – just from what is contained in the bag and the boxes – we can compute the probability of (1) that a given number will be drawn from the bag; let these probabilities be $\mu(1)$, $\mu(2)$, $\mu(3)$, $\mu(4)$; (2) that a given letter will be drawn from the box U_t, where t stands for one of the numbers 1, 2, 3, 4; let these probabilities be $v_t(a)$, $v_t(b)$, $v_t(c)$, $v_t(d)$. These latter are conditional probabilities. To compute the probabilities of drawing the letter $x(=a, b, c, d)$ in the whole experiment we must resort to the formula

$$(7)\qquad P(x)=\mu(1)\,v_1(x)+\mu(2)\,v_2(x)+\mu(3)\,v_3(x)+\mu(4)\,v_4(x)$$

which is analogous with the formula (3).

The probability of drawing a or b is given by the formula:

$$P(a \text{ or } b)=\mu(1)\left[v_1(a)+v_1(b)\right]+\mu(2)\left[v_2(a)+v_2(b)\right]+ \mu(3)\left[v_3(a)+v_3(b)\right]+\mu(4)\left[v_4(a)+v_4(b)\right].$$

Let it be noted that nothing changes in these considerations if it is assumed that some boxes contain only balls marked with the letters a and b, and the others, only balls marked with the letters c and d. In such a case for some boxes $v_i(a)+v_i(b)=1$ and $v_i(c)+v_i(d)=0$, and for the others $v_i(a)+v_i(b)=0$ and $v_i(c)+v_i(d)=1$. The probability $P(a$ or $b)$ is then the sum of those $\mu(i)$ for which $v_i(a)+v_i(b)=1$. It need be neither 0 nor 1, although the relative probability: that of drawing a or b from a given box always is 0 or 1.

We have considered the determination of the probabilities of drawing balls with appropriate numbers on the basis of the knowledge of the principles of drawing and the knowledge of what the bag and the boxes contain. Let us now consider a problem which is reverse in a sense: can the bag and the boxes be made to contain such cubes and balls, respectively, that the probabilities of drawing balls marked with the various letters should have the values determined in advance: $P(a)=w_1, P(b)=w_2, P(c)=w_3, P(d)=w_4$? Of course, $w_1+w_2+w_3+w_4$ must be 1, and $w_i \geqslant 0$, for $i=1, 2, 3, 4$, but if it is so, the problem can be solved without difficulty. It suffices so to adjust the balls contained in the boxes that the balls marked with the letter a should be w_1, the balls marked with the letter b should be w_2, etc. (w_i must be rational numbers, for otherwise only an arbitrarily exact approximation can be obtained – this difficulty will be disregarded here as inessential for further considerations).

As result we obtain what we want to have, regardless of what the bag contains during the first draw.

Since the problem is easy to solve, let us make it a little more complicated. Let be required that the boxes 1 and 3 contain only balls marked with the letters a and b, and the boxes 2 and 4, only balls marked with the letters c and d; in other words, let $v_i(a)+v_i(b)$ be 1 for $i=1, 3$, and let it be 0 for $i=2, 4$.

This can be done, too, even so that the numbers 3 and 4 may not be represented at all in the bag, so that $\mu(3)=\mu(4)=0$. Let it be noted that under such conditions it follows that $P(1)=P(a)+P(b)=w_1+w_2$ and $P(2)=P(c)+P(d)=w_3+w_4$. From the formula (7) we obtain $w_1=P(a)=\sum_{i=1}^{4}\mu(i)\, v_i(a)=\mu(1)\, v_1(a)=(w_1+w_2)\, v_1(a)$. Hence $v_1(a)=w_1/(w_1+w_2)$ is the sufficient and necessary condition. Other probabilities are computed in an analogous manner.

When computing probabilities of formulas we have to do with a procedure which is quite similar to the box scheme described above, although it is much more complicated. But like in the box scheme we have double drawing, first of the model – which corresponds to drawing a cube from the bag – and then of a sequence of elements from the model drawn by lot – which in turn corresponds to drawing a ball from a box. As in the case of the box scheme we may here pose the question: given the probability P for the expression S of a certain theory, can we so select the models $\{\mathcal{M}_t\}_{t\in T}$, the probability μ in the set T, and the probabilities v_t for sequences of elements from the models $\mathcal{M}_t$, that the formula (0) should hold, i.e., that

$$P(\alpha)=E_\mu(v_t(\alpha))$$

should hold for every formula α from S?

Note that we are in a similar situation as in the case of the box scheme, when it was required that certain boxes should contain exclusively balls marked with the letters a and b, and the others, only balls marked with the letters c and d. However the models $\mathcal{M}_i$ be selected, sentences (formulas without free variables) will be true or false in those models, so that for the sentences α we shall have either $v_t(\alpha)=1$ or $v_t(\alpha)=0$.

The answer to the question posed above is in the affirmative: it is possible so to select the models $\mathcal{M}_t$ and the corresponding probabilities μ and v_t.

But before that answer in the affirmative is formulated as a theorem, let the formal properties of formulas and probability be examined. Let S be the set of formulas. There is no need to assume that the formulas belonging to S are elementary, that is, that they include only individual variables. It must be assumed, however, that all the formulas in S are of a certain fixed type, but only such in which the variables of higher types are quantifier-bound.

There is no need to explain the first assumption: the point is that it should be permitted to connect formulas belonging to S by sentential connectives without going outside the set S. The second assumption makes us possible to confine ourselves to drawing a sequence of elements of models (the probability v_t) without the need of drawing, for instance, subsets. The latter would hold if the formulas belonging to S would include variables ranging over sets. It seems that the question: what is the probability of the formula $\alpha(A)$?, where A is a set, means the same as the question: what is the probability of drawing the set A which satisfies α?

Now let Z stand for the set of sentences belonging to S (i.e., formulas without free variables), and let $Cn(X)$ stand for the set of consequences that can be deduced from the subset X of S. Remember that (1) Both S and Z are closed under sentential connectives, in particular alternation, conjunction and negation; (2) We call a system such a set X for which $Cn(X)=X$; (3) Every system which includes a formula also includes all the substitutions of that formula; (4) For every system X holds the formula $Cn(X \cap Z)=X$.

Let now X_0 be a consistent system ($X_0 \neq S$). By probability in S, related to X_0, we mean the function $P(\alpha)$, defined for α belonging to S, with values ranging between 0 and 1, equal to 1 for the formulas belonging to X_0, and such that if the conjunction of two formulas, α and β, is contradictory in the sense that its negation belongs to X_0 ($(\alpha \wedge \beta)' \in X_0$), then for their disjunction we have $P(\alpha \vee \beta)=P(\alpha)+P(\beta)$.

It can be seen from the above that if P is a probability related to X_0, and X_1 is a system contained in X_0, then P also is a probability related to X_1.

Now let P be a probability related to X_0, and let F stand for the set of those formulas α belonging to S for which $P(\alpha)=1$. Obviously, X_0 is included in F, but although F need not be a system (for instance, F need not be closed under substitutions, it may be so that $P(\alpha(x_1))=1$ and $P(\alpha(x_2))<1$), yet there is a greatest system included in F, namely the system $Cn(F \cap Z)$. Of course, P is a probability related to that system.

As can be seen, from the formal point of view nearly every probability may be treated as related to the various systems (unless $Cn(F \cap Z)$ is the system of tautologies); if it is so, what is the meaning of relating P to X_0 and not to some other X_1, included in X_0?

The meaning of relating it precisely in such a way will be explained by an example. Let X_1 be included in X_0, which is in turn included in F, and let α be a sentence that belongs to X_0 but not to X_1. If we treat P as related to X_0, then we think that whatever model be drawn, α will be true in it. But if we treat P as related to X_1, we do not exclude the falsehood of α in a model of the class from which we draw; the only provision is that the subclass of those models in which α is false must have probability equal to 0.

Thus, the relating of probability to a certain definite system X_0 restricts its interpretation by drawing of models and from models more than does the computation of that probability.

If $S(\mathcal{M}_t)$ stands for the set of the true formulas in the model $\mathcal{M}_t$ then if we want to interpret, in the class of models $\{\mathcal{M}_t\}_{t \in T}$, the probability P related to X_0, we should require that X_0 be included in every $S(\mathcal{M}_t)$, but not more, so that the intersection of all $S(\mathcal{M}_t)$ should give exactly X_0. In the case of elementary theories such a class of models can always be found for a given system. This is confirmed by Gödel's theorem on the completeness of the first-order functional calculus, which – in a formulation that suits our purpose well – says that: for every consistent system X there is a model $\mathcal{M}$ such that $X \subset S(\mathcal{M})$.

Gödel's theorem does not hold for non-elementary theories and therefore if we want to obtain, for a given probability P related to X_0, a class of models in which that probability can be interpreted properly, we must accept the condition that there exists for X_0 such a class of models $\{\mathcal{M}_t\}_{t \in T}$ that X_0 is exactly the common part of all the $S(\mathcal{M}_t)$. A system which has that property will be called ω-regular. Note that a system X_0 for which there exists no such model $\mathcal{M}$ that $X_0 \subset S(\mathcal{M})$ certainly is not ω-regular (such a system is called ω-inconsistent), but the existence of one such model does not ensure ω-regularity; hence ω-regularity is a stronger property than ω-consistency.

Let now X stand for an arbitrary system including X_0, to which the probability P is related (X may include F or not). As we know, we have $Cn(X \cap Z) = X$. Let the set $X \cap Z$ of all the sentences from X be arranged as the sequence $\zeta_1, \zeta_2, \ldots$, and let z_n stand for the conjunction of the first n sentences of that sequence. The sequence of probabilities, $P(z_1)$, $P(z_2)$,

$P(z_3), \ldots$, is certainly non-increasing, and hence it may converge to zero or to some number greater than zero. In the latter case we have $P(z_n) \geqslant \varepsilon > 0$ for all n. But at the same time we have

$$\bigcup_{n=1}^{\infty} Cn(z_n) = Cn(X \cap Z) = X.$$

It seems therefore that should we like to extend the probability P so that it should cover systems as well, we would have to impart to the system X the probability ε. Here the difficulty emerges for if that probability is understood as a drawing of models then this would have to mean that the drawing of such a model $\mathcal{M}_t$ in which every formula belonging to X would be true, i.e., $X \subset S(\mathcal{M}_t)$, has the probability $\varepsilon > 0$; yet, if the given theory is not elementary, it may happen that such models do not exist at all, in other words, that the system X is ω-inconsistent. We must protect ourselves against such a possibility by imposing upon P a condition which, in the case of $P(z_n) \geqslant \varepsilon > 0$, would guarantee the existence of an appropriate model. In order to avoid the formation of the sequence ζ_i and the conjunctions z_n, let us formulate that condition as follows:

(C) If there exists such an $\varepsilon > 0$ that for every sentence ζ belonging to the system X we have $P(\zeta) \geqslant \varepsilon$, then X is not an ω-inconsistent system.

A probability that satisfies the condition (C) will be called continuous. Note again that the condition (C) is essential only in the case of non-elementary theories. In elementary theories there are no ω-inconsistent systems, and hence every probability is continuous.

These introductory considerations may now be followed by the final formulation of the theorem on the semantic representation of probability [1].

THEOREM. For every continuous probability in S, related to an ω-regular system X_0, and for every family of models $\{\mathcal{M}_t\}_{t \in T}$ such that for every system Y, complete and ω-consistent and including X_0, there is such a t in T that $Y = S(\mathcal{M}_t)$, there exist

(1) denumerably additive probability μ defined for the subsets of T;

(2) probabilities ν_t, each defined for the subsets of the set $A_t^{\omega_0}$ which

consists of sequences of elements of the model $\mathcal{M}_t$ such that if we symbolize by $\sigma_t(\alpha)$ the set of those sequences from $A_t^{\omega_0}$ which satisfy α we have for every α from S:

$$P(\alpha)=E_\mu(v_t(\sigma_t(\alpha)))=\int_T v_t(\sigma_t(\alpha))\,\mu(dt).$$

The proof of that theorem does not involve difficulties from the mathematical point of view. Given the family $\{\mathcal{M}_t\}_{t\in T}$, we first define the measure for subsets so that to every sentence ζ from Z we ascribe the set $T(\zeta)$ of those t for which ζ is true in $\mathcal{M}_t$. Next we put $\mu(T(\zeta))=P(\zeta)$ and in this way the probability μ for the sets $T(\zeta)$ is given. Availing ourselves of the continuity of P and of the assumption concerning the family $\{\mathcal{M}_t\}_{t\in T}$ we extend μ as to become denumerably additive probability (Kolmogorov's theorem on the extension of measure). Finally, μ being already given, we determine v_t by means of the Radon-Nikodym theorem.

This outline does not tantamount to a proof, but the mathematical apparatus to be used in this connection is unfortunately too complicated for a complete proof to be given here.

Two important remarks must be made here.

First, the probability μ is unique, the probabilities v_t are not unique, but the family $\{v_t\}_{t\in T}$ is unique "almost everywhere". This means that if both μ, $\{v_t\}_{t\in T}$ and μ', $\{v_t'\}_{t\in T}$ satisfy the thesis of the new theorem, then $\mu=\mu'$, and by drawing from the set T according to the probability μ we have the probability 0 of drawing such a t that $v_t\neq v_t'$ (in other words: $\mu\{t\in T: v_t\neq v_t'\}=0$).

Second, throughout all these considerations we have been making the tacit assumption that we have to do with an ordinary theory, in which the set S is denumerable. This is a very natural assumption, which was not questioned for many years, but in recent times investigations have, for various reasons, covered theories in which the set S is non-denumerable.

The reservation must be made in advance that for such theories our theorem cannot be demonstrated with the methods outlined above. The Radon-Nikodym theorem simply does not suffice to determine the probabilities v_t. Still worse, I am inclined to believe that for non-denumerable theories the theorem given above is just not true.

The theorem given above amounts to what is stated in the title of the present paper: a semantic representation of the probability of formulas in formalized theories. It also enables us to make a natural transition from probability defined over formulas to a probability connected with a random drawing of a point from a set. In that case it is sometimes said that we have to do with "Kolmogorov's scheme"; this refers to such an interpretation of probability which fall under Kolmogorov's well-known axiom system.

Thus, this is a path from logical probability to Kolmogorov's scheme. It has long been known that that path can be covered in various ways. The way shown in the present paper is confined to semantics, and hence to the conceptual apparatus most closely connected with logic. This fact seems to indicate that this is a proper path to follow.

NOTE

[1] The full proof of this theorem may be found in J. Łoś: Remarks on Foundations of Probability, Semantical Interpretation of the Probability of Formulas, *Proceedings of the International Congress of Mathematicians 1962*, Stockholm, 1963, pp. 225–229.

J. ŁOŚ

SEMANTYCZNA REPREZENTACJA PRAWDOPODOBIEŃSTWA WYRAŻEŃ W TEORIACH SFORMALIZOWANYCH

(Streszczenie)

W pracy przedstawiona jest ogólna postać prawdopodobieństwa na ciele wyrażeń (formuł) elementarnej teorii sformalizowanej.

Jeśli S jest zbiorem wszystkich takich wyrażeń, $\{\mathcal{M}_t\}_{t\in T}$ (gdzie $\mathcal{M}_t = \langle A_t, R_1^{(t)}, R_2^{(t)}, \ldots \rangle$) klasą modeli, $\bar{\nu}_t$ prawdopodobieństwem w zbiorze ciągów (nieskończonych) utworzonych z elementów zbioru A_t, μ prawdopodobieństwem w zbiorze T, zaś dla α w S (które może, choć nie musi zawierać zmienne indywiduowe) $\sigma_t(\alpha)$ oznacza zbiór tych ciągów (nieskończonych), które spełniają α, wówczas

1° $\quad P_t(\alpha) = \bar{\nu}_t(\sigma_t(\alpha))$

jest dla każdego t w T prawdopodobieństwem w S wyrażającym szanse spełnienia wyrażenia α przez ciąg elementów modelu $\mathcal{M}_t$ wybrany zgodnie z prawdopodobieóstwem $\bar{v}_t$.

Jeśli funkcja $P_t(\alpha)$, przy każdym α, jako funkcja zmiennej t przebiegającej zbiór T, jest mierzalna względem prawdopodobieństwa v_t, czyli posiada wartość oczekiwaną względem tego prawdopodobieństwa, to

$$2^\circ \qquad P(\alpha) = \int_T P_t(\alpha)\,\mu(dt) = \int_T \bar{v}_t(\sigma_t(\alpha))\,\mu(dt)$$

jest prawdopodobieństwem w S.

Całka po prawej stronie wzoru 2° wyraża właśnie wartość oczekiwaną względem μ funkcji określonej wzorem 1°.

Zachodzi następujące twierdzenie (którego pełny dowód może czytelnik znaleźć w mojej pracy cytowanej na s. 192):

Dla każdego prawdopodobieństwa P w S istnieją: klasa modeli $\{\mathcal{M}_t\}_{t \in T}$, prawdopodobieństwa v_t w zbiorze ciągów nieskończonych, utworzonych z elementów zbiorów modeli $\mathcal{M}_t$, wreszcie prawdopodobieństwo μ w zbiorze T takie, że zachodzi wzór 2°.

Twierdzenie to daje interpretację semantyczną prawdopodobieństwa w teoriach elementarnych. Każde takie prawdopodobieństwo $P(\alpha)$ wyraża szanse znalezienia ciągu elementów spełniającego α, przez podwójne losowanie. Najpierw (zgodnie z μ) losujemy jakiś model z danego zbioru modeli, a następnie z tego modelu losujemy ciąg elementów (zgodnie z v_t). Należy zauważyć, że o P nie zakładamy niczego poza addytywnością i unormowaniem ($P(\alpha \vee \beta) = P(\alpha) + P(\beta)$ dla α i β sprzecznych, $P(\alpha) = 1$ dla tautologii), natomiast w tezie możemy stwierdzić σ-addytywność (przeliczalną addytywność) zarówno μ (co jest właściwie konieczne dla istnienia całki we wzorze 2°), jak też każdego v_t. Istotnym założeniem jest przeliczalność zbioru S. Przy specjalnych założeniach co do P twierdzenie to daje się rozszerzyć na teorie nieelementarne.

FRANZ KAMBER

THE STRUCTURE OF THE PROPOSITIONAL CALCULUS OF A PHYSICAL THEORY*

*(Translated by Herbert Korte**)*

INTRODUCTION

In content this note follows the work [5]. Like the latter, this note was inspired by Birkhoff and von Neumann's investigations of the propositional calculus in quantum theory [2], [6].

Starting with postulates of measurements which are performable on a physical system, a propositional calculus (P, R, N, K) is imposed on the system under consideration (Section 6). In the physical interpretation P denotes the set of propositions, R the implication relation, N negation and K the commensurability relation. These operations and relations suffice for the axioms of a semiboolean algebra (Sections 3.2 and 6.3). Two-place operations which correspond to the disjunction and conjunction of propositions are defined for commensurable pairs of propositions. In contrast to [2], [4], and [5] no operations on incommensurable propositions will be assumed here (Section 6.4).

The propositional calculus (P, R, N, K) of a physical system depends therefore on the family of measurements performable on the system.

The most important example for physics of such a propositional calculus was already given by von Neumann in [6], Ch. III: the set of projection operators on a Hilbert Space possesses in a natural way the structure of a semi-boolean algebra (it is isomorphic to $P(C\omega_0)$, cf. Section 5.3).

General characteristics of semiboolean algebras and related structures will be investigated in Sections 1–3.

Finally, a statistical description of a physical system is given through a semiboolean algebra P (set of propositions) and a set Σ of probability functions μ on P, which shall represent the statistical states of ensembles (Section 6.5).

The question of the existence of 'hidden parameters' shall be discussed in the framework sketched above (Section 7). If a physical system which is described by (P, Σ) allows the introduction of a 'hidden parameter',

C. A. Hooker (ed.), *The Logico-Algebraic Approach to Quantum Mechanics*, 221–245.

then a classical extension belongs in a natural way to (P, Σ), i.e. there is an embedding of (P, Σ) into a classical system (B, Σ'), B a boolean algebra. The question of the existence of the 'hidden parameters' leads therefore essentially to the problem of the embeddability of semiboolean algebras into boolean algebras.

Examples of systems will be given, which are important for quantum theory and which do not allow the introduction of 'hidden parameters' (Sections 5.3 and 7.2).

For certain simple quantum theoretical systems an introduction of 'hidden variables' exist. In these cases nothing more can be said mathematically. One can, however, ask under which conditions the corresponding classical extension is interpretable through measurements in the sense of Section 6. It becomes clear that the question of the interpretability of the classical extension can be related to the feasability of certain experiments which contradict the quantum mechanical description of the physical system (Sections 7.3–4; also comp. [4]).

The material dealt with is divided in the following way: In Part I (Sections 1–5) the necessary mathematical tools are presented. Part II (Sections 6–7) contains the construction of the propositional calculus and the discussion of the question of 'hidden variables'.

I heartily thank Professor Dr B. L. van der Waerden at Zürich as well as Dr H. D. Dombrowski and Dr K. Horneffer in Göttingen for several discussions and critical remarks. Furthermore, the author thanks the Swiss National Fund for Scientific Research for financial support.

PART I

1. A SYSTEM OF POSTULATES FOR A PARTIAL BOOLEAN ALGEBRA

1.1. *Notations*

A relation ρ between two sets M and N is understood as a subset $\rho \subseteq M \times N$. For $(x, y) \in \rho$ we also write $x \rho y$. For $M = N$ and $n > 1$, a relation $\rho \subseteq M \times M$ determines in the n-fold product M^n, n-place relations $\rho^{(n-1)}$ and $\rho_{(n-1)}$ through

$$\rho^{(n-1)} = \{(x_1, \ldots, x_n) \in M^n / x_i \rho x_j, \quad i, j = 1, \ldots, n\},$$
$$\rho_{(n-1)} = \{(x_1, \ldots, x_n) \in M^n / x_i \rho x_{i+1}, \quad i = 1, \ldots, n-1\}.$$

For mappings we write $\varphi: A \to B$, and for a constructive mapping $A \dashrightarrow B$. Relations $\rho \subseteq M \times N$ will often be denoted by $M \leftrightarrow N$.

1.2. *Postulates for Boolean Algebras*

The following system of postulates for a boolean algebra (BA), which may be found e.g. in Birkhoff [1], Chapt. X, p. 157, appears to us especially suitable for a generalization of partial BA, since it only contains simple operations, with the exception of the distributive laws:

Let a set A be given as well as

$$\begin{array}{lll} 2 & \text{0-place operations} & 0, 1 \in A, \\ 1 & \text{1-place operations} & N: A \to A, \\ 2 & \text{2-place operations} & D_j: A \times A \to A \quad (j=1,2). \end{array}$$

The axioms BA, 1–5) are then a complete system of postulates for boolean algebras:

BA, 1) $D_2(x, x) = x,\ x \in A.$
BA, 2) $D_2(0, x) = D_2(x, 0) = x,\ x \in A.$
BA, 3) $D_1(1, x) = x,\ x \in A.$
BA, 4) $D_1(x, Nx) = 0,\ D_2(x, Nx) = 1,\ x \in A.$
BA, 5) $D_i(x, D_j(y, z)) = D_j(D_i(x, y), D_i(x, z)),\ x, y, z \in A,\ i \neq j.$

1.3. *Postulates for Partial* BA

For a partial BA the 2-place operations D_j $(j=1,2)$ shall not be given any more on $A \times A$, but only on a 2-place relation $K \subseteq A \times A$. The system of postulates given in Section 1.2 is easily transferred to partial structures; every postulate demands in addition a postulate about the relation K.

DEFINITION. A partial BA is given through (A, K, D_j, N), $(j=1, 2)$: a set A, a relation $K \subseteq A \times A$,

$$\begin{array}{lll} 2 & \text{0-place operations} & 0, 1 \in A, \\ 1 & \text{1-place operations} & N: A \to A, \\ 2 & \text{2-place operations} & D_j: K \to A \ (j=1,2), \end{array}$$

with the following axioms:

PBA, 1) K is reflexive and symmetric;
$D_2(x, x) = x,\ x \in A.$

PBA, 2) $0 K x$ for $x \in A$;
$D_2(0, x) = D_2(x, 0) = x,\ x \in A.$

PBA, 3) $1 K x$ for $x \in A$;
$D_1(1, x) = x,\ x \in A.$

PBA, 4) $x K N y$ for $x K y$;
$D_1(x, Nx) = 0,\ D_2(x, Nx) = 1,\ x \in A.$

PBA, 5) $(x, y, z) \in K^{(2)} \Rightarrow x K D_j(y, z),\ (j = 1, 2)$;
$D_j(x, D_i(y, z)) = D_i(D_j(x, y), D_j(x, z))$ for $(x, y, z) \in K^{(2)},\ i \neq j.$

Remark. This system of postulates implies the uniqueness of the units $0, 1 \in A$.

A subset $B \subset A$ with $B \times B \subseteq K$, which is closed with respect to all operations is according to Section 1.2 a boolean algebra with respect to the induced operations: B then signifies a boolean algebra in A.

LEMMA 1.1. *It follows:* $D_1(x, x) = x$ for $x \in A$; $N^2 = id_A$.

Proof. Comp. [1], p. 156.

THEOREM 1.2. *For a non-empty set $S \subseteq A$, with $S \times S \subseteq K$, there exists a boolean algebra B in A with $S \subseteq B$.*

COROLLARY. (1) *A has a covering of* BA's $B \subseteq A$.

(2) *For D_j and N it follows:*

D_j *is commutative;* $D_1(x, y) = x \Leftrightarrow D_2(x, y) = y$ for $x K y$;
$N D_j(x, y) = D_i(Nx, Ny),\ i \neq j,\ x K y$;
$D_j(x, D_j(y, z)) = D_j(D_j(x, y), z),\ (x, y, z) \in K^{(2)},\ j = 1, 2$;
$D_1(0, x) = 0,\ D_2(1, x) = 1,\ x \in A.$

Proof. (1) We recursively define n-place functions $\Psi^{(n)}: K^{(n-1)} \to A$ by:

$$\Psi^{(1)}: A \to A, \quad \Psi^{(1)} = N_\varepsilon, \quad N_\varepsilon = \begin{cases} N, & \varepsilon = 1. \\ \mathrm{id}_A, & \varepsilon = 0. \end{cases}$$

$$\begin{array}{ccccccc} \Psi^{(n)}: K^{(n-1)} & \dashrightarrow & K^{(n-1)} & \xrightarrow[\mathrm{id} \times \cdots \times D_j \times \cdots \times \mathrm{id}]{} & K^{(n-2)} & \xrightarrow{\Psi^{(n-1)}} & A \\ \cap & & \cap & \searrow & \cap & & \\ A^n & \xrightarrow{N_{\varepsilon_1} \times \cdots \times N_{\varepsilon_n}} & A^n & & A^{n-1} & & \end{array}$$

The mapping $K^{(n-1)} \dashrightarrow K^{(n-2)}$ exists because of PBA, 4, 5).

(2) Let $S \subseteq A$ and $S \times S \subseteq K$. Then $S^n \subseteq K^{(n-1)} \subseteq A^n$. Let $B^{(n)} \subseteq A$ be the images of $S^n \subseteq K^{(n-1)}$ by means of n-place functions. Then $B^{(n)} \subseteq B^{(m)}$ for $n \leqslant m$ and for $B = \bigcup_{n \geqslant 1} B^{(n)}$ it follows that:

> $B \times B \subseteq K$, $B \supseteq S$ and B is closed with respect to all operations. $S \subseteq B$, 0, $1 \in B$, $N: B \to B$ follows directly from the postulates and Lemma 1.1.

For $B \times B \subseteq K$ and $D_j: B \times B \to B$ it is sufficient to show that:

$$B^{(n)} \times B^{(n)} \subseteq K \quad \text{and} \quad B^{(n)} \times B^{(n)} \xrightarrow{D_j} B^{(2n)}.$$

This, however, follows directly from Lemma 1.3 and the inclusion $S^{2n} \subseteq K^{(2n-1)}$. Q.E.D.

LEMMA 1.3. *Let $\Psi^{(n)}$, $\Phi^{(n)}$ be n-place functions $K^{(n-1)} \to A$. Then it follows that:*

$$\begin{array}{ccc} K^{(n-1)} \times K^{(n-1)} & \xrightarrow{\Psi^{(n)} \times \Phi^{(n)}} & A \times A \\ \cup & & \cup \\ K^{(2n-1)} & \dashrightarrow & K \end{array}$$

and $\chi_j: K^{(2n-1)} \xrightarrow{\Psi^{(n)} \times \Phi^{(n)}} K \xrightarrow{D_j} A$ is a 2n-place function ($j = 1, 2$).

Proof. Induction on n. Q.E.D.

2. THE UNIVERSAL MAPPING OF A PARTIAL BA IN A BA

2.1. *Homomorphisms of Partial* BA

A mapping $\varphi: A \to B$ of a partial BA is called a homomorphism in case

$$\begin{array}{ccc} A \times A & \xrightarrow{\varphi \times \varphi} & B \times B \\ \cup & & \cup \\ K_A & \dashrightarrow & K_B \end{array}, \qquad \begin{array}{ccc} A & \xrightarrow{\varphi} & B \\ \downarrow N_A & & \downarrow N_B \\ A & \xrightarrow{\varphi} & B \end{array}, \qquad \begin{array}{ccc} K_A & \xrightarrow{\varphi \times \varphi} & K_B \\ \downarrow D_j & & \downarrow D_j \\ A & \xrightarrow{\varphi} & B \end{array}$$

are commutative. With this we get the category $\mathfrak{P}$ of the part. BA with homomorphisms of partial BA. According to the definition, a BA is also a partial BA. More exactly: the category $\mathfrak{B}$ of the BA with BA-homomorphisms is a full subcategory of $\mathfrak{P}$:

$$\mathfrak{B} \underset{i}{\subset} \mathfrak{P}.$$

In Section 2.2 it will be shown that a functor $\mathfrak{r}:\mathfrak{B}\to\mathfrak{B}$ exists with $\mathfrak{r}\circ\mathfrak{i}\cong \mathrm{id}_{\mathfrak{B}}$. But first we need some concepts.

DEFINITION. $j\subseteq A$ is called an ideal, in case

(1) $xKy,\ x, y\in j\Rightarrow D_2(x, y)\in j.$
(2) $xKy,\ x\in j,\ y\in A\Rightarrow D_1(x, y)\in j.$

It follows: $j\subset A$ exactly when $1\notin j$ or $j\cap N(j)=\emptyset$.

DEFINITION. A proper ideal $p\subset A$ is called a prime-ideal, in case

$$xKy,\ D_1(x, y)\in p\Rightarrow x\in p \quad \text{or} \quad y\in p.$$

LEMMA 2.1. *Equivalent conditions are:*

(i) $p\subset A$ prime,
(ii) $p\cup N(p)=A$ ($\cup$ = union),
(iii) $\exists\beta_p\in \mathrm{Hom}_{\mathfrak{P}}(A, \mathbf{Z}_2)$ with $\ker\beta_p=p$.

Moreover, $\mathbf{Z}_2$ is furnished with a trivial BA structure.
Proof. Comp. [5], (Section 3).

2.2. *The Universal Mapping* $A\xrightarrow{u} B_A$

We consider for $A\in\mathfrak{P}$ the following covariant functor $F_A:\mathfrak{B}\to\mathfrak{Ens}$ (category of sets):

$X\in\mathfrak{B}: F_AX=\mathrm{Hom}_{\mathfrak{P}}(A, X)$, partial homomorphisms of A to X.

$$\varphi: X\to Y: \qquad F_A\varphi: F_AX\to F_AY$$
$$\qquad\qquad\qquad\quad\ \ \ni \qquad\quad \ni$$
$$\qquad\qquad\qquad\quad\ \ f \ \ \to \varphi\circ f$$

THEOREM 2.2. (i) *For all $A\in\mathfrak{P}$ there exists a 'universal pair' (B_A, u) with respect to F_A, i.e. $B_A\in\mathfrak{B}$, $u\in \mathrm{Hom}_{\mathfrak{P}}(A, B_A)$ with:*

for $\varphi\in \mathrm{Hom}_{\mathfrak{P}}(A, X)$, $X\in\mathfrak{B}$ $\exists$ exactly one $\tilde\varphi\in \mathrm{Hom}_{\mathfrak{B}}(B_A, X)$ such that

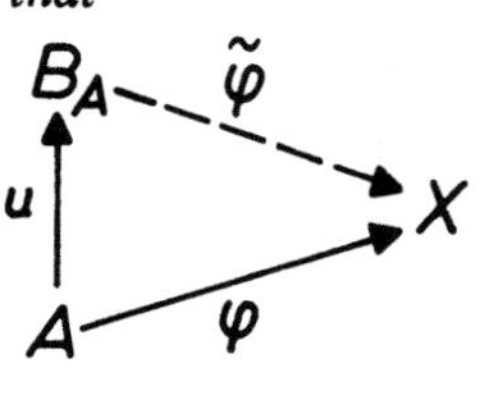

is commutative. Through this property (B_A, u) *is determined up to equivalence.*

(ii) *With respect to* $\mathfrak{r}(A)=B_A$, $\mathfrak{P}\xrightarrow{\mathfrak{r}}\mathfrak{B}$ *becomes in a natural way a functor with* $\mathfrak{r}\circ\mathfrak{i}\cong id_{\mathfrak{B}}$.

Proof. (i) Let $(A_\iota)_{\iota\in I}$ be a family of BA in A with $\bigcup_{\iota\in I} A_\iota=A$, $\bigcup_{\iota\in I} A_\iota\times\times A_\iota=K$. Such families exist according to Theorem 1.2.

A homomorphism $\varphi: A\rightarrow X$ for $X\in\mathfrak{B}$ is given through BA-homomorphisms $\varphi_\iota: A_\iota\rightarrow X$ with $\varphi_\iota \mid A_\iota\cap A_\lambda=\varphi_\lambda \mid A_\iota\cap A_\lambda$ for ι, $\lambda\in I$. The direct sum[1] $\bar{A}=\bigoplus_{\iota\in I} A_\iota \xleftarrow{v_\iota} A_\iota$ is universal for mappings $\varphi: A_\iota\rightarrow X$: There exists exactly one $\bar{\varphi}:\bar{A}\rightarrow X$ with

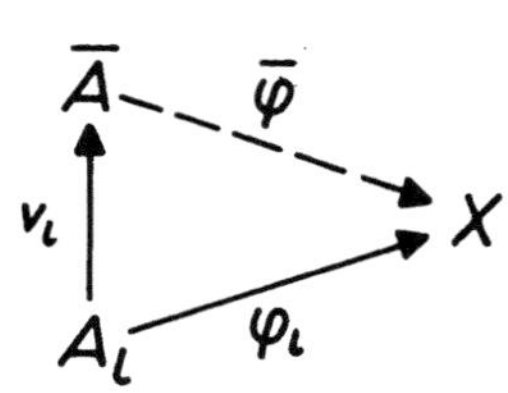

which is commutative for $\iota\in I$.

If we consider in $\bar{A}$ the ideal j, generated from the elements of the form:

$$v_\iota(x)\triangle v_\lambda(x),\ x\in A_\iota\cap A_\lambda, \quad (\triangle=\text{symm. difference})$$

and the surjective homomorphism $\pi:\bar{A}\rightarrow\bar{A}/j=B_A$, then $\pi\circ v_\iota: A_\iota\rightarrow B_A$ define, according to construction, a partial homomorphism $u: A\rightarrow B_A$ and by assumption $\bar{\varphi}$ vanishes on j and induces therewith $\tilde{\varphi}: B_A\rightarrow X$ with $\tilde{\varphi}\circ u=\varphi$.

(ii) Let A_1, $A_2\in\mathfrak{P}$, $\varphi\in\mathrm{Hom}_{\mathfrak{P}}(A_1, A_2)$. Then it follows that:

$$\begin{array}{ccc} B_{A_1} & \overset{\tilde{\varphi}}{\dashrightarrow} & B_{A_2} \\ {\scriptstyle u_1}\uparrow & & \uparrow{\scriptstyle u_2} \\ A_1 & \xrightarrow{\varphi} & A_2 \end{array}$$

because $u_2\circ\varphi\in\mathrm{Hom}_{\mathfrak{P}}(A_1, B_{A_2})$ and $\tilde{\varphi}$ exists uniquely because of the universality of u_1.

With $\mathfrak{r}(A)=B_A$, $\mathfrak{r}(\varphi)=\tilde{\varphi}$, $\mathfrak{r}:\mathfrak{P}\rightarrow\mathfrak{B}$ becomes a covariant functor, which obviously induces an identity on $\mathfrak{B}$, i.e. for $B\in\mathfrak{B}\subset\mathfrak{P}$, (B, id_B) is F_B – universal. Q.E.D.

2.3. *Characterization of B_A through prime-ideals in A*

Let π be the set of prime-ideals in A for $A \in \mathfrak{P}$. Then the Power-set (Potenzmenge) is in a natural way $P(\Pi) \cong \times_{p \in \Pi} A/p$.

We consider the mapping:

$$\beta : A \to \times_{p \in \Pi} A/p, \quad \text{given by} \quad \beta = (\beta_p)_{p \in \Pi},$$
$$\beta_p : A \to \mathbf{Z}_2 \quad \text{with} \quad \ker \beta_p = p \quad \text{(Lemma 2.1).}$$

Then $\beta \in \mathrm{Hom}_{\mathfrak{P}}(A, \mathring{B}_A)$, where $\mathring{B}_A$ is a BA generated from $\beta(A)$ in $P(\Pi)$.

THEOREM 2.3. *$(\mathring{B}_A, \beta)$ is F_A-universal, i.e. there exists an isomorphism $\gamma : B_A \to \mathring{B}_A$ with $\gamma \circ u = \beta$.*

Proof. (B_A, u) F_A-universal $\Rightarrow \gamma$ with $\gamma \circ u = \beta$ exists uniquely. If $x \neq 0$, $x \in B_A$, then there exists a prime-ideal $p' \subset B_A$ with $x \notin p'$. To that corresponds uniquely, according to the universality of (B_A, u) a prime-ideal $p \subset A$, which according to the construction of $\mathring{B}_A$ determines a prime-ideal $p'' \subset \mathring{B}_A$ with $\beta_p = \beta_{p''} \circ \beta$.

But then it follows:

$$\beta_{p'} \circ u = \beta_p = \beta_{p''} \circ \beta = (\beta_{p''} \circ \gamma) \circ u.$$

u universal $\Rightarrow \beta_{p'} = \beta_{p''} \circ \gamma$.

With this we have $0 \neq \beta_{p'}(x) = \beta_{p''}(\gamma(x)) \Rightarrow x \notin \ker \gamma$, i.e. $\ker \gamma = 0$. γ however is surjective since $\beta(A)$ generates $\mathring{B}_A$. Q.E.D.

3. PARTIAL-ORDERING RELATIONS IN PARTIAL BOOLEAN ALGEBRAS

3.1. A partial ordering $R \subseteq A \times A$ in a partial BA (A, K, D_j, N) is called *compatible* with the partial structure of an algebra, in case

(1) $R \subseteq K$,

(2) $x K y \Rightarrow D_1(x, y) = x \cap y(R)$ (Infinum with respect to R in A)

holds.

LEMMA 3.1. *A compatible partial-ordering R in (A, K, D_j, N) fulfills the following conditions:*

(i) $0, 1 \in A$ *are the smallest and largest elements in* A *with respect to* R,

(ii) N *is* R*-dual, i.e.* $xRy \Rightarrow NyRNx$ *and* $xRNx \Rightarrow x=0$,

(iii) $xKy \Rightarrow D_2(x, y) = x \cup y(R)$ *(Supremum with respect to* R *in* A*)*,

(iv) *If* $B \subseteq A$ *is a* BA *with respect to* R *in* A*, i.e.*

(1) $Nx \in B$ *for* $x \in B$,

(2) $x, y \in B \Rightarrow x \cap y(R)$ *exists with respect to* R *in* A *and* $x \cap y(R) \in B$,

(3) $(B, N, \cap)$ *is a Boolean-Algebra*,

then it follows that $B \times B \subseteq K$ *and that* B *is a* BA *in* (A, K, D_j, N).

Proof. Through computation with the help of Conditions 1, 2, about R.

What can be said now about the existence and uniqueness of a compatible relation R? This question will be answered in Theorem 3.2. First of all the following relations can be introduced in an arbitrary partial BA:

$$\rho_1 \subseteq A \times A : x\rho_1 y \Leftrightarrow xKy \text{ and } D_1(x, y) = x \text{ (or } D_2(x, y) = y).$$
$$\rho_2 \subseteq A \times A : x\rho_2 y \Leftrightarrow \exists x = x_1, \ldots, x_i, \ldots, x_n = y \text{ with } x_i \rho_1 x_{i+1} \ (i = 1, \ldots, n-1).$$

ρ_1 is reflexive, and it follows that $\rho_1 \subseteq K$ and $x\rho_1 y$, $y\rho_1 x \Rightarrow x = y$. ρ_2 is reflexive and transitive, and therefore $\rho_1 \subseteq \rho_2$.

THEOREM 3.2. (i) *Uniqueness of* R: R *is a compatible partial-ordering in* $A \Rightarrow R = \rho_1 = \rho_2$.

(ii) *Existence of* R: $\rho_1 = \rho_2 \Rightarrow R \equiv \rho_1 = \rho_2$ *is a compatible partial-ordering in* A.

(iii) *For* $\rho_1 = \rho_2$ *it is necessary and sufficient:*

R) $(x, y, z) \in \rho_{1(2)} \Rightarrow xKz$.

Proof. (i) and (iii) are simple to prove.

(ii) Let $\rho_1 = \rho_2$. Then $R = \rho_1 = \rho_2$ is a partial-ordering in A according to previously determined characteristics of ρ_1 and ρ_2.

Furthermore: $R = \rho_1 \subseteq K$.

We still have to prove $xKy \Rightarrow D_1(x, y) = x \cap y(R)$.

Let xKy be given. Then $D_1(D_1(x, y), x) = D_1(x, y)$ since D_1 is commutative, associative and $D_1(x, x) = x$; likewise for y. Also: $D_1(x, y)\ Rx, y$ according to definition of ρ_1.

Given zRx, y. Then it follows because of $R \subseteq K:(x, y, z) \in K^{(2)}$. According to Theorem 1.2 BA $B \subseteq A$, $\{x, y, z\} \subseteq B$ exists. Since according to assumption $D_1(z, x) = z$, $D_1(z, y) = z$, it follows that $D_1(z, D_1(x, y)) = D_1(D_1(z, x), y) = D_1(z, y) = z$, i.e. $zRD_1(x, y)$. From that it follows that $D_1(x, y) = x \cap y(R)$. Q.E.D.

3.2. *Semi-Boolean Algebras*

A partial boolean algebra based on a partially-ordered set with an involution shall be constructed in Section 6.

The connection of such sets with partial BA shall be presented in this section.

A partial-ordered set with involution (P, R, N) is given through: $0, 1 \in P$, partial-ordering $R \subseteq P \times P$, involution $N: P \to P$, $N^2 = \mathrm{id}_P$ with:

HJ, 1) $0Rx$, $xR1$, $x \in P$,
HJ, 2) $xRy \Rightarrow NyRNx$ i.e. N is R-dual,
HJ, 3) $xRNx \Rightarrow x = 0$.

A BA B with respect to R in P is defined as in Lemma 3.1, iv.

A relation $K \subseteq P \times P$ in a partially-ordered set (P, R, N) with involution is called a *commensurability-relation* (K-R), in case the following conditions are satisfied:

KR, 1) K symmetric,
KR, 2) $R \subseteq K$,
KR, 3) If $B \subseteq P$ is a BA with respect to R in $P \Rightarrow B \times B \subseteq K$,
KR, 4) $S \subseteq P$, $S \times S \subseteq K \Rightarrow$ there exists a BA $B \subseteq P$ with $S \subseteq B$.

LEMMA 3.3. (i) *If there exists in* (P, R, N) *a* K-R, *then it is uniquely determined and given through:* $K = \bigcup B \times B$, *where* B BA *are in* P *with respect to* R.

(ii) *If there exists in* (P, R, N) *a* K-R $K \subseteq P \times P$, *then the quasi-modular law follows in* (P, R, N) (comp. [5], Section 2): $x, y, z \in P$, xRz, $yRNz \Rightarrow z \cap (x \cup y) = x$.

Proof. (i) follows from KR 1, 3, 4).

(ii) For $x, y, z \in P$, with xRz, $yRNz$ it follows because of KR, 1, 2, 3, 4) that $(x, y, z) \in K^{(2)}$. Then there exists according to KR, 4) a BA $B \subseteq P$ with $\{x, y, z\} \subseteq B$ and the quasi-modular law is satisfied in B. Q.E.D.

A partially-ordered set with involution (P, R, N) is called a *semi-boolean algebra* (SBA) in case there exists in (P, R, N) one and according to Lemma 3.3 only one K-R $K \subseteq P \times P$.

The following theorem completely leads the structure of SBA back to the structure of the partial BA with condition R).

THEOREM 3.4. (i) *Let (P, R, N, K) be a* SBA.

Because of KR, 1–4) *R determines two functions:*

$$D_j: K \to P,\ D_1(x, y) = x \cap y(R),\ D_2(x, y) = x \cup y(R) \text{ for } xKy.$$

(P, K, D_j, N) *is then a partial* BA *which satisfies condition R) and R is the uniquely determined partial-ordering in P compatible with the partial* BA-*structure.*

(ii) *Let (A, K, D_j, N) be a partial* BA *which satisfies condition* R). *Let $R = \rho_1 = \rho_2$ be the compatible partial-ordering in (A, K, D_j, N)* (Comp. Theorem 3.2). *Then (A, R, N, K) is a* SBA *and it follows that: $x \cap y(R) = D_1(x, y)$, $x \cup y(R) = D_2(x, y)$ for xKy.*

Proof. (i) is clear.

(ii)

HJ, 1–3) follow from Lemma 3.1, (i–ii).

KR, 2) follows from Theorem 3.2.

KR, 3) follows from Lemma 3.1, (iv).

KR, 4) follows from Theorem 1.2. Q.E.D.

COROLLARY. PBA, 1–5 and R) is a complete system of postulates for semi-boolean algebras.

Remark. The defining characteristics KR, 1–4 of a K-R are exactly those which one would expect from a physical commensurability-concept.

4. REPRESENTABLE PROBABILITY FUNCTIONS OF A SBA

4.1. We define a probability function (W-F) on a SBA P as a mapping $\mu: P \to \mathbf{R}$ with

W, 1) $\mu(x \cup y) = \mu(x) + \mu(y)$ for $xRNy$ (or equivalently $yRNx$).

W, 2) $\mu(x) \geqslant 0$, $x \in P$; $\mu(1) = 1$.

A W-F is a monotonically increasing function. $j_\mu = \{x \in P / \mu(x) = 0\}$ is an ideal in P.

2-valued W-F, i.e. $\mu(x) = 0, 1$ for $x \in P$, are exactly the homomorphisms $\mathrm{Hom}_{\mathfrak{P}}(P, \mathbf{Z}_2)$.

On boolean-algebras this definition agrees with the usual one.

We are interested here in especially those W-F, which can be represented as mean values by two-valued W-F.

More exactly: A W-F on P is called *representable* (through prime-ideals $p \in \Pi$) in case there exists on $\mathring{B}_P$ a W-F v, such that $\mu = \beta^* v$ (comp. Section 2.3).

Every normalized linear combination of 2-valued W-F is representable. If $L(\Pi)$ is a σ-setalgebra (σ-Mengenalgebra) in $P(\Pi)$ and $L(\Pi) \supseteq \mathring{B}_P$, v a normalized measure on $L(\Pi)$, then

$$\mu(x) = \int_{\Pi} \beta_p(x)\, dv(p)$$

is a representable W-F on P.

4.2. We shall now consider whole systems (P, Σ): P SBA, Σ a set of W-F on P, which satisfy the following condition:

For $x \neq y$, $x, y \in P$ there exists $\mu \in \Sigma$ with $\mu(x) \neq \mu(y)$. A system (P, Σ) is called *representable*, in case all $\mu \in \Sigma$ are representable.

It shall be called *classical*, in case $P = B$ is a boolean-algebra. A classical system (i, B, Σ') is called a *classical extension* of (P, Σ), in case there exists a mapping $i: P \to B$ with the characteristic:

for $\mu \in \Sigma$ there exists a $\mu' \in \Sigma'$ such that $\mu = i^* \mu'$. On the basis of the condition on (P, Σ), i must be one-to-one, i.e. $x \neq y \Rightarrow i(x) \neq i(y)$.

A representable system naturally possesses a classical extension. But the converse also holds.

LEMMA 4.1. *If* (P, Σ) *possesses a classical extension* (i, B, Σ') *then* (P, Σ) *is representable.*

Proof. On the basis of the universality of β (Theorem 2.3) there exists the map j:

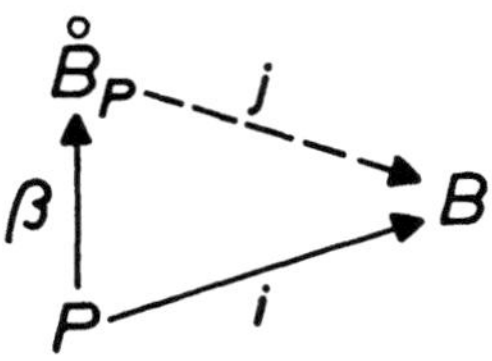

Whence for $\mu = i^*\mu'$, $\mu' \in \Sigma'$ and $\nu = j^*u' : \beta^*\nu = \beta^*(j^*\mu') = (j \circ \beta)^*\mu' = i^*\mu' = \mu$, i.e. μ is representable for $\mu \in \Sigma$. Q.E.D.

THEOREM 4.2. *Let (P, Σ) be given; let $L(\Omega)$ be an algebra of sets, $\tilde{\Sigma}$ a set of* W-F *on $L(\Omega)$ and $\rho \subseteq L(\Omega) \times P$ a consistent structural (strukturtreue) relation between $L(\Omega)$ and P with conditions:*

(1) $\rho_x = \{\tilde{x} \in L(\Omega) / \tilde{x}\, \rho\, x\}$ *is not empty for* $x \in P$.
(2) *For* $\mu \in \Sigma$ *there exists* $\tilde{\mu} \in \tilde{\Sigma}$ *with:*
(a) $\tilde{\mu}(\tilde{x} \triangle \tilde{y}) = 0$ for $\tilde{x}, \tilde{y} \in \rho_x$, $x \in P$.
(b) $\tilde{\mu}(\tilde{x}) = \mu(x)$ for $\tilde{x} \in \rho_x$.

To this representation of (P, Σ) there belongs then in a natural way a classical extension (i, B, Σ') such that $B \cong L(\Omega)/j$ and

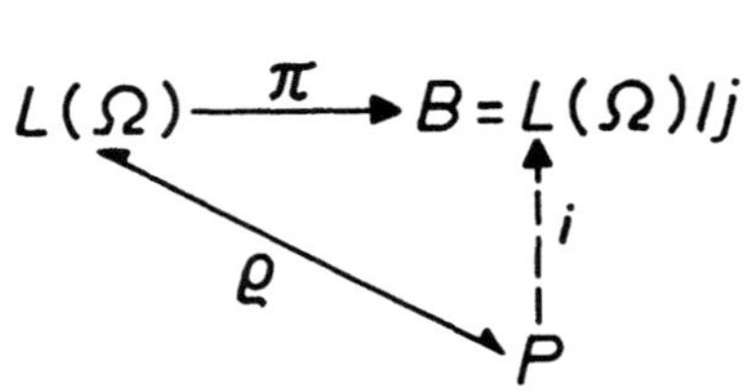

is commutative. π is surjective and i is one-to-one.

Proof. Let j be generated from elements of the form $\tilde{x} \triangle \tilde{y}$, $\tilde{x}, \tilde{y} \in \rho_x$, $x \in P$. Let $B = L(\Omega)/j$ and π be the natural mapping. $i(x) = \pi(\tilde{x})$ for $\tilde{x} \in \rho_x$, is then well-defined. Let $\tilde{\mu} \in \tilde{\Sigma}$ and $\tilde{\mu}(\tilde{x}) = \mu(x)$ for $\tilde{x} \in \rho_x$, $x \in P$. Then because of 2, a) $\tilde{\mu}(\tilde{x} \triangle \tilde{y}) = 0$ for $\tilde{x}, \tilde{y} \in \rho_x$, $x \in P$, d. h. $j_{\tilde{\mu}} \supseteq \{\tilde{x} \triangle \tilde{y}\}$.

Since $j_{\tilde{\mu}}$ is an ideal in $L(\Omega)$ it follows that $j_{\tilde{\mu}} \supseteq \ker \pi = j$ and $\tilde{\mu}$ induces μ' on B with $\pi^*\mu' = \tilde{\mu}$. Hence: $\mu(x) = \tilde{\mu}(\tilde{x}) = \mu'(\pi(\tilde{x})) = \mu'(i(x)) = i^*\mu'(x)$ for $\tilde{x} \in \rho_x$, $x \in P$, i.e. (P, Σ) possesses a classical extension: (i, B, Σ'), $\Sigma' = \{\mu'$ on $B, \pi^*\mu' = \tilde{\mu}, \tilde{\mu} \in \tilde{\Sigma}\}$. Q.E.D.

5. The Embedding Problem for Partial BA

5.1. The representability of systems (P, Σ) led us in a natural way to embeddable SBA.

Several statements about partial BA shall be formulated here.

A partial BA A is *embeddable*, in case there exists a one-to-one homomorphism $i: A \to B$ of A into a BA B.

The presence of such a homomorphism $i: A \to B$ is called an *embedding* of A in B in case $i(A) \subseteq B$ generates the BA B.

THEOREM 5.1. *Every embedding of a partial* BA *is a quotient of the universal mapping* $u: A \to B_A$:

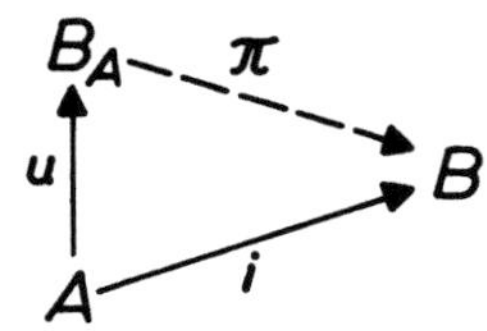

There exists therefore essentially at most one embedding of A into a Boolean algebra, which is, in case it exists, given through $A \xrightarrow{u} B_A$.

Proof. π exists uniquely, since u is universal. $i(A)$ generates B, $u(A)$ generates $B_A \Rightarrow \pi$ surjective. Q.E.D.

5.2. *Finite Character of Embeddability-Property*

THEOREM 5.2. *A partial* BA *A is embeddable in case all finite unions* $\bigcup_{i=1}^{n} B_i$ *of finite* BA *in A are structurepreservingly embeddable in* BA.

Proof. This follows from the study of the universal mapping $u: A \to B_A$, in case one covers up A through the class of finite BA in A. This class of BA in A satisfies the condition in the proof of Theorem 2.2(i). Q.E.D.

COROLLARY. *If A is not embeddable, then a finite union* $\bigcup_{i=1}^{n} B_i$ *of finite* BA B_i *in A is not embeddable.*

5.3. *Examples of Non-Embeddable Partial* BA

Let $P(\Lambda, n-1)$ be the SBA of the closed linear subspaces in the n-dimensional Hilbert Space $H^n(\Lambda)$ over the field (Körper)

$$\Lambda = \mathbf{R}, \mathbf{C}; \qquad n = 3 \ldots \omega_0.$$

Then there exists a corollary from a theorem by Gleason [3] which holds for the universal mapping:

$$P(\Lambda, n-1) \overset{u}{\to} B_{P(\Lambda, n-1)} = 0 \quad \text{for} \quad n > 2,$$

and for $n = \omega_0$

$$P(\Lambda, \omega_0) \overset{u}{\to} B_{P(\Lambda, \omega_0)}$$

there exists no embedding.

Moreover, if for the continuous geometry $P_c(\Lambda)$ over a field Λ

$$P_c(\Lambda) \overset{u}{\to} B_{P_c(\Lambda)} = 0,$$

then $P_c(\Lambda)$ is the metric termination (Abschluss) of the direct limit of a directed sequence of projective geometries:

$$P(\Lambda, 1) \overset{\Delta}{\to} P(\Lambda, 3) \overset{\Delta}{\to} \cdots \overset{\Delta}{\to} P(\Lambda, 2^{n-1} - 1) \overset{\Delta}{\to} P(\Lambda, 2^n - 1) \to \cdots,$$

where Δ are the injective diagonal mappings (comp. [1], Chap. VIII, p. 125). Hence:

THEOREM 5.3. *A system (P, Σ) is not representable (and hence also does not possess a classical extension) in case P contains a segment $P(o, a)$ $a \in P$ which is isomorphic to* SBA *$P(\Lambda, n-1)$, $P_c(\Lambda)$, $n = 3, 4, \ldots, \omega_0$.*

If $P \cong P(\Lambda, n-1)$, $P_c(\Lambda)$, $n = 3 \cdots < \omega_0$ then $Hom_{\mathfrak{P}}(P, \mathbf{Z}_2)$ is the empty set, i.e. there are no 2-valued W-F *on P.*

PART II

6. THE PROPOSITIONAL CALCULUS IN A PHYSICAL THEORY

6.1. Let $\mu_1, \ldots, \mu_n$ be compatible measurements on a physical system. After a simultaneous performance of these measurements one gets a measurement result $(x_1, \ldots, x_n) \in \mathbf{R}^n$.

The most general proposition, which can be made about the physical system under consideration with respect to the measurement μ_j $(j = 1, \ldots, n)$ is of the form: the measurement result $(x_1, \ldots, x_n)$ lies in the subset $\xi \subseteq \mathbf{R}^n$.

We shall call the elements $\xi, \eta, \ldots$ of a suitable class $A = L(\mathbf{R}^n)$ of subsets in $\mathbf{R}^n$ which generates an algebra of sets, the *experimental states of affairs* with respect to the measurements μ_j $(j=1,\ldots,n)$ (in [2]; Section 2: experimental propositions).

If after performing the measurements μ_j we have: $(x_1,\ldots,x_n)\in\xi$ (respectively $(x_1,\ldots,x_n)\notin\xi$,) then we say: The measurements determine the existence (respectively the non-existence) of the state of affairs ξ.

6.2. We consider now an algebra of sets $A_\nu = L_\nu(\mathbf{R}^{n_\nu})$ belonging to all possible collections of compatible measurements. Let the following relation R' be introduced implicitly in $A = \dot{\bigcup}_\nu A_\nu \Big/ \begin{matrix} 0_\nu \sim 0 \\ 1_\nu \sim 1 \end{matrix}$ (disjoint union of A_ν with identification of all null and single elements): $\xi R'\eta$ holds exactly, when η always exists just in case ξ exists, $\xi, \eta\in A$. Let R'_s be the correlated symmetrized relation: $\xi R'_s\eta \Leftrightarrow \xi R'\eta$ and $\eta R'\xi$. States of affairs ξ, η with $\xi R'_s\eta$ cannot be differentiated through measurements.

In order to free ourselves from the implicit definition we demand that R' satisfies the following postulates which immediately permit a physical interpretation:

(A) (a) R' is reflexive and transitive (R'_s is therewith an equivalence relation)

(b) $0R'\xi$ and $\xi R'1$ for $\xi\in A$.

(c) $\xi, \eta\in A_\nu$, $\zeta\in A_\lambda$, $\xi, \eta R'\zeta \Rightarrow \xi\cup\eta R'\zeta$.

(d) $\xi, \eta\in A_\nu$, $\xi\subseteq\eta\Rightarrow\xi R'\eta$.

(e) $\xi R'\eta\Rightarrow\eta^\perp R'\xi^\perp$ ($\perp$: Complement in the A_ν).

(f) $\xi R'\eta\Rightarrow\exists\xi', \eta'\in A_\nu$, $\xi R'_s\xi'$, $\eta R'_s\eta'$.

(g) $\xi_1, \eta_1\in A_\lambda$, $\xi_2, \zeta_1\in A_\mu$, $\eta_2, \zeta_2\in A_\nu$, $\xi_1 R'_s\xi_2$, $\eta_1 R'_s\eta_2$, $\zeta_1 R'_s\zeta_2$ $\Rightarrow\exists\xi', \eta', \zeta'\in A_\chi$ with $\xi' R'_s\xi_i$, $\eta' R'_s\eta_i$, $\zeta' R'_s\zeta_i$ $(i=1, 2)$.

According to (a) R' shall possess the characteristics of an implication relation;

(b)–(e) connect R' with the algebraic structure of A_ν.

(f) says that states of affairs implied by states of affairs can be replaced through equivalent states of affairs in a boolean algebra A_ν.

For the postulate (g) comp. Section 6.3).

We consider now the set $P = A/R'_s$ of classes of equivalent states of affairs; let $A\xrightarrow{\pi}P$ be the natural mapping of A on P.

We call P the *proposition set* which belongs to the physical system. Let the propositions be designated by $x, y\ldots,\in P$.

From the demands (A)–(a–e) the following theorem results:

THEOREM 6.1. (i) *R' induces in P a partial-ordering relation R. $\pi(0)=0$ and $\pi(1)=1$ are the smallest and the largest element of P with respect to R. The complement-structure $\perp$ in the A_v determines an involutory R-dualautomorphism $N: P \to P$, which satisfies the relation*

$$xRNx \Rightarrow x=0$$

(P, R, N) is therewith a partial-ordered set with involution.

(ii) *$B_v = \pi(A_v)$ is a boolean algebra in (P, R, N) (comp.* Section 3.1) *and π_v; $A_v \to B_v \subseteq P$ is a surjective algebraic homomorphism.*

The B_v cover P since π is surjective.

Remark. R and N in P can be interpreted according to construction as implication and negation of propositions; $\cap$ and $\cup$ as conjunction and disjunction in the boolean algebras B_v.

6.3. *Commensurability*

We call two propositions $x, y \in P$ *physically commensurable*, in case they can be represented through the states of affairs ξ, η which belong to the same group of compatible measurements:

$$xKy \Leftrightarrow \exists A_v, \quad \xi, \eta \in A, \quad \pi(\xi)=x, \quad \pi(\eta)=y.$$

The relation K is thus given by: $K = \bigcup_v B_v \times B_v \subseteq P \times P$. The postulate (A)-(g) can now be formulated as follows:

3 pairwise commensurable propositions can be represented through states of affairs which belong to the same group of compatible measurements:

$$xKy, yKz, xKz \Rightarrow \exists B_v, \qquad x, y, z \in B_v.$$

THEOREM 6.2. *There exist in (P, R, N) at least one (and according to Lemma* 3.3 *at most one) commensurability relation which satisfies conditions* KR, 1–4). *It is given by the physical commensurability $K = \bigcup_v B \times B_v \subseteq P \times P$. (P, R, N, K) is thus a* SBA.

COROLLARY. *The quasi-modular law (Lemma* 3.3*) holds in (P, R, N, K).*

Proof. Since B_v are boolean algebras in P, we can define the operations $D_j: K \to P$ through:

$$D_1(x, y) = x \cap y(R) \text{ (Infinum with respect to } R,$$
$$D_2(x, y) = x \cup y(R) \text{ (Supremum with respect to } R) \text{ for } xKy.$$

(P, K, D_j, N) becomes a partial BA because of (A)-(g) and Theorem 6.1. R satisfies the condition $R \subseteq K$ because of (A)-(f), and according to the definition $D_1(x, y) = x \cap y(R)$ for xKy. According to Theorem 3.2 and 3.4, (P, R, N, K) is a SBA, i.e. K satisfies the conditions, KR 1–4). Q.E.D.

Remarks (1) The commensurability relation in quantum theory satisfies conditions KR, 1–4):

Let P be a set of projection operators on a Hilbert space with:

(1) $E, F \in P, [E, F] = 0 \Rightarrow E \circ F \in P.$
(2) $E \in P \Rightarrow 1 - E \in P.$

With ERF for $E \leqslant F$, $NE = 1 - E$, P is then a SBA and EKF is characterized by $[E, F] = 0$. This calculus was already presented by J. von Neumann in [6], Chapter III, Section 5.

(2) All classical theories presuppose that $K = P \times P$. P must necessarily be a boolean algebra. In classical mechanics there even exists a B_v with $B_v = P: B_v$ belongs to the $2f$ measurements of the positions q_j and momenta p_j $(j = 1 \ldots f)$. In quantum theory this is no longer the case.

(3) The quasi-modular law appears here as a consequence of (A)-(f, g). It was used already in [5] as a replacement for the modular law introduced in [2], Section 11. In orthocomplemented lattices it extensively replaces the modular law and has the advantage that it only contains operations on commensurable elements.

Summing up we can say:

On the basis of A, a–g) the proposition set P of a physical system possesses the structure of a semi-boolean algebra (P, R, N, K).

6.4. In [2], Section 8, an additional postulate is demanded:

(B) For arbitrary x, $y \in P$ there exists the infinum $x \cap y(R)$ w.r.t. R.

Thus P becomes an orthocomplemented quasi-modular lattice, because then the supremum $x \cup y(R)$ exists as well and is given by $x \cup y = N(Nx \cap Ny)$ and N automatically becomes an orthocomplement of the lattice structure.

Such lattices were investigated in [5] and the results were applied to the propositional calculus of quantum theory under the presupposition of the above lattice-postulate (B).

For the discussion of the question of 'hidden parameters' (Section 7) we shall limit ourselves, however, to the semi-boolean structure of P for the following reason:

There exist SBA P, which are embeddable in a BA B, as well as in a non-boolean orthocomplemented lattice V. Therefore the extension of the partial structure in P to a lattice structure is in general not uniquely determined.

The giving up of postulate (B) therefore means in a certain sense that one understands the operations $x \cap y$, $x \cup y$ for $(x, y) \notin K$ not to be fixed. Physically this means: For $(x, y) \notin K$, $x \cap y$, $x \cup y$ are dependent on the experimental information about the physical system under consideration. For example let x, y be propositions about the position and momentum of a 'particle': Then classically: $x \cap y \neq 0$ and quantum theoretically: $x \cap y = 0$.

The operations for propositions $(x, y) \notin K$ are determined permanently by postulate (B), and the latter may very well find its justification in a solid theory, – e.g. quantum theory.

For the discussion of the question of the 'hidden parameter' we shall freely admit here, that the operations of incommensurable propositions (in case they exist at all) can change through an eventual introduction of 'hidden parameters'.

This is indeed a necessary condition for quantum theoretical systems (comp. Section 7, 3–4).

We shall base the following considerations only on the semi-boolean structure of the propositional calculus which is permitted through the postulates (A, a–g);

6.5. *States*

If we consider the ensemble of physical systems which are all described through the same SBA P (in the sense of Section 6.1) then a measurement of a proposition $x \in P$ on all single systems determines a probability $0 \leqslant \mu(x) \leqslant 1$ for the existence of $x \in P$.

The function $\mu: P \to [0, 1]$ depends on the 'physical state' of the ensemble. Two ensembles, for which $\mu = \mu'$ holds, cannot be statistically differentiated. We shall consider such ensembles as statistically equivalent and represent their *statistical state* through the function $\mu : P \to [0, 1]$.

For example, in statistical mechanics the state of an ensemble is characterized through definite values of temperature and volume. In quantum mechanics the state is determined by concrete measurements of a maximal class of compatible magnitudes.

The following postulate about the states μ of an ensemble agrees with the implicit definition of the relation R in P:

(C) $\mu \mid B_\nu$ is a probability function on the BA $B_\nu \subseteq P$ for all B_ν. But this means exactly that μ is a W-F on P, i.e. μ satisfies the conditions W 1.2 of Section 4.1.

With this we can say:

A physical system is described through a pair (P, Σ):

P: Proposition set, which possess the structure of a SBA and which, by means of the relation R, is connected with the measurements performed on the system.

Σ: Set of W-F μ on P which represent the states of statistical ensembles.

A presupposition shall finally be made about the set Σ of states:

(D)-(a) For $x \neq 0$, $x \in P$, there exists a $\mu \in \Sigma$ with $\mu(x) = 1$,

(b) If $\mu(y) = 1$ always follows from $\mu(x) = 1$, then $x R y$.

Except in Section 7, 3–4, we only use the weaker postulate.

(D′) For $x \neq y$, $x, y \in P$ there exists a $\mu \in \Sigma$ with $\mu(x) \neq \mu(y)$.

Postulate (D) is likewise in agreement with the implicit definition of P and R.

7. Hidden parameters

7.1. Let us consider the standard example of a theory with 'hidden parameters', viz. statistical mechanics:

There is e.g. a phase space $\Omega = \mathbf{R}^f \times \dot{\mathbf{R}}^f$ ($\dot{\mathbf{R}}^f$ dualspace to $\mathbf{R}^f$) given, and $A = L(\Omega)$ (σ-algebra of L-measurable sets), $P = L(\Omega)/N$, (N ideal of the L-null sets), and the states $\mu \in \Sigma$ are normalized measures on P (comp. [2], Section 5).

If $\pi : A \to P$ is the canonical mapping of A on P, then the W-F $\mu' = \pi^* \mu$, $\mu \in \Sigma$ are measures on the σ-set algebra $L(\Omega)$ or, what comes to the same: $\pi^* \mu$ is representable through prime ideals p_ω, $\omega \in \Omega$ of the form $p_\omega = \{x \in L(\Omega) / \omega \notin x\}$.

Moreover, the p_ω represent exactly the deterministic two-valued states of classical mechanics: $\mu_\omega(x)=\begin{cases}1, \omega\in x\\ 0, \omega\notin x\end{cases}$, which are determined by $\omega\in\Omega$, i.e. by determined values of position and momentum.

The states of statistical mechanics are thus simply average values concerning the two-valued states of classical mechanics.

Since the parameters $\omega\in\Omega$ do not occur anymore in the description by (P, Σ), one designates them as 'hidden parameters' of the system (P, Σ).

If an arbitrary physical system is now described through the pair (P, Σ), then P admits a covering by the boolean-algebras $B_\nu: P=\bigcup_\nu B_\nu$.

The B_ν however, are quotients of the set algebras $L_\nu(\mathbf{R}^{n_\nu})$ (comp. Section 6) and hence what was said above can be repeated word for word for the subsystems $(B_\nu, \Sigma \mid B_\nu)$: 'Hidden parameters' can always be introduced on the boolean parts $(B_\nu, \Sigma \mid B_\nu)$ of (P, Σ).

The main question is now: When can 'hidden parameters' be introduced for the whole system (P, Σ)?

We must first formulate precisely what this means. But according to the above, the generalization is apparent.

DEFINITION. An *introduction* of *'hidden parameters'* in (P, Σ) is given by a set Ω, a set algebra $L(\Omega)\subseteq P(\Omega)$, a set $\tilde{\Sigma}$ of W-F on $L(\Omega)$ and a structure-preserving relation $\rho\subseteq L(\Omega)\times P$ with the conditions:

(1) $\rho_x=\{\tilde{x}\in L(\Omega)/\tilde{x}\,\rho\, x\}$ is non-empty for all $x\in P$,

(2) for $\mu\in\Sigma$ there exists $\tilde{\mu}\in\tilde{\Sigma}$ with:

(a) $\tilde{\mu}(\tilde{x}\triangle\tilde{y})=0$ for $\tilde{x}, \tilde{y}\in\rho_x$, $x\in P$ ($\triangle$: symmetric difference in $L(\Omega)$).
From this it follows that $\tilde{\mu}=$constant on ρ_x for $x\in P$.

(b) $\tilde{\mu}(\tilde{x})=\mu(x)$ for $\tilde{x}\in\rho_x$, $x\in P$. (For this we likewise set: $\mu=\rho^*\tilde{\mu}$).

According to Theorem 4.2 there exists a classical extension (i, B, Σ') of (P, Σ) which corresponds naturally to the presentation given above:

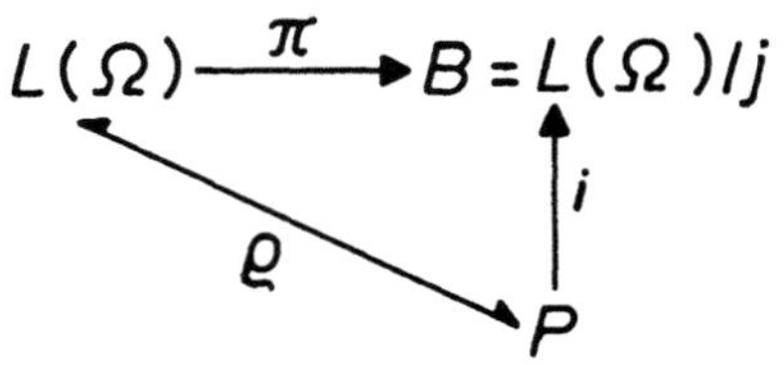

i is an embedding and π a surjection; for $\mu \in \Sigma$ there exists $\tilde{\mu} \in \tilde{\Sigma}$ and $\mu' \in \Sigma'$ with: $\mu = i^* \mu'$, $\tilde{\mu} = \pi^* \mu'$, $\mu = \rho^* \tilde{\mu}$.

Hence an introduction of 'hidden parameters' in (P, Σ) is divided into two steps:

(1) a classical extension (i, B, Σ') to (P, Σ) must be constructed.

(2) Introduction of 'hidden parameters' in (i, B, Σ') analogous to statistical mechanics.

The second step is always realizable; especially every classical system (B, Σ) permits an introduction of 'hidden parameters'. If B is a σ-boolean algebra, then according to a theorem by Loomis, $L(\Omega)$ can be chosen as a σ-set algebra and $\pi: L(\Omega) \to B$ can be chosen as a σ-homomorphism ([1], Chapt. X, p. 167).

Remark. In this formulation an introduction of 'hidden parameters' is not necessarily uniquely determined.

According to Lemma 4.1 and Theorem 4.2, a system (P, Σ) allows an introduction of 'hidden parameters' exactly, when (P, Σ) is representable, i.e. in case $\mu \in \Sigma$ are average values of two-valued W-F on P. The property of representability thus characterizes the systems which permit the introduction of 'hidden parameters'.

7.2. *Systems without 'Hidden Parameters'*

The existence of a classical extension (i, B, Σ') of (P, Σ) is acc. to Section 7.1 necessary and sufficient for the possibility of introducing 'hidden parameters' in (P, Σ). Then by Theorem 5.1 the universal mapping $u: P \to B_P$ is necessarily an embedding of P in B_P. If this is not the case then the system (P, Σ) does not possess 'hidden parameters'.

Since according to Section 5.3 no embedding exists for $P = P(\Lambda, n-1)$, $\Lambda = \mathbf{R}, \mathbf{C}$, $n = 3, \ldots, \omega_0$ and for $P = P_c(\Lambda)\,u$ it follows:

THEOREM 7.1. *If a physical system is described by (P, Σ) and if P contains a sequent $P(0, a)$, $a \in P$, which is isomorphic to one of the* SBA $P(\Lambda, n-1)$, $P_c(\Lambda)$, $\Lambda = \mathbf{R}, \mathbf{C}$ $n = 3, \ldots, \omega_0$, *then no introduction of 'hidden parameters' exists for (P, Σ).*

This is especially the case, if P is a direct product of SBA of the above form.

Remark. Such SBA do occur in a quantum theoretic description of physical systems: E.g. $P=P(\mathbf{C}, \omega_0)$ in quantum mechanics; SBA of type $P_c(\Lambda)$ were considered by von Neumann in [2].

7.3. *Quantum Theoretical Systems*

A physical system described through (P, Σ) is called *quantum theoretic* in case the following conditions are satisfied:

(Q) There exists propositions $\mathring{x}$, $\mathring{y} \in P$ such that the greatest lower and least upper bounds of pairs of propositions $\mathring{x}$, $\mathring{y}$, $N\mathring{x}$, $N\mathring{y}$ are given by the following schema:

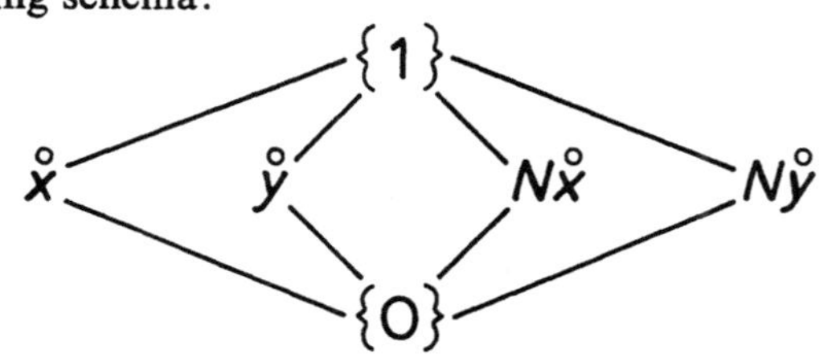

It follows furthermore: $\mu(\mathring{x})=0, 1 \Rightarrow 0<\mu(\mathring{y})<1$ for $\mu \in \Sigma$ and conversely.

Such propositions are characteristic of quantum theory: The propositions of the example given in [2], p. 831, as well as propositions about momentum and position of a particle or about the polarization-direction of an electron satisfy this condition.

For such systems: If $K \subset P \times P$, then $(\mathring{x}, \mathring{y}) \notin K$.

There exist quantum theoretic systems which allow the introduction of 'hidden parameters':

(1) The quantum theoretic description of a 1-dimensional particle by position and momentum.

(2) The quantum theoretic description of the spin of an electron.[2] In this case $P=P(\mathbf{C}, 1)$.

In both examples the embeddability of P rests on the fact that the commensurability relation K has a very simple structure:

> There exists a covering of P through BA $A_\iota \subseteq P$, $\iota \in I$ with $\bigcup_{\iota \in I} A_\iota \times A_\iota = K$, $A_\iota \cap A_\lambda = \{0, 1\}$, $\iota \neq \lambda$, ι, $\lambda \in I$.

7.4. We now consider a quantum theoretical system (P, Σ) which permits the introduction of 'hidden parameters'.

Let $(P, \Sigma) \xrightarrow{i} (B, \Sigma')$ be the naturally related classical extension of (P, Σ) according to Section 7.1.

On the basis of the postulates (D) and (Q) the images $\hat{x}=i(\mathring{x})$, $\hat{y}=i(\mathring{y})$ generate a subalgebra C in B which is isomorphic to the free algebra BA generated by two elements: $C \cong FB(\hat{x}, \hat{y})$. Further, there exists for $z \in C$, $z \neq 0$ a $\mu' \in \Sigma'$ with $\mu'(z) > 0$.

The propositions $\mathring{x}$, $\mathring{y}$ therefore appear in B as *independent* elements.

We therefore can say: As long as no experiments can be specified which allow the independent determination of propositions $\mathring{x}$, $\mathring{y}$ [which satisfy condition (Q) in (P, Σ)], the classical extension (B, Σ') of the quantum theoretic system (P, Σ) cannot be interpreted physically.

The existence of 'hidden parameters' [with whose help one obtains a refined description of the physical system under consideration that agrees with experiment] cannot, in contrast to examples in Section 7.2, be excluded here mathematically (comp. [6], p. 108). But it can be excluded empirically, as long as the quantum theoretic description of a physical system agrees with experiment (comp. [4]).

7.5. *Closing Remarks*

In summary we can therefore say the following about the question of 'hidden parameters':

On the basis of the postulates (A), (C), (D) and the definition in Section 7.1, 'hidden parameters' can exactly then be introduced into physical systems described by (P, Σ), when (P, Σ) is representable.

To an introduction of 'hidden parameters' there corresponds in a natural way a classical extension

$$i: (P, \Sigma) \to (B, \Sigma')$$

of (P, Σ).

P must necessarily then be embeddable in a BA B.

Whence:

(1) 'Hidden parameters' can always be introduced in a classical system (B, Σ).

(2) If P is not embeddable in a BA B, then an introduction of 'hidden parameters' is mathematically excluded. (comp. the examples in Section 7.2).

(3) If the system (P, Σ) allows an introduction of 'hidden parameters' and in addition (P, Σ) satisfies condition (Q) in Section 7.3, then nothing has as yet been said about the physical meaning of these 'hidden parameters'.

It becomes apparent however: As long as the propositions $\mathring{x}$, $\mathring{y} \in P$, which satisfy the condition (Q) in (P, Σ), are not measurable *independently* from each other, the corresponding classical statistical extension (B, Σ') itself is not interpretable through measurements in the sense of Section 6.1.

In conclusion it is to be pointed out again that *no* operations of incommensurable propositions go into the postulates (A), (C), (D).

Institute for Theoretical Physics at the University of Zürich

NOTES

* Communicated by W. Maak at the Meeting of February 21, 1964.

** The University of Western Ontario. I should like to thank Prof. W. Demopoulos and Prof. C. A. Hooker for directing my attention to this paper.

[1] In the category $\mathfrak{B}$ there exist arbitrary direct sums.

[2] The second example which is more interesting, mathematically and physically than the first, I owe by way of oral communication to Mr. Kochen.

BIBLIOGRAPHY

[1] Birkhoff, G.: 'Lattice Theory', *Amer. Math. Soc. Coll. Publ.* **XXV** (rev. Ed.), 1961.

[2] Birkhoff, G. and von Neumann, J.: 'The Logic of Quantum Mechanics', *Ann. of Math.* **37** (1936), 823.

[3] Gleason, A. M.: 'Measures on the Closed Subspaces of a Hilbert-Space', *Journ. of Math. and Mech.* **6** No. 6 (1957).

[4] Jauch, J. M. and Piron, C.: 'Can Hidden Variables be Excluded in Quantum Mechanics?', *Helv. Phys. Acta* **36** (1963), 827.

[5] Kamber, F.: '2-wertige Wahrscheinlichkeitsfunktionen auf orthokomplementären Verbänden', Diss., Universität Zürich (1963), see also *Math. Annalen* **158** (1965), 158.

[6] von Neumann, J.: 'Mathematische Grundlagen der Quantenmechanik', *Grundlehren Math. Wiss.* **XXXVIII**, Springer 1932.

NEAL ZIERLER AND MICHAEL SCHLESSINGER

BOOLEAN EMBEDDINGS OF ORTHOMODULAR SETS AND QUANTUM LOGIC

0. INTRODUCTION

By a "quantum logic" we mean a pair F, P where P is a set and F is a set of functions from P to the closed real unit interval satisfying three postulates which we describe in intuitive terms here. Cf. [2], [4], [7]. P may be interpreted as the set of events and F the set of states of a "physical system", and $f(x)$ then becomes the probability of occurrence of the event x in the state f. Since the outcome of an experiment is an estimate for some $f(x)$, or a collection of such estimates, it is natural to identify events which cannot be distinguished by experiment. Thus, we assume first:

E1. Two events which have the same probability of occurrence in all states are the same event.

Second, the existence of the event "not x" for each event x is postulated:

E2. If x is an event, then so is its negation x': for all f in F, $f(x')=1-f(x)$.

A (weak) implication relation "$\leqslant$" is introduced in P by defining $x\leqslant y$ to mean that $f(x)\leqslant f(y)$ for all f. The "mutual exclusiveness" of pairs of events may be formalized as follows: x and y are mutually exclusive if $x\leqslant y'$. This is symmetric, for if $x\leqslant y'$, then $f(x)\leqslant f(y')=1-f(y)$ or $f(y)\leqslant 1-f(x)=f(x')$ for all f, so $y\leqslant x'$ too. Note that if $x\leqslant y'$, then in any state in which x is certain to occur, y is certain not to occur, and conversely. The notion of simultaneous measurability (or "commutativity") will be formalized in the next section, but it seems clear that mutually exclusive events should be simultaneously measurable. Thus, our third and final postulate is:

E3. For a finite number $x_1,\ldots, x_n$ of pairwise mutually exclusive events, "x_1 or x_2 or ... or x_n" is an event denoted $\bigvee x_i$ such that $f(\bigvee x_i)=\sum f(x_i)$.

If $n=2$ and $x_2=x_1'$, then $f(\bigvee x_i)=\sum f(x_i)=f(x_1)+f(x_2)=f(x_1)+1-f(x_1)=1$ for all f. This event, characterized in accordance with (E1)

C. A. Hooker (ed.), The Logico-Algebraic Approach to Quantum Mechanics, 247–262.

by the fact that it has probability one of occurring in all states, is given the name "1". Then $1'$ is called "0", and we observe that for all x, $0 \leqslant x$ and $x \leqslant 1$.

The relation "$\leqslant$" is a partial ordering for P and, in this ordering, 0 and 1 are the least and greatest elements of P. Further, the negation mapping $x \to x'$ is an orthocomplementation in P (for details, see Section I), and such a partially ordered set is said to be an "orthocomplemented set". If the greatest lower bound or least upper bound of two elements x and y exists in P for the partial ordering defined above, it is denoted xy or $x \vee y$ respectively. P is said to be "weakly modular" or an "orthomodular set" if it satisfies

(W). Whenever $x \leqslant y$, then yx' exists and $y = yx' \vee x$.

It is shown in [7] that (W) holds automatically for the orthocomplemented set P of a quantum logic F, P.

On the other hand, if P is merely an orthocomplemented set (that is, a partially ordered set with a least element 0, a greatest element 1 and an orthocomplementation $x \to x'$), a "state" for P is a function f from P to the closed real unit interval such that $f(0) = 0$, $f(1) = 1$ and $f(\bigvee x_i) = \sum f(x_i)$ whenever $x_1, \ldots, x_n$ are a finite number of elements of P such that $x_i \leqslant x_j'$ for $1 \leqslant i < j \leqslant n$ for which the least upper bound $\bigvee x_i$ exists in P. A set F of states for P is said to be "full" if it determines the order in P; i.e., $x \leqslant y$ if and only if $f(x) \leqslant f(y)$ for all f. It is shown in [7] that if F is a full set of states for an orthomodular set P, then the pair F, P is a quantum logic. Thus, the set of quantum logics coincides with the set of ordered pairs F, P where P is an orthomodular set and F is a full set of states for P.

"Simultaneous measurability" may be defined for pairs of elements in an arbitrary orthomodular set P. It is shown in Section I that if x and y are simultaneously measurable, then $x \vee y$ and xy belong to P, and that in the center C of P (which consists of all those elements of P which are simultaneously measurable with every element of P) the distributive law $x(y \vee z) = xy \vee xz$ holds; i.e., C is a Boolean algebra. Thus, if P is deterministic (every pair of elements is simultaneously measurable), then $P = C$, and C is a Boolean algebra. Conversely, if P is a Boolean algebra, then (see [7] and below) every element is simultaneously measurable with every other, and P is deterministic. Accordingly, a quantum logic F, P

in which P is a Boolean algebra is said to be "deterministic" or "classical". Indeed, such a P may be realized as the field of Borel measurable subsets of a phase space, and then F becomes a family of probability measures for this space.

If the quantum logic F, P is not deterministic, it seems natural to ask if it is possible to adjoin new events to P and define the states of F on them in such a way that the enlarged pair is a deterministic logic. (Because of the dualism between events and observables (see [4], [8]), this may be restated as: Do there exist additional observables which eliminate the indeterminacy from a given system?) This is the same as asking if there exists a deterministic logic G, Q and a mapping α of P in Q satisfying

(I1) if $x \leqslant y$ then $\alpha(x) \leqslant \alpha(y)$

(I2) $\alpha(x') = \alpha(x)'$

(I3) if $\alpha(x) \leqslant \alpha(y)$, then $x \leqslant y$

such that for every f in F there exists g in G with $f = g \circ \alpha$.

In Section II, an affirmative answer is obtained for the question: "Can every orthomodular set P be embedded in a Boolean algebra (does there exist a Boolean algebra B and a mapping α of P in B satisfying (I1)–(I3))?" by constructing the minimal Boolean extension of P. This is a pair φ, A where A is a Boolean algebra and $\varphi : P \to A$ satisfies (I1)–(I3) such that each $a \in A$ is of the form $\bigvee \bigwedge \varphi(x_{ij})$, $x_{ij} \in P$, and such that whenever f is a homomorphism of P (i.e., a mapping satisfying (I1) and (I2)) in a Boolean algebra B, then there exists a unique homomorphism h of A in B such that $f = h \circ \varphi$. This construction turns out to be a useful tool for answering the original question, and it is used in Section IV to show that no embedding in a deterministic logic exists for a large class of quantum logics which includes the logic of non-relativistic quantum mechanics. The last conclusion was reached by von Neumann [5; IV, 2], but for a much stronger definition of embedding than the one we use here.

Section III is devoted chiefly to a study of the extent to which an embedding α of P in a Boolean algebra B can preserve the lattice operations. (3) Because of (I2), only the preservation of joins need be considered, for if $\alpha(x \vee y) = \alpha(x) \vee \alpha(y)$, then $\alpha(x'y') = \alpha((x \vee y)') = \alpha(x \vee y)' = (\alpha(x) \vee \alpha(y))' =$

$\alpha(x')\,\alpha(y')$, and conversely. The most innocent requirement is:

(I4) if x and y are in the center of P, then $\alpha(x \vee y) = \alpha(x) \vee \alpha(y)$.

With obvious modifications, the construction works just as well with (I4) as without. Next we consider the following two equivalent conditions.

(I5) if x and y commute, then $\alpha(x \vee y) = \alpha(x) \vee \alpha(y)$,

(I5′) if x and y are mutually exclusive, then $\alpha(x \vee y) = \alpha(x) \vee \alpha(y)$.

It is shown by a class of examples which includes non-relativistic quantum mechanics that, in general, no embedding exists which satisfies (I5). A stronger conclusion is reached for

(I6) if $x \vee y$ exists in P, then $\alpha(x \vee y) = \alpha(x) \vee \alpha(y)$.

Theorem 3.1 asserts that if there exists an embedding α of P in a Boolean algebra that satisfies (I6), then $x \vee y$ exists in P only when x commutes with y.

I. PRELIMINARIES

Let P be a partially ordered set with least and greatest elements 0, 1 respectively. If the greatest lower bound or least upper bound of two elements x and y exists in P, we denote it by xy or $x \vee y$, respectively.

DEFINITION 1.1. An *orthocomplementation* in P is a mapping $x \to x'$ of P into P such that

(i) $(x')' = x$

(ii) $x < y$ if and only if $y' < x'$

(iii) $x' \vee x$ and $x'x$ exist and equal 1 and 0 respectively.

A partially ordered set with orthocomplementation is called an *orthocomplemented set*. Two elements x, y of P are called *orthogonal* (written $x \perp y$) if $x \leqslant y'$. It is clear that De Morgan's Law holds in P in the sense that if either $x \vee y$ or $x'y'$ exists, then so does the other, and $(x \vee y)' = x'y'$.

DEFINITION 1.2. An orthocomplemented set P is called *weakly modular*, or an *orthomodular set*, if for any $x, y \in P$ such that $x \leqslant y$, we have

(i) $x'y$ exists,

(ii) $y = x'y \vee x$.

Note that by (i) the join of two orthogonal elements exists, so the right-hand side of (ii) exists a priori.

DEFINITION 1.3. Let P be an orthocomplemented set. For $x, y \in P$, we say x *commutes with* y if xy and xy' exist and x is their least upper bound:

$$x = xy \vee xy'.$$

If x commutes with y, then, clearly, x commutes with y' too.

Throughout the remainder of this note we assume that P (and sometimes Q) is an orthomodular set and refer to its members as $p, q, \ldots, z$. Occasional results apply to a general orthocomplemented set, and these are noted by referring to P as an orthocomplemented set.

The following two lemmas may be found in [7, Section I].

LEMMA 1.1. *If* $p \leqslant r$ *and* $q \perp r$, *then* $(p \vee q)\, r = p$.

LEMMA 1.2. *If* x *commutes with* y, *then* y *commutes with* x.

COROLLARY. *If* x *commutes with* y, *then* $x \vee y$, *as well as* xy, *exists in* P.

By virtue of Lemma 1.2, we may now say "x and y commute" instead of "x commutes with y" or the like. We now prove

LEMMA 1.3. *Suppose that* x *commutes with* y *and with* z, *and that* $y \vee z$, $xy \vee xz$, *and* $x'y \vee x'z$ *exist. Then*

$$x(y \vee z) = xy \vee xz.$$

Proof. Apply Lemma 1.1 with $p = xy \vee xz$, $q = x'y \vee x'z$, $r = x$. Now $p \vee q = (xy \vee x'y) \vee (xz \vee x'z) = y \vee z$ since x commutes with y and z, so that $(p \vee q)\, r = (y \vee z)\, x = p = xy \vee xz$.

COROLLARY. x *commutes with* $y \vee z$.

Proof. Since x' also commutes with y and z, $x'(y \vee z) = x'y \vee x'z$ by Lemma 1.3. Then $x(y \vee z) \vee x'(y \vee z) = xy \vee xz \vee x'y \vee x'z = xy \vee x'y \vee xz \vee x'z = y \vee z$.

Let C be the set of all $p \in P$ such that p commutes with q for all $q \in P$. C is called the *center* of P.

PROPOSITION 1.1. *C is a Boolean algebra.*

Proof. Suppose x and y belong to C. Then $x \vee y$ belongs to C by the corollary to Lemma 1.3. Since C is clearly closed to orthocomplementation, $xy = (x' \vee y')'$ belongs to C too. Clearly 0 and 1 belong to C, and C is a complemented lattice. Since the distributive law holds in C by Lemma 1.3, it is a Boolean algebra.

If $C = P$, we call P *deterministic*. Clearly P is deterministic if and only if it is a Boolean algebra. It is not difficult to show that if P is a lattice, it is deterministic if and only if $xy = 0$ implies $x \perp y$. (See [7; Section I]). By virtue of Lemma 1.3 we see that the center of a lattice may be characterized as the set of all x such that $x(y \vee z) = xy \vee xz$ for all y and z (Cf. [1]).

DEFINITION 1.4. By a *homomorphism* $f: P \to Q$ of orthocomplemented sets we mean a function such that

(i) f is isotone: $x \leqslant y$ implies $f(x) \leqslant f(y)$

(ii) $f(x') = f(x)'$

(iii) If $x, y \in C$ and $f(x) \vee f(y)$ exists in Q, then $f(x \vee y) = f(x) \vee f(y)$.

f is called an *embedding* if in addition

(iv) $f(x) \leqslant f(y)$ implies $x \leqslant y$.

Note that if f is a homomorphism, then $f(0) = 0$ and $f(1) = 1$, because $0 \leqslant 1$ implies $f(0) \leqslant f(1) = f(0') = f(0)'$. Thus $f(0) \leqslant f(0)'$ so that $f(0) = 0$, and $f(1) = f(0)' = 1$.

DEFINITION 1.5. An *ideal* in P is a non-empty subset I of P such that

(i) if $x \in I$ and $y \leqslant x$, then $y \in I$

(ii) if $x, y \in C \cap I$, then $x \vee y \in I$.

An ideal I is said to be *proper* if $x \in I$ implies $x' \notin I$. The kernel of a homomorphism is clearly a proper ideal. An ideal I is *maximal* if it is proper and if it is contained in no other proper ideal. The union of a chain of proper ideals is clearly a proper ideal, so, by Zorn's lemma, every proper ideal is contained in a maximal ideal.

We adopt the following notation (where I, J are subsets of P):

$$I_c = I \cap C$$
$$\langle I \rangle = \{x \in P \mid \exists y \in I \quad \text{with} \quad x \leqslant y\}$$
$$I \vee J = \{x \in P \mid x = y \vee z,\ y \in I,\ z \in J\}$$

(I, J) = ideal Generated by I, J
$= \bigcap K$, where the intersection runs over all ideals K of P such that $I \subset K$ and $J \subset K$.

$$I' = \{x \in P \mid x' \in I\}$$

(x) = the principle ideal generated by x; $(x) = \{y \in P \mid y \leqslant x\}$.
$(x_1 \ldots, x_n)$ = the ideal generated by $x_1 \ldots x_n$.

LEMMA 1.4. *Let I and J be ideals in P. Then $(I, J) = I \cup J \cup \langle I_c \vee J_c \rangle$.*

Proof. Let $K = I \cup J \cup \langle I_c \vee J_c \rangle$. Since $I \subset K$, $J \subset K$ and $K \subset (I, J)$ it suffices to show that K is an ideal. Clearly property (i) of Definition 1.5 is satisfied: If $y \in K$ and $x \leqslant y$, then $x \in K$. Suppose that x, $y \in K_c$; say $x \leqslant p \vee q$, $y \leqslant r \vee s$, with p, $r \in I_c$ and q, $s \in J_c$. Then $x \vee y \leqslant p \vee q \vee r \vee s = p \vee r \vee q \vee s \in I_c \vee I_c \vee J_c = I_c \vee J_c \subset K$, and it follows that property (ii) of Definition 1.5 is satisfied.

PROPOSITION 1.2. *Let I and J be proper ideals in P. Then (I, J) is proper if and only if $I' \cap J = \phi$.*

Proof. Suppose (I, J) is improper. Then there exists $x \in P$ such that $x \in (I, J)$ and $x' \in (I, J)$.

Case 1. $x \in \langle I_c \vee J_c \rangle$. Then x and x' belong to $(I, J)_c$ so $1 = x \vee x'$ does too. Hence there exist $y_1 \in I_c$ and $y_2 \in J_c$ such that $y_1 \vee y_2 = 1$. By the distributive law in C we get $y_2' = y_2' y_1$ or $y_2' \leqslant y_1$. Thus $y_2 \in I' \cap J \neq \phi$.

Case 2. $x \in I$, $x' \in J$. Then $x' \in I' \cap J \neq \phi$. Thus we have shown that if (I, J) is improper, then $I' \cap J \neq 0$. The converse is obvious.

COROLLARY. *A proper ideal I is maximal if and only if $x \notin I$ implies $x' \in I$.*

Proof. Suppose I is a maximal and $x \notin I$. If $x = 1$, then $0 = x' \in I$, so assume $x \neq 1$. Then $J = (x)$ is a proper ideal properly contained in the ideal (I, J) which must therefore be improper by the maximality of I. Hence $I' \cap J \neq \phi$ by Proposition 1.2, and it follows that $x \in I'$.

Conversely suppose $x \notin I$ implies $x' \in I$. Let K be an ideal containing I. If $K \neq I$, there exists $x \in K$ with $x \notin I$. Then $x' \in I \subset K$, so that K is improper.

Let S denote the set of all homomorphisms of P in 2 (the two-element Boolean algebra) and let M denote the set of all maximal ideals. A one-

one mapping of S on M is clearly given by assigning its kernel to each homomorphism.

In the second example of Section III it is shown that there is in general no embedding f of an orthocomplemented set P in a Boolean algebra which satisfies

(iii)* $f(x \vee y) = f(x) \vee f(y)$ whenever x commutes with y.

Consider also the apparently weaker

(iii)** $f(x \vee y) = f(x) \vee f(y)$ whenever $x \perp y$.

LEMMA 1.5. *Let f be a homomorphism satisfying* (iii)** *of P in an orthomodular set Q. Then f also satisfies* (iii)*.

Proof. Suppose x commutes with y. Then $x = xy' \vee xy$ and $y = xy \vee x'y$ so $f(x \vee y) = f(xy' \vee xy \vee xy \vee x'y) = f(xy' \vee xy \vee x'y)$

$= f(xy') \vee f xy) \vee f(x'y)$ by two applications of (iii)**

$= f(xy') \vee f(xy) \vee f(xy) \vee f(x'y)$

$= f(xy' \vee xy) \vee f(xy \vee x'y)$ by two applications of (iii)**

$= f(x) \vee f(y)$.

II. THE EMBEDDING

Let P be an orthocomplemented set. By a *minimal Boolean extension* of P we mean a Boolean algebra A together with a homomorphism $\varphi: P \to A$, such that for any homomorphism $g: P \to B$ of P into a Boolean algebra B, there exists a unique homomorphism $h: A \to B$ such that $h \circ \varphi = g$.

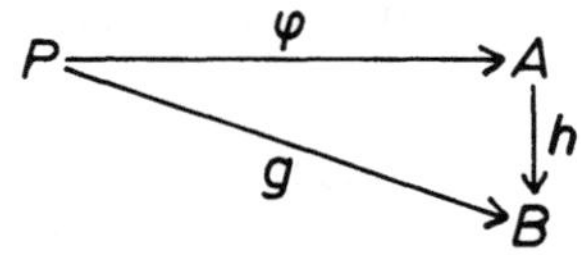

If a minimal Boolean extension exists, it is evidently unique up to a cononical isomorphism; i.e., given two minimal Boolean extensions (φ_1, A_1), (φ_2, A_2), there exist unique homomorphisms $h_1: A_1 \to A_2$ and $h_2: A_2 \to A_1$ such that $h_1 \circ \varphi_1 = \varphi_2$, $h_2 \circ \varphi_2 = \varphi_1$ and $h_1 \circ h_2$ and $h_2 \circ h_1$ are the identity mappings. If P is a Boolean algebra, we may evidently take $P = A$, $\varphi =$ identity.

THEOREM 2.1. *Let P be an orthomodular set. Then P has a minimal*

Boolean extension (φ, A). *Moreover,* φ *is an embedding, and its image generates the Boolean algebra* A.

Our construction is analagous to that of M. H. Stone [6] for representing Boolean algebras as fields of sets (and degenerates into that representation if P is a Boolean algebra).

Let S be the set of all homomorphisms from P to 2. In the induced topology of S (as a subset of 2^P) let A be the Boolean algebra of closed and open subsets. For $x \in P$ define

$$\varphi(x) = \{\alpha \in S \mid \alpha(x) = 1\}.$$

Then $\varphi(x) \in A$, and the sets $\varphi(x)$ separate the points of S and thus generate A as a Boolean algebra. We omit proofs of these facts, which are identical with those given in [6]. Clearly φ is a homomorphism.

LEMMA 2.1. φ *is an embedding.*

Proof. Suppose $x \leqslant y$ is false. We must show that there is an $\alpha \in S$ such that $\alpha(y) < \alpha(x)$ so that $\varphi(x) \leqslant \varphi(y)$ is false. Now the ideal (x', y) is proper by Proposition 1.2, so there is a maximal ideal containing (x', y), and thus an $\alpha \in S$ such that $\alpha(x') = \alpha(y) = 0$. Then $0 = \alpha(y) < 1 = \alpha(x)$.

LEMMA 2.2. *Universal Mapping Property. Let* $g: P \to B$ *be a homomorphism of* P *into a Boolean algebra* B. *Then there exists a unique homomorphism* $h: A \to B$ *such that* $g = h \circ \varphi$.

Proof. Let

$$a = \bigvee_i \bigwedge_j \varphi(x_{ij}) \in A.$$

Define

$$h(a) = \bigvee_i \bigwedge_j g(x_{ij}) \in B.$$

If also

$$a = \bigvee_p \bigwedge_q \varphi(y_{pq})$$

then

$$\begin{aligned} aa' = 0 &= \Big(\bigvee_i \bigwedge_j \varphi(x_{ij})\Big) \wedge \Big(\bigwedge_p \bigvee_q \varphi(y'_{pq})\Big) \\ &= \bigvee_{i,p} \bigwedge_{j,q} \varphi(x_{ij})\, \varphi(y'_{pq}). \end{aligned}$$

Hence h will be well defined if we can show that $\bigwedge_j \varphi(x_j)=0$ implies $\bigwedge_j g(x_j)=0$.

Suppose then that $b=\bigwedge_j g(x_j)\neq 0$; then (b') is a proper ideal in B, so that there is a maximal ideal containing it, and thus a homomorphism $\beta:B\to 2$ such that $\beta(b')=0$. Since $g(x_j)\geqslant b$, we have $\beta\circ g(x_j)=1$. Let $\alpha=\beta\circ g\in S$. Then $\alpha(x_j)=1$ for all j so that $\alpha\in\bigwedge_j \varphi(x_j)$ and $\bigwedge_j \varphi(x_j)\neq 0$. Thus h is well defined, and it is clearly a homomorphism. Uniqueness follows from the fact that A is generated by the image of φ. This completes the proof of Lemma 2.2 and Theorem 2.1.

Remark 1. If P is a Boolean algebra, then (φ, A) as constructed above is the Stone representation of P as a field of sets (see [6]).

Remark 2. If we drop the requirement that P be weakly modular, and define a "homomorphism" as an isotone map commuting with orthocomplementation, then the theorem remains true in the following sense: there exists an embedding $\varphi:P\to A$ of P in a Boolean algebra A (with $\varphi(P)$ generating A) such that for any "homomorphism" $g:P\to B$ of P into a Boolean algebra B, there exists unique Boolean algebra homomorphism $h:A\to B$ such that $g=\varphi\circ h$. However, Remark 1 becomes false, as may be seen by taking $P=2^3$ (the Boolean algebra with 8 elements) in which case $A=2^4$.

III. EXAMPLES; ADDITIVITY

Example 3.1. This example shows that in the diagram

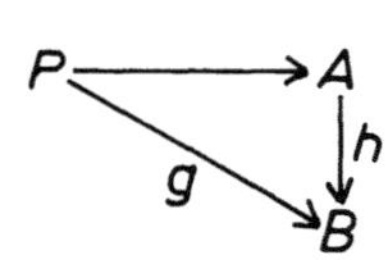

h need not be an embedding, even when g is. Let P be the lattice of closed subspaces of a real or complex Hilbert space of dimension $\geqslant 2$. Here the center is trivial, i.e. consists of 0 and 1. (In fact, if $x\in P$, $x\neq 0, 1$, then there exists y such that $xy=0$ and $x \not\leftrightarrow y$. Then $x\neq xy\vee xy'$, and x does not commute with y.) Take three lines x_1, x_2, $x_3\in P$, mutually non-orthogonal. Then $I=(x_1', x_2', x_3')$ is a proper ideal (for in this case $I=(x_1')\cup(x_2')\,(x_3')$; if there were a $c\in P$ such that, c, $c'\in I$, say $c\leqslant x_1'$, $c'\leqslant x_2'$, we would have $x_2\leqslant c\leqslant x_1'$, i.e. $x_1\perp x_2$). Thus $a=\varphi(x_1)\,\varphi(x_2)\,\varphi(x_3)\neq 0$.

Let $B = A/(a)$, h the canonical projection from A to B and $g = h \circ \varphi : P \to B$. We claim that g is an embedding. Since h clearly is not, this will prove the statement at the beginning of the paragraph.

Suppose for some $p, q \in P$, we have $g(p) \leqslant g(q')$. Then $g(p)\, g(q) = 0$, so $\varphi(p)\, \varphi(q) \leqslant a$. We must show that $p \leqslant q'$. We may assume that neither p nor q is 0. Then we must have, for some i, $x_i \ngeqslant p$ and $x_i \ngeqslant q$. For if not, then we would have, say $p \leqslant x_1$ and $p \leqslant x_3$ so that $p \leqslant x_1 \wedge x_3 = 0$, and $p = 0$. Suppose then that $p \leqslant x_1$ and $q \nleqslant x_1$. Then if $p \nleqslant q'$; the ideal (q', q', x_1) is proper by the argument used above, so there exists $\alpha \in S$ for which $\alpha(p') = \alpha(q') = \alpha(x_1) = 0$. Thus $\alpha \in \varphi(p)\, \varphi(q)$, while $\alpha \notin \varphi(x_1)\, \varphi(x_2)\, \varphi(x_3) = a$ so that $\varphi(p)\, \varphi(q) \leqslant a$ is false, unless $p \leqslant q'$, and the proof is complete.

Remark. The above procedure will provide a counter-example to the statement (*) g is an embedding implies h is an embedding in any orthocomplemented set with the following properties:

(1) An ideal (y_1, y_2, y_3) is proper whenever each (y_i, y_j) is proper.

(2) There exists $x_1, x_2, x_3 \in P$ such that (x_1', x_2', x_3') is proper and $x_i x_j = 0$ if $i \neq j$.

The statement (*) is equivalent to this:

(**) for any $a \in A$, $a \neq 0$, there exists $x, y \in P$ such that $0 < \varphi(x)\, \varphi(y) \leqslant a$.

It is conjectured that (*) holds only when P is a Boolean algebra (The converse is obvious).

The above example shows that there does not in general exist a strictly minimal extension of P; i.e. an embedding $\varphi_1 : P \to A_1$ where A_1 is a Boolean algebra, such that for any embedding $g : P \to B$ of P into a Boolean algebra B, there exists a unique embedding $h : A \to B$ such that $h \circ \varphi_1 = g$. Indeed it is easily seen that (φ_1, A_1) is isomorphic to (φ, A), while it is known that (φ, A) does not in general enjoy this strictly minimal property.

DEFINITION 3.1. A function f from the orthomodular set P to the closed real unit interval is said to be a *state* for P if $f(0) = 0$, $f(1) = 1$ and $f(\bigvee a_i) = \sum f(a_i)$ whenever n is a positive integer and $a_1, \cdots, a_n$ are pair-wise orthogonal.

Example 3.2. This example shows that if the definition of homomorphism is strengthened in a certain natural way, then it becomes impossible to embed P in a Boolean algebra. (This is equivalent to showing that when the new minimal extension (φ, A) is constructed, φ is not an embedding). Suppose that in the definition of homomorphism (Definition 1.4) we replace (iii) by (orthogonal) additivity

(iii)* If x and y commute, then $f(x \vee y) = f(x) \vee f(y)$, or, as noted in Lemma 1.5, the equivalent

(iii)** If x and y are orthogonal, then $f(x \vee y) = f(x) \vee f(y)$.

Let f be an additive homomorphism of P in a Boolean algebra B. B has plenty of homomorphisms onto 2 by the Stone representation (Theorem 2.1), and if g is any one of these, $g \circ f$ is a two-valued state. Our example consists of a class of orthomodular sets without two-valued states. Let P be the lattice of closed subspaces of a separable real or complex Hilbert space H of dimension $d \geqslant 3$. Gleason [3] has shown that for each state s for P there exists a positive semi-definite self-adjoint operator T of the trace class such that for each a in P, $s(a) = \text{trace}\,(TP_a)$ where P_a is the orthogonal projection of H onto a. It follows readily that every state takes on at least $d+1$ values. It should also be mentioned that, in fact, the non-existence of two-valued states is an elementary geometric fact contained quite explicitly in [3, Paragraph 2.8].

The preceding example shows that there is in general no embedding of P in a Boolean algebra that has the additivity property (iii)*. If it is replaced by strong additivity:

(iii)*** $\theta(x \vee y) = \theta(x) \vee \theta(y)$ whenever $x \vee y$ exists, the conclusion can be strengthened as follows.

THEOREM 3.1. *Suppose there exists a strongly additive embedding θ of P in a Boolean algebra B. Then $x \vee y$ exists in P only when x commutes with y.*

Proof. Let x and y be elements of P for which $x \vee y$ exists in P. Then $0 = \theta(x \vee y)\,(\theta(x \vee y))' = \theta(x \vee y)\,(\theta(x) \vee \theta(y))' = \theta(x \vee y)\,\theta(x)'\,\theta(y)' = \theta(x \vee y)\,\theta(x')\,\theta(y)'$. Hence, for such x, y, $\theta(x \vee y)\,\theta(x') \leqslant \theta(y)$. Now $(x \vee y)\,x'$ exists by weak modularity, and θ also preserves joins. Thus, $\theta((x \vee y)\,x') \leqslant \theta(y)$ and so $(x \vee y)\,x' \leqslant y$. Suppose $z \leqslant y$ and $z \leqslant x'$. Then $z \leqslant x \vee y$ so $z \leqslant (x \vee y)\,x'$ and it follows that $(x \vee y)\,x'$ is the greatest lower bound $x'y$ of x' and y. Now suppose $w \geqslant x'y$ and $w \geqslant x'y'$. Then $w \geqslant (x \vee y)\,x'$ and $w \geqslant x'y' = (x \vee y)' \geqslant (x \vee y)'\,x'$ (which exists since x' commutes with $x \vee y$). Hence $w \geqslant (x \vee y)\,x' \vee (x \vee y)'\,x' = x'$, and it follows that x' is the least upper bound of $x'y$ and $x'y'$. Therefore x', and hence x, commutes with y as was to be proved.

COROLLARY. *Suppose there exists a strongly additive embedding of the orthomodular lattice P in a Boolean algebra. Then P is a Boolean algebra (and conversely).*

Proof. According to the theorem, P now coincides with its center, and so is a Boolean algebra by Proposition 1.1.

IV. EXTENSIONS OF STATES

A state g for Q is an *extension to* Q by α of a state f for P if α is a homomorphism of P in Q and $f = g \circ \alpha$. A set T of states for Q is *an extension to* Q by α of a set S of states for P if every member of S has an extension to Q by α in T.

LEMMA 4.1. *If a state can be extended from P to a Boolean algebra D, then it can be extended to A by φ.*

Proof. Suppose g is an extension to D by α of a state for P, and let h be the homomorphism of Lemma 2.2 of A in D such that $\alpha = h \circ \varphi$. Evidently $s = g \circ h$ is a state for A, and is an extension to A by φ of the state $g \circ \alpha$ for $P: g \circ \alpha = g \circ (h \circ \varphi) = (g \circ h) \circ \varphi = s \circ \varphi$.

Let I denote the ideal in A generated by all elements of the form $\varphi(a \vee b)\, \varphi(a')\, \varphi(b')$ for $a \perp b$. Let $B = A/I$, let β denote the canonical homomorphism of A on B and let $\alpha = \beta \circ \varphi$.

LEMMA 4.2. *Suppose $I \neq A$. Then α is an additive homomorphism of P in the Boolean algebra B.*

Proof. It is clearly a homomorphism. To prove additivity, suppose $a \perp b$ and observe that $\varphi(a \vee b) = \varphi(a \vee b)\, \varphi(a')\, \varphi(b') \vee \varphi(a) \vee \varphi(b)$ since $\varphi(a) \vee \varphi(b) \leqslant \varphi(a \vee b)$. Then $\beta \circ \varphi(a \vee b) = \beta(\varphi(a) \vee \varphi(b)) = \beta \circ \varphi(a) \vee \beta \circ \varphi(b)$, the latter equality coming from (iii) of Definition 1.5, since A coincides with its center.

LEMMA 4.3. *Let g be an extension to A by φ of a state for P. Then g vanishes on I.*

Proof. If $a \perp b$, then $g \circ \varphi(a \vee b) = g \circ \varphi(a) \vee g \circ \varphi(b) = g(\varphi(a) \vee \varphi(b))$, the first equality because $g \circ \varphi$ is a state for P and the second because g is a state for A. Then $g(\varphi(a \vee b)\, \varphi(a')\, \varphi(b')) = 0$ and $g \equiv 0$ on I.

THEOREM 4.1. *Suppose some state for P has an extension to a Boolean algebra. Then P has a 2-valued state.*

Proof. $I \neq A$ by Lemma 4.3, so B is a Boolean algebra. If g is any state for B, $g \circ \alpha$ is a state for P since α is additive by Lemma 4.2. The result now follows from the fact that B has plenty of two-valued states (=homomorphisms in 2).

COROLLARY 4.1. *Let P be the set of events of nonrelativistic quantum mechanics, i.e., the lattice of closed subspaces of complex separable Hilbert space. Then no state for P has an extension to a Boolean algebra.*

Proof. P has no two-valued states (see Example 3.2).

LEMMA 4.4. *If a state for P has an extension to a Boolean algebra, then it can be extended to B by α.*

Proof. By Lemma 4.1 a state s for P with an extension to a Boolean algebra has an extension g to A by φ. Now if a and b are elements of A for which $\beta(a)=\beta(b)$, then $a=b\vee c$ with $c\in I$ and $c\perp b$. Hence $g(a)=g(b)+g(c)=g(b)$ since g is a state and $g(c)=0$ by Lemma 4.3. Hence $f(\beta(a))=g(a)$ defines a function f on B, and $f\circ\alpha=f\circ\beta\circ\varphi=g\circ\varphi=s$.

Let S and T be sets of states for the orthomodular sets P and Q respectively. An *extension* of S, P in T, Q is a homomorphism of P in Q extending S to T and is an *embedding* of S, P in T, Q if it embeds P in Q. S is *full* for P if $s(a)\leqslant s(b)$ for all s in S implies $a\leqslant b$. The pair S, P is said to be a *quantum logic* if S is full for P and *deterministic* if P is a Boolean algebra.

LEMMA 4.5. *Suppose a quantum S, P is extended by δ to T, Q. Then δ is an embedding of P in Q.*

Proof. If $x\leqslant y$ is false, there exists s in S with $s(y)<s(x)$ since S is full for P. Then $S=f\circ\delta$, f a state for Q, and so $\delta(x)\leqslant\delta(y)$ would imply $s(x)=f(\delta(x))\leqslant f(\delta(y))=s(y)$ contrary to our choice of s.

THEOREM 4.2. *Suppose a quantum logic S, P has an extension in a deterministic logic. Then the additive homomorphism α of P in B is an embedding and extends S to a set of states for B.*

Proof. S, P may be extended to B by α in accordance with Lemma 4.4, so α is an embedding by Lemma 4.5.

COROLLARY 4.2. *Under the hypothesis, P has a full set of two-valued states.*

Proof. If $x \leqslant y$ is false, so is $\alpha(x) \leqslant \alpha(y)$, since α is an embedding. Since B is a Boolean algebra, it has a two-valued state f such that $f(\alpha(x)) \leqslant f(\alpha(y))$ is false, and $f \circ \alpha$ is a state for P by the additivity of α.

Remark 1. The preceding corollary shows that if an orthomodular set P may be additively embedded in a Boolean algebra, it has a full set of two-valued states. The converse is also true, for suppose the set S of all two-valued states for P is full. Then the Stone construction ($\delta(x) = \{s \in S \mid s(x) = 1\}$) provides an additive embedding of P in the Boolean algebra of all closed and open subsets of S.

Remark 2. Let δ be an additive homomorphism of P in a Boolean algebra D and let h be the homomorphism of A in D such that $\delta = h \circ \varphi$. Then if $a \perp b$, $h \circ \varphi(a) \vee h \circ \varphi(b) = h \circ \varphi(a \vee b) = h(\varphi(a \vee b)\ \varphi(a')\ \varphi(b') \vee h \circ \varphi(a) \vee h \circ \varphi(b)$; (since h is additive as a homomorphism of Boolean algebras) so h vanishes on I. Let η denote the resulting canonical homomorphism of B on D; then $\eta \circ \beta = h$ and $\delta = \eta \circ \beta \circ \varphi = \eta \circ \alpha$. In summary: P has additive Boolean extension if and only if $I \neq A$; if $I \neq A$, α, B is the minimal additive Boolean extension of P.

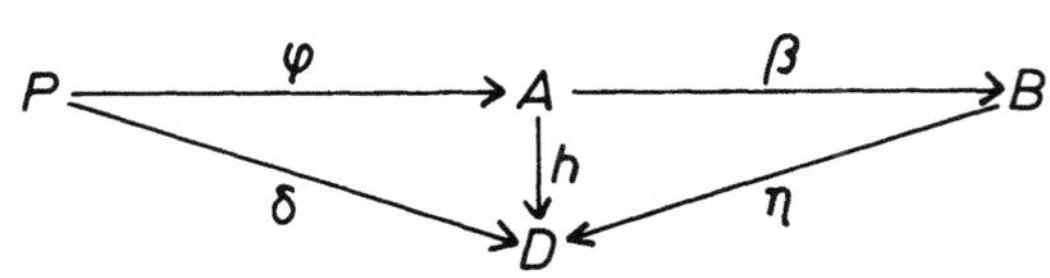

The MITRE Corporation
Bedford, Massachusetts
and
The Institute for Defense Analyses
Princeton, New Jersey

BIBLIOGRAPHY

[1] Birkhoff, G., 'Neutral Elements in General Lattices', *Bulletin of the American Mathematical Society* **46** (1940), 702–705.

[2] Birkhoff, G. and Neumann, J. von, 'The Logic of Quantum Mechanics', *Annals of Mathematics* **37** (1936), pp. 823–843.

[3] Gleason, A. M., 'Measures on the Closed Subspaces of a Hilbert Space', *Journal of Rational Mechanics and Analysis* **6** (1957), pp. 885–894.

[4] Mackey, G. W., 'Quantum Mechanics and Hilbert Space', *American Mathematical Monthly* **64** (1957), Part II, 45–57.

[5] Neumann, J. von, *Mathematische Grundlagen der Quantenmechanik*, Berlin 1931.

[6] Stone, M. H., 'The Theory of Representations for Boolean Algebras', *Transactions of the American Mathematical Society* **40** (1936), 37–111.
[7] Zierler, N., 'Axioms for Non-Relativistic Quantum Mechanics', *Pacific Journal of Mathematics* **11** (1961), 1151–1169.
[8] Zierler, N., 'Order Properties of Bounded Observables', *Proceedings of the American Mathematical Society* **14** (1963), 346–351.

SIMON KOCHEN AND E. P. SPECKER

LOGICAL STRUCTURES ARISING IN QUANTUM THEORY*

I

The logical structures studied in this paper are generalizations of the propositional calculus. The classical propositional calculus is essentially Boolean algebra or, alternatively, the theory of functions on an arbitrary set S with values in a two-element set. The generalization consists in allowing partial functions on the set S, i.e., functions defined on certain subsets of S, and defining an equivalence relation among these functions such that any two constant functions with the same constant value belong to the same equivalence class. The generalization is equally natural for functions with values in the field of real numbers and we shall consider this case first.

The admissible partial functions and the equivalence relation are determined by a given structure on the set S. In Section II, we shall introduce and discuss in some detail the simplest such structure, viz. graphs of a certain type; for, though rather removed from applications, they are our most fruitful source of examples. A system of partial functions closer to applications is the following: Let S be the set of unit vectors in unitary n-space U^n. Functions of the system are real-valued functions whose domain of definition is a unitary basis $(\beta_1, \ldots, \beta_n)$. With such a function f we associate a linear map $f^*: U^n \to U^n$ by defining

$$f^*(\sum x_i \beta_i) = \sum x_i f(\beta_i)\, \beta_i.$$

Two functions f_1, f_2 are equivalent if and only if the corresponding maps f_1^*, f_2^* are equal.

Systems of partial functions on a set S can be correlated to physical theories in the following way: Elements of S correspond to (pure) states, equivalence classes of functions correspond to observables. (The term observable will therefore be used for such classes.) If $a \in S$ and if f is an element of the observable q, then $f(a)$ is the value of the observable q for the physical system in state a. In classical theories, every observable

C. A. Hooker (ed.), The Logico-Algebraic Approach to Quantum Mechanics, 263–276.

has a value for all states – the functions are defined for the whole set S; in quantum theory, an observable has a (fixed) value only for certain states – the functions are partial functions. (Probability distributions associated with observables and states will not be considered in this paper.)

Sum and product of two observables q_1, q_2 are only defined if there exist functions $f_i \in q_i$ $(i=1, 2)$ having the same domain (for which we say "q_1 and q_2 are commeasurable"). In this case, the sum q_1+q_2 is the equivalence class of the functions f_1+f_2 $(f_i \in q_i, i=1, 2)$ and similarly for the product. The set Q of observables is thus made into a "partial algebra". Partial algebras will also be defined independently of a system of functions. They are structures $\langle A; \between; +, \cdot, \cdot', 1\rangle$; $\between$ is a binary relation (commeasurability); $+$, $\cdot$ are partial binary operations (q_1+q_2, $q_1 \cdot q_2$ being defined iff $q_1 \between q_2$); $\cdot'$ is the multiplication of an element of A by a real number; 1 is the unit element of A.

The subset of idempotent elements of a partial algebra forms a "partial Boolean algebra" $\mathfrak{B}$. The operations in $\mathfrak{B}$ are defined in the usual way: $q_1 \vee q_2 = (q_1+q_2) - (q_1 \cdot q_2)$ etc. The notion of validity of a formula α of the propositional calculus is defined by associating a mapping with α. Consider, e.g., the associative law α

$$[(x_1 \vee x_2) \vee x_3] \leftrightarrow [x_1 \vee (x_2 \vee x_3)].$$

A triple $\langle q_1, q_2, q_3\rangle$ of elements of $\mathfrak{B}$ is in the domain of α if all the operations in α can be performed for q_1, q_2, q_3 (this requires five commeassurabilities). α is said to be "Q-valid" if the element

$$[(q_1 \vee q_2) \vee q_3] \leftrightarrow [q_1 \vee (q_2 \vee q_3)]$$

is the unit element 1 of $\mathfrak{B}$ for all triples in the domain of α and all partial Boolean algebras $\mathfrak{B}$.

It may well be that all formulas of propositional calculus in Whitehead-Russell [3], (1.2 to 5.75), are Q-valid. The simplest formula (known to us) which is classically valid but not Q-valid is

$$[(x_1 \leftrightarrow x_2) \leftrightarrow (x_3 \leftrightarrow x_4)] \leftrightarrow [(x_1 \leftrightarrow x_4) \leftrightarrow (x_2 \leftrightarrow x_3)].$$

We shall axiomatize the notion of Q-validity and outline the corresponding completeness proof. We do not know whether the set of Q-valid formulas is recursive.

The notion of validity considered in this paper is based on the class of all partial Boolean algebras. It is equally natural to base corresponding notions on certain subclasses, e.g., the class of transitive partial Boolean algebras. (A partial Boolean algebra is transitive iff $a \subseteq b$, i.e., $a \wedge b = a$ and $b \subseteq c$ implies $a \between c$ and therefore $a \subseteq c$). Another natural subclass is the class of partial Boolean algebras associated with n-dimensional Euclidean or unitary space ($\mathfrak{B}(E^n)$, $\mathfrak{B}(U^n)$ as defined in Section 5, Example (1)) or with Hilbert space. It has been shown in Specker [2] that $\mathfrak{B}(E^3)$ cannot be imbedded into a Boolean algebra. This is an immediate consequence of the theorem – not stated in [2] – that some classically valid formula does not hold in $\mathfrak{B}(E^3)$. The relation of the notions of validity, imbeddability, and the connection of these notions with the problem of hidden variables will be discussed in another paper.

II

Let $\mathfrak{G}$ be a graph, i.e., a structure $\langle G, R\rangle$ on an underlying nonempty set G where R is a binary symmetric and irreflexive relation. Elements of G are called "vertices". $R(a, b)$ is read as "a and b are connected". A graph $\mathfrak{G}$ satisfies condition C iff it has the following properties:

(1) Any two connected vertices belong to exactly one triangle. Formally: For all a, b if $R(a, b)$ then there exists exactly one c such that $R(a, c)$ and $R(b, c)$.

(2) G contains at least one pair of connected vertices.

Examples of graphs satisfying condition C:

(a) $G=(a, b, c)$, $R(x, y)$ iff $x \neq y$ ($\mathfrak{G}$ is a triangle).

(b) Graph in Figure 1.

(c) G is the set of all lines through the origin of 3-dimensional Euclidean [alternatively: unitary] space; two lines are connected iff they are [alternatively: unitarily] orthogonal.

We associate a class F of functions with a graph satisfying condition C: $f \in F$ iff the values of f are real numbers and the domain of f – dom f – is a set of three vertices of $\mathfrak{G}$ any two of which are connected.

We define a relation E on $F \times F$: $E(f, g)$ holds iff one of the following conditions is satisfied:

(1) $f = g$.

(2) The sets dom f and dom g have one element in common, say dom $f=(a, b, c)$, dom $g=(a, b', c')$ and we have $f(a)=g(a)$ and

$$f(b)=f(c)=g(b')=g(c').$$

(3) $f(x_1)=g(x_2)$ for all $x_1 \in \text{dom } f$, $x_2 \in \text{dom } g$. (f and g are both constant functions with the same constant value.)

The equivalence classes of the relation E are called "observables"; Q is the set of all observables.

Two observables q_1, q_2 are said to be "commeasurable" ($q_1 \mathbin{\rotatebox[origin=c]{180}{$\lor$}} q_2$) if there exist functions $f_i \in q_i$ ($i=1, 2$) such that dom f_1 = dom f_2. Sum and product of commeasurable observables are defined as follows: q_1+q_2 is the equivalence class of the functions f_1+f_2, $q_1 \cdot q_2$ is the equivalence class of the functions $f_1 \cdot f_2$, where $f_i \in q_i$, $i=1, 2$, and dom f_1 = dom f_2. (One verifies that the equivalence classes do not depend on the choice of the functions.) If q is an observable and a is a real number then all the functions af for $f \in q$ belong to the same equivalence class which is by definition the class $a \cdot' q$.

III

With these definitions, the set Q of observables is made into what we shall call a "partial algebra": $\mathfrak{A} = \langle A; \circlearrowleft; +, \cdot, \cdot'; 1\rangle$. A partial algebra $\mathfrak{A}$ is given by a nonempty set A, a binary relation denoted by $\circlearrowleft$, two binary partial functions from A into A (sum $+$, product $\cdot$), a function from $R \times A$ into A (R field of reals) and an element 1 of A. The properties are as follows:

(1) The relation $\circlearrowleft$ is symmetric and reflexive.

(2) For all q in A, $q \circlearrowleft 1$. (The constant function 1 is commeasurable with all observables.)

(3) The partial functions sum and product are defined exactly for those pairs $\langle q_1, q_2\rangle$ of $A \times A$ for which $q_1 \circlearrowleft q_2$.

(4) If any two of the observables q_1, q_2, q_3 are commeasurable (i.e., $q_i \circlearrowleft q_j$ for $i, j=1, 2, 3$) then $(q_1+q_2) \circlearrowleft q_3$, $(q_1 \cdot q_2) \circlearrowleft q_3$ and $a \cdot' q_1 \circlearrowleft q_3$ (a a real number).

(5) If any two of q_1, q_2, q_3 are commeasurable then the polynomials in q_1, q_2, q_3 form a commutative algebra over the field of real numbers.

(This condition is equivalent to a longer but more elementary one: If $q_1 \mathrel{\underset{\circ}{|}} q_2$ then $q_1+q_2=q_2+q_1$ etc.)

Remarks

(1) If any two of $q_1, \ldots, q_n$ are commeasurable then the polynomials in $q_1, \ldots, q_n$ form a commutative algebra over the field of real numbers.

(2) If $\mathfrak{G}$ is the graph considered in example (c) then the associated partial algebra is isomorphic to the following algebra $\mathfrak{A}$: A is the set of 3×3 real symmetric [alternatively: Hermitian] matrices; $M_1 \mathrel{\underset{\circ}{|}} M_2$ iff $M_1M_2=M_2M_1$ (i.e., if the matrices commute); sum and product are the usual sum and product of matrices.

IV

Let $\mathfrak{A}$ be a partial algebra and let P_n be the set of polynomials (with real coefficients) containing no other variables than $x_1, \ldots, x_n$. We define recursively the domain $D_{\varphi,n}$ of a polynomial $\varphi \in P_n$ and a map φ^* corresponding to φ. $D_{\varphi,n}$ will be a subset of the n-fold Cartesian product A^n of A, φ^* will be a map from $D_{\varphi,n}$ into A; we put $\langle q_1, \ldots, q_n\rangle = \mathbf{q}$.

(1) If φ is the polynomial 1 then $D_{\varphi,n}=A^n$ and $\varphi^*(\mathbf{q})=1$.

(2) If φ is the polynomial x_i $(i=1, \ldots, n)$ then $D_{\varphi,n}=A^n$ and $\varphi^*(\mathbf{q})=\varphi^*(q_1, \ldots, q_n)=q_i$.

(3) If $\varphi=a\psi$ (a a real number) then $D_{\varphi,n}=D_{\psi,n}$ and $\varphi^*(\mathbf{q})=a\cdot'\psi^*(\mathbf{q})$.

(4) If $\varphi=\psi\otimes\chi$ (where $\otimes$ is either $+$ or $\cdot$) then $\mathbf{q}\in D_{\varphi,n}$ iff $\mathbf{q}\in D_{\psi,n}\cap D_{\chi,n}$ and $\psi^*(\mathbf{q}) \mathrel{\underset{\circ}{|}} \chi^*(\mathbf{q})$; $\varphi^*(\mathbf{q})=\psi^*(\mathbf{q})\otimes\chi^*(\mathbf{q})$.

We say "φ is identically 1 on $\mathfrak{A}$" or "the identity $\varphi=1$ holds in $\mathfrak{A}$" iff $\varphi^*(\mathbf{q})=1$ for all $\mathbf{q}\in D_{\varphi,n}$.

Roughly speaking, the identity $\varphi=1$ means that the corresponding function on A is 1 whenever it is defined.

An identity

$$\varphi(x_1, \ldots, x_n)=\psi(x_1, \ldots, x_n)$$

can be interpreted in two ways.

(1) Whenever both φ and ψ are defined then they are equal: If $\langle q_1, \ldots, q_n\rangle=\mathbf{q}\in D_{\varphi,n}\cap D_{\psi,n}$ then $\varphi^*(\mathbf{q})=\psi^*(\mathbf{q})$.

(2) Whenever both φ and ψ are defined and $\varphi^*(\mathbf{q}) \mathrel{\underset{\circ}{|}} \psi^*(\mathbf{q})$ then $\varphi^*(\mathbf{q})=\psi^*(\mathbf{q})$.

The following are examples of identities holding in all partial algebras (in the sense of (1) and therefore also in the sense of (2)):

$$x_1 + x_2 = x_2 + x_1$$
$$x_1 + (x_2 + x_3) = (x_1 + x_2) + x_3.$$

The following identity does not hold in all partial algebras (not even in the sense of (2)):

$$(x_1 + x_2) + (x_3 + x_4) = (x_1 + x_4) + (x_2 + x_3).$$

We construct a partial algebra in which this identity does not hold; the algebra is given by a graph of 11 vertices (v_1 and v_2 are represented twice in the diagram).

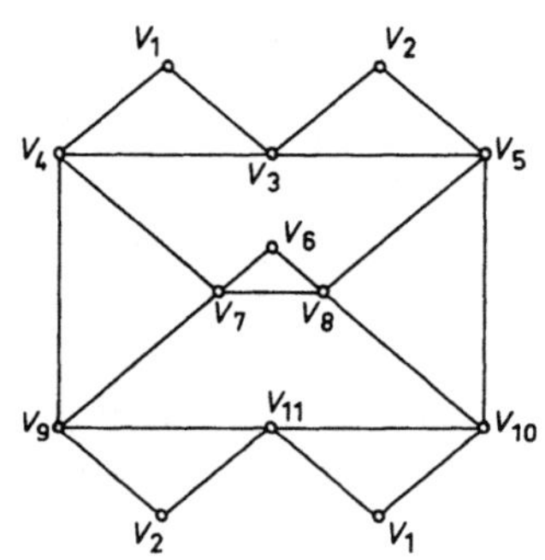

Fig. 1

In order to define observables we use the following notation: $[i, j, k; a_i, a_j, a_k]$ is the function f whose domain is set (v_i, v_j, v_k) of vertices and for which we have $f(v_h) = a_h$, $h = i, j, k$. The observables are defined by the functions following on the same line, both functions being equivalent.

$$\begin{aligned}
q_1&:[\ 1,\ 3,\ 4;\ 1,\ 0,\ 0], &&[\ 1,\ 11,\ 10;\ 1,\ 0,\ 0]\\
q_2&:[\ 3,\ 1,\ 4;\ 1,\ 0,\ 0], &&[\ 3,\ 2,\ 5;\ 1,\ 0,\ 0]\\
q_3&:[\ 2,\ 3,\ 5;\ 1,\ 0,\ 0], &&[\ 2,\ 9,\ 11;\ 1,\ 0,\ 0]\\
q_4&:[11,\ 2,\ 9;\ 1,\ 0,\ 0], &&[11,\ 1,\ 10;\ 1,\ 0,\ 0]
\end{aligned}$$

We then have

$$\begin{aligned}
q_1 + q_2&:[1,\ 3,\ 4;\ 1,\ 1,\ 0], &&[4,\ 7,\ 9;\ 0,\ 1,\ 1]\\
q_3 + q_4&:[2,\ 9,\ 11;\ 1,\ 0,\ 1], &&[4,\ 7,\ 9;\ 1,\ 1,\ 0]
\end{aligned}$$

$$\begin{array}{rl}
q_1+q_4: & [1,\ 10,\ 11;\ 1,\ 0,\ 1],\quad [5,\ 8,\ 10;\ 1,\ 1,\ 0] \\
q_2+q_3: & [2,\ 3,\ 5;\ 1,\ 1,\ 0],\quad [5,\ 8,\ 10;\ 0,\ 1,\ 1] \\
(q_1+q_2)+(q_3+q_4): & [4,\ 7,\ 9;\ 1,\ 2,\ 1],\quad [6,\ 7,\ 8;\ 1,\ 2,\ 1] \\
(q_1+q_4)+(q_2+q_3): & [6,\ 8,\ 10;\ 1,\ 2,\ 1],\quad [6,\ 7,\ 8;\ 1,\ 1,\ 2]
\end{array}$$

The observables $(q_1+q_2)+(q_3+q_4)$ and $(q_1+q_4)+(q_2+q_3)$ are commeasurable as they are both represented on the triangle (v_6, v_7, v_8); they are different because they are represented there by different functions.

V

As in the case of ordinary commutative algebras, the subset B of idempotent elements of a partial algebra forms a partial Boolean algebra. In detail: Let $\langle A; \mathrel{\underset{\circ}{|}}; +, \cdot, \cdot'; 1\rangle$ be a partial algebra. Let B be the subset of elements $a \in A$ such that $a \cdot a = a$. Define $a \mathrel{\underset{\circ}{|}} b$ for $a, b \in B$ iff $a \mathrel{\underset{\circ}{|}} b$ in A; $a \vee b = (a+b) - a \cdot b$ (where $c-d$ is to be understood in the obvious way), $\neg a = 1 - a$, 1 same element as in A, $0 = 0 \cdot' 1$ (0 number zero). The partial Boolean algebra $\langle B; \mathrel{\underset{\circ}{|}}, \vee, \neg; 1, 0\rangle$ then satisfies the following conditions:

(1) The relation $\mathrel{\underset{\circ}{|}}$ is symmetric and reflexive.

(2) For all $q \in B: q \mathrel{\underset{\circ}{|}} 1$ and $q \mathrel{\underset{\circ}{|}} 0$.

(3) The partial function $\vee$ is defined exactly for those pairs $\langle q_1, q_2\rangle$ of $B \times B$ for which $q_1 \mathrel{\underset{\circ}{|}} q_2$.

(4) If any two of q_1, q_2, q_3 are commeasurable then $(q_1 \vee q_2) \mathrel{\underset{\circ}{|}} q_3$ and $\neg q_1 \mathrel{\underset{\circ}{|}} q_2$.

(5) If any two of q_1, q_2, q_3 are commeasurable then the Boolean polynomials in q_1, q_2, q_3 form a Boolean algebra. (As in the case of algebras this condition can be replaced by a more elementary one.)

Properties (1)–(5) define the notion of partial Boolean algebra independently of the notion of partial algebra.

Examples of partial Boolean algebras:

(1) Let U^n be the n-dimensional unitary vector space, B the set of linear subspaces of U^n. For $a, b \in B$, $a \mathrel{\underset{\circ}{|}} b$ holds iff there exists a unitary basis of U^n such that some subset of this basis is a basis of a and some subset is a basis of b. $a \vee b$ is the span of a and b; $\neg a$ is the complement of a, 0 is the 0-dimensional subspace, 1 is the whole space U^n.

(2) Let $\mathfrak{B}_i$, $i \in I$, be a nonempty family of Boolean algebras such that the following conditions are satisfied:

(a) For $i, j \in I$ there exists $k \in I$ such that $B_i \cap B_j = B_k$. (The intersection of two algebras of the family is an algebra of the family; all algebras have therefore the same element 0 and the same element 1.)

(b) If $B = \bigcup B_i$ (union over I) and if a, b, c are elements of B such that any of them lie in some common algebra B_i, then there exists $k \in I$ such that $a, b, c \in B_k$.

The algebra $\mathfrak{B}$ on the set B is then defined as follows: $a \between b$ iff there exists $i \in I$ such that $a, b \in B_i$; $a \vee b = c$ in $\mathfrak{B}$ iff there exists $i \in I$ such that $a \vee b = c$ in $\mathfrak{B}_i$; $\neg a = b$ in $\mathfrak{B}$ iff there exists $i \in I$ such that $\neg a = b$ in $\mathfrak{B}_i$; 1 and 0 are the common unit and zero elements of the algebras $\mathfrak{B}_i$.

It can be shown that every partial Boolean algebra is isomorphic to an algebra of this type.

VI

We now define the "logic" associated with a partial Boolean algebra $\mathfrak{B}$. Let α be a formula of the propositional calculus (in the connectives $\vee$ and $\neg$) and φ the corresponding Boolean polynomial. α is said to be valid in $\mathfrak{B}$ iff $\varphi = 1$ is an identity of $\mathfrak{B}$ in the sense defined in Section IV. A formula is Q-valid if it holds in all partial Boolean algebras, it is C-valid if it holds in all Boolean algebras (i.e., is an identity of the propositional calculus). Clearly, every Q-valid formula is also C-valid.

THEOREM. *Let α be a formula in $x_1, \ldots, x_n$ whose only subformulas in x_i alone are x_i or $\neg x_i$ $(i = 1, \ldots, n)$ and such that for all i, j $(1 \leqslant i < j \leqslant n)$ there exists a subformula $\alpha_{i,j}$ in x_i, x_j alone. Then α is Q-valid if it is C-valid.*

Proof. Let $\mathfrak{B}$ be a partial Boolean algebra, let φ be the polynomial corresponding to α and let $\langle q_1, \ldots, q_n \rangle = \mathbf{q}$ be a sequence of elements in $\mathfrak{B}$ belonging to $D_{\varphi, n}$. Let $\varphi_{i,j}$ be a subpolynomial of φ such that no subpolynomial of $\varphi_{i,j}$ is a polynomial in x_i and x_j; $\varphi_{i,j}$ is then one of the four polynomials $x_i \vee x_j$, $\neg x_i \vee x_j$, $x_i \vee \neg x_j$, $\neg x_i \vee \neg x_j$; from this follows $q_i \between q_j$ and the theorem by Remark (1) in Section III.

COROLLARY. *A formula in one or two variables is Q-valid if it is C-valid.*

Examples

(1) The distributive law

$$[(x_1 \vee x_2) \wedge x_3] \leftrightarrow [(x_1 \wedge x_3) \vee (x_2 \wedge x_3)]$$

satisfies the hypothesis of the theorem and is therefore Q-valid. ($\wedge$, $\rightarrow$, $\leftrightarrow$ are here and in the following understood as being defined in terms of $\vee$ and $\neg$.)

(2) The associative law

$$[(x_1 \vee x_2) \vee x_3] \leftrightarrow [x_1 \vee (x_2 \vee x_3)]$$

does not satisfy the hypothesis of the theorem as there is no subformula containing x_1 and x_3 but not x_2; it is nevertheless Q-valid. For let $\langle q_1, q_2, q_3 \rangle = \mathbf{q}$ be a triple of elements in the partial Boolean algebra belonging to $D_{\varphi, 3}$ (φ being the polynomial corresponding to the formula). We then have $q_1 \between q_2$, $q_2 \between q_3$, $(q_1 \vee q_2) \between q_3$, $q_1 \between (q_2 \vee q_3)$; from this we obtain $(q_1 \vee q_2) \between (q_2 \vee q_3)$. Any two of the three elements q_1, q_2, $q_4 = q_2 \vee q_3$ are commeasurable and we have $(q_1 \vee q_2) \vee q_4 = q_1 \vee (q_2 \vee q_4)$. As we also have $q_2 \vee q_4 = q_2 \vee q_3$ we obtain

$$(q_1 \vee q_2) \vee (q_2 \vee q_3) = q_1 \vee (q_2 \vee q_3).$$

In exactly the same way we prove

$$(q_1 \vee q_2) \vee (q_2 \vee q_3) = (q_1 \vee q_2) \vee q_3.$$

The simplest example (known to us) of a C-valid formula which is not Q-valid is

$$[(x_1 \leftrightarrow x_2) \leftrightarrow (x_3 \leftrightarrow x_4)] \leftrightarrow [(x_1 \leftrightarrow x_4) \leftrightarrow (x_2 \leftrightarrow x_3)].$$

The proof is by considering the same algebra and the same observables as in Section IV for the corresponding formula with $+$ instead of $\leftrightarrow$.

VII

We shall now axiomatize the set of Q-valid formulas. Most axiom systems of the classical propositional calculus (e.g., Principia Mathematica) consist of Q-valid formulas. Modus ponens, however, does not hold for the notion of Q-validity: There are Q-valid formulas α, $\alpha \rightarrow \beta$ such that β is not Q-valid. (A refutation of β is given by a sequency of observables for which α cannot be evaluated.)

Let Σ be the set of formulas of the propositional calculus in the variables $x_1, x_2, \ldots$ and the connectives $\vee$ and $\neg$. (As before, $\wedge$, $\rightarrow$ and $\leftrightarrow$ are thought of as being defined.) Let Σ^* be the set of formulas of Σ and the formulas of the type $\delta(\alpha_1, \ldots, \alpha_n)$ where n is a positive integer and $\alpha_1, \ldots, \alpha_n$ are formulas of Σ. We define the following notion: "The sequence $\gamma_1, \ldots, \gamma_m$ of formulas of Σ^* is a Q-proof of the formula α of Σ." The notion of Q-proof is reduced to the following notion: "The sequence $\gamma_1, \ldots, \gamma_m$ of formulas of Σ^* is Φ-admissible." (Φ will be subset of Σ.) A sequence is then a Q-proof if it is (α)-admissible and contains α as one of its formulas. ((α) is the unit set of α.)

The notion of admissibility is based on rules of inference.

R$_1$. $$\frac{\delta(\alpha_1, \ldots, \alpha_n)}{\delta(\alpha_i, \alpha_j)} \quad (\text{where } 1 \leqslant i \leqslant n,\ 1 \leqslant j \leqslant n).$$

R$_2$. $$\frac{\delta(\alpha_1, \alpha_1), \delta(\alpha_1, \alpha_2), \ldots, \delta(\alpha_i, \alpha_j) \ldots, \delta(\alpha_n, \alpha_n)}{\delta(\alpha_1, \ldots, \alpha_n)}.$$

(The premiss consists of the n^2 formulas $\delta(\alpha_i, \alpha_j)$ such that $1 \leqslant i \leqslant n$, $1 \leqslant j \leqslant n$.)

R$_3$. $$\frac{\delta(\alpha_1, \alpha_2), \alpha_2 \leftrightarrow \alpha_3}{\delta(\alpha_1, \alpha_3)}.$$

R$_4$. $$\frac{\delta(\neg \alpha_1, \alpha_2)}{\delta(\alpha_1, \alpha_2)}.$$

R$_5$. $$\frac{\delta(\alpha_1, \alpha_2, \alpha_3)}{\delta(\alpha_1 \vee \alpha_2, \alpha_3)}.$$

S$_1$. $$\frac{\delta(\alpha_1, \ldots, \alpha_n)}{\beta(\alpha_1, \ldots, \alpha_n)}.$$

where $\beta(x_1, \ldots, x_n)$ is a C-valid formula.

S$_2$. $$\frac{\alpha_1, \alpha_1 \rightarrow \alpha_2}{\alpha_2}.$$

(This form of modus ponens is of course different from the one mentioned at the beginning of Section VII.) S$_1$ is a scheme of schemes; it can be replaced by a finite number of ordinary schemes (e.g., schemes corresponding to the axioms of Principia Mathematica).

A sequence $\gamma_1, \ldots, \gamma_n$ of formulas of Σ^* is Φ-admissible iff the following conditions are satisfied:

(1) Φ is a subset of Σ.

(2) For all i, $1 \leqslant i \leqslant n$, γ_i is either of the type $\text{⫰}(\beta_1, \beta_1)$ or $\text{⫰}(\beta_1, \beta_2)$ (where β_1, $\beta_1 \wedge \beta_2$ are subformulas of a formula $\alpha \in \Phi$) or there exist indices $i_1, \ldots, i_m$ such that $1 \leqslant i_k < i$ $(k = 1, \ldots, m)$ and γ_i follows from $\gamma_{i_1}, \ldots, \gamma_{i_m}$ by one of the above rules.

Remarks

(1) The notion $\text{⫰}(\alpha_1, \ldots, \alpha_n)$ can be eliminated from our axiom system by replacing the formula $\text{⫰}(\alpha_1, \ldots, \alpha_n)$ by a Q-identity in $\alpha_1, \ldots, \alpha_n$ containing all formulas $\alpha_i \vee \alpha_j$ $(1 \leqslant i < j \leqslant n)$ as subformulas.

(2) In the rules of inference and in deductions the "variables" $x_1, x_2, \ldots$ are considered as parameters and hold fixed throughout the argument. So they should perhaps rather be called "constants".

THEOREM. *There exists a Q-proof for a formula α of Σ if and only if α holds in all partial Boolean algebras.*

(1) Assume that there exists a Q-proof $\gamma_1, \ldots, \gamma_m$ of α (formula in $x_1, \ldots, x_n$), let $\mathfrak{B}$ be a partial Boolean algebra, let $\langle q_1, \ldots, q_n \rangle$ be a sequence of elements in the domain of definition of φ (the polynomial associated with α). We show by induction on i: If γ_i is a formula of Σ, then $\langle q_i, \ldots, q_n \rangle$ is in the domain of definition of the corresponding Boolean polynomial χ_i and $\chi_i^*(q_1, \ldots, q_n) = 1$ (γ_i has therefore no other variables than $x_1, \ldots, x_n$!); if γ_i is a formula $\text{⫰}(\beta_1, \ldots, \beta_k)$, then $\langle q_1, \ldots, q_n \rangle$ is in the domain of the corresponding polynomials $\psi_1, \ldots, \psi_k$ and the elements $\psi_i^*(q_1, \ldots, q_n)$ are all in relation ⫰. Clearly the statement is true if γ_i is $\text{⫰}(\alpha_1, \alpha_1)$ or $\text{⫰}(\alpha_1, \alpha_2)$, where α_1, $\alpha_1 \vee \alpha_2$ are subformulas of α. If the statement is true for the formulas in the premiss of a rule, it is also true for the conclusion. Let us verify as an example modus ponens (rule S_2): If φ_1, φ_2 are the polynomials associated with the formulas α_1, α_2, the induction hypothesis applied to $\alpha_1, \alpha_1 \rightarrow \alpha_2$ gives

$$\varphi_1^*(q_1, \ldots, q_n) = 1,$$
$$\neg \varphi_1^*(q_1, \ldots, q_n) \cup \varphi_2^*(q_1, \ldots, q_n) = 1.$$

Clearly

$$\neg \varphi_1^*(q_1, \ldots, q_n) = 0,$$
$$0 \cup \varphi_2^*(q_1, \ldots, q_n) = \varphi_2^*(q_1, \ldots, q_n) = 1.$$

(2) Assume that there does not exist a Q-proof of the formula α of Σ.

Our aim is to construct a partial Boolean algebra in which α does not hold.

Throughout the remainder of this section α is a fixed formula of Σ; we assume that α contains exactly the variables $x_1, \ldots, x_n$. A formula of Σ^* is called "α-provable" iff there exists an (α)-admissible sequence containing it. We state some simple lemmas on the notion of α-provability.

(a) *If α_1 is α-provable, so is $\between(\alpha_1, \alpha_1)$.*

(b) *Let Ω be the subset of formulas β of Σ such that $\between(\beta, \beta)$ is α-provable. Formulas of Ω contain no other variables than $x_1, \ldots, x_n$. $x_1, \ldots, x_n$ and α are formulas of Ω.*

(c) *The relation "$\alpha_1 \leftrightarrow \alpha_2$ is α-provable" is an equivalence relation on the set Ω. This equivalence relation is compatible with the operations $\vee$, $\neg$ and the relation $\between$.*

(d) *If α_1 and $\alpha_1 \leftrightarrow \alpha_2$ are α-provable so is α_2.*

(e) *If α_1 and α_2 are α-provable, so is $\alpha_1 \leftrightarrow \alpha_2$.*

We now define a partial Boolean algebra $\langle B; \between; \vee, \neg; 1, 0\rangle$ associated with the formula α. Elements of B are the equivalence classes of the relation "$\alpha_1 \leftrightarrow \alpha_2$ is α-provable". The equivalence classes are composed in the obvious way: $[\alpha_1] \vee [\alpha_2]$ is the class of $\alpha_1 \vee \alpha_2$, etc. $[\alpha_1] \between [\alpha_2]$ if and only if $\between(\alpha_1, \alpha_2)$ is α-provable. 1 is the class of α-provable formulas. We have $q \between 1$ for all q as for every formula β of Ω the formula $\between(\beta, \beta \leftrightarrow \beta)$ is α-provable. 0 is the class $\neg 1$. The axioms of partial Boolean algebras are easily verified.

Let q_i be the class of the formula x_i $(i = 1, \ldots, n)$; if β is any formula of Ω, ψ the Boolean polynomial associated with β, the class of β is the element $\psi^*(q_1, \ldots, q_n)$, i.e., β is α-provable if and only if $\psi^*(q_1, \ldots, q_n) = 1$. The formula α is therefore Q-provable if and only if $\varphi^*(q_1, \ldots, q_n) = 1$ in the partial Boolean algebra just defined.

Remark

The completeness theorem can easily be extended to a theorem on the completeness of the rules $R_1, \ldots, R_5$, S_1, S_2. Let Γ be a subset of Σ^*, $\gamma \in \Sigma^*$; then γ follows from Γ by the rules $R_1, \ldots, R_5$, S_1, S_2 and $\between(x_i, x_i)$ $(i = 1, 2, \ldots)$ $(\Gamma \vdash \gamma)$ iff γ is a consequence of Γ in all partial Boolean algebras $\mathfrak{B}$. (γ is a consequence of Γ in B if γ is defined and true for all sequences $\langle q_1, \ldots, q_n\rangle$ of elements of B for which all the formulas of Γ are defined and true.) A formula $\alpha \in \Sigma$ is therefore Q-provable iff $\Gamma(\alpha) \vdash \alpha$, where $\Gamma(\alpha)$ is the set of commeasurability relations of the subformulas of α.

IX

Let S be a set, S_0 a subset of S, let $\vee$ be a binary operation $S^2 \rightarrow S$ and let $\neg$ be a unary operation $S \rightarrow S$. The algebra $\mathfrak{S} = \langle S, S_0; \vee, \neg \rangle$ is called a "truth-table"; S_0 is the set of designated elements. A formula α in the variables $x_1, \ldots, x_n$ defines a map $\alpha^*: S^n \rightarrow S$; α holds in $\mathfrak{S}$ iff α^* maps S^n into S_0.

THEOREM. *There exists a truth-table $\mathfrak{S}$ with a two-element S_0 such that a formula is Q-valid iff it holds in $\mathfrak{S}$.*

Proof (in outline). (1) We construct (e.g., by an infinite direct product) a partial Boolean algebra $\mathfrak{B}$ such that a formula is Q-valid iff it holds in $\mathfrak{B}$. (2) We define the truth-table $\mathfrak{S}$ as follows: $S = B \cup (u)$ (where $u \notin B$), $S_0 = (1, u)$; the operations $\neg'$ and $\vee'$ in $\mathfrak{S}$ are defined by putting $\neg' q = \neg q$ for $q \in B$, $\neg u = u$; $q_1 \vee' q_2 = q_1 \vee q_2$ for $q_1, q_2 \in B$ and $q_1 \mathrel{\text{\d{$\smile$}}} q_2$, $q_1 \vee' q_2 = u$ otherwise. One verifies that a formula holds in $\mathfrak{S}$ iff it holds in $\mathfrak{B}$.

Remark

There does not exist a truth-table $\mathfrak{S}$ with only one designated element such that a formula is Q-valid iff it holds in $\mathfrak{S}$.

THEOREM. *If all Q-valid formulas hold in the truth-table $\mathfrak{S}$ having three (or less) elements then all C-valid formulas hold in $\mathfrak{S}$.*

By a theorem in Section VI the above theorem (suggested by a question of Leon Henkin) is an immediate consequence of the following:

THEOREM. *If all C-valid formulas in one or two variables hold in the truth-table $\mathfrak{S}$ having three (or less) elements then all C-valid formulas hold in $\mathfrak{S}$.*

Proof. (1) Assume that the algebra $\mathfrak{S}$ is generated by a two-element subset: There exist elements $a, b \in S$ and a Boolean polynomial φ (corresponding to a formula α) such that $S = (a, b, \varphi^*(a, b))$. Let $\beta(x_1, \ldots, x_n)$ be a formula in the n variables $x_1, \ldots, x_n$ which does not hold in $\mathfrak{S}$. Let $\langle q_1, \ldots, q_n \rangle$ be a sequence such that $\beta^*(q_1, \ldots, q_n) \notin S_0$. Define a formula

$\gamma(x_1, x_2)$ as follows: $\gamma(x_1, x_2)$ is $\beta(\xi_1, \ldots, \xi_n)$ where ξ_i is x_1, x_2, or $\alpha(x_1, x_2)$ according to whether q_i is a, b or $\varphi^*(a, b)$. γ does not hold in S because $\gamma^*(a, b) = \beta^*(q_1, \ldots, q_n) \notin S_0$. The formula γ is therefore not C-valid and neither is β.

(2) Assume that the algebra $\mathfrak{S}$ is not generated by a two-element subset of S. Then $\neg x = x$ for all $x \in S$ and $x \vee y = x$ or $x \vee y = y$ for all $x, y \in S$. Every element $x \vee y$ is designated; for if, e.g., $x \vee y = x$, x is a value of $(x_1 \vee x_2) \vee \neg x_2$ which is C-valid. Every C-valid formula is of the form $\neg \ldots \neg (\beta_1 \vee \beta_2)$ and holds therefore in $\mathfrak{S}$.

The above theorem does not contradict Reichenbach [1] which connects "quantum-logic" and three-valued logic because the notions of validity involved are different.

Cornell University, Ithaca
Eidgenössische Technische Hochschule, Zürich

NOTE

* This work was supported in part by a U.S. National Science Foundation grant.

BIBLIOGRAPHY

[1] Reichenbach, Hans, *Philosophic Foundations of Quantum Mechanics*, Berkeley and Los Angeles, 1944.
[2] Specker, Ernst, "Die Logik nicht gleichzeitig entscheidbarer Aussagen", *Dialectica*, **14** (1960), pp. 239–246.
[3] Whitehead, Alfred North, and Russell, Bertrand, *Principia Mathematica*, vol. 1, 2nd ed., Cambridge, 1925.

SIMON KOCHEN AND E. P. SPECKER

THE CALCULUS OF PARTIAL PROPOSITIONAL FUNCTIONS

I

The calculus of partial propositional functions has been introduced in [2]. It is a variant of the classical propositional calculus, a variant which takes into account that pairs of propositions may be "incompatible" and cannot therefore be connected. As is well known, such pairs are considered in Quantum Theory; but they may also be said to occur in natural languages. Difficulties arising from propositions of the type "If two times two are five, then there exist centaurs" seem to be due as much to incompatibility as to material implication.

The calculus has been based in [2] on the connectives $\rightharpoondown$, $\vee$ and a relation ♂ (called "commeasurability"). A method of eliminating ♂ has been sketched; if carried out, this elimination leads to a rather complicated system.

The choice of connectives being as free in the new calculus as in the classical one, we choose falsity (f) and implication ($\rightarrow$) as new basic connectives. For commeasurability of ϕ, ψ is most naturally expressed as $f \rightarrow (\phi \rightarrow \psi)$. "If ϕ, ψ are commeasurable, then $\phi \rightarrow \psi$ makes sense; whatever makes sense is implied by f. Conversely, what is implied by anything makes sense; if $\phi \rightarrow \psi$ makes then sense, ϕ, ψ are commeasurable."

Presented this way, the calculus PP_1 of partial propositional functions has the same set of formulas as the classical propositional calculus; it differs from it by the notion of validity. The formal notion of validity in PP_1 (called "Q-validity" in [2]) is based on the notion of partial Boolean algebra. In order to make this paper somewhat independent from [2], we assume familiarity with this notion only in the last section. Whenever partial Boolean algebras are mentioned in earlier sections, the reader may think of the partial algebra of linear subspaces of the 3-dimensional orthogonal space (as defined in Section III, example 2) or of the partial algebra of closed linear subspaces of Hilbert space. These algebras are the most interesting examples and are at the origin of the notion of partial Boolean

C. A. Hooker (ed.), The Logico-Algebraic Approach to Quantum Mechanics, 277–292.

algebras. Their relation to Quantum Theory has been considered in [2], [3].

The notion of validity in PP_1 may also be explained somewhat informally. Let S be a set of propositions. Assume that there is defined on S a binary relation $\mathfrak{J}$ of commeasurability and a partial function $\rightarrow$ from $S \times S$ to S, $s_1 \rightarrow s_2$ being defined if and only if $\mathfrak{J}(s_1, s_2)$; assume furthermore that the set S_1 of true sentences of S is given. (In general, propositions depend on parameters and are therefore neither true nor false.) Let ϕ be a formula of PP_1, e.g. $x_1 \rightarrow (x_2 \rightarrow x_1)$. ϕ can be evaluated for a pair if and only if $\mathfrak{J}(s_2, s_1)$ and $\mathfrak{J}(s_1, s_2 \rightarrow s_1)$ hold: $s_2 \rightarrow s_1$ has to be defined and putting $s_3 = s_2 \rightarrow s_1$ also $s_1 \rightarrow s_3$. If these conditions are satisfied, the value assigned to ϕ for $\langle s_1, s_2 \rangle$ is $s_1 \rightarrow (s_2 \rightarrow s_1)$. The formula ϕ holds in the structure $\langle S, \mathfrak{J}, \rightarrow, S_1 \rangle$ if the assigned value is an element of S_1 for all such pairs $\langle s_1, s_2 \rangle$: "ϕ holds iff it is true whenever it makes sense." Our axiom system is based on the assumption that ϕ makes sense if and only if $f \rightarrow \phi$ holds. The question of validity of ϕ is thereby reduced to the question whether ϕ is derivable from $f \rightarrow \phi$. The notion of derivation will be formalized by rules of inference $R_1, \ldots, R_7$ (given in 7.1). The rules are adopted from a system of Wajsberg [4] for the propositional calculus. We have learnt from Wajsberg's work also in an other respect; indeed, the main idea behind the series of derived rules in Section VIII is due to him. We shall prove completeness, i.e. we show that ϕ holds in all partial Boolean algebras iff ϕ is derivable from $f \rightarrow \phi$. The formulas in such a derivation all make sense provided ϕ does; a proof of ϕ in the system PP_1 is therefore essentially a proof based on subformulas of ϕ.

As pointed out in [2], most formulas of *Principia Mathematica* hold in the calculus of partial propositional functions. Contrary to a conjecture of [2], it is not true for all formulas, the "praeclarum theorema" of Leibniz (PM 3.47) being a counter-example. The formula PM 3.47 holds however in the partial Boolean algebra $B(E^{\omega})$ associated with Hilbert space E^{ω} and a fortiori in $B(E^3)$. Axiomatizations of the sets of formulas holding in $B(E^{\alpha})$ $(\alpha = 3, \ldots, \omega)$ and relations between these sets will be given in another paper.

In some of the following sections, we write $\phi\psi$ instead of $\phi \rightarrow \psi$, $\phi\psi\chi$ instead of $\phi(\psi\chi)$. There is no danger of misunderstanding since conjunction does nor occur explicitly.

II

Let P_1 be the system of the classical propositional calculus as defined e.g. in [1]: Symbols of P_1 are

$$(\rightarrow)\, f x_0 \;\; x_1 \;\; x_2 \;\; \ldots$$

f and $x_0, x_1, x_2, \ldots$ are formulas; if ϕ, ψ are formulas, then $(\phi \rightarrow \psi)$ is a formula. Outermost parentheses in formulas may be omitted.

III

A structure $B = \langle B, ♂, 0, 1, \rightharpoondown, \vee \rangle$ is of type *PB* (partial Boolean) if it satisfies the following conditions

(a) B is a non-empty set;

(b) ♂ is a binary relation on B(♂(a, b) is read: "a and b are commeasurable");

(c) 0 and 1 are elements of B;

(d) $\rightharpoondown$ is a unary function from B to B;

(e) $\vee$ is a binary function. The domain of $\vee$ is the set of those ordered pairs $\langle a, b \rangle$ of $B \times B$ for which ♂(a, b); the co-domain of $\vee$ is the set B.

The notion of partial Boolean algebra is defined by imposing restrictions on structures of type *PB*. An example of such a restriction is: $\rightharpoondown\rightharpoondown a = a$ for all $a \in B$.

We define two structures of type *PB* which are partial Boolean algebras:

(1) The Boolean algebra of two elements. B is the set $(0, 1)$; $0 \neq 1$. ♂(a, b) holds for all a, b in B. $\rightharpoondown 0 = 1$, $\rightharpoondown 1 = 0$. $0 \vee 0 = 0$, $0 \vee 1 = 1$, $1 \vee 0 = 1$, $1 \vee 1 = 1$.

(2) The partial algebra $B(E^3)$ of linear subspaces of E^3 (3-dimensional orthogonal space).

(a) B is the set of linear subspaces of E^3;

(b) ♂(a, b) for subspaces a, b iff a and b are orthogonal in the sense of elementary geometry, i.e. if there exists a basis of E^3 containing a basis of a and of b; (if a is a subspace of b, ♂(a, b) holds.)

(c) 0 is the 0-dimensional, 1 is the 3-dimensional subspace of E^3;

(d) $\rightharpoondown a$ is the orthogonal complement of a;

(e) $a \vee b$ is the union (span) of a and b, defined only for those pairs $\langle a, b \rangle$ for which ♂(a, b) holds.

IV

We state some properties of the structure $B(E^3)$ defined in example (2) of Section III. These properties hold in all partial Boolean algebras as defined in [2]; it will follow from the completeness theorem in Section X that they form an axiom system for partial Boolean algebras.

For all elements a, b, c of B:

4.1. $\neg 0=1, \neg 1=0$

4.2. $\neg\neg a=a$

4.3. $\mathfrak{d}(1, a)$

4.4. *If* $\mathfrak{d}(a, b)$, *then* $\mathfrak{d}(b, a)$

4.5. *If* $\mathfrak{d}(\neg a, b)$, *then* $\mathfrak{d}(a, b)$

4.6. $1 \vee a=1, a \vee 1=1$

4.7. $0 \vee a=a, a \vee 0=a$ ($\mathfrak{d}(0, a)$ *holds by* 4.1, 4.3, 4.5)

4.8. *If* $\neg a \vee b=1$ *and* $\neg b \vee a=1$, *then* $a=b$

4.9. *If* $\mathfrak{d}(a, b)$, *then* $\mathfrak{d}(\neg b, a)$, $\mathfrak{d}(\neg a, \neg b \vee a)$ *and* $\neg a \vee \vee(\neg b \vee a)=1$

4.10. *If* $\mathfrak{d}(a, \neg b)$, $\mathfrak{d}(a, c)$, $\mathfrak{d}(b, c)$ *then* $\mathfrak{d}(\neg a, b)$, $\mathfrak{d}(\neg a, c)$, $\mathfrak{d}(\neg b, c)$, $\mathfrak{d}(\neg a, \neg b \vee c)$, $\mathfrak{d}(\neg(\neg a \vee b), \neg a \vee c)$, $\mathfrak{d}(\neg(\neg a \vee (\neg b \vee c)), \neg(\neg a \vee b) \vee (\neg a \vee c))$, *and* $\neg(\neg a \vee (\neg b \vee \vee c)) \vee (\neg(\neg a \vee b) \vee (\neg a \vee c))=1$. (All operations are defined by the hypotheses.)

The Theorems 4.9, 4.10 are special cases of the following: *If* $\mathfrak{d}(a, b)$, $\mathfrak{d}(a, c)$, $\mathfrak{d}(b, c)$, *then all Boolean identities in a, b, c hold.*

V

Let $B=\langle B, \mathfrak{d}, 0, 1, \neg, \vee\rangle$ be a structure of type *PB* as defined in Section III and let *N* be the set of natural numbers. We associate functions with formulas ϕ of P_1 (defined in Section II). The domain D_ϕ of the function $[\phi]$ associated with ϕ is a subset of B^N (the set of functions from *N* to *B*), the codomain of $[\phi]$ is *B*. The functions $[\phi]$ and their domains D_ϕ are defined simultaneously by recursion (with respect to the length of ϕ).

(1) $D_f=B^N$ and $[f](q)=0$ for all $q \in D \cdot ([f]$ is the constant function 0 defined for all sequences.)

(2) $D_{x_0}=D_{x_1}=D_{x_2}=\cdots=B^N$ and $[x_0](q)=q(0)$, $[x_1](q)=q(1)$, $[x_2](q)=q(2),\ldots$ ($[x_1]$ *is the* projection of B^N on its coordinate 1.)

(3) $q\in D_{\phi\to\psi}$ if and only if $q\in D_\phi$ and $q\in D_\psi$ and $\eth(\neg[\phi](q), [\psi](q))$. The function

$$[\phi\to\psi]:D_{\phi\to\psi}\Rightarrow B$$

is defined as follows:

$$[\phi\to\psi](q)=\neg([\phi](q))\vee[\psi](q) \text{ for } q\in D_{\phi\to\psi}.$$

Example.
The set $D_{x_0\to x_1}$ consists of those sequences $\langle q(0), q(1),\ldots\rangle$ for which $\eth(\neg q(0), q(1))$ and $[x_0\to x_1](q)=\neg q(0)\vee q(1)$. Roughly speaking, D_ϕ is the set of those sequences in B^N for which ϕ can be evaluated and $[\phi](q)$ is the result of the evaluation.

DEFINITION of validity in a structure of type *PB*: A formula ϕ of P_1 holds (is valid) in the structure $\langle B, \eth, 0, 1, \neg, \vee\rangle$ of type *PB* if and only if $[\phi](q)=1$ for all $q\in D_\phi$.

Remarks
(1) If $\langle B, \eth, 0, 1, \neg, \vee\rangle$ is the two element Boolean algebra defined in example (1) of Section III, D_ϕ is equal to B^N for all formulas ϕ and the above construction is the one given by Tarski for the notion of satisfaction.

(2) A formula valid in all partial Boolean algebras has been called "*Q*-valid" in [2].

DEFINITION of (semantic) consequence in the structure of type *PB*: The formula ψ of *P* is a semantic consequence of the formulas $\phi_1, \phi_2,\ldots,\phi_n$ of P_1 in the structure $\langle B, \eth, 0, 1, \neg, \vee\rangle$ of type *PB* if and only if the following condition is satisfied for all q in B^N:
If $q\in D_{\phi_i}$ and $[\phi_i](q)=1$ for all i, $1\leqq i\leqq n$, then $q\in D_\psi$ and $[\psi](q)=1$.
Semantic consequence is expressed as follows:

$$\phi_1,\ldots,\phi_n \Vdash \psi$$

VI

We introduce a shorter notation: Instead of $\phi \rightarrow \psi$ we write $\phi\psi$; association is to the right, i.e. $\phi\psi\chi$ is $\phi(\psi\chi)$. The formula $(\phi \rightarrow (\psi \rightarrow \chi)) \rightarrow ((\phi \rightarrow \psi) \rightarrow (\phi \rightarrow \chi))$ is therefore written $(\phi\psi\chi)\,(\phi\psi)\,(\phi\chi)$. Throughout this section, validity and semantic consequence is with respect to a fixed structure $B = \langle B, \mathfrak{J}, 0, 1, \neg, \vee \rangle$ of type PB satisfying 4.1–4.10 (i.e. a partial Boolean algebra). Formulas are formulas of P_1.

6.1. *$q \in D_{f\phi}$ iff $q \in D_\phi$*

Proof. $q \in D_{f\phi}$ iff $q \in D_f$, $q \in D_\phi$ and $\mathfrak{J}(\neg[f](q), [\phi](q))$. Therefore, if $q \in D_{f\phi}$ then $q \in D_\phi$. Assume $q \in D_\phi$. By definition, $D_f = B^N$, $[f](q) = 0$; by 4.1, $\neg 0 = 1$; by 4.3 $\mathfrak{J}(1, [\phi](q))$, i.e. $q \in D_{f\phi}$.

6.2. *$f\phi$ is valid for all formulas ϕ of P_1.*

Proof. Assume $q \in D_{f\phi}$; then $[f\phi](q) = \neg[f](q) \vee [\phi](q) = 1 \vee a = 1$ (*by* 4.1, 4.6).

6.3. *ϕ is valid iff $f\phi \Vdash \phi$.*

Proof. (1) Assume ϕ valid and $q \in D_{f\phi}$; then $q \in D_\phi$ by 6.1 and $[\phi](q) = 1$ by validity. (2) Assume $f\phi \Vdash \phi$ and $q \in D_\phi$; then $q \in D_{f\phi}$ by 6.1, $[f\phi](q) = 1$ by 6.2, and therefore $[\phi](q) = 1$ by $f\phi \Vdash \phi$.

6.4. $\phi \Vdash f\phi$

Proof. If $q \in D_\phi$, then $q \in D_{f\phi}$ by 6.1; $[f\phi](q) = 1$ by 6.2.

6.5. $f\phi\psi \Vdash f\psi$

Proof. Assume $q \in D_{f\phi\psi}$; then $q \in D_{\phi\psi}$, $q \in D_\psi$; by 6.1, $q \in D_{f\psi}$; by 6.2 $[f\psi](q) = 1$.

6.6. $(\phi f)f \Vdash \phi$

Proof. Assume $q \in D_{(\phi f)f}$; then $q \in D_\phi$. Putting $[\phi](q) = a$ and assuming $[(\phi f)f](q) = 1$, we have $\neg(\neg a \vee 0) \vee 0 = 1$. By 4.7, $b \vee 0 = b$ for all b; therefore $\neg\neg a = 1$; by 4.2, $\neg\neg a = a$, i.e. $[\phi](q) = 1$.

6.7. $f\phi\psi \Vdash \phi\psi\phi$

Proof. Assume $q \in D_{f\phi\psi}$; then $q \in D_{\phi\psi}$, $q \in D_\phi$, $q \in D_\psi$. Putting $[\phi](q) = a$

$[\psi](q)=b$, we have $\mathfrak{d}(\rightharpoondown a, b)$. Therefore by 4.5, $\mathfrak{d}(a, b)$; by 4.9, $\mathfrak{d}(\rightharpoondown b, a)$, $\mathfrak{d}(\rightharpoondown a, \rightharpoondown b \vee a)$ and $\rightharpoondown a \vee (\rightharpoondown b \vee a)=1$. Hence $q \in D_{\psi\phi}$, $q \in D_{\phi\psi\phi}$, and $[\phi\psi\phi](q)=1$.

6.8. $f\psi\chi, f\phi\psi, f\phi\chi \Vdash (\phi\psi\chi)(\phi\psi)(\phi\chi)$

Proof. Assume $q \in D_{f\psi\chi}$, $q \in D_{f\phi\psi}$, and $q \in D_{f\phi\chi}$; then $q \in D_{\psi\chi}$, $q \in D_{\phi\psi}$, $q \in D_{\phi\chi}$, $q \in D_{\phi}$, $q \in D_{\psi}$, $q \in D_{\chi}$. Putting $[\phi](q)=a$, $[\psi](q)=b$, $[\chi](q)=c$, we have $\mathfrak{d}(\rightharpoondown b, c)$, $\mathfrak{d}(\rightharpoondown a, b)$, $\mathfrak{d}(\rightharpoondown a, c)$. Therefore, by 4.5, $\mathfrak{d}(a, b)$, $\mathfrak{d}(b, c)$, $\mathfrak{d}(a, c)$ and, by 4.10, $\mathfrak{d}(\rightharpoondown(\rightharpoondown a \vee b), \rightharpoondown a \vee c)$; hence $q \in D_{(\phi\psi)(\phi\chi)}$. Furthermore by 4.10, $\mathfrak{d}(\rightharpoondown(\rightharpoondown a \vee (\rightharpoondown b \vee c)), \rightharpoondown(\rightharpoondown a \vee b) \vee (\rightharpoondown a \vee c))$, i.e. $q \in D_{(\phi\psi\chi)(\phi\psi)(\phi\chi)}$. Again by 4.10, $\rightharpoondown(\rightharpoondown a \vee (\rightharpoondown b \vee c)) \vee (\rightharpoondown(\rightharpoondown a \vee b) \vee (\rightharpoondown a \vee c))=1$, i.e. $[(\phi\psi\chi)(\phi\psi)(\phi\chi)](q)=1$.

6.9. $\phi, \phi\psi \Vdash \psi$

Proof. Assume $q \in D_{\phi\psi}$; then $q \in D_{\psi}$. Assuming $[\phi](q)=1$ and putting $[\psi](q)=a$, we have $[\phi\psi](q)=\rightharpoondown[\phi](q) \vee [\psi](q)=\rightharpoondown 1 \vee a=0 \vee a=a$ (by 4.1, 4.7). Assuming $[\phi\psi](q)=1$, we have $1=a$, i.e. $[\psi](q)=1$.

6.10. $f\phi\psi, \psi\chi, \chi\psi \Vdash f\phi\chi$

Proof. Assume $q \in D_{f\phi\psi}$, $q \in D_{\psi\chi}$, $q \in D_{\chi\psi}$. Then $q \in D_{\phi}$, $q \in D_{\psi}$, $q \in D_{\chi}$. Putting $[\phi](q)=a$, $[\psi](q)=b$, $[\chi](q)=c$, we have $\mathfrak{d}(\rightharpoondown a, b)$; $[\psi\chi](q)= \rightharpoondown b \vee c$, $[\chi\psi](q)=\rightharpoondown c \vee b$. Assuming $[\psi\chi](q)=1$ and $[\chi\psi](q)=1$, we have $\rightharpoondown b \vee c=1$ and $\rightharpoondown c \vee b=1$. Therefore by 4.8, $b=c$. Hence $\mathfrak{d}(\rightharpoondown a, c)$, $q \in D_{\phi\chi}$; by 6.1, 6.2, $q \in D_{f\phi\chi}$, $[f\phi\chi](q)=1$.

6.11. *Remark*

All rules $\phi_1, \ldots, \phi_m \Vdash \psi$ in 6.1–6.9 have the property that $q \in D_{\psi}$ provided $q \in D_{\phi_i}$, $i=1, \ldots, m$. The rule 6.10 does not have this property as can be shown by an example in $B(E^3)$.

7.1. DEFINITION of the calculus PP_1 of partial propositional functions.

(1) Formulas of PP_1 are the formulas of P_1.

(2) PP_1 has the following rules of inference

R_1: $\phi \vdash f \rightarrow \phi$

R_2: $f \rightarrow (\phi \rightarrow \psi) \vdash f \rightarrow \psi$

R_3: $(\phi \rightarrow f) \rightarrow f \vdash \phi$

R_4: $f \rightarrow (\phi \rightarrow \psi) \vdash \phi \rightarrow (\psi \rightarrow \phi)$

R_5: $f\to(\psi\to\chi),\ f\to(\phi\to\psi),\ f\to(\phi\to\chi)\vdash(\phi\to(\psi\to\chi))\to((\phi\to\psi)\to(\phi\to\chi))$

R_6: $\phi,\ \phi\to\psi\vdash\psi$

R_7: $f\to(\phi\to\psi),\ \psi\to\chi,\ \chi\to\psi\vdash f\to(\phi\to\chi)$

(3) A rule

$$\phi_1,\dots,\phi_m\vdash\gamma_n$$

is a derivable rule of PP_1 iff there exists a sequence $\gamma_1\dots\gamma_n$ of formulas of PP_1 such that each γ_i ($i\leq n$) is either one of the formulas $\phi_1,\dots,\phi_m$ or follows from formulas $\gamma_{i_1},\dots,\gamma_{i_k}$ ($i_j<i$, $j=1,\dots,k$) by one of the rules $R_1,\dots,R_7$.

(4) A formula ϕ of PP_1 is provable in PP_1 iff $f\to\phi\vdash\phi$ is a derivable rule of PP_1.

7.2. THEOREM. *If $\phi_1,\dots,\phi_m\vdash\psi$ is a derivable rule of PP_1, then $\phi_1,\dots,\phi_m\Vdash\psi$ holds in every partial Boolean algebra.*

Proof. Let $\langle\gamma_1,\dots,\gamma_n\rangle$ be a sequence as defined in (3) of 7.1 and assume $q\in D_{\phi i}$, $[\phi_i](q)=1$ for $i=1,\dots,m$. We prove by induction with respect to j: $q\in D_{\gamma j}$ and $[\gamma_j](q)=1$. The inductive step is provided for each of the rules R_i by $6.3+i$ ($i=1,\dots,7$).

THEOREM. *A provable formula of PP_1 holds in all partial Boolean algebras (is "Q-valid").*

Proof. Assume $f\to\phi\vdash\phi$; then $f\to\phi\Vdash\phi$ by the preceding theorem. By 6.1, ϕ is valid iff $f\to\phi\Vdash\phi$ holds.

7.3. The rest of the paper is devoted to the proof of the converse: If ϕ holds in all partial Boolean algebras, then ϕ is provable in PP_1. By 6.1, it suffices to show: If $f\to\phi\Vdash\phi$ holds in all partial Boolean algebras, then $f\to\phi\vdash\phi$ is a derivable rule of PP_1.

7.4. It might be suspected that $\phi_1,\dots,\phi_m\vdash\psi$ follows generally from $\phi_1,\dots,\phi_m\Vdash\psi$. This is not so as shown by the following counterexample. Clearly $f\Vdash x_0$ holds in all partial Boolean algebras as there is no q such that $[f](q)=1$. However, $f\vdash x_0$ is not a derivable rule. For, if the variable x_0 does not occur in the premise of the rules $R_1,\dots,R_7$, neither does it occur in the conclusion. $f \vdash \phi$ is therefore derivable only for formulas ϕ not containing x_0. The system PP_1 can be made complete in the above strong sense by adjoining the infinite list of axioms $fx_0, fx_1, \ldots$

7.5. We shall state a series of derivable rules, numbered $S_1, S_2, \ldots$ For clarity, they will be included in brackets:

$$S_i\colon [\phi_1, \ldots, \phi_m \vdash \psi].$$

Proofs of such rules will be given in the following form

$$[\gamma_1; \gamma_2; \ldots; R_2\colon\gamma_5; \ldots; S_2\colon\gamma_7; \ldots; D\colon\gamma_9; \ldots; \gamma_n]$$

$\gamma_1, \ldots, \gamma_n$ will be formulas of P_1, γ_n the formula ψ. A formula γ_k $(1 \leqq k \leqq n)$ not preceded by some R_i, S_i or D is one of the formulas $\phi_1, \ldots, \phi_m$, $\gamma_1, \ldots, \gamma_{k-1}$. If γ_k is preceded by R_i (or: by S_i), it follows from $\gamma_{k-t_i}, \ldots, \gamma_{k-1}$ by the rule R_i (or: S_i), where t_i is the number of formulas in the premise of R_i (or: S_i). If γ_k is preceded by D, it is obtained from γ_{k-1} by substituting w (truth) for the subformula ff or by substituting ff for w.

VIII

Derivable rules:

S_1: $[f\phi\psi \vdash f\psi\phi]$
$[f\phi\psi; R_4\colon\phi\psi\phi; R_1\colon f\phi\psi\phi; R_2\colon f\psi\phi]$

S_2: $[f\phi \vdash f\phi f]$
$[f\phi; R_1\colon ff\phi; S_1\colon f\phi f]$

S_3: $[f\phi\psi \vdash f\phi]$
$[f\phi\psi; S_1\colon f\psi\phi; R_2\colon f\phi]$

S_4: $[f\phi\psi\chi \vdash f\chi]$
$[f\phi\psi\chi; R_2\colon f\psi\chi; R_2\colon f\chi]$

S_5: $[f\phi\psi\chi \vdash f\psi]$
$[f\phi\psi\chi; R_2\colon f\psi\chi; S_3\colon f\psi]$

S_6: $[\phi \vdash ff]$
$[\phi; R_1\colon f\phi; R_1\colon ff\phi; S_3\colon ff]$

DEFINITION: w is ff

S_7: $[\phi \vdash w]$

$[\phi;\ S_6{:}ff;\ D{:}w]$

S_8: $[f\phi \vdash f\phi w]$

$[f\phi;\ S_7{:}w;\ S_6{:}ff;\ R_1{:}fff;\ f\phi;\ S_2{:}f\phi f;\ fff;\ f\phi f;\ f\phi f;\ R_5{:}\ (\phi ff)(\phi f)(\phi f);\ R_1{:}f(\phi ff)(\phi f)(\phi f);\ S_3{:}f\phi ff;\ D{:}f\phi w]$

S_9: $[f\phi \vdash \phi w]$

$[f\phi;\ S_8{:}f\phi w;\ S_1{:}fw\phi;\ R_4{:}w\phi w;\ S_7{:}w;\ w\phi w;\ R_6{:}\phi w]$

S_{10}: $[f\phi \vdash \phi w\phi]$

$[f\phi;\ S_8{:}f\phi w;\ R_4{:}\phi w\phi]$

S_{11}: $[f\phi\psi \vdash f\phi\psi\phi]$

$[f\phi\psi;\ R_4{:}\phi\psi\phi;\ R_1{:}f\phi\psi\phi]$

S_{12}: $[f\phi\phi,\ f\phi\psi \vdash f\phi\phi\psi]$

$[f\phi\psi; f\phi\phi; f\phi\psi;\ R_5{:}(\phi\phi\psi)(\phi\phi)(\phi\psi);\ R_1{:}f(\phi\phi\psi)(\phi\phi)(\phi\psi);\ S_3{:}f\phi\phi\psi]$

S_{13}: $[f\phi\psi,\ f\phi\chi,\ f\psi\chi \vdash f\phi\psi\chi]$

$[f\psi\chi;\ f\phi\psi;\ f\phi\chi;\ R_5{:}(\phi\psi\chi)(\phi\psi)(\phi\chi);\ R_1{:}f(\phi\psi\chi)(\phi\psi)(\phi\chi);\ S_3{:}f\phi\psi\chi]$

S_{14}: $[f\phi\psi,\ f\phi\chi,\ f\psi\chi \vdash f(\phi\psi)\chi]$

$[f\psi\chi;\ S_1{:}f\chi\psi;\ f\phi\chi;\ S_1{:}f\chi\phi;\ f\chi\psi;\ f\phi\psi;\ S_{13}{:}f\chi\phi\psi;\ S_1{:}f(\phi\psi)\ S_1{:}f(\phi\psi)\chi]$

S_{15}: $[f\phi\psi,\ f\phi\chi,\ \psi\chi \vdash \phi\psi\chi]$

$[f\phi\psi;\ S_1{:}f\psi\phi;\ f\phi\chi;\ S_1{:}f\chi\phi;\ \psi\chi;\ R_1{:}f\psi\chi;\ f\psi\phi;\ f\chi\phi;\ S_{14}{:}f(\psi\chi)\phi;\ R_4{:}(\psi\chi)\phi(\psi\chi);\ \psi\chi;\ (\psi\chi)\phi(\psi\chi);\ R_6{:}\phi(\psi\chi)]$

S_{16}: $[f\phi\psi,\ f\phi\chi,\ \psi\chi \vdash (\phi\psi)(\phi\chi)]$

$[f\psi\chi;\ f\phi\psi;\ f\phi\chi;\ R_5{:}(\phi\psi\chi)(\phi\psi)(\phi\chi);\ f\phi\psi;\ f\phi\chi;\ \phi\chi;\ S_{15}{:}\phi\psi\chi;\ (\phi\psi\chi)(\phi\psi)(\phi\chi){:}\ R_6{:}(\phi\psi)(\phi\chi)]$

S_{17}: $[f\phi\phi,\ f\phi\psi,\ f\phi\chi,\ \phi\psi\chi \vdash \psi\phi\chi]$

$[\phi\psi\chi;\ R_1{:}f\phi\psi\chi;\ R_2{:}f\psi\chi;\ f\phi\psi;\ f\phi\chi;\ R_5{:}(\phi\psi\chi)(\phi\psi)(\phi\chi);\ \phi\psi\chi;\ (\phi\psi\chi)(\phi\psi)(\phi\chi);\ R_6{:}(\phi\psi)(\phi\chi);\ R_1{:}f(\phi\psi)(\phi\chi);\ \phi\psi\chi;\ R_1{:}f\phi\psi\chi;\ R_2{:}f\psi\chi;\ f\phi\psi;\ S_1{:}f\psi\phi;\ f\psi\chi;\ f\phi\chi;\ S_{13}{:}f\psi\phi\chi;\ f\psi\phi;\ R_4{:}\psi\phi\psi;\ R_1{:}f\psi\phi\psi;\ f\psi\phi\chi;\ (\phi\psi)(\phi\chi);\ S_{16}{:}(\psi\phi\psi)(\psi\phi\chi);\ \psi\phi\psi;\ (\psi\phi\psi)(\psi\phi\chi);\ R_6{:}\psi\phi\chi]$

S_{18}: $[f\phi\phi,\ \phi\psi\phi \vdash \psi\phi\phi]$

$[\phi\psi\phi;\ R_1{:}f\phi\psi\phi;\ R_2{:}f\psi\phi;\ S_1{:}f\phi\psi;\ f\phi\phi;\ f\phi\psi;\ f\phi\phi;\ \phi\psi\phi;\ S_{17}{:}\psi\phi\phi]$

S_{19}: $[f\phi \vdash f\phi w]$

$[f\phi;\ S_9{:}\phi w;\ R_1{:}f\phi w]$

S_{20}: $[f\phi \vdash fw\phi]$

$[f\phi;\ S_{19}{:}f\phi w;\ S_1{:}fw\phi]$

S_{21}: $[f\phi \vdash f(w\phi)\,w]$

$[f\phi;\ S_{20}{:}fw\phi;\ S_{20}{:}fww\phi;\ S_{21}{:}f(w\phi)\,w]$

S_{22}: $[f\phi \vdash f\phi w\phi]$

$[f\phi;\ S_{10}{:}\phi w\phi;\ R_1{:}f\phi w\phi]$

S_{23}: $[f\phi \vdash f(w\phi)\,\phi]$

$[f\phi;\ S_{22}{:}f\phi w\phi;\ S_1{:}f(w\phi)\,\phi]$

S_{24}: $[f\phi \vdash f(w\phi)(w\phi)]$

$[f\phi;\ S_{20}{:}fw\phi;\ f\phi;\ S_{21}{:}f(w\phi)\,w;\ f\phi;\ S_{23}{:}f(w\phi)\,\phi;\ f(w\phi)\,w;$
$fw\phi;\ S_{13}{:}f(w\phi)(w\phi)]$

S_{25}: $[f\phi \vdash (w\phi)(w\phi)]$

$[f\phi;\ S_{21}{:}f(w\phi)\,w;\ R_4{:}(w\phi)\,w(w\phi);\ f\phi;\ S_{24}{:}f(w\phi)(w\phi);$
$(w\phi)\,w(w\phi);\ S_{18}{:}w(w\phi)(w\phi);\ S_7{:}w;\ w(w\phi)(w\phi);$
$R_6{:}(w\phi)(w\phi)]$

S_{26}: $[f\phi \vdash (w\phi)\,\phi]$

$[f\phi;\ S_{25}{:}(w\phi)(w\phi);\ f\phi;\ S_{23}{:}f(w\phi)\,\phi;\ f\phi;\ S_{21}{:}f(w\phi)\,w;\ f\phi;$
$S_{24}{:}f(w\phi)(w\phi);\ f(w\phi)\,w;\ f(w\phi)\,\phi;\ (w\phi)(w\phi);\ S_{17}{:}w(w\phi)\,\phi;$
$S_7{:}w;\ w(w\phi)\,\phi;\ R_6{:}(w\phi)\,\phi]$

S_{27}: $[f\phi \vdash f\phi\phi]$

$[f\phi;S_{10}{:}\phi w\phi;f\phi;S_{26}{:}(w\phi)\,\phi;f\phi;S_{22}{:}f\phi w\phi;(w\phi)\,\phi;\phi w\phi;$
$R_7{:}f\phi\phi]$

S_{28}: $[f\phi\psi \vdash f\phi\phi]$

$[f\phi\psi;\ S_3{:}f\phi;\ S_{27}{:}f\phi\phi]$

S_{29}: $[f\phi\psi \vdash f\psi\psi]$

$[f\phi\psi;\ S_1{:}f\psi\phi;\ S_{28}{:}f\psi\psi]$

S_{30}: $[f\phi\psi \vdash f\phi\phi\psi]$

$[f\phi\psi;\ S_{28}{:}f\phi\phi;\ f\phi\psi;\ S_{12}{:}f\phi\phi\psi]$

S_{31}: $[f\phi\psi,\ f\phi\chi,\ \phi\psi\chi \vdash \psi\phi\chi]$

$[f\phi\psi;\ S_{28}{:}f\phi\phi;\ f\phi\psi;\ f\phi\chi;\ \phi\psi\chi;\ S_{17}{:}\psi\phi\chi]$

S_{32}: $[\phi\psi\phi \vdash \psi\phi\phi]$

$[\phi\psi\phi;\ R_1{:}f\phi\psi\phi;\ S_{28}{:}f\phi\phi;\ \phi\psi\phi;\ S_{18}{:}\psi\phi\phi]$

S_{33}: $[f\phi \vdash \phi\phi]$

$[f\phi;\ S_{10}{:}\phi w\phi;\ S_{32}{:}w\phi\phi;\ S_7{:}w;\ w\phi\phi;\ R_6{:}\phi\phi]$

S_{34}: $[f\phi\psi \vdash f(\phi f)\,\psi]$

$[f\phi\psi;\ R_2{:}f\psi;\ R_1{:}ff\psi;\ f\phi\psi;\ S_3{:}f\phi;\ S_2{:}f\phi f;\ f\phi\psi;\ ff\psi;$
$S_{13}{:}f(\phi f)\,\psi)]$

S_{35}: $[f\phi\psi \vdash f\phi\psi f]$

$[f\phi\psi;\ S_1{:}f\psi\phi;\ S_{34}{:}f(\psi f)\,\phi;\ S_1{:}f\phi\psi f]$

S_{36}: $[f\phi\psi, f\phi\chi, f\psi\chi \vdash f((\phi f)\,\psi)\,\chi]$

$[f\phi\chi;\ S_{34}{:}f(\phi f)\,\chi;\ f\phi\psi;\ S_{34}{:}f(\phi f)\,\psi;\ f(\phi f)\,\chi;\ f\psi\chi;$
$S_{13}{:}f((\phi f)\,\psi)\,\chi]$

IX

9.1. Let $P_1^n(y)$ be the system P_1 introduced in Section II in which the series $x_0, x_1, \ldots$ is replaced by $y_0, y_1, \ldots$ and where no other variables than $y_0, \ldots, y_n$ occur. If γ is a formula of $P_1(y)$ and $*$ is an n-sequence $\langle\phi_0, \ldots, \phi_n\rangle$ of formulas of P_1 then γ^* is the result of substituting ϕ_i for y_i $(i=1, \ldots, n)$. We have $f^* = f$, $y_0^* = \phi_0, \ldots; (\gamma_1\gamma_2)^* = \gamma_1^*\gamma_2^*$. If $*$ is the sequence $\langle\phi_0, \ldots, \phi_n\rangle$ of formulas of P_1, then f^{**} is the following sequence of formulas: It is $f\phi_0\phi_0$ in case $n=0$; *it is* $f\phi_0\phi_1, \ldots, f\phi_i\phi_j$ $(i<j), \ldots, f\phi_{n-1}\phi_n$ in case $n \geqq 1$. We state a metarule

M_1: *If* γ_1, γ_2 *are formulas of* $P_1^n(y)$ *and if* $*$ *is an n-sequence of formulas of* P_1, *then* $[f^{**} \vdash f\gamma_1^*\gamma_2^*]$ *is a derivable rule.*

The proof (by induction) follows from the rules R_2, S_3, S_{13}, S_{27}.

9.2. M_2: *If the formula* γ *of* $P_1^n(y)$ *is an identity of the classical propositional calculus and if* $*$ *is an n-sequence of formulas of* P_1, *then*

$$[f^{**} \vdash \gamma^*]$$

is a derivable rule.

Proof. γ being an identity, there exists by Wajsberg [4], p. 138, a sequence $\langle\gamma_1, \ldots, \gamma_m\rangle$, $\gamma_m = \gamma$, of formulas of $P_1^n(y)$ having the following property:

For each i, $1 \leqq i \leqq m$, one of the following alternatives hold:

(a) there exist formulas ϕ, ψ, χ of $P_1^n(y)$ such that γ_i is one of the following formulas ("γ_i is an axiom")

(a_1) $f\phi$

(a_2) $\phi\psi\phi$

(a_3) $(\phi\psi\chi)\,(\phi\psi)\,(\phi\chi)$.

(b) There exists j, $j<i$, such that γ_j is $(\gamma_i f)\,f$.

(c) There exist j, k, $j<i$, $j<k$ such that γ_k is $\gamma_j\gamma_i$.

We describe a modification of the sequence $\langle\gamma_1^*,\ldots,\gamma_m^*\rangle$ which transforms it into a proof of $[f^{**}\vdash\gamma_m^*]$. The formula γ_i^* will be replaced by one of the following sequences (the last formula being γ_i^* itself):

(a$_1$) $M_1{:}f\phi^*\phi^*;\ R_2{:}f\phi^*$

(a$_2$) $M_1{:}f\phi^*\psi^*;\ R_4{:}\phi^*\psi^*\phi^*$

(a$_3$) $M_1{:}f\psi^*\chi^*;\ M_1{:}f\phi^*\psi^*;\ M_1{:}f\phi^*\chi^*;\ R_5{:}(\phi^*\psi^*\chi^*)(\phi^*\psi^*)(\phi^*\chi^*)$

(b) $(\gamma_i^* f)\,f;\ R_3{:}\gamma_i^*$

(c) $\gamma_j^*;\ \gamma_j^*\gamma_i^*;\ R_6{:}\gamma_i^*$

9.3. The following rules S_{37}, S_{38}, S_{39} are special cases of the metarule M_2:

S_{37}: $[f\phi\phi\vdash((\phi f)f)\,\phi]$

S_{38}: $[f\phi\phi\vdash\phi(\phi f)f]$

S_{39}: $[f\phi_1\phi_2,\ f\phi_1\psi_1,\ f\phi_1\psi_2,\ f\phi_2\psi_1,\ f\phi_2\psi_2,\ f\psi_1\psi_2\vdash(\psi_1\psi_2)(\phi_2\phi_1)(\phi_1\psi_1)(\phi_2\psi_2)]$

Rule S_{40} follows easily from R_6, R_7, S_1, S_{39}.

S_{40}: $[f\phi_1\psi_1,\ \phi_1\phi_2,\ \phi_2\phi_1,\ \psi_1\psi_2,\ \psi_2\psi_1\vdash(\phi_1\psi_1)(\phi_2\psi_2)]$

9.4. We proceed to prove the substitutivity property of equivalence. Let γ be a formula of $P_1''(y)$, let $\langle\phi_0',\phi_1,\ldots,\phi_n\rangle$ and $\langle\phi_0'',\phi_1,\ldots,\phi_n\rangle$ be n-sequences of formulas of P_1^*; let γ_1^* be the formula corresponding to the first, γ_2^* the formula corresponding to the second sequence. We then have the two following metarules:

M_3: $[f\gamma_1^*,\ \phi_0'\phi_0'',\ \phi_0''\phi_0'\vdash\gamma_1^*\gamma_2^*]$

M_4: $[\gamma_1^*,\ \phi_0'\phi_0'',\ \phi_0''\phi_0'\vdash\gamma_2^*]$

The proof of M_3 is by induction with respect to the length of γ; the inductive step is provided by S_{40}.

Proof. of M_4: $[\gamma_1^*;\ R_1{:}f\gamma_1^*;\ \phi_0'\phi_0'';\ \phi_0''\phi_0';\ M_3{:}\gamma_1^*\gamma_2^*;\ \gamma_1^*;\ \gamma_1^*\gamma_2^*;\ R_6{:}\gamma_2^*]$.

X

THEOREM. *If the formula ϕ of PP_1 holds in all partial Boolean algebras* (as defined in [2]), *then $f\to\phi\vdash\phi$ is a derivable rule of PP_1.*

Instead of giving the (rather tedious) reduction of this completeness theorem to the one given in [2], we outline the adaptation of the proof in [2] to the present case.

10.1. With each formula ϕ of P_1 we associate a partial Boolean algebra B_ϕ such that $f \to \phi \vdash \phi$ is a derivable rule of PP_1, if ϕ holds in B_ϕ. Let Ω be the set of formulas ψ of P_1 such that $[f\phi \vdash f\psi]$ is a derivable rule and let the relation $\simeq$ on $\Omega \times \Omega$ be defined as follows: $\psi_1 \simeq \psi_2$ iff the rules $[f\phi \vdash \psi_1\psi_2]$ and $[f\phi \vdash \psi_2\psi_1]$ are derivable. Ω and $\simeq$ have the following properties:

(1) $f \in \Omega$, $w \in \Omega$, $\phi \in \Omega$ (S_6, S_7, R_1)

(2) *If* $\psi_1\psi_2 \in \Omega$, *then* $\psi_i \in \Omega$ $(i=1, 2$; *by* $R_2, S_3)$

(3) $\simeq$ *is an equivalence relation* (S_{33}, M_4)

(4) *If* $\psi \simeq \psi'$, *then* $\psi f \simeq \psi' f$ (S_{40})

(5) *If* $\psi_1\psi_2 \in \Omega$, *then* $((\psi_1 f) f)\psi_2 \in \Omega$ *and* $((\psi_1 f) f)\psi_2 \simeq \psi_1\psi_2$ (S_{37}, S_{38})

(6) *If* $\psi_1 \simeq \psi'_1$ *and* $\psi_2 \simeq \psi'_2$, *then* $\psi_1\psi_2 \in \Omega$ *iff* $\psi'_1\psi'_2 \in \Omega (M_4)$

(7) *If* $\psi_1 \simeq \psi'_1$, $\psi_2 \simeq \psi'_2$ *and* $\psi_1\psi_2 \in \Omega$, *then* $(\psi_1 f)\psi_2 \simeq (\psi'_1 f)\psi'_2 (M_4)$

(8) $\psi \simeq w$ *iff* $[f\phi \vdash \psi]$ *is derivable.*

Proof. Assume $[f\phi \vdash \psi]$; then $[f\phi \vdash \psi w]$ by S_9, $[f\phi \vdash w\psi]$ by S_{10}, R_6. If $[f\phi \vdash w\psi]$, then $[f\phi \vdash \psi]$ by R_6.

10.2. We define a structure $B_\phi = \langle B, \delta, 0, 1, \rightharpoondown, \vee \rangle$ of type *PB* (cf. Section III):

(a) B is the set of equivalence classes of the relation $\simeq$ on Ω.

(b) $\delta(a_1, a_2)$ holds for $a_i \in B$ $(i=1, 2)$ iff there exist formulas $\psi_1 \in a_i$ $(i=1, 2)$ such that $[f\phi \vdash f\psi_1\psi_2]$ is derivable. By (6), $\delta(a_1, a_2)$ iff $[f\phi \vdash f\psi_1\psi_2]$ is derivable for all formulas $\psi_i \in a_i$ $(i=1, 2)$.

(c) 0 is the class of f, 1 is the class of w (f, $w \in \Omega$ by (1)).

(d) By (4), there exists for every class $a \in B$ a class $b \in B$ such that $(\psi f) \in b$ if $\psi \in a$; let this class b be $\rightharpoondown a$.

(e) Assume $a_i \in B$, $\psi_i \in a_i$, $\psi'_i \in a_i$ $(i=1, 2)$ and $\delta(a_1, a_2)$. Then $\psi_1\psi_2 \in \Omega$, $\psi'_1\psi'_2 \in \Omega$ and the formulas $(\psi_1 f)\psi_2$, $(\psi'_1 f)\psi'_2$ belong to the same class b: let $a_1 \vee a_2$ be this class b.

10.3. *The structure* B_ϕ *defined in* 10.2 *is a partial Boolean algebra*, i.e. it satisfies the following 5 axioms of [2]:

(A1) The relation $\mathfrak{d}$ is symmetric and reflexive (symmetry by S_1, reflexivity by S_{28}).

(A2) For all $b \in B$: $\mathfrak{d}(b, 1)$, $\mathfrak{d}(b, 0)$ (S_2 and S_8).

(A3) The partial function $\vee$ is defined exactly for those pairs $\langle b_1, b_2 \rangle$ for which $\mathfrak{d}(b_1, b_2)$ (by definition).

(A4) If $\mathfrak{d}(b_1, b_2)$, $\mathfrak{d}(b_1, b_3)$ and $\mathfrak{d}(b_2, b_3)$, then $\mathfrak{d}(b_1 \vee b_2, b_3)$, $\mathfrak{d}(\rightharpoondown b_1, b_2)$ (the first conclusion by S_{36}, the second by S_{34}).

(A5) For all $b_0, b_1, b_2 \in B$: If $\mathfrak{d}(b_0, b_1)$, $\mathfrak{d}(b_0, b_2)$ and $\mathfrak{d}(b_1, b_2)$, then the Boolean polynomials in b_0, b_1, b_2 form a Boolean algebra.

Proof. By 4.8, it suffices to show: If P is a Boolean polynomial such that $P(y_0, y_1, y_2) = 1$ in the Boolean sense, then $P(b_0, b_1, b_2) = 1$ in B_ϕ. Let γ be the formula of $P_1^2(y)$ translating the polynomial P (the translation of $y_0 \vee y_1$ being $(y_0 f)\, y_1$ etc.); $P = 1$ being a Boolean identity, γ is an identity of the classical propositional calculus. Assume $\psi_i \in b_i$ $(i = 0, 1, 2)$, $* = \langle \psi_0, \psi_1, \psi_2 \rangle$ and $\mathfrak{d}(b_0, b_1)$, $\mathfrak{d}(b_0, b_2)$, $\mathfrak{d}(b_1, b_2)$. We then have f^{**} and by M_4: $[f^{**} \vdash \gamma^*]$, i.e. $[f\phi \vdash \gamma^*]$; by (8) of 10.1 therefore $\gamma^* \simeq w$, i.e. $\gamma^* \in 1$. The formula γ^* is an element of $P(b_0, b_1, b_2)$: The class of $(\phi_0 f)\, \phi_1$ is by definition $\{(\psi_0 f) f\} \vee \{\psi_1\}$ which is the same as $\{\psi_0\} \vee \{\psi_1\}$, i.e. $b_0 \vee b_1$. We therefore have $\gamma^* \in 1$, $\gamma^* \in P(b_0, b_1, b_2)$, i.e. $P(b_0, b_1, b_2) = 1$.

10.4. *There exists a sequence $q \in B^N$ such that for all formulas ψ of Ω, $q \in D_\psi$ and $[\psi](q) = \{\psi\}$* (equivalence class of ψ).

Proof. q is defined as follows: If the variable x_n is an element of Ω, then $q(n) = \{x_n\}$; otherwise $q(n) = 0$. The theorem is then proved by induction with respect to the length of ψ. If ψ is a variable or f, it holds by definition. Assume therefore $\psi = \psi_1 \psi_2$; then $\psi_1, \psi_2 \in \Omega$ ((2) of 10.1) and $q \in D_{\Psi_i}$, $[\psi_i](q) = \{\psi_i\}$ $(i = 1, 2)$ by the hypothesis of the induction. In order to prove $q \in D\psi_1\psi_2$, we have to show: $\mathfrak{d}(\rightharpoondown[\psi_1](q), [\psi_2](q))$. We have $\rightharpoondown[\psi_1](q) = \rightharpoondown\{\psi_1\} = \{\psi_1 f\}$; $[\psi_2](q) = \{\psi_2\}$; we therefore have to show $[f\phi \vdash f(\psi_1 f)\, \psi_2]$, which follows immediately from S_{34}. Furthermore $[\psi_1\psi_2](q) = \rightharpoondown(\psi_1](q) \vee [\psi_2](q) = \rightharpoondown\{\psi_1\} \vee \{\psi_2\} = \{\psi_1 f\} \vee \{\psi_2\} = \{((\psi_1 f) f)\, \psi_2\}$; by (5) of 10.1, $((\psi_1 f) f)\, \psi_2 \simeq \psi_1 \psi_2$ and therefore $[\psi_1\psi_2](q) = \{\psi_1\psi_2\}$.

10.5. *If the formula ψ of Ω holds in the partial Boolean algebra B_ϕ, then $[f\phi \vdash \psi]$ is a derivable rule of PP_1.*

Proof. Let q be the sequence defined in 10.4; then $[\psi](q) = \{\psi\}$. If ψ

holds in B_ϕ, then $[\psi](q)=1$, i.e. $\{\psi\}=1$, $\psi \simeq w$. By (8) of 10.1, $\psi \simeq w$ iff the rule $[f\phi \vdash \psi]$ is derivable.

By (1) of 10.1, ϕ is a formula of Ω. Therefore:

ϕ holds in the partial Boolean algebra B_ϕ iff $[f \rightarrow \phi \vdash \phi]$ is a derivable rule of PP_1.

Cornell University, Ithaca
Eidgenössische Technische Hochschule, Zürich

BIBLIOGRAPHY

[1] Church, A., *Introduction to Mathematical Logic*, vol. 1, Princeton University Press, Princeton, 1956.
[2] Kochen, S. and Specker, E. P., 'Logical Structures Arising in Quantum Theory', to appear in the *Proc. of the Model Theory Symp.* held in Berkeley, June–July 1963.
[3] Specker, E., 'Die Logik nicht gleichzeitig entscheidbarer Aussagen', *Dialectica* **13** (1960), 239–246.
[4] Wajsberg, M., 'Metalogische Beiträge II', *Wiadomości Matematyczne* **47** (1939), 119–139.

SIMON KOCHEN AND E. P. SPECKER

THE PROBLEM OF HIDDEN VARIABLES IN QUANTUM MECHANICS

(Communicated by A. M. Gleason)

0. INTRODUCTION

Forty years after the advent of quantum mechanics the problem of hidden variables, that is, the possibility of imbedding quantum theory into a classical theory, remains a controversial and obscure subject. Whereas to most physicists the possibility of a classical reinterpretation of quantum mechanics remains remote and perhaps irrelevant to current problems, a minority have kept the issue alive throughout this period. (See Freistadt [5] for a review of the problem and a comprehensive bibliography up to 1957.) As far as results are concerned there are on the one hand purported proofs of the non-existence of hidden variables, most notably von Neumann's proof, and on the other, various attempts to introduce hidden variables such as de Broglie [4] and Bohm [1] and [2]. One of the difficulties in evaluating these contradictory results is that no exact mathematical criterion is given to enable one to judge the degree of success of these proposals.

The main aim of this paper is to give a proof of the nonexistence of hidden variables. This requires that we give at least a precise necessary condition for their existence. This is carried out in Sections I and II. The proposals in the literature for a classical reinterpretation usually introduce a phase space of hidden pure states in a manner reminiscent of statistical mechanics. The attempt is then shown to succeed in the sense that the quantum mechanical average of an observable is equal to the phase space average. However, this statistical condition does not take into account the algebraic structure of the quantum mechanical observables. A minimum such structure is given by the fact that some observables are functions of others. This structure is independent of the particular theory under consideration and should be preserved in a classical reinterpretation. That this is not provided for by the above statistical condition is easily shown by constructing a phase space in which the statistical condition is satisfied but the quantum mechanical observables

C. A. Hooker (ed.), The Logico-Algebraic Approach to Quantum Mechanics, 293–328.

become interpreted as independent random variables over the space.

The algebraic structure to be preserved is formalized in Section II in the concept of a partial algebra. The set of quantum mechanical observables viewed as operators on Hilbert space form a partial algebra if we restrict the operations of sum and product to be defined only when the operators commute. A necessary condition then for the existence of hidden variables is that this partial algebra be imbeddable in a commutative algebra (such as the algebra of all real-valued functions on a phase space). In Sections III and IV it is shown that there exists a finite partial algebra of quantum mechanical observables for which no such imbedding exists. The physical description of this result may be understood in an intuitive fashion quite independently of the formal machinery introduced. An electric field of rhombic symmetry may be applied to an atom of orthohelium in its lowest energy state in any one of a specified finite number of directions. The proposed classical interpretation must then predict the resulting change in the energy state of the atom in every one of these directions. For each such prediction there exists a direction in this specified set in which the field may be applied such that the predicted value is contradicted by the experimentally measured value.

The last section deals with the logic of quantum mechanics. It is proved there that the imbedding problem we considered earlier is equivalent to the question of whether the logic of quantum mechanics is essentially the same as classical logic. The precise meaning of this statement is given in that section. Roughly speaking a propositional formula $\psi(x_1, \ldots, x_n)$ is valid in quantum mechanics if for every "meaningful" substitution of quantum mechanical propositions P_i for the variables x_i this formula is true, where a meaningful substitution is one such that the propositions P_i are only conjoined by the logical connectives in $\psi(P_1, \ldots, P_n)$ if they are simultaneously measurable. It then follows from our results that there is a formula $\varphi(x_1, \ldots, x_{86})$ which is a classical tautology but is false for some meaningful substitution of quantum mechanical propositions. In this sense the logic of quantum mechanics differs from classical logic. The positive problem of describing quantum logic has been studied in Kochen and Specker [10] and [11].

In Section V the present proof has been compared with von Neumann's well-known proof of the non-existence of hidden variables. Von Neumann's proof is essentially based on the non-existence of a real-valued function on the set of quantum mechanical observables which is multiplicative on commuting observables and linear. In our proof we show the non-existence of a real-valued function which is both multiplicative and linear only on commuting observables. Thus, in a formal sense our result is stronger than von Neumann's. In Section V we attempt to show that this difference is essential. We show that von Neumann's criterion applies to a single particle of spin $\frac{1}{2}$, implying that there is no classical description of this system. On the other hand, we contradict this conclusion by constructing a classical model of a spin $\frac{1}{2}$ particle. This is done by imbedding the partial algebra of self-adjoint operators on a two-dimensional complex Hilbert space into the algebra of real-valued functions on a suitable phase space in such a way that the statistical condition is satisfied.

I. DISCUSSION OF THE PROBLEM

For our purposes it is convenient to describe a physical theory within the following framework. We are given a set $\mathcal{O}$ called the set of observables and a set S called the set of states. In addition, we have a function P which assigns to each observable A and each state ψ a probability measure $P_{A\psi}$ on the real line $\mathbf{R}$. Physically speaking, if U is a subset of $\mathbf{R}$ which is measurable with respect to $P_{A\psi}$, then $P_{A\psi}(U)$ denotes the probability that the measurement of A for a system in the state ψ yields a value lying in U. From this we obtain in the usual manner the expectation of the observable A for the state ψ,

$$\mathrm{Exp}_{\psi}(A) = \int_{-\infty}^{\infty} \lambda \, dP_{A\psi}(\lambda).$$

States are generally divided into two kinds, pure states and mixed states. Roughly speaking, the pure states describe a maximal possible amount of knowledge available in the theory about the physical system in question; the mixed states give only incomplete information and describe our ignorance of the exact pure state the system is actually in.

We illustrate these remarks with an example from Newtonian mechanics. Suppose we are given a system of N particles. Then each pure state ψ of the system is given by a $6N$-tuple $(q_1, \ldots, q_{3N}, p_1, \ldots, p_{3N})$ of real numbers denoting the coordinates of position and momentum of the particles. In this case, the probability $P_{A\psi}$ assigned to each observable is an atomic measure, concentrated on a single real number a. That is, $P_{A\psi}(U)=1$ if $a\in U$ and $P_{A\psi}(U)=0$ if $a\notin U$. Thus, if we introduce the phase space Ω of pure states, which we may here identify with a subset of $6N$-dimensional Euclidean space, then each observable A becomes associated with a real-valued function $f_A:\Omega\to\mathbf{R}$ given by $f_A(\psi)=a$.

If N is large it is not feasible to determine the precise pure state the system may be in. We resort in this case to the notion of a mixed state which gives only the probability that the system is in a pure state which lies in a region of Ω. More precisely, a mixed state ψ is described by a probability measure μ_ψ on the space Ω, so that, for each measurable subset Γ of Ω, $\mu_\psi(\Gamma)$ is the probability that the system is in a pure state lying in Γ. It follows immediately that the probability measure $P_{A\psi}$ assigned to an observable A and mixed state ψ is given by the formula

$$(1) \qquad P_{A\psi}(U)=\mu_\psi(f_A^{-1}(U)).$$

Thus, we have

$$(2) \qquad \mathrm{Exp}_\psi(A)=\int_\Omega f_A(\omega)\,d\mu_\psi(\omega).$$

In the case of quantum mechanics the set $\mathcal{O}$ of observables is represented by self-adjoint operators on a separable Hilbert space $\mathcal{H}$. The pure states are given by the one-dimensional linear subspaces of $\mathcal{H}$. The probability $P_{A\psi}$ is defined by taking the spectral resolution of A:

$$A=\int_{-\infty}^{\infty}\lambda\,dE_A(\lambda),$$

where E_A is the spectral measure corresponding to A. Then

$$P_{A\psi}(U)=\langle E_A(U)\,\psi,\psi\rangle,$$

where ψ is any unit vector in the one-dimensional linear subspace corresponding to the pure state ψ. Hence, by the spectral theorem

$$\mathrm{Exp}_\psi(A)=\int_{-\infty}^{\infty}\lambda\,d\langle E_A(U)\,\psi,\psi\rangle=\langle A\psi,\psi\rangle.$$

Although there may be states ψ for each observable A in this theory such that $P_{A\psi}$ is atomic, there are no longer, as in classical mechanics, states ψ such that $P_{A\psi}$ is atomic simultaneously for all observables.

The problem of hidden variables may be described within the preceding framework. Let us recall that the hidden variables problem was successfully solved in a classical case, namely, the theory of thermodynamics. The theory of macroscopic thermodynamics is a discipline which is independent of classical mechanics. This theory has its own set of observables such as pressure, volume, temperature, energy, and entropy and its own set of states. This theory shares with quantum mechanics the property that the probability $P_{A\psi}$ is not atomic even for pure states ψ. In most cases this probability is sufficiently concentrated about a single point so that it is in practice replaced by an atomic measure. However, there are cases where distinct macroscopic phenomena (such as critical opalescence) depend upon these fluctuations.

It proves possible in this case to introduce an underlying theory of classical mechanics on which thermodynamics may be based. In terms of the preceding description a phase space Ω of "hidden" pure states is introduced. In physical terms the system is assumed to consist of a large number of molecules and Ω is the space of the coordinates of position and momentum of all the molecules. Every pure state ψ of the original theory of thermodynamics is now interpreted as a mixed state of the new theory, *i.e.*, as a probability measure μ_ψ over the space Ω. Every observable A of thermodynamics is interpreted as a function $f_A:\Omega\rightarrow\mathbf{R}$, and it is assumed that condition (1) and hence (2) holds. It is in this way that the laws of thermodynamics become consequences of classical Newtonian mechanics *via* statistical mechanics. The formula (2) is the familiar statistical mechanical averaging process. This example has been considered as the classic case of a successful introduction of hidden variables into a theory.

The problem of hidden variables for quantum mechanics may be interpreted in a similar fashion as introducing a phase space Ω of hidden states for which condition (1) is true. This statistical condition (1) has in fact been taken as a proof of the success of various attempts to introduce a phase space into quantum mechanics. Now, in fact the condition (1) can hardly be the only requirement for the existence of hidden variables. For we may always introduce, at least mathematically, a phase space Ω

into a theory so that (1) is satisfied. To see this, let

$$\Omega=\mathbf{R}^{\mathcal{O}}=\{\omega \mid \omega:\mathcal{O}\to\mathbf{R}\}.$$

If $A\in\mathcal{O}$, let $f_A:\Omega\to\mathbf{R}$ be defined by $f_A(\omega)=\omega(A)$. If $\psi\in S$, let

$$\mu_\psi=\prod_{A\in\mathcal{O}} P_{A\psi},$$

the product measure of the probabilities $P_{A\psi}$. Then,

$$\mu_\psi f_A^{-1}(U)=\mu_\psi(\{\omega \mid \omega(A)\in U\})=P_{A\psi}(U).$$

We have two reasons for mentioning this somewhat trivial construction. First, in the various attempts to introduce hidden variables into quantum mechanics, the only explicitly stated requirement that is to be fulfilled is the condition (1). (See Bohm [1] and [2], Bopp [3], Siegel and Wiener [16], and especially the review of [16] in Schwartz [15].) Of course, the above space Ω is far more artificial than the spaces proposed in these papers, but the only purpose here was to point out the insufficiency of the condition (1) as a test for the adequacy of the solution of the problem.

Our second reason for introducing the space $\mathbf{R}^{\mathcal{O}}$ is that it indicates the direction in which the condition (1) is inadequate. For each state ψ, as interpreted in the space $\mathbf{R}^{\mathcal{O}}$, the functions f_A are easily seen to be measurable functions with respect to the probability measure μ_ψ. In the language of probability theory the observables are thus interpreted as random variables for each state ψ. It is not hard to show furthermore that in this representation the observables appear as independent random variables.

Now it is clear that the observables of a theory are in fact not independent. The observable A^2 is a function of the observable A and is certainly not independent of A. In any theory, one way of measuring A^2 consists in measuring A and squaring the resulting value. In fact, this may be used as the *definition* of a function of an observable. Namely, we define the observable $g(A)$ for every observable A and Borel function $g:\mathbf{R}\to\mathbf{R}$ by the formula

(3) $$P_{g(A)\psi}(U)=P_{A\psi}(g^{-1}(U))$$

for each state ψ. If we assume that every observable is determined by the function P, *i.e.*, $P_{A\psi}=P_{B\psi}$ for every state ψ implies that $A=B$, then the formula (3) defines the observable $g(A)$. This definition coincides with the definition of a function of an observable in both quantum and classical mechanics.

Thus the measurement of a function $g(A)$ of an observable A is independent of the theory considered – one merely writes $g(a)$ for the value of $g(A)$ if a is the measured value of A. The set of observables of a theory thereby acquires an algebraic structure, and the introduction of hidden variables into a theory should preserve this structure. In more detail, we require for the successful introduction of hidden variables that a space Ω be constructed such that condition (1) is satisfied and also that

$$f_{g(A)}=g(f_A) \tag{4}$$

for every Borel function g and observable A of the theory. Note that this condition is satisfied in the statistical mechanical description of thermodynamics.

Our aim is to show that for quantum mechanics no such construction satisfying condition (4) is possible. However, condition (4) as it stands proves too unwieldy and we shall first replace it by a more tractable condition.

II. PARTIAL ALGEBRAS

We shall say that the observables A_i, $i\in I$, in a theory are *commeasurable* if there exists an observable B and (Borel) functions f_i, $i\in I$, such that $A_i=f_i(B)$ for all $i\in I$. Clearly in this case it is possible to measure the observables A_i, $i\in I$, simultaneously for it is only necessary to measure B and apply the function f_i to the measured value to obtain the value of A_i. In quantum mechanics a set $\{A_i \mid i\in I\}$ of observables is said to be simultaneously measurable if as operators they pairwise commute. A classical theorem on operators shows that this coincides with the above definition (see, *e.g.*, Neumark [12, Thm. 6]). (Note that as a result in the case of quantum mechanics the A_i, $i\in I$, are commeasurable if they are pairwise commeasurable.)

If A_1 and A_2 are commeasurable then we may define the observables $\mu_1A_1+\mu_2A_2$ and A_1A_2 for all real μ_1, μ_2. For then $A_1=f_1(B)$ and $A_2=$

$=f_2(B)$ for some observable B and functions f_1 and f_2. Hence we have

(5) $$\mu_1 A_1 + \mu_2 A_2 = (\mu_1 f_1 + \mu_2 f_2)(B),$$ $$A_1 A_2 = (f_1 f_2)(B).$$

With linear combinations and products of commeasurable observables defined the set of observables acquires the structure of a *partial algebra.* Note that condition (4) implies that the partial operations defined in (5) are preserved under the map f. These ideas will now be formalized in the following definitions.

DEFINITION. A set A forms a *partial algebra* over a field K if there is a binary relation ♀ (commeasurability) on A, (*i.e.*, $♀ \subseteq A \times A$), operations of addition and multiplication from ♀ to A, scalar multiplication from $K \times A$ to A, and an element 1 of A, satisfying the following properties:

(1) The relation ♀ is reflexive and symmetric, *i.e.*, $a ♀ a$ and $a ♀ b$ implies $b ♀ a$ for all $a, b \in A$.

(2) For all $a \in A$, $a ♀ 1$.

(3) The relation ♀ is closed under the operations, *i.e.*, if $a_i ♀ a_j$ for all $1 \leqq i, j \leqq 3$ then $(a_1 + a_2) ♀ a_3$, $a_1 a_2 ♀ a_3$ and $\lambda a_1 ♀ a_3$, for all $\lambda \in K$.

(4) If $a_i ♀ a_j$ for all $1 \leqq i, j \leqq 3$, then the values of the polynomials in a_1, a_2, a_3 form a commutative algebra over the field K.

It follows immediately from the definition of a partial algebra that if D is a set of pairwise commeasurable elements of A then the set D generates a commutative algebra in A.

We have defined the notion of a partial algebra over an arbitrary field K but there are two cases which are of interest to us. The first is the field $\mathbf{R}$ of real numbers and the second is the field Z_2 of two elements. For the case of a partial algebra over Z_2 we may define the Boolean operations in terms of the ring operations in the usual manner: $a \cap b = ab$, $a \cup b = a + b - ab$, $a' = 1 - a$. It follows that if $a_i ♀ a_j$, $1 \leqq i, j \leqq 3$, then the polynomials in a_1, a_2, a_3 form a Boolean algebra. We shall call a partial algebra over Z_2 a *partial Boolean algebra.* It is clear how we may define this notion directly in terms of the operations $\cap$, $\cup$, $'$. What makes a partial Boolean algebra important for our purposes is that the set of idempotent elements of a partial algebra $\mathcal{N}$ forms a partial Boolean algebra. This is a counterpart of the familiar fact that the set of idempotents of a commutative algebra forms a Boolean algebra.

We consider some examples of partial algebras. Let $H(U^\alpha)$ be the set of all self-adjoint operators on a complex Hilbert space U^α of dimension α. If we take the relation ♀ to be the relation of commutativity then $H(U^\alpha)$ forms a partial algebra over the field $\mathbf{R}$ of reals. In this case the idempotents are the projections of U^α. Thus the set $\mathbf{B}(U^\alpha)$ of projections forms a partial Boolean algebra. Because every projection corresponds uniquely to a closed linear subspace of U^α, we may alternatively consider $\mathbf{B}(U^\alpha)$ as the partial Boolean algebra of closed linear subspaces of U^α. The direct definition of the relation ♀ in this interpretation of $\mathbf{B}(U^\alpha)$ is: a ♀ b if there exists elements c, d, e in $\mathbf{B}(U^\alpha)$ which are mutually orthogonal with $a=c\oplus d$ and $b=d\oplus e$. Furthermore $a\cap b$ denotes the intersection of the two subspaces a and b, $a\cup b$ denotes the space spanned by a and b, and a' denotes the orthogonal complement of a.

We have seen that the set $\mathcal{O}$ of observables of a physical theory forms a partial algebra over $\mathbf{R}$ if we take ♀ to be the relation of commeasurability. If A is an idempotent in $\mathcal{O}$, then it follows from the definition of A^2, that the measured values of the observable A can only be 1 or 0. By identifying these values with truth and falsity we may consider each such idempotent observable as a proposition of the theory. (See von Neumann [19, Ch. III.5] for a more detailed discussion of this point.) Thus, the set of propositions of a physical theory form a partial Boolean algebra. It is a basic tenet of quantum theory that the set of its observables may be identified with a partial sub-algebra Q of $H(U^\omega)$, the partial algebra of self-adjoint operators on a separable complex Hilbert space. This implies then that the propositions of quantum mechanics form a partial Boolean sub-algebra $\mathcal{B}$ of $\mathbf{B}(U^\omega)$.

Every commutative algebra A forms a partial algebra if we take the relation ♀ to be $A\times A$. The following construction of a partial algebra is of interest because it gives us an alternative way of viewing partial algebras. Let C_i, $i\in I$, be a non-empty family of commutative algebras over a fixed field K which satisfy the following conditions:

(a) For every $i, j\in I$ there is a $k\in I$ such that $C_i\cap C_j=C_k$.

(b) If $a_1,\ldots, a_n$ are elements of $C=\bigcup_{i\in I} C_i$ such that any two of them lie in a common algebra C_i, then there is a $k\in I$ such that $a_1,\ldots, a_n\in C_k$.

The set C forms a partial algebra over K if we define the relations (i) a ♀ b, (ii) $ab=c$, and (iii) $a+b=c$ in C by the condition that there exist an $i\in I$ such that (i) $a, b\in C_i$, (ii) $ab=c$ in C_i and (iii) $a+b=c$ in C_i respectively. It is not difficult to show that every partial algebra is isomorphic to an algebra of this type. (We may thus view a partial algebra as a category in which the objects are commutative algebras and the maps are imbeddings.)

DEFINITION. A map $h: U \to V$ between two partial algebras over a common field K is a *homomorphism* if for all $a, b \in U$ such that $a ♀ b$ and all $\mu, \lambda \in K$,

$$h(a) ♀ h(b),$$
$$h(\mu a + \lambda b) = \mu h(a) + \lambda h(b),$$
$$h(ab) = h(a)\, h(b),$$
$$h(1) = 1.$$

Given this definition we may state what our condition (4) of Section I on the existence of hidden variables implies for the partial algebra Q of observables of quantum mechanics. The set $\mathbf{R}^{\Omega}$ of all functions $f: \Omega \to \mathbf{R}$ from a space Ω of hidden states into the reals forms a commutative algebra over $\mathbf{R}$. From the way in which the partial operations on the set of observables of a theory are defined (Equation (5)), condition (4) implies that there is an imbedding of the partial algebra into the algebra $\mathbf{R}^{\Omega}$. Our conclusion of this discussion is then the following:

A necessary condition for the existence of hidden variables for quantum mechanics is the existence of an imbedding of the partial algebra Q of quantum mechanical observables into a commutative algebra.

A possible objection to this conclusion is that the map of Q into the commutative algebra C need not be single-valued since a given quantum-mechanical observable may split into several observables in C. Thus, Q might be a homomorphic image of C. We shall meet this objection in Section V by showing that even such a many-valued map of Q into C does not exist.

Now if $\varphi: \mathcal{N} \to C$ is an imbedding of a partial algebra $\mathcal{N}$ into a commutative algebra, it follows immediately that φ restricted to the partial Boolean algebra of idempotents of $\mathcal{N}$ is an imbedding into the Boolean algebra of idempotents of C. Thus, the existence of hidden variables implies the existence of an imbedding of the partial Boolean algebra of propositions of quantum mechanics into a Boolean algebra. We may justify the last statement independently of the previous discussion. For the set of propositions of a classical reinterpretation of quantum mechanics must form a Boolean algebra. But the conjunction of two commeasurable propositions has the same meaning in quantum mechanics as in classical physics and so should be preserved in the classical interpretation.

Let $h: Q \to \mathbf{R}$ be a homomorphism of the partial algebra Q of quantum mechanical observables into $\mathbf{R}$. Physically speaking h may be considered as a *prediction* function which simultaneously assigns to every observable a predicted measured value. If we assume the existence of a hidden state space Ω, so that Q is imbeddable by a map f into the algebra $\mathbf{R}^{\Omega}$, then each hidden state $\omega \in \Omega$ defines such a homomorphism $h: Q \to \mathbf{R}$, namely $h(A) = f_A(\omega)$. Thus, the existence of hidden variables implies the existence of a large number of prediction functions. Every homomorphism $h: \mathcal{N} \to \mathbf{R}$ is by restriction a homomorphism of the partial Boolean algebra of idempotents onto Z_2. The following theorem characterizes the imbedding of a partial Boolean algebra into a Boolean algebra in terms of its homomorphisms onto Z_2.

THEOREM 0. *Let $\mathcal{N}$ be a partial Boolean algebra. A necessary and sufficient condition that $\mathcal{N}$ is imbeddable in a Boolean algebra B is that for every pair of distinct elements a, b in $\mathcal{N}$ there is a homomorphism $h: \mathcal{N} \to Z_2$ such that $h(a) \neq h(b)$.*

Proof. Suppose $\varphi: \mathcal{N} \to B$ is an imbedding. Since $\varphi(a) \neq \varphi(b)$ if $a \neq b$, there exists by the semi-simplicity property of Boolean algebras (see *e.g.*, Halmos [8, sect. 18, Lemma 1]), a homomorphism $h: B \to Z_2$ such that $h\varphi(a) \neq h\varphi(b)$. Hence $k = h\varphi$ is the required homomorphism of $\mathcal{N}$ onto Z_2.

To prove the converse, let S be the set of all non-trivial homomorphisms of $\mathcal{N}$ into Z_2. Define the map $\varphi: \mathcal{N} \to Z_2^S$ by letting $\varphi(a)$ be the function $g: S \to Z_2$ such that $g(h) = h(a)$ for every $h \in S$. Then it is easily checked that φ is an imbedding of $\mathcal{N}$ into the Boolean algebra Z_2^S.

The next two sections are devoted to showing that there does not exist even a single homomorphism of the partial Boolean algebra $\mathcal{B}$ of the propositions of quantum mechanics onto Z_2.

III. THE PARTIAL BOOLEAN ALGEBRA $\mathbf{B}(E^3)$

Let $\mathbf{B}(E^\alpha)$ denote the partial Boolean algebra of linear subspaces of α dimensional Euclidean space E^α. Our aim in this section is to show that there is a finite partial Boolean subalgebra D of $\mathbf{B}(E^3)$ such that there is no homomorphism $h:D\to Z_2$. In the next section we shall show that the elements of D in fact correspond to quantum mechanical observables.

Let D be a partial Boolean subalgebra of $\mathbf{B}(E^3)$ with a homomorphism $h:D\to Z_2$. If s_1, s_2, s_3 are mutually orthogonal one-dimensional linear subspaces of D, then

$$(6)\qquad \begin{aligned} &h(s_1)\cup h(s_2)\cup h(s_3)=h(s_1\cup s_2\cup s_3)=h(E^3)=1 \quad \text{and}\\ &h(s_i)\cap h(s_j)=h(s_i\cap s_j)=h(0)=0 \end{aligned}$$

for $1\leqq i\neq j\leqq 3$. Hence, exactly one of every three mutually orthogonal lines is mapped by h onto 1. If we replace the lines by lines of unit length then h induces a map $h^*:T\to\{0, 1\}$ from a subset T of the unit sphere S into $\{0, 1\}$ such that for any three mutually orthogonal points in T exactly one is mapped by h^* into 1.

It will be convenient in what follows to represent points on S by the vertices of a graph. Two vertices which are joined by an edge in the graph represent orthogonal points on S. When we say that a graph Γ is *realizable* on S we mean that there is an assignment of points of S to the vertices of Γ, distinct points for distinct vertices, with the orthogonality relations as indicated in Γ.

LEMMA 1. *The following graph Γ_1 is realizable on S.*

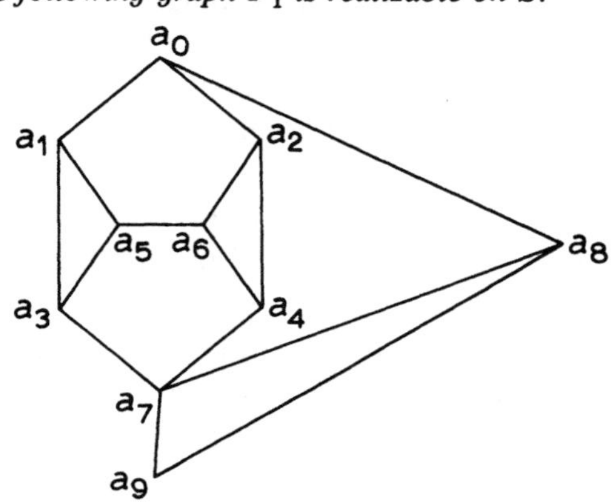

In fact, if p and q are points on S such that $0 \leqq \sin\theta \leqq \frac{1}{3}$ *where* θ *is the angle subtended by p and q at the center of S, then there exists a map* $u: \Gamma_1 \to S$ *such that* $u(a_0) = p$ *and* $u(a_9) = q$.

Proof. Since $u(a_8)$ is orthogonal to $u(a_0) \cup u(a_9)$ and $u(a_7)$ is orthogonal to $u(a_8)$, $u(a_7)$ lies in the plane $u(a_0) \cup u(a_9)$. Also since $u(a_7)$ is orthogonal to $u(a_9)$, we have that $\varphi = \pi/2 - \theta$, where φ is the angle subtended at the center of S by $u(a_0)$ and $u(a_7)$. Let $u(a_5) = \bar{i}$ and $u(a_6) = \bar{k}$. Then we may take

$$u(a_1) = (\bar{j} + x\bar{k})(1 + x^2)^{-1/2} \quad \text{and} \quad u(a_2) = (\bar{i} + y\bar{j})(1 + y^2)^{-1/2}.$$

The orthogonality conditions then force

$$u(a_3) = (x\bar{j} - \bar{k})(1 + x^2)^{-1/2},$$
$$u(a_4) = (y\bar{i} - \bar{j})(1 + y^2)^{-1/2},$$

and hence,

$$u(a_0) = (xy\bar{i} - x\bar{j} + \bar{k})(1 + x^2 + x^2y^2)^{-1/2},$$
$$u(a_7) = (\bar{i} + y\bar{j} + xy\bar{k})(1 + y^2 + x^2y^2)^{-1/2}.$$

Thus

$$\cos\varphi = \frac{xy}{((1 + x^2 + x^2y^2)(1 + y^2 + x^2y^2))^{1/2}}.$$

By elementary calculus the maximum value of this expression is $\frac{1}{3}$. Hence Γ_1 is realizable if $0 \leqq \cos\varphi \leqq \frac{1}{3}$, *i.e.*, $0 \leqq \sin\theta \leqq \frac{1}{3}$.

LEMMA 2. *The following graph* Γ_2 *is realizable on S.*

The graph Γ_2 *is obtained from the above diagram by identifying the points* p_0 *and a,* q_0 *and b, and* r_0 *and c. The vertices of* Γ_2 *are the points on the rim of this diagram.*

Proof. For $0 \leqq k \leqq 4$, let

$$P_k = \cos\frac{\pi k}{10}\bar{i} + \sin\frac{\pi k}{10}\bar{j},$$

$$Q_k = \cos\frac{\pi k}{10}\bar{j} + \sin\frac{\pi k}{10}\bar{k},$$

$$R_k = \sin\frac{\pi k}{10}\bar{i} + \cos\frac{\pi k}{10}\bar{k}.$$

Let $u(p_k)=P_k$, $u(q_k)=Q_k$, $u(r_k)=R_k$, for $0 \leqq k \leqq 4$. Since the subgraph of Γ_2 contained between the points p_0, p_1, and r_0 is a copy of Γ_1 and the angle subtended by P_0, P_1 is $\pi/10$ ($\sin \pi/10 < \frac{1}{3}$), we may extend u to a realization of this subgraph on S. A realization of the subgraph of Γ_2 contained between the points p_1, p_2, and r_0 is then obtained by rotating

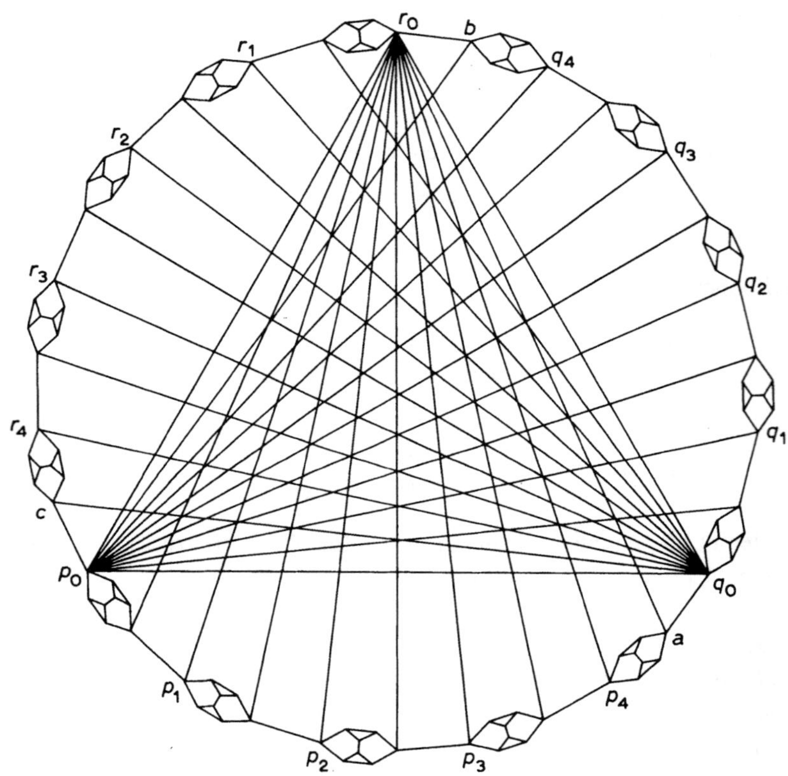

P_0 to P_1 about R_0. The remainder of the realization u is obtained by similar rotations about R_0, P_0, and Q_0.

Let T be the image of Γ_2 under u, consisting of 117 points on S. Let D be the partial Boolean subalgebra generated by T in $\mathbf{B}(E^3)$. (This corresponds to completing the graph Γ_2 so that every edge lies in a triangle. In the resulting graph the points and edges correspond to one and two-dimensional linear subspaces of $\mathbf{B}(E^3)$ respectively.)

THEOREM 1. *The finite partial Boolean algebra D has no homomorphism onto Z_2.*

Proof. As we have seen, such a homomorphism $h:D \rightarrow Z_2$ induces a map $h^*:T \rightarrow \{0, 1\}$ satisfying condition (6). Reverting to the graph Γ_2, we shall assume that there is a map $k:\Gamma_2 \rightarrow \{0, 1\}$ satisfying condition (6). Let us consider the action of k on a copy in Γ_2 of the graph Γ_1. Suppose that $k(a_0)=1$, then it follows that $k(a_9)=1$. For if $k(a_9)=0$, then since $k(a_8)=0$ we must have $k(a_7)=1$. Hence, $k(a_1)=k(a_2)=k(a_3)=k(a_4)=0$; so that $k(a_5)=k(a_6)=1$, a contradiction.

Now since p_0, q_0, and r_0 lie in a triangle in Γ_2, exactly one of these points is mapped by k onto 1, say $k(p_0)=1$. Hence, by the above argument $k(p_1)=1$. Continuing in this manner in Γ_2 we find $k(p_2)=k(p_3)= =k(p_4)=k(q_0)=1$. But $k(q_0)=1$ contradicts the condition that $k(p_0)=1$, and proves the theorem.

Remark

Theorem 1 implies that there is no map of the sphere S onto $\{0, 1\}$ satisfying condition (4), and hence no homomorphism from $\mathbf{B}(E^3)$ onto Z_2. This result, first stated in Specker [17], can be obtained more simply either by a direct topological argument or by applying a theorem of Gleason [6]. However, it seems to us important in the demonstration of the non-existence of hidden variables that we deal with a small finite partial Boolean algebra. For otherwise a reasonable objection can be raised that in fact it is not physically meaningful to assume that there are a continuum number of quantum mechanical propositions.

To obtain a partial Boolean subalgebra of $\mathbf{B}(E^3)$ which is not imbeddable in a Boolean algebra a far smaller graph than Γ_2 suffices. The

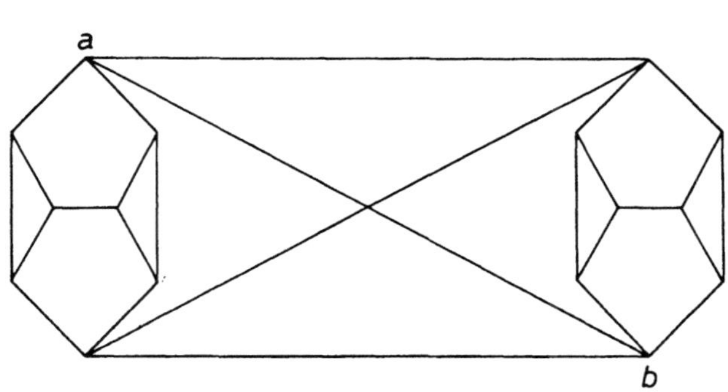

following graph Γ_3 may be shown to be realizable on S in similar fashion to the proof of Lemma 2. Let F be the partial Boolean algebra generated by the set of 17 points on S corresponding to Γ_3. If $h:F\to Z_2$ is a homomorphism then as we have seen in the proof of Theorem 1, if $h(a)=1$ then $h(b)=1$; by symmetry also $h(b)=1$ implies $h(a)=1$. That is, $h(a)=$ $=h(b)$ in every homomorphism $h:F\to Z_2$. If $\varphi:F\to B$ is an imbedding of F into a Boolean algebra, then by the semi-simplicity of B there exists a homomorphism $h':B\to Z_2$ such that $h'(\varphi(a))\neq h'(\varphi(b))$. Hence, $h=h'\varphi$ is a homomorphism from F onto Z_2 such that $h(a)\neq h(b)$, a contradiction.

IV. THE OPERATORS AS OBSERVABLES

Let us consider a system in which the total angular momentum operator J commutes with the Hamiltonian operator H, so that $\bar{J}$ is a constant of the motion. We assume further that the system is in a state for which the principal quantum number $n=2$ and the azimuthal quantum number $j=1$, so that the total angular momentum is $\sqrt{2}\hbar$. The eigenspace N corresponding to the eigenvalue $2\hbar^2$ of J^2 is three-dimensional. We adopt the convention that $\hbar=1$.

Let J_x, J_y, and J_z be the components of $\bar{J}$ in three mutually orthogonal directions x, y, and z. We shall show that in the three dimensional representation given by $n=2, j=1$ the following relations hold.

$$(7) \qquad [J_x^2, J_y^2]=[J_y^2, J_z^2]=[J_z^2, J_x^2]=0.$$

In the usual representation in which J^2 and J_z are diagonal we have (see Schiff [14], p. 146)

$$J_z=\frac{1}{\sqrt{2}}\begin{pmatrix}1&0&0\\0&0&0\\0&0&-1\end{pmatrix},\quad J_x=\frac{1}{\sqrt{2}}\begin{pmatrix}0&1&0\\1&0&1\\0&1&0\end{pmatrix},\quad J_y=\frac{1}{\sqrt{2}}\begin{pmatrix}0&-i&0\\i&0&-i\\0&i&0\end{pmatrix}.$$

It is now easily checked that the relations (7) follow. It may be of some interest to give a coordinate-free proof of these relations. The following proof was suggested to us by J. Chaiken. Let $J_\pm=J_x\pm iJ_y$. From the commutation relations $[J_x, J_y]=iJ_z$, *etc.*, for J_x, J_y, and J_z it follows that

$$[J_x^2, J_y^2]=(J_z-I)\,J_+^2-(J_z+I)\,J_-^2.$$

Now if $J_z\psi = m\psi$ then

$$J_z J_+ \psi = \begin{cases} (m+1)\, J_+ \psi & \text{if} \quad -j \leqq m < j \\ 0 & \text{if} \quad m = j. \end{cases}$$

Hence, if φ is any vector in the three-dimensional representation ($n=2$, $j=1$), then $J_+^2 \varphi$ is either zero or an eigenvector of J_z with eigenvalue $+1$. In either case, $(J_z - I)\, J_+^2 \varphi = 0$. Hence $(J_z - I)\, J_+^2 = 0$ in this representation. Similarly, $(J_z + I)\, J_-^2 = 0$, so that $[J_x^2, J_y^2] = 0$. This establishes (7). Note that these relations do not hold in any higher dimensional representation.

We now show that there is an imbedding ψ of the partial Boolean algebra $\mathbf{B}(E^3)$ into the partial Boolean algebra $\mathscr{B}$ of quantum mechanical propositions. Let P be the projection operator belonging to the three-dimensional eigenspace N. To each one-dimensional linear subspace α of E^3 there corresponds an operator J_α, the component of angular momentum in the direction in physical space defined by α. Let $\psi(\alpha) = PJ_\alpha^2$. If β is a two-dimensional linear subspace of E^3 let α be the orthogonal complement of β in E^3. We define $\psi(\beta) = P - PJ_\alpha^2$. Finally we let $\psi(E^3) = P$ and $\psi(0) = 0$. This defines the map ψ. To show that ψ is an imbedding it clearly suffices to prove that if α and β are orthogonal one-dimensional linear subspaces of E^3, then $[PJ_\alpha^2, PJ_\beta^2] = 0$. But this is precisely the relation (7) which we have established. Note that the projection operator PJ_α^2 is an element of $\mathscr{B}$; it corresponds to the proposition P_α: "For the system in energy state $n=2$ and total angular momentum state $j=1$, the component of angular momentum in the direction α is not 0."

Since then the finite partial Boolean algebra D has been imbedded in $\mathscr{B}$, it follows by Section III that there is no homomorphism of $\mathscr{B}$ onto Z_2.

In the above argument we have assumed that in the three-dimensional representation the observables J_x^2, J_y^2 and J_z^2 are commeasurable. This remains to be justified. Of course, we have seen that these operators commute and it is a generally accepted assumption of quantum mechanics that commuting operators correspond to commeasurable observables. A rationale for this assumption, as we pointed out in Section II, is that if A_i, $i \in I$, is a set of mutually pairwise commuting self-adjoint operators, then there exists a self-adjoint operator B and Borel functions f_i, $i \in I$ such that $A_i = f_i(B)$. However this justification hinges on the existence of a physical observable which corresponds to the operator B.

We shall now show that there is in this case an operator H_J of which J_x^2, J_y^2, and J_z^2 are functions and which corresponds to an observable.

Let a, b, and c be distinct real numbers and define

$$H_J = aJ_x^2 + bJ_y^2 + cJ_z^2.$$

Then it is easily checked that in the three dimensional representation

$$\begin{aligned} J_x^2 &= (a-b)^{-1}(c-a)^{-1}(H_J-(b+c))(H_J-2a),\\ J_y^2 &= (b-c)^{-1}(a-b)^{-1}(H_J-(c+a))(H_J-2b),\\ J_z^2 &= (c-a)^{-1}(b-c)^{-1}(H_J-(a+b))(H_J-2c). \end{aligned} \tag{8}$$

Consider now a physical system the total angular momentum of which is spin angular momentum S, with S having the constant value $\sqrt{2}\hbar$. An example of such a system is an atom of orthohelium in the 2^3S_1 state, *i.e.*, the lowest triplet state of helium, with the principal quantum number $n=2$, the orbital quantum number $l=0$, and spin $s=1$. (Note that this is a stable state for the atom even though it is not the ground state. (It is called a metastable state.) The reason for the stability is that the ground state ($n=1$) of the atom occurs only for parahelium, *i.e.*, the singlet state of helium with $s=0$; and transitions are forbidden between the singlet and the triplet states of helium).

We now apply to the system in this state a small electric field E which has rhombic symmetry about the atom. (Such a field, for instance, results from placing point charges at the points $(\pm u, 0, 0)$, $(0, \pm v, 0)$ $(0, 0, \pm w)$, with u, v, and w distinct, the atom being at the origin.) By perturbation methods it may be shown that the Hamiltonian H of the system is perturbed to a new Hamiltonian $H+H_S$, where, from the rhombic symmetry of the field, the additional term H_S, called the spin-Hamiltonian, has the form $H_S = aS_x^2 + bS_y^2 + cS_z^2$ with a, b, and c distinct in the three-dimensional representation. (See *e.g.*, Stevens [18] and Pryce [13] for a proof.)

Thus the operator $H_S = aS_x^2 + bS_y^2 + cS_z^2$ corresponds to a physical observable – the change in the energy of the lowest orbital state of orthohelium resulting from the application of a small electric field with rhombic symmetry. The change in energy levels may be measured by studying the spectrum of the helium atom after the field is applied. The possible measured values in the change in energy levels is either $a+b$, $b+c$, or $c+a$, since these are the eigenvalues of H_S in the three-dimensional representation. Since a, b, and c are distinct, so are $a+b$, $b+c$, and $c+a$. Thus, a measurement of H_S leads immediately to the simultaneous measurement of S_x^2, S_y^2 and S_z^2. If, for instance, the measured value of H_S is $a+b$, then we infer that the values of S_x^2 and S_y^2 are each 1 and the value of S_z^2 is 0. (This is equivalent to applying the relations (8) to H_S.)

We remark that although such an experiment has probably not been carried out on the helium atom, related experiments are described in the literature. For instance Griffith and Owen [7] investigated in paramagnetic resonance experiments a nickel Tulton salt, nickel fluosilicate. This salt consists of a nickel ion surrounded by an octahedron of water molecules and it occurs in the state $J^2 = S^2 = 2\hbar^2$. The water molecules form a crystalline electric field with rhombic symmetry about the nickel ion. The resulting spin-Hamiltonian H_S takes the form $aS_x^2 + bS_y^2 + cS_z^2$ with a, b, and c distinct. This is in all respects similar to the situation we have discussed above. Of course, in this case the electric field is supplied by the crystal and cannot be switched on and off or rotated at will to measure S_x^2, S_y^2, and S_z^2 in any three prescribed orthogonal directions. Nevertheless, the experimental agreement with the quantum mechanical predictions here suggests a similar agreement for the case of an external electric field applied to a helium atom.

To sum up the last two sections we shall recapitulate our case against the existence of hidden variables for quantum mechanics. We have used the formal technique of introducing the concept of a partial algebra to discuss this question but we may now give a direct intuitive argument. If a physicist X believes in hidden variables he should be able to predict (in theory) the measured value of every quantum mechanical observable. We now confront X with the problem of simultaneously answering the question:

"Is the component of spin angular momentum in the direction α equal to zero for the lowest orbital state of orthohelium ($n=2$, $l=0$, $s=1$)" where α varies over the 117 directions provided in the proof of Theorem 1. For each such prediction by X we can find, by Theorem 1, three orthogonal directions x, y, z among the 117 for which this prediction contradicts the statement

"Exactly one of the three components of spin angular momentum S_x, S_y, S_z of the lowest orbital state of orthohelium is zero."

This statement is what is predicted by quantum mechanics since

$$S_x^2+S_y^2+S_z^2=S^2=2\hbar^2$$

and each of S_x^2, S_y^2, S_z^2 thus has the value 0 or $\hbar^2$. Thus the prediction of X contradicts the prediction of quantum mechanics. Furthermore as we have seen in this section this prediction may be experimentally verified by simultaneously measuring S_x^2, S_y^2, and S_z^2. Our conclusion is that every prediction by physicist X may be contradicted by experiment. (It has been argued (See Bohm [2, Sect. 9]) that with the introduction of a hidden state space Ω the present quantum mechanical observables such as spin will not be the fundamental observables of the new theory. Certainly, many new possible observables are thereby introduced (namely, functions $f:\Omega\to\mathbf{R}$). The quantum observables represent not true observables of the system itself which is under study, but reflect rather properties of the disturbed system and the apparatus. This is nevertheless no argument against the above proof. For in a classical interpretation of quantum mechanics observables such as spin will still be functions on the phase space of the combined apparatus and system and as such should be simultaneously predictable).

V. HOMOMORPHIC RELATIONS

In Section I we reduced the question of hidden variables to the existence of an imbedding of Q into a commutative algebra C. We discuss here a possible objection to this reduction. It may be argued that in a classical reinterpretation of quantum mechanics a given observable may split into several new observables. Thus, the correspondence between Q and C may take the form of a homomorphism $\psi:C\to Q$ from C onto Q. This possibility is provided for in the following theorem.

DEFINITION. Let $\mathscr{N}$ and $\mathscr{L}$ be partial algebras over a common field K. A relation $R\subseteq\mathscr{N}\times\mathscr{L}$ is called a *homomorphic relation* between $\mathscr{N}$ and $\mathscr{L}$ if, for all $x \,♀\, y$ in $\mathscr{N}$ and $\alpha \,♀\, \beta$ in $\mathscr{L}$, $R(x,\alpha)$ and $R(y,\beta)$ imply that $R(\lambda x+\mu y, \lambda\alpha+\mu\beta)$ and $R(xy,\alpha\beta)$ for every λ, $\mu\in K$ and also $R(1,1)$.

The homomorphic relation $R\subseteq\mathscr{N}\times\mathscr{L}$ has *domain* $\mathscr{N}$ if for all $x\in\mathscr{N}$ there is an $\alpha\in\mathscr{L}$ such that $R(x,\alpha)$. The relation R is *non-trivial* if not $R(1,0)$.

If $\varphi:\mathcal{N}\to\mathcal{L}$ is a homomorphism then the graph of φ *i.e.*, the relation $R(x, \alpha)$ defined by $\varphi(x)=\alpha$, is a non-trivial homomorphic relation with domain $\mathcal{N}$. Similarly a homomorphism $\psi:\mathcal{L}\to\mathcal{N}$ of $\mathcal{L}$ onto $\mathcal{N}$ defines the non-trivial homomorphic relation R with domain $\mathcal{N}$ by taking $R(x, \alpha)$ if $\psi(\alpha)=x$.

THEOREM 2. *Let $\mathcal{N}$ be a partial algebra and assume that there exists a non-trivial homomorphic relation R with domain $\mathcal{N}$ between $\mathcal{N}$ and a commutative algebra C. Then there exists a commumative algebra C' and a homomosphism $h:\mathcal{N}\to C'$ from $\mathcal{N}$ onto C'.*

Proof. Let S be the set of all elements α in C such that $R(x, \alpha)$ for some $x\in\mathcal{N}$. Let $\bar{S}$ be the subalgebra generated by S in C. Define I to be the set of all $\alpha\in C$ such that $R(0, \alpha)$. Then I is clearly closed under linear combinations. Next let $\beta\in\bar{S}$, so that

$$\beta=\sum_i \lambda_i\beta_{i1}\beta_{i2}\cdots\beta_{in_i}$$

for some $\lambda_i\in K$, and $\beta_{ij}\in S$.

If $\alpha\in I$, then $\alpha\beta_{ij}\in I$. Hence $\alpha\beta=\sum_i \lambda_i\,\alpha\beta_{i1}\dots\beta_{in_i}\in I$. Finally, $1\notin I$. We have shown that I is a proper ideal of the algebra $\bar{S}$. Let $C'=\bar{S}/I$ and let $\bar{S}\to C'$ be the canonical homomorphism. Define $h:\mathcal{N}\to C'$ by $h(x)= =\varphi(\alpha)$ where $\alpha\in S$ is such that $R(x, \alpha)$. Then it is easily checked that h is well-defined and a homomorphism.

If we now take $\mathcal{N}$ to be the partial algebra Q, it follows from this theorem that there is no non-trivial homomorphic relation with domain Q between Q and a commutative algebra.

VI. A CLASSICAL MODEL OF ELECTRON SPIN

We prove here that the problem of hidden variables as we have formulated it in Section I has a positive solution for a restricted part of quantum mechanics. The portion of quantum mechanics with which we deal is obtained by restricting our Hilbert space to be two-dimensional. Thus, the state vectors are assumed to range over two-dimensional unitary space U^2, and the observables to range over the set H_2 of two-dimensional self-adjoint operators.

As will be seen, the problem reduces to considering the case of spin operators. Thus, our problem becomes essentially that of constructing a classical model for a single particle of spin $\frac{1}{2}$, say an electron. Needless to say, we do not maintain that this classical model of electron spin remains valid in the general context of quantum mechanics. In fact, as was shown in Section IV, there exists a system of two electrons in a suitable external field such that there is no classical model for the spin of the system.

Our aim in constructing a classical model for electron spin is two-fold. In the first place, we wish to exhibit a classical interpretation of a part of quantum mechanics so that it may be compared with various attempts to introduce hidden variables into quantum mechanics. We believe these attempts to be unsuccessful, so it would be as well if we could give an example of what is for us a successful introduction of hidden variables into a theory. In the second place, we shall use this model in discussing von Neumann's proof in [19] of the non-existence of hidden variables.

As formulated in Section I, our problem is to define a "phase" space Ω such that for each operator $A \in H_2$ there is a real-valued function $f_A : \Omega \to \mathbf{R}$ and for each vector $\psi \in U^2$ there exists a probability measure μ_ψ on Ω such that

(I) $f_{u(A)} = u(f_A)$ for each (Borel) function u; and
(II) the quantum mechanical expectation

$$\langle A\psi, \psi \rangle = \int_\Omega f_A(\omega)\, d\mu_\psi(\omega).$$

Let V be the set of operators in H_2 of trace zero. V forms a three-dimensional vector space over $\mathbf{R}$. This is easily seen by noting that the Pauli spin matrices

$$\sigma_x = \begin{pmatrix} 0 & 1 \\ 1 & 0 \end{pmatrix}, \quad \sigma_y = \begin{pmatrix} 0 & -i \\ i & 0 \end{pmatrix}, \quad \sigma_z = \begin{pmatrix} 1 & 0 \\ 0 & -1 \end{pmatrix}$$

form an orthonormal basis for V. If we assign to $(\sigma_x, \sigma_y, \sigma_z)$ an orthonormal basis (i, j, k) in three-dimensional Euclidean space E^3, we obtain a vector isomorphism $P: V \to E^3$. To every spin matrix σ, *i.e.*, a matrix σ in V with eigenvalues ± 1, there corresponds under the map P a point P_σ on the unit sphere S^2 in E^3. Physically, one speaks of the spin matrix σ as corresponding to the observable "the spin angular momentum of the electron (say) in the direction $0P_\sigma$," where 0 is the origin in E^3.

Now let A be any matrix in H_2 with distinct eigenvalues λ_1, λ_2. We let

$$\sigma(A)=\left(\frac{2}{\lambda_1-\lambda_2}\right)A-\left(\frac{\lambda_1+\lambda_2}{\lambda_1-\lambda_2}\right)I.$$

Then $\sigma(A)$ is a spin matrix such that the eigenvectors of $\sigma(A)$ corresponding to $+1$ and -1 are the same as the eigenvectors of A corresponding to λ_1 and λ_2 respectively.

We are now ready to choose the appropriate space Ω and functions f_A. For Ω we choose S^2. If $A\in H_2$ with distinct eigenvalues λ_1 and λ_2, we let

$$f_A(p)=\begin{cases}\lambda_1 & \text{for} \quad p\in S^+_{P_{\sigma(A)}}\\ \lambda_2 & \text{otherwise.}\end{cases}$$

Here $S^+_{P_{\sigma(A)}}$ denotes the upper hemisphere of S^2 with the North Pole at $P_{\sigma(A)}$.

If the eigenvalues of A are equal, so that $A=\lambda I$, say, then we let

$$f_A(p)=\lambda, \quad \text{for all} \quad p\in S^2.$$

With this definition, it is a simple matter to check that the condition (I): $f_{u(A)}=u(f_A)$ holds. We need only note that for two-dimensional operators it is sufficient to consider linear functions: $u(A)=\alpha A+\beta I$, with α, $\beta\in\mathbf{R}$. Then condition (I) follows immediately from the fact that $\sigma_{\alpha A+\beta I}=\sigma_A$.

Next we wish to assign a probability measure μ_ψ to each vector $\psi\in U^2$. Let σ_ψ denote the spin matrix for which ψ is the eigenvector belonging to the eigenvalue $+1$. We may thus assign to each $\psi\in U^2$ a point $P_{\sigma\psi}$ of S^2. We shall write P_ψ for $P_{\sigma\psi}$. Physically, if ψ is the state vector of an electron, then the electron is said to have "spin in the direction $0P_\psi$."

To delimit the problem and at the same time to obtain a solution with natural isotropy properties, we shall assume that the probability measures μ_ψ satisfy the following conditions:

(a) For each $\psi\in U^2$, the measure μ_ψ arises from a continuous probability density $u_\psi(p)$ on S^2, so that

$$\mu_\psi(E)=\int\limits_E u_\psi(p)\,dp$$

for every measurable subset E of S^2.

(b) The probability density $u_\psi(p)$ is a function only of the angle θ subtended at 0 by the points p and P_ψ on S^2. We may thus write $u_\psi(\theta)$ for the function $u_\psi(p)$.

(c) Let $u(\theta)\,(=u_{\psi_0}(\theta))$ be the probability density assigned to the state vector

$$\psi_0=\begin{pmatrix}1\\0\end{pmatrix}.$$

(Note that $\sigma_{\psi_0}=\sigma_z$, so that $P_{\psi_0}=(0,0,1)$.) Let $\psi\in U^2$. If α is the polar angle of the point P_ψ on S^2, then we assume that $u_\psi(\theta)=u(\theta+\alpha)$. Thus, the probability takes the same functional form for all states ψ.

(d) We assume that $u(\theta)=0$ for $\theta>\pi/2$.

An examination of the problem shows that these are natural properties to assign to the quantum states considered as probability distributions over the hidden states. We shall show that there do exist measures μ_ψ satisfying the above conditions as well as condition (II). In fact, we shall see that these conditions determine the density functions u_ψ uniquely.

Using these assumptions we may simplify the problem of finding measures μ_ψ which satisfy condition (II) as follows. Since f_A is a linear function of A, the integral $\int_\Omega f_A(\omega)\,d\mu_\psi(\omega)$ is a linear function of A. On the other hand the expectation function $\langle A\psi,\psi\rangle$ is also a linear function of A. Since every matrix A in H_2 is a linear function of a projection matrix, it is sufficient to verify condition (II) for projection matrices. Next, by condition (c) we may assume that

$$\psi=\begin{pmatrix}1\\0\end{pmatrix},$$

so that $P_\psi=(0,0,1)$. Furthermore, by condition (b), it is sufficient to consider the case where $P_{\sigma(A)}$ has azimuthal angle equal to zero. In what follows we shall make the above assumptions on A and ψ.

It is now necessary to express the expectation $\langle A\psi,\psi\rangle$ as a function of the angle subtended at 0 by the points $P_\psi=(0,0,1)$ and $P_{\sigma(A)}$, *i.e.*, as a function of the polar angle ρ of $P_{\sigma(A)}$.

In spherical polar coordinates we may write

$$P_{\sigma(A)}=(\sin\rho,0,\cos\rho).$$

Hence,

$$\sigma(A)=\sigma_x \sin\rho+\sigma_z \cos\rho$$
$$=\begin{pmatrix}\cos\rho & \sin\rho\\ \sin\rho & -\cos\rho\end{pmatrix}.$$

The eigenvector η of $\sigma(A)$ belonging to the eigenvalue $+1$ is

$$\eta=\begin{pmatrix}\cos(\rho/2)\\ \sin(\rho/2)\end{pmatrix}.$$

Since A was assumed to be a projection matrix, η is also the eigenvector of A belonging to the eigenvalue $+1$. Thus,

$$\langle A\psi, \psi\rangle=\langle\langle\psi, \eta\rangle\,\eta, \psi\rangle$$
$$=|\langle\psi, \eta\rangle|^2$$
$$=\cos^2(\rho/2).$$

Our problem is thus reduced to solving for $u(\theta)$ the integral equation

$$\cos^2(\rho/2)=\int\limits_{S^2} f_A(p)\,u(\theta)\,dp.$$

Since

$$f_A(p)=\begin{cases}1 & \text{on}\quad S^+_{P_{\sigma(A)}}\\ 0 & \text{otherwise}\end{cases}$$

this equation becomes

$$\cos^2(\rho/2)=\int\limits_T u(\theta)\,dp$$

where $T=S^+_{P_{\sigma(A)}}\cap S^+_{P_\psi}$. Thus

$$\cos^2(\rho/2)=\int\limits_{\rho-\pi/2}^{\pi/2}\int\limits_{-\varphi\theta}^{\varphi\theta} u(\theta)\sin\theta\,d\varphi\,d\theta$$

where φ_θ is the azimuthal angle of the point $Q=(\sin\theta\cos\varphi_\theta, \sin\theta\sin\varphi_\theta, \cos\theta)$ with polar angle θ which lies on the great circle C perpendicular to the point $P_{\sigma(A)}=(\sin\rho, 0, \cos\rho)$.

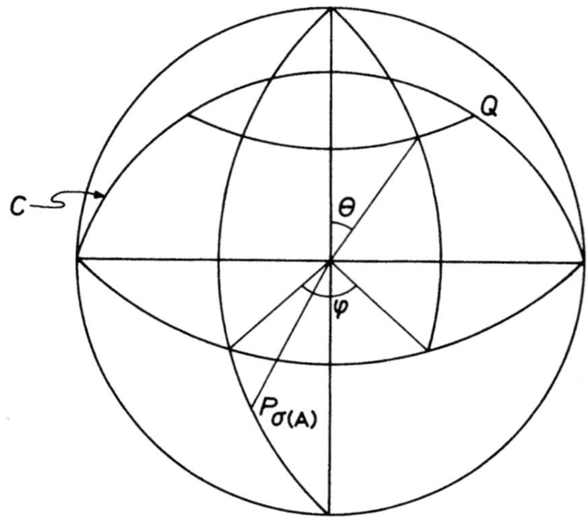

Using the orthogonality of Q and $P_{\sigma(A)}$, we have

$$\sin\rho \sin\theta \cos\varphi_\theta + \cos\rho \cos\theta = 0$$

or

$$\varphi_\theta = \cos^{-1}(-\cot\rho \cot\theta).$$

Thus

$$\tfrac{1}{2}(1+\cos\rho) = 2 \int_{\rho-\pi/2}^{\pi/2} u(\theta) \sin\theta \cos^{-1}(-\cot\rho \cot\theta)\, d\theta.$$

Letting $x = \rho - \pi/2$, we have

$$\tfrac{1}{2}(1-\sin x) = -2 \int_{\pi/2}^{x} u(\theta) \sin\theta \cos^{-1}(\cot\theta \tan x)\, d\theta.$$

Now, differentiating both sides with respect to x, we obtain

$$-\tfrac{1}{2}\cos x = -2u(x) \sin x \cos^{-1}(\cot x \tan x) + \int_{\pi/2}^{x} \frac{u(\theta) \sin\theta \cot\theta \sec^2 x}{(1-\cot^2\theta \tan^2 x)^{1/2}}\, d\theta$$

or,

$$\cos^3 x = -4 \int_{\pi/2}^{x} \frac{u(\theta)\cos\theta}{(1-\cot^2\theta\tan^2 x)^{1/2}}\, d\theta.$$

If we set $z=\cos^2 x$, $s=\cos^2\theta$, and $w(s)=u(\theta)$, we find

$$z = \int_0^z \frac{2w(s)}{(z-s)^{1/2}}\, ds.$$

This is a special case of Abel's integral equation, and is easily solved by Laplace transforms. Namely, if $*$ denotes convolution and $L(f) = \int_0^\infty f(x)\, e^{-tx}\, dx$, the Laplace transform, then

$$z = w * 2z^{-1/2}.$$

Hence,

$$L(z) = L(w)\, L(2z^{-1/2}),$$

or

$$L(w) = L(z)/L(2z^{-1/2}) = \frac{1}{2\sqrt{\pi}}\, t^{-3/2} = L((1/\pi)\, s^{1/2}),$$

so that

$$w(s) = (1/\pi)\, s^{1/2}.$$

We thus have shown that

$$u(\theta) = \begin{cases} (1/\pi)\cos\theta & \text{if } 0 \leqq \theta \leqq \pi/2 \\ 0 & \text{otherwise.} \end{cases}$$

On the basis of this mathematical solution, we may construct a simple classical model of electron spin. The same model then serves (by linearity) for the more general case of operators in H_2.

We start with a sphere with fixed center 0. A point P on the sphere represents the quantum state "spin in the direction $0P$". If the sphere is in such a quantum state it is at the same time in a hidden state which is represented by another point $T \in S_P^+$. The point T has been determined as follows. A disk D of the same radius as the sphere is placed perpendicular to the $0P$ axis with center directly above P. A particle is placed on the disk and the disk shaken "randomly". That is the disk is so shaken that the probability of the particle being in a region U in D is proportional to the area of U (*i.e.*, the probability is uniformly distributed). The point T is the orthogonal projection of the particle (after shaking) onto the sphere. It is easily seen that the probability density function for the projection is given by

$$u(T)=\begin{cases}(1/\pi)\cos\theta & 0\leqq\theta\leqq\pi/2\\ 0 & \text{otherwise},\end{cases}$$

where θ is the angle subtended by T and P at 0.

Suppose we now wish to measure the spin angular momentum in a direction $0Q$. This is determined as follows. If $T\in S_Q^+$, then the spin angular momentum is $+\hbar/2$, if $T\notin S_Q^+$ then the spin is $-\hbar/2$. The sphere is now in the new quantum state of spin in the direction $0Q$ if $T\in S_Q^+$ or spin in the direction $0Q^*$ (where Q^* is the antipodal point of Q) if $T\notin S_Q^+$. The new hidden state of the sphere is now determined as before, by shaking the particle on the disk D, the disk being placed with center above Q if $T\in S_Q^+$ or with center above Q^* if $T\notin S_Q^+$.

It should be clear from the preceding analysis that the probabilities and expectations that arise from this model are precisely the same as those arising from quantum mechanical calculations for free electron spin. In the model the disk D, the particle, and its projection are to be considered as the hidden apparatus. The probabilities arise through the ignorance of the observer of the sphere of the actual location of the particle on the disk. To an observer of the complete system of sphere and disk the model is a deterministic classical system.

Note that in the above model we could keep the disk fixed vertically above the sphere and instead rotate the sphere to determine each new hidden state. If we now further replace the shaking disk by a random vertically falling water drop, we may say that rain falling on a ball forms a classical model of electron spin.

We remark finally that the conditions (I) and (II) say nothing about the propagation of the probabilities in time. That is to say, although these conditions give the probabilities arising at each experiment, they do not deal with the change of probabilities during the time between experiments. However, in the situation we are examining of free electron spin this causes no difficulty since every state is in this case stationary, and the probabilities remain constant in the time between experiments.

We now consider the bearing of this model on von Neumann's discussion of the hidden variables problem given in [19, Ch. IV]. In that chapter von Neumann gives what he considers to be a necessary condition for the existence of hidden variables for quantum mechanics. This condition is the existence of a function

$$\mathscr{E}: H \to \mathbf{R},$$

where H is the set of self-adjoint operators, such that

(1) $\mathscr{E}(I)=1$.

(2) $\mathscr{E}(aA)=a\mathscr{E}(A)$, for all $a \in \mathbf{R}$, $A \in H$.

(3) $\mathscr{E}(A^2)=\mathscr{E}^2(A)$, for all $A \in H$.

(4) $\mathscr{E}(A+B)=\mathscr{E}(A)+\mathscr{E}(B)$, for all $A, B \in H$.

In [19] it is then shown that there does not exist a function satisfying these conditions. (In [19] a further condition is added on $\mathscr{E}$: (5) If A is "essentially positive" then $\mathscr{E}(A) \geqq 0$. But we shall not require this condition in our proof.) We present another proof below. This is done for two reasons. First, our proof is simpler, and is in fact trivial. Second, this proof shows that there is even no function $\mathscr{E}: H_2 \to \mathbf{R}$ satisfying conditions (1)–(4), a result we require for our later discussion.

LEMMA. *If the function* $\mathscr{E}: H \to \mathbf{R}$ *satisfies* (1)–(3) *together with condition*

(4)′ $\mathscr{E}(A+B)=\mathscr{E}(A)+\mathscr{E}(B)$, *for all* $A, B \in H$ *such that* $AB=BA$, *then* $\mathscr{E}(AB)=\mathscr{E}(A)\,\mathscr{E}(B)$, *for all* $A, B \in H$ *such that* $AB=BA$. (In the terminology of Section II, $\mathscr{E}$ is thus a homomorphism of the partial algebra H into $\mathbf{R}$.)

Proof. Assume $AB=BA$. Then

$$\begin{aligned}\mathscr{E}^2(A)+2\mathscr{E}(A)\,\mathscr{E}(B)+\mathscr{E}^2(B)&=(\mathscr{E}(A)+\mathscr{E}(B))^2\\&=\mathscr{E}^2(A+B)\\&=\mathscr{E}((A+B)^2)\\&=\mathscr{E}(A^2+2AB+B^2)\\&=\mathscr{E}(A^2)+\mathscr{E}(2AB)+\mathscr{E}(B^2)\\&=\mathscr{E}^2(A)+2\mathscr{E}(AB)+\mathscr{E}^2(B).\end{aligned}$$

Hence, $\mathscr{E}(A)\,\mathscr{E}(B)=\mathscr{E}(AB)$.

COROLLARY. *If the function $\mathscr{E}$ satisfies conditions* (1), (2), (3), (4)′, *then* $\mathscr{E}(A)$ *lies in the spectrum of* A.

Proof. Suppose to the contrary that $A-\mathscr{E}(A)$ has an inverse B. Then by the Lemma,

$$\begin{aligned} 1 &= \mathscr{E}(I) \\ &= \mathscr{E}((A-\mathscr{E}(A))\,B) \\ &= \mathscr{E}(A-\mathscr{E}(A))\,\mathscr{E}(B) \\ &= (\mathscr{E}(A)-\mathscr{E}(\mathscr{E}(A))\,\mathscr{E}(B) \\ &= (\mathscr{E}(A)-\mathscr{E}(A))\,\mathscr{E}(B) \\ &= 0. \end{aligned}$$

THEOREM 3. *There is no function* $\mathscr{E}:H\to\mathbf{R}$ *satisfying conditions* (1)–(4).

Proof. Consider the two matrices

$$A=\begin{pmatrix}1 & 0\\ 0 & 0\end{pmatrix}, \qquad B=\tfrac{1}{2}\begin{pmatrix}1 & 1\\ 1 & 1\end{pmatrix}.$$

The matrices A and B are projection matrices and hence have eigenvalues 0 and 1. The matrix $A+B$ has eigenvalues $1\pm\frac{1}{2}(2)^{1/2}$. Hence, $\mathscr{E}(A+B)\neq\mathscr{E}(A)+\mathscr{E}(B)$, by the above corollary.

As the proof shows, there is no function $\mathscr{E}$ with properties (1)–(4) even when the domain of $\mathscr{E}$ is restricted to H_2.

Now, von Neumann's criterion has been criticized in the literature in requiring the additivity of $\mathscr{E}$ even for non-commuting operators, *i.e.*, in requiring condition (4) rather than (4)′. (See for example Hermann [9, pp. 99–104].) As the above Lemma shows, it is precisely on this point that von Neumann's criterion differs from our point of view. For we showed that there does not exist a function satisfying (1), (2), (3), and (4)′. We may now go further. We have here constructed a classical system C (the sphere and the disk). From this system we obtained a new system Q (the sphere without the disk) such that the pure states of Q are certain mixed states of C and the observables of Q are among the observables of C. The pure states of Q may then be described by vectors in U^2 and the observables of Q by operators in H_2, just as in quantum mechanics. If we now accept von Neumann's criterion, we must conclude that we cannot introduce hidden variables into the system Q. But this can hardly be a reasonable conclusion, since we may reintroduce into Q the states and observables of C which we ignored in forming Q, to recover the classical system C.

VII. THE LOGIC OF QUANTUM MECHANICS

In this section we discuss the non-existence of an imbedding of $\mathscr{B}$ into a Boolean algebra from a different point of view. It will turn out that a consequence of this result is that the logic of quantum mechanics is different from classical logic. Since the set of propositions of a classical physical theory forms a Boolean algebra B it follows that the propositions valid in such a theory are precisely the classical tautologies. This means that if we are given a classical tautology such as

$$(9) \qquad x_1 \wedge (x_2 \wedge x_3) \equiv (x_1 \wedge x_2) \wedge x_3$$

then every substitution of elements of B for x_1, x_2, x_3 yields the element 1 of B. In the case of a theory such as quantum mechanics where the set of propositions form a partial Boolean algebra $\mathscr{B}$ it is not clear what it means for a proposition to be valid. To take the preceding proposition (9) as an example, it is not possible to substitute arbitrary elements of a_1, a_2, a_3 of $\mathscr{B}$ for x_1, x_2, x_3. It is necessary in this case that the commeasurability relations $a_2 \,♀\, a_3$, $a_1 \,♀\, a_2$, $a_1 \,♀\, a_2 \wedge a_3$, $a_1 \wedge a_2 \,♀\, a_3$, and $a_1 \wedge (a_2 \wedge a_3) \,♀\, (a_1 \wedge a_2) \wedge a_3$ be satisfied, to allow an application of the partial operations in $\mathscr{B}$. A proposition is then valid in $\mathscr{B}$ if every such "meaningful" substitution of elements yields the element 1 of $\mathscr{B}$.

A Boolean function $\varphi(x_1, \ldots, x_n)$ such as (9) may be considered as a polynomial over Z_2. We shall now give a formal definition for a polynomial $\varphi(x_1, \ldots, x_n)$ over a field K to be identically 1 in a partial algebra $\mathscr{N}$ over K. We first recursively define the *domain* D_φ of $\varphi(x_1, \ldots, x_n)$ in $\mathscr{N}$. We simultaneously define a map φ^* corresponding to $\varphi(x_1, \ldots, x_n)$. D_φ is a subset of the n-fold Cartesian product $\mathscr{N}^n$ of $\mathscr{N}$ and φ^* is a map from D_φ into $\mathscr{N}$. Let $a = \langle a_1, \ldots, a_n \rangle$ be an arbitrary element of $\mathscr{N}^n$.

1. If φ is the polynomial 1, then $D\varphi = \mathscr{N}^n$ and $\varphi^*(a) = 1$.
2. If φ is the polynomial x_i $(i = 1, 2, \ldots, n)$, then $D\varphi = \mathscr{N}^n$ and $\varphi^*(a) = a_i$.
3. If $\varphi = k\psi$ with $k \in K$, then $D_\varphi = D_\psi$ and $\varphi^*(a) = k\psi^*(a)$.
4. If $\varphi = \psi \otimes \chi$ (where $\otimes$ is either $+$ or $\cdot$), then $a \in D_\varphi$ if and only if $a \in D_\psi \cap D_\chi$ and $\psi^*(a) \,♀\, \chi^*(a)$; $\varphi^*(a) = \psi^*(a) \otimes \chi^*(a)$.

We say that the identity $\varphi(x_1, \ldots, x_n) = 1$ holds in $\mathscr{N}$ if $\varphi^*(a) = 1$ for all $a \in D_\varphi$. More generally, if $\varphi(x_1, \ldots, x_n)$ and $\psi(x_1, \ldots, x_n)$ are two polynomials over K, we shall say that the identity $\varphi(x_1, \ldots, x_n) = \psi(x_1, \ldots, x_n)$ holds in $\mathcal{N}$ if $\varphi^*(a) = \psi^*(a)$ for all $a \in D_\varphi \cap D_\psi$.

Let $\varphi(x_1, \ldots, x_n)$ be a propositional (*i.e.*, a Boolean) function. Then $\varphi(x_1, \ldots, x_n)$ may be considered as a polynomial over Z_2. Let $\mathcal{N}$ be a partial Boolean algebra. Then φ is *valid* in $\mathcal{N}$ if the identity $\varphi = 1$ holds in $\mathcal{N}$. If for some $a \in D_\varphi$ we have $\varphi^*(a) = 0$, then φ is *refutable* in $\mathcal{N}$. If φ and ψ are two propositional functions, then $\varphi = \psi$ is *valid* in $\mathcal{N}$ if the identity $\varphi = \psi$ holds in $\mathcal{N}$. We illustrate these definitions with an example. We shall show that the tautology (9) is valid in every partial Boolean algebra $\mathcal{N}$. In fact, we show that the identity $x_1 \wedge (x_2 \wedge x_3) = (x_1 \wedge x_2) \wedge x_3$ is valid in $\mathcal{N}$; this means that we do not require that $a_1 \wedge (a_2 \wedge a_3) ♀ (a_1 \wedge a_2) \wedge a_3$. To see this note that if $a_2 ♀ a_3$, $a_1 ♀ a_2$, $a_1 ♀ a_2 \wedge a_3$, $a_1 \wedge a_2 ♀ a_3$ then

$$
\begin{aligned}
a_1 \wedge (a_2 \wedge a_3) &= a_1 \wedge (a_2 \wedge (a_2 \wedge a_3)) \\
&= (a_1 \wedge a_2) \wedge (a_2 \wedge a_3).
\end{aligned}
$$

The last equality holds because the elements a_1, a_2 and $a_2 \wedge a_3$ are pairwise commeasurable and hence by the definition of a partial algebra generate a Boolean algebra in $\mathcal{N}$. Similarly, $(a_1 \wedge a_2) \wedge a_3 = (a_1 \wedge a_2) \wedge (a_2 \wedge a_3)$, proving the result.

In the case of quantum mechanics these considerations are more than theoretical possibilities, they occur in ordinary reasoning about physical systems. For instance, the orbital angular momentum $\bar{L}$ of an atom is commeasurable with the spin angular momentum $\bar{S}$. If the system has spherical symmetry then a component of $\bar{L} + \bar{S}$ (= total angular momentum $\bar{J}$) is commeasurable with the Hamiltonian H, although components of $\bar{L}$ and $\bar{S}$ are separately not commeasurable with H. Thus a statement specifying H and a component of $\bar{L} + \bar{S}$ is of the type considered here.

If $\mathcal{N}$ is a Boolean algebra this definition of validity coincides with the usual definition. In that case the set of valid propositional functions coincides with the classical tautologies, *i.e.*, those propositional functions which are valid in Z_2. In the following theorem we connect the validity of classical tautologies in a partial Boolean algebra $\mathcal{N}$ with the imbeddability of $\mathcal{N}$ into a Boolean algebra.

For the sake of obtaining a complete correspondence in this theorem we introduce the following weakening of the notion of imbedding.

DEFINITION. Let $\mathcal{N}$, $\mathcal{L}$ be partial Boolean algebras. A homomorphism $\varphi:\mathcal{N}\to\mathcal{L}$ is a *weak imbedding* of $\mathcal{N}$ into $\mathcal{L}$ if $\varphi(a)\neq\varphi(b)$ whenever $a\between b$ and $a\neq b$ in $\mathcal{N}$. Thus a weak imbedding is a homomorphism which is an embedding on Boolean subalgebras of $\mathcal{N}$.

The counterpart of Theorem 0 of Section II is that $\mathcal{N}$ is weakly imbeddable in a Boolean algebra if and only if for every non-zero element a in $\mathcal{N}$ there is a homomorphism $h:\mathcal{N}\to Z_2$ such that $h(a)\neq 0$.

THEOREM 4. *Let $\mathcal{N}$ be a partial Boolean algebra.*

(1) *$\mathcal{N}$ is imbeddable into a Boolean algebra if and only if, for every classical tautology of the form $\varphi\equiv\psi$, $\varphi=\psi$ is valid in $\mathcal{N}$.*

(2) *$\mathcal{N}$ is weakly imbeddable into a Boolean algebra if and only if every classical tautology φ is valid in $\mathcal{N}$.*

(3) *$\mathcal{N}$ may be mapped homomorphically into a Boolean algebra if and only if every classical tautology φ is not refutable in $\mathcal{N}$.*

Proof. The necessity of the condition in each case is clear. We shall give a uniform proof of sufficiency for the three cases where $\mathcal{N}$ satisfies the condition that $\mathcal{N}$ is (1) imbeddable, (2) weakly imbeddable or (3) mapped homomorphically into a Boolean algebra. Let

$$s_i(x)=\begin{cases}x & i=1,2\\ 1 & i=3,\end{cases}\qquad t_i(y)=\begin{cases}y & i=1\\ 0 & i=2,3.\end{cases}$$

Let K_1 be the set of all equations of the form $\alpha+\beta=\gamma$ or $\xi\eta=\zeta$ which subsist among elements of $\mathcal{N}$. (In the language of model theory, K_1 denotes the positive statements from the diagram of $\mathcal{N}$.) Let K_2 be the elementary axioms describing the class of Boolean algebras. Write $K=$ $=K_1\cup K_2$. Then the class of all models of K consist precisely of the homomorphic images of $\mathcal{N}$ which are Boolean algebras.

Suppose now that $\mathcal{N}$ does not satisfy condition (i) ($i=1$, 2, or 3). Then by Theorem 0 and its counterpart for weak imbeddings there exist two distinct elements a, b in $\mathcal{N}$ such that for every Boolean algebra B and every homomorphism $h:\mathcal{N}\to B$ we have $h(s_i(a))=h(t_i(b))$. Since then $s_i(a)$ and $t_i(b)$ are identified in every model of K, we have by the Completeness Theorem for the Predicate Calculus that

$$K\vdash s_i(a)=t_i(b).$$

Hence, there is a finite subset

$$L=\{\alpha_j+\beta_j=\gamma_j,\ \xi_k\eta_k=\zeta_k \mid 1\leqq j\leqq n,\ 1\leqq k\leqq m\}$$

of K_1 such that

$$K_2\cup L\vdash s_i(a)=t_i(b)$$

so that

$$K_2\vdash(\bigwedge_j(\alpha_j+\beta_j+\gamma_j=0)\wedge\bigwedge_k(\xi_k\eta_k+\zeta_k=0))\rightarrow s_i(a)=t_i(b)$$

or

$$K_2\vdash(\bigvee_{j,k}(\alpha_j+\beta_j+\gamma_j)(\xi_k\eta_k+\zeta_k)=0)\rightarrow s_i(a)=t_i(b),$$

i.e., $K_2\vdash\rho(\alpha_1,\ldots,\zeta_m)=0\rightarrow s_i(a)=t_i(b)$ where

$$\rho(\alpha_1,\ldots,\zeta_m)=\bigvee_{j,k}(\alpha_j+\beta_j+\gamma_j)(\xi_k\eta_k+\zeta_k).$$

Since the constants $\alpha_1,\ldots,\zeta_m$, a, b do not occur in K_2, we may replace them by variables $x_1,\ldots,x_n$, x, y to obtain

$$(10)\qquad K_2\vdash\rho(x_1,\ldots,x_n)=0\rightarrow s_i(x)=t_i(y).$$

Hence, the implication $\rho(x_1,\ldots,x_n)=0\rightarrow s_i(x)=t_i(y)$ is valid in all Boolean algebras. Let

$$\varphi \text{ denote } s_i(x)\rightarrow\rho$$

and

$$\psi \text{ denote } t_i(y)\rightarrow\rho.$$

Then it follows from (10) that $\varphi=\psi$ is Boolean identity, *i.e.*, $\varphi\equiv\psi$ is a classical tautology. (Note that for $i=2$, 3, $\psi=1$ so that $\varphi\equiv\psi$ reduces to φ.) On the other hand the substitution of the elements $\alpha_1,\ldots,\zeta_m$, $s_i(a)$, $t_i(b)$ from $\mathcal{N}$ for the variables $x_1,\ldots,x_n$, $s_i(x)$, $t_i(y)$ yields a value 0 for ρ, and hence a value $s_i(a)'$ for φ and $t_i(b)'$ for ψ. (Here u' denotes $1-u$.) Hence, under this valuation of φ and ψ in $\mathcal{N}$, we have

$$\begin{aligned}\varphi=a',\ \psi=b',\quad &\text{so that}\quad \varphi\neq\psi,\quad \text{for}\quad i=1\\ \varphi=a',\quad &\text{so that}\quad \varphi\neq 1,\quad \text{for}\quad i=2\\ &\qquad\qquad\ \ \varphi=0,\quad \text{for}\quad i=3,\end{aligned}$$

proving the theorem.

Since in the case of quantum mechanics there is, by Theorem 1, no homomorphism of D onto Z_2, we obtain the following consequence of Theorem 4.

COROLLARY. *There is a propositional formula φ which is a classical tautology but which is false under a (meaningful) substitution of quantum mechanical propositions for the propositional variables of φ.*

It is in fact not difficult to construct such a formula. Assign to each one-dimensional linear subspace L_i of D a distinct propositional variable x_i. To each orthogonal triple L_i, L_j, L_k of D assign the Boolean function

$$x_i+x_j+x_k+x_ix_jx_k.$$

Note that classically this formula is valid if and only if exactly one of x_i, x_j, x_k is valid. Hence the formula

$$\varphi=1-\prod(x_i+x_j+x_k+x_ix_jx_k),$$

where the product extends over all orthogonal triples of D, is classically valid, by Theorem 1. On the other hand, the substitution of the quantum mechanical statement P_i of Section IV for each x_i makes φ false since each factor of the product takes the value 1. Thus, the formula φ is the formal counterpart of the argument given at the end of Section IV. Actually, the formula φ is uneconomical in the number of variables used. A more judicious choice of variables corresponding to the graph Γ_2 yields a formula in 86 variables which is classically valid and quantum mechanically refutable.

This way of viewing the results of Sections III and IV, seems to us to display a new feature of quantum mechanics in its departure from classical mechanics. It is of course true that the Uncertainty Principle, say, already marks a departure from classical physics. However, the statement of the Uncertainty Principle involves two observables which are not commeasurable, and so may be refuted in the future with the addition of new states. This is the view of those who believe in hidden variables. Thus, the Uncertainly Principle as applied to the two-dimensional situation described in Section VI becomes inapplicable once the system is imbedded in the classical one. The statement $\varphi(P_1,\ldots,P_n)$ we have constructed deals only in each of the steps of its construction with commeasurable observables, and so cannot be refuted at a later date.

BIBLIOGRAPHY

[1] Bohm, D., 'Quantum Theory in Terms of "Hidden" Variables I', *Phys. Rev.* **85** (1952), 166–179.

[2] Bohm, D., 'A Suggested Interpretation of the Quantum Theory in Terms of "Hidden" Variables II', *Phys. Rev.* **85** (1952), 180–193.

[3] Bopp, F., 'La méchanique quantique est-elle une méchanique statistique classique particulière?', *Ann. L'Inst. H. Poincaré* **15** II (1956), 81–112.

[4] Broglie, L. de, *Non-Linear Wave Mechanics*, Elsevier, 1960.

[5] Freistadt, H., 'The Causal Formulation of the Quantum Mechanics of Particles', *Nuovo Cimento Suppl.*, Ser. 10, **5** (1957), 1–70.

[6] Gleason, A., 'Measures on Closed Subspaces of Hilbert Space', *J. of Math. and Mech.* **6** (1957), 885–893.

[7] Griffith, J. H. E. and Owen, J., 'Paramagnetic Resonance in the Nickel Tutton Salts', *Proc. Royal Society of London*, Ser. A, **213** (1952), 459–473.

[8] Halmos, P., *Lectures on Boolean Algebras*, Van Nostrand Studies, 1963.

[9] Hermann, G., *Die naturphilosophischen Grundlagen der Quantenmechanik*, Abhandlungen der Fries'schen Schule, 1935.

[10] Kochen, S. and Specker, E., *Logical Structures Arising in Quantum Theory*, The Theory of Models, 1963 Symposium at Berkeley, pp. 177–189.

[11] Kochen, S. and Specker, E., *The Calculus of Partial Propositional Functions*, Logic, Methodology and Philosophy of Science, 1964 Congress at Jerusalem, pp. 45–57.

[12] Neumark, M. A., 'Operatorenalgebren im Hilbertschen Raum', in *Sowjetische Arbeiten zur Funktionalen Analyse*, Verlag Kultur and Fortschritt, Berlin, 1954.

[13] Pryce, M. H. L., 'A Modified Perturbation Method for a Problem in Paramagnetism', *Phys. Soc. Proc. A*, **63** (1950), 25–29.

[14] Schiff, L., *Quantum Mechanics*, 2nd ed., McGraw-Hill, 1955.

[15] Schwartz, J., 'The Wiener-Siegel Causal Theory of Quantum Mechanics', in *Integration of Functionals*, New York University, 1957.

[16] Siegel, A., and Wiener, N., 'The Differential Space of Quantum Theory', *Phys. Rev.* **101** (1956).

[17] Specker, E., 'Die Logik nicht gleichzeitig entscheidbarer Aussagen', *Dialectica* **14** (1960), 239–246.

[18] Stevens, K. W. H., 'The Spin-Hamiltonian and Line Widths in Nickel Tutton Salts', *Proc. Roy. Soc. of London*, Ser. A. **214** (1952), 237–244.

[19] Neumann, J. von, *Mathematical Foundations of Quantum Mechanics*, P.V.P., 1955.

Cornell University
and
Eidgenössische Technische Hochschule, Zürich

PATRICK SUPPES

LOGICS APPROPRIATE TO EMPIRICAL THEORIES

I. INTRODUCTION

To those like myself who are mainly concerned with the methodology of the empirical sciences, the present symposium is both sobering and encouraging. It is sobering as one thinks of the scientific contrast between the majority of papers read here and the standard sources in the methodology and philosophy of science. Yet it is encouraging, because the hope is engendered that many of the methods, and perhaps above all, the intellectual standards of these papers, will extend themselves in a natural way to logical investigations of the empirical sciences. The logical and philosophical foundations of physics, for example, seem to be at about the stage where the foundations of mathematics were during most of the nineteenth century. Nearly any physicist and a large number of philosophers are prepared to deliver at a moment's notice a lecture on the foundations of quantum mechanics. The situation is far different with respect to the foundations of mathematics. With an ever-increasing volume of deep and rigorous results, mathematicians unacquainted with the literature are not prone to deliver casually-put-together *obiter dicta* on logic and related topics.

The present paper is meant to be a quite modest contribution to the foundations of physics conceived of as the same sort of discipline as the foundations of mathematics. My objective is to give two sorts of examples of empirical theories for which it is natural and convenient to introduce a logic that is deviant in some respect. The first class of examples is derived from theories of measurement and the more general consideration of physical principles of invariance. The second class is drawn from quantum mechanics.

II. PRINCIPLES OF INVARIANCE AND THREE-VALUED LOGIC

In theories of measurement or in more general physical theories prin-

C. A. Hooker (ed.), The Logico-Algebraic Approach to Quantum Mechanics, 329–340.

ciples of invariance arise in a natural way. For example, the statement that Venus has greater mass than Earth is true or false, independent of what unit of mass is used. Put another way, the truth value of the statement is invariant under an arbitrary (positive) similarity transformation of the units of mass used in attempting to verify or falsify the statement. On the other hand, the statement that the mass of Venus is greater than 10^{10} has no such invariance, for its truth or falsity will vary with the units of mass selected.

Let us generalize on this example by looking at a language in which we can express the results of n distinct measurements on physical objects. To avoid many repetitious details, I shall use as a basis the formal language of Tarski's monograph [51] enriched by individual variables "a", "b", "c",..., which take as values physical objects and n function symbols "m_1",..., "m_n". Terms are constructed from these function symbols and individual variables in the expected fashion; in particular, it is intended that $m_i(a)$ be a real number for $1 \leqslant i \leqslant n$. Let us call this language for a fixed n, $\mathscr{L}_n$. This formalization of $\mathscr{L}_n$ is necessary only to give definiteness to some of the logical results and problems it seems of interest to formulate. With inessential modifications for what is to be stated in this paper, $\mathscr{L}_n$ could easily be adapted to express in canonical form the results of a wide class of physical experiments.

The general semantical notion of a model of a language I shall assume is familiar. By a *model* of $\mathscr{L}_n$, I mean an ordered $(n+2)$-tuple $\mathfrak{M} = \langle \mathfrak{S}, A, m_1, \ldots, m_n \rangle$, where $\mathfrak{S}$ is the usual system of real numbers under the operations of addition and multiplication and the relation less than (technical details about $\mathfrak{S}$ are not critical here), A is a finite, non-empty set, and $m_1, \ldots, m_n$ are real-valued functions on A. The intended interpretation should be obvious. The set A is meant to be a set of physical objects whose various properties are measured by the numerical functions $m_1, \ldots, m_n$.

Associated with each function m_i expressing the results of a particular kind of measurement is a group G_i of (numerical) transformations. The group G_i characterizes the degree of uniqueness of the measurement. If m_i is a mass measurement, then G_i is the group of similarity transformations. If m_i is a classical position measurement on the x-axis, then G_i is the group of linear transformations, and so forth.

We first define when two models of $\mathscr{L}_n$ are appropriately related by these groups of transformations.

DEFINITION 1. *Let* $\mathfrak{M}_1 = \langle \mathfrak{S}, A_1, m_1, \ldots, m_n \rangle$ *and* $\mathfrak{M}_2 = \langle \mathfrak{S}, A_2, m'_1, \ldots, m' \rangle$ *be two models of* $\mathscr{L}_n$, *and let* $\mathfrak{G} = \langle G_1, \ldots, G_n \rangle$ *be an n-tuple of groups of transformations. Then* $\mathfrak{M}_1$ *is* $\mathfrak{G}$-related to $\mathfrak{M}_2$ *if and only if*

(i) $A_1 = A_2$;

(ii) *for* $1 \leqslant i \leqslant n$, *there is a* φ *in* $\mathfrak{G}_i$ *such that*

$$\varphi \circ m_i = m'_i,$$

where $\circ$ *is functional composition.*

The intuitive idea of truth-value invariance under the appropriate groups of transformations is formalized in the next definition.

DEFINITION 2. *An atomic formula F of* $\mathscr{L}_n$ *is* $\mathfrak{G}$-empirically meaningful *if and only if F is satisfied in a model* $\mathfrak{M}$ *of* $\mathscr{L}_n$ *when and only when it is satisfied in every model* $\mathfrak{G}$*-related to* $\mathfrak{M}$.

In other words, an atomic formula is said to be empirically meaningful only when it has the appropriate invariance properties – that the condition expressed in Definition 2 is always sufficient and not merely necessary is not a point of argument here. For a more extensive discussion of these matters see Suppes [59].

DEFINITION 3. *A formula of* $\mathscr{L}_n$ *is* $\mathfrak{G}$-empirically meaningful *if and only if each of its atomic formulas is such.*

The reason for requiring that each atomic formula of a formula be meaningful is to obtain truth-functionality for the three-valued logic characterized in the next theorem. The substance of this logic is to assign the ordinary values truth and falsity to meaningful formulas, and the single value *meaninglessness* (abbreviated μ) to the remainder. The proof of the theorem is straightforward and is omitted.

THEOREM 1. *Assign the truth value* μ *to* $\mathfrak{G}$*-empirically meaningless formulas of* $\mathscr{L}_n$. *Then the formulas of* $\mathscr{L}_n$ *are truth-functional in the three-valued logic T, F and* μ *defined by the following tables for negation and conjunction.*

S	$\neg S$
T	F
F	T
μ	μ

$\wedge$	T	F	μ
T	T	F	μ
F	F	F	μ
μ	μ	μ	μ

The ordinary definitions of disjunction, implication and so forth in terms of negation and conjunction are assumed in working out additional tables. The resulting truth tables are precisely those which Kleene [52a, p. 334] has called the *weak tables*, and they do seem to be about the weakest possible extension of classical two-valued logic.

A simple example will show why meaningfulness needs to be initially defined in terms of the atomic formulas rather than arbitrary formulas in order to have the truth-functional result of Theorem 1. Suppose we had proceeded the other way, and let m_1 be a mass measurement function with the associated group of similarity transformations. Then the sentence

$$(\exists a)(m_1(a)=1) \vee \neg(\exists a)(m_1(a)=1)$$

would have the value T although its single component atomic formula has the value μ. On the other hand, the sentence

$$(\exists a)(m_1(a)=1) \vee (\exists a)(m_1(a)=2)$$

would have the value μ as does its component atomic formulas, and thus the logic would not be truth-functional.

When we turn our attention to what happens to the set of meaningful formulas under inferences involving quantifiers the results are not as satisfactory as the sentential ones of Theorem 1. In formulating the theorem it is useful to introduce the notion of a *non-trivial* n-tuple $\mathfrak{G}$ of groups for a model $\mathfrak{M}$. By this I mean at least one G_i is such that every model $\mathfrak{G}$-related to $\mathfrak{M}$ may not be obtained by the identity transformation on the positive integers (a weaker notion of non-triviality will suffice to prove the theorem, but is not important for the purposes at hand).

THEOREM 2. *Let $\mathfrak{M}$ be a model of $\mathscr{L}_n$ and let $\mathfrak{G}$ be non-trivial for $\mathfrak{M}$. Then the set of $\mathfrak{G}$-meaningful formulas of $\mathscr{L}_n$ whose closures are true in $\mathfrak{M}$ is not a deductive system (i.e., the set is not closed under logical consequence).*

Proof. (I sketch the proof without too great concern for use-mention distinctions.) Let G_i be non-trivial for $\mathfrak{M}=\langle\mathfrak{S}, A, m_1,\ldots, m_n\rangle$. Then there is an atomic formula of $\mathscr{L}_n$, a positive integer k, a model $\mathfrak{M}'$ $\mathfrak{G}$-related to $\mathfrak{M}$ and an a in A such that

$$m_i(a)=k$$
$$m_i'(a)\neq k.$$

Now the integer k may be described in $\mathscr{L}_n$ by an expression of the form "$1+\cdots+1$", and thus there is a $\mathfrak{G}$-meaningful sentence of $\mathscr{L}_n$ which is true in $\mathfrak{M}$ and which asserts that for every number x if $x=k$ then $x<k+1$. For the variable in the sentence of $\mathscr{L}_n$ corresponding to "x" we now specify a term of the form "$m_i(a)$", and we obtain a formula which is not $\mathfrak{G}$-meaningful and thus not in the original set.

The difficulties posed by Theorem 2 are real. It certainly does not seem satisfactory to have a notion of meaningfulness that is not deductively closed. In Suppes [59] I proposed taking the bull by the horns and redefining the concept of logical consequence. We may say that a formula is a *$\mathfrak{G}$-meaningful logical consequence* of a set A of formulas of $\mathscr{L}_n$ if and only if S is a logical consequence of A and S is $\mathfrak{G}$-meaningful whenever every formula in A is also.

Two problems beset this new notion of meaningful logical consequence. First, it will not in general be possible to give finitary rules of inference, because for many n-tuples of groups the notion of meaningfulness will not be decidable. Secondly, there is a problem of completeness. Putting aside the finitary problem we may ask this question. If a standard set of rules of inference is used, augmented by the requirement that each step of an inference be $\mathfrak{G}$-meaningful, does being a meaningful logical consequence imply derivability by the rules of inference? This problem has not been solved for any significant n-tuples of groups.

The discussion of empirical meaningfulness and the related three-value logic has been restricted to the simple languages $\mathscr{L}_n$, but if we are willing to move from talk about formulas and sentences to talk about propositions or relations, it is easy to generalize the results to richer physical theories. The three-valued logic of T, F and μ applies in a wholly natural way to the system-of-classical-particle-mechanics functions introduced in McKinsey and Suppes [55], for example. In this case the groups of transformations are no longer restricted to one-to-one functions from the real numbers to the real numbers, but include the geometrical and Galilean transformations familiar in classical mechanics. Other examples from classical and relativistic physics are easily constructed.

In the past thirty or forty years a large number of different multi-valued logics have been proposed, and the technical literature on the formal structure of these logics has reached sizable proportions. The philosophical defense of the significance of multi-valued logics has been much less substantial. The interest of the discussion in this paper is meant to be in the intuitively direct philosophical defense possible for the three-valued logic introduced. The extension of classical two-valued logic developed here is too weak to raise any interesting technical problems at the level of sentential logic, and perhaps too complicated at the level of predicate logic, for which the problem of completeness in terms of meaningful logical consequence is central. But the empirical or methodological defense of the logic is clear. The data of experiments, or the empirical predictions of physical theories, are best expressed not by a particular model of the theory but by an appropriate equivalence class of models. Sentences or propositions of the theory which do not have an invariant truth value (in the two-valued sense) over these equivalence classes simply do not have any clear empirical meaning. A three-valued logic, with the third value being meaninglessness, provides a simple device for segregating these non-invariant propositions that do not have a definite empirical content, and for calling attention to the otiose features of any one model of a theory.

III. FAMILIES OF BOOLEAN ALGEBRAS IN QUANTUM MECHANICS[1]

Turning now to the logic of quantum mechanics, I try to develop an approach that hews somewhat closer to classical lines than many discussions. I also frankly admit that many of the proposals made in the literature are hard for me to fathom from an empirical or experimental standpoint. It is a simple enough matter formally to define orthocomplemented modular lattices and relate them to the structure of subspaces that arise in quantum mechanics, but exactly how this logic corresponds to the set of experimental propositions or events is ordinarily not elaborated in any detail. For example, the much-cited article of Birkhoff and von Neumann [36] restricts the consideration of the decisive supporting experimental consequences of quantum-mechanical theory to a few lines (p. 831).

My objective here is to give a simpler and more classical logic for the experimental propositions or events. I mean by an experimental event an event that may at least in principle be observed by some configuration of experimental apparatus. In this sense the Heisenberg uncertainty principle is not an experimental proposition expressing the observation of a single event, for it is in general not possible to measure simultaneously position and momentum. In probabilistic terms, as I have emphasized elsewhere (Suppes [61a]), there is no joint probability distribution of position and momentum. It is not a matter of measuring position and probability with zero variance. It is rather that there is no possibility of measuring them jointly at all, because their joint distribution does not exist.

To simplify developments and thereby to avoid a number of technical issues, I shall make the restrictive assumption that all observables may be expressed as functions of position and momentum. Thus for a system of n particles the probability space is the $6n$-dimensional Euclidean space $\mathscr{E}_{6n}$, and I shall represent any point by $6n$ real numbers $(q_1, \ldots, q_{3n}, p_{3n+1}, \ldots, p_{6n})$, where it is understood that the first $3n$ coordinates are the position measurements and the last $3n$ the momentum measurements, with q_1, q_2, q_3 the position coordinates of the first particle, p_{3n+1}, p_{3n+2}, p_{3n+3} its momentum coordinates, etc. A one-dimensional event is just a measurable subset of points of $\mathscr{E}_{6n}$ restricted in some one dimension, and in general, an event is a measurable cylinder set $C_S \subseteq \mathscr{E}_{6n}$, where S is the set of dimensions of $\mathscr{E}_{6n}$ on which C_S is restricted, i.e., $S \subseteq \{1, \ldots, 6n\}$. The following theorem about cylinder sets is easily proved.

THEOREM 3. *For any non-empty set* $R \subseteq \{1, \ldots, 6n\}$, *the collection of measurable cylinder sets* C_S *such that* $S \subseteq R$ *forms a Boolean algebra, which we designate* $\mathfrak{B}_R$.

The following theorem can be proved on the basis of the standard formalism of classical quantum mechanics.

THEOREM 4. *Let* R *be a non-empty subset of* $\{1, \ldots, 6n\}$. *Then there exists a quantum-mechanical probability distribution on* $\mathfrak{B}_R$ *if and only if there is no number* m *such that* m *and* $3n+m$ *are both in* R.

In other words, a distribution exists on $\mathfrak{B}_R$ whenever no two coordinates representing conjugate observables are in R. It is obvious, of course, that all kinds of proper distributions exist on any $\mathfrak{B}_R$ apart from quantum mechanics. The restriction of the theorem applies only to those distributions derived from quantum-mechanical considerations. I say that a subset R of $\{1, \ldots, 6n\}$ is *admissible* whenever it satisfies the restriction stated in Theorem 4. Analogous to the definitions of Section II, I then define a measurable cylinder set C_R as (*experimentally*) *meaningful* if and only if R is admissible. A collection $\{C_S\}$ of measurable cylinder sets is *simultaneously meaningful* if and only if there is a subset R of $\{1,\ldots, 6n\}$ such that for each cylinder set C_S in the collection $S \subseteq R$ and R is admissible.

The departures from classical logic that follow from these definitions are evident. For example, the intersection or union of two meaningful events is not necessarily meaningful. The truth tables given in Theorem 1 certainly do not hold for this logic of meaningfulness.

My own view is that the most natural way to look at this situation is in terms of a family of logics defined in the following manner. An admissible set R of coordinates is *admissibly complete* iff it is a proper subset of no admissible set. A Boolean algebra $\mathfrak{B}_R$ of measurable cylinder sets of $\mathscr{E}_{6n}$ is a *maximal, meaningful* Boolean algebra iff R is admissibly complete. From Theorem 4 it is evident there is no unique maximal Boolean algebra. Moreover, the meaningful Boolean algebras do not form a lattice, because two maximal such algebras, of which there are 2^{3n} in the present setup, do not have a meaningful greatest lower bound or meet. On the other hand, it is evident that any two maximal Boolean algebras as defined here are isomorphic, although it would be most surprising if for any Hamiltonian the quantum-mechanical probability distributions on the two algebras were also isomorphic.

I thus come to the following thesis.

The logic of the experimental propositions of the quantum mechanics of n particles for which all observables are defined in terms of position and momentum, is the family $\mathfrak{F}(\mathscr{E}_{6n})$ *of* 2^{3n} *maximal meaningful Boolean algebras* $\mathfrak{B}_R$ *and their subalgebras* $\mathfrak{B}_S$, *with* $S \subseteq R$.

At least one of these maximal algebras is always appropriate for the logic of a given experiment, and the intersection of all those that are is the unique subalgebra of the experiment.

It may be useful to compare the logic defined here with other proposals that have been made. I first mention the three-valued logic of Reichenbach [44]. His table for conjunction is the following

$\wedge$	T	F	I
T	T	F	I
F	F	F	F
I	I	F	I

which is Kleene's [52a, p. 334] *strong* table for conjunction. But Reichenbach's three-valued logic is highly misleading, for the structure of experimental propositions or events proposed by him is not constructed so as to be truth-functional in this logic, and the same is true of the family $\mathfrak{F}(\mathscr{E}_{6n})$ of Boolean algebras with $\wedge$ interpreted as intersection. The actual structure Reichenbach seems to be after in terms of what can and cannot be measured is rather close to what has been suggested in this paper.

By adopting a more complicated set of values, a truth-functional scheme for $\mathfrak{F}(\mathscr{E}_{6n})$ rather similar to the three-valued logic proposed in the first part of this paper may be set up. The idea is that the truth values are now not simply T, F and μ, but are ordered pairs (T, R), (F, R) and (μ, R), where R is the set of coordinates of the cylinder set corresponding to the proposition in question. To get truth-functionality, we must, as in the previous case, pay particular attention to the atomic formulas. What I mean is easily illustrated by an example. Let

$$A = \{x : x \in \mathscr{E}_{6n} \;\&\; x_1 \in (0, 1) \;\&\; x_{3n+1} \in (0, 1)\},$$
$$B = \{x : x \in \mathscr{E}_{6n} \;\&\; x_1 \notin (0, 1) \;\&\; x_{3n+1} \notin (0, 1)\},$$

where x_i is the ith coordinate of point x in $\mathscr{E}_{6n}$. Then neither A nor B is meaningful, but $A \cup B$ is, being just $\mathscr{E}_{6n}$ itself. The disjunction corresponding to $A \cup B$ is declared meaningless, for the same reasons that a similar decision was made about '$x = 1 \vee x \neq 1$' in the earlier discussion of invariance. Holding these remarks in mind, the intuitive basis for the following truth tables for negation and conjunction is evident. (I shall not formalize an elementary language to make their application precise, for it is clear from the earlier discussion how this may be done.) Because meaningfulness is preserved under complementation of cylinder sets, negation is simple.

P	$\neg P$
(T, R)	(F, R)
(F, R)	(T, R)
(μ, R)	(μ, R)

In the case of conjunction, two natural cases arise according to whether or not the set of coordinates $R \cup S$ is admissible.

Case 1. $R \cup S$ admissible.

$\wedge$	(T, S)	(F, S)
(T, R)	$(T, R \cup S)$	$(F, R \cup S)$
(F, R)	$(F, R \cup S)$	$(F, R \cup S)$

Case 2. $R \cup S$ not admissible.

$\wedge$	(T, S)	(F, S)	(μ, S)
(T, R)			
(F, R)	$(\mu, R \cup S)$ for all entries		
(μ, R)			

A theorem like Theorem 1 may be proved for these tables. We obtain truth-functionality by widening the usual net of considerations that count in determining the truth value of a molecular sentence to include, not the full factual content of the proposition, but the set of coordinates relevant to the factual content of its atomic constituents. This inclusion marks a wider departure from classical logic than is true of the tables of Theorem 1.

The central point of comparison with the logic of quantum mechanics proposed by Birkhoff and von Neumann [36] has already been mentioned. They require that the structure of experimental propositions be a lattice, and thus that the *conjunction* of two meaningful propositions in the lattice also be in the lattice, but for reasons already stated this requirement seems too strict when one proposition expresses a possible result of measuring position and the other of measuring momentum. The fact that the closed linear subspaces of a complex separable Hilbert space form such a lattice is not sufficient, in my view, to maintain that it expresses the more restricted logic of experimental propositions, in spite of the central importance of this lattice of subspaces in the formulation of the theory.

A generalization of the logic of von Neumann and Birkhoff has been given by Varadarajan [62] and also by Mackey [63]. They define a *logic* as an orthocomplemented partially ordered set, which concept is easily made explicit. Let A be a non-empty set, let $\leqslant$ be a binary relation on A, let $'$ be a unary operation on A, and let 0 and 1 be elements of A. Let us also define another binary relation and a binary operation on A. Elements a and b are *disjoint* (in symbols: $a \perp b$) if and only if $a \leqslant b'$. And, if $a \perp b$ we define the sum $a \dot{+} b$ in the expected fashion: $a \dot{+} b = c$ if and only if $a \leqslant c$, $b \leqslant c$ and for every d in A, if $a \leqslant d$ and $b \leqslant d$ then $c \leqslant d$. In terms of these notions we then have:

DEFINITION 4. *A structure* $\mathscr{A} = \langle A, \leqslant, ', 0, 1 \rangle$ *is an orthocomplemented* *partial ordering if and only if the following axioms are satisfied for a, b and c in A:*

1. $a \leqslant a$.
2. *If* $a \leqslant b$ *and* $b \leqslant a$ *then* $a = b$.
3. *If* $a \leqslant b$ *and* $b \leqslant c$ *then* $a \leqslant c$.
4. $0 \leqslant a$.
5. $a \leqslant 1$.
6. *If* $a \leqslant b$ *then* $b' \leqslant a'$.
7. $(a')' = a$.
8. $a \dot{+} a' = 1$.
9. *If* $a \leqslant b$ *then there is a,* c *in* A *such that* $c \perp a$ *and* $a \dot{+} c = b$.
10. *For any sequence of pairwise disjoint elements* $a_1, \ldots, a_n, \ldots$ *of* A *there is a* c *in* A *such that for all* n, $a_n \leqslant c$ *and for every* d *in* A, *if for every* n, $a_n \leqslant d$, *then* $c \leqslant d$.

The purpose of Axiom 10 from the standpoint of probability distributions on logics is obvious; it guarantees the necessary underlying structure for a countably additive measure.

Although an orthocomplemented partial ordering is not a lattice, the requirements are still too strict for the collection of experimentally meaningful events in the family $\mathfrak{F}(\mathscr{E}_{6n})$ of Boolean algebras. The intended interpretation is that A be $\bigcup \mathfrak{F}(\mathscr{E}_{6n})$, $\leqslant$ be set inclusion $\subseteq$, $'$ be set complementation, 0 be the empty set and 1 be $\mathscr{E}_{6n}$. But this structure does not satisfy Axiom 10, because the union of two disjoint meaningful events is not itself meaningful. For example, let

$$C = \{x : x \in \mathscr{E}_{6n} \;\&\; x_1 \in (0, 1) \;\&\; x_2 \in (0, 1)\},$$
$$D = \{x : x \in \mathscr{E}_{6n} \;\&\; x_1 \in (1, 2) \;\&\; x_{3n+2} \in (0, 1)\},$$

where x_i is the $i$the coordinate of point x in $\mathscr{E}_{6n}$. Clearly both C and D are meaningful, disjoint events, but $C \cup D$ is not meaningful, and is thus not in $\bigcup \mathfrak{F}(\mathscr{E}_{6n})$.

The results stated for $\mathfrak{F}(\mathscr{E}_{6n})$ thus far have been rather negative, but there is a positive result related to the work of Kochen-Specker [*], namely, $\mathfrak{F}(\mathscr{E}_{6n})$ with the appropriate structure is a partial Boolean algebra. Roughly speaking, every partial Boolean algebra may be represented by a collection of Boolean algebras satisfying certain consistency conditions. For details see the article of Kochen-Specker [*].

In this sense the family of Boolean algebras constructed here originally from probabilistic considerations provides yet another intuitive argument for Kochen and Specker's concept of a partial Boolean algebra. However, the issue is not entirely simple. From a probability standpoint it is also convenient to have well defined the probability of the union of any two disjoint events, as in the case of $C \cup D$ above. No conflict with Theorem 4 exists, for the theorem is concerned with the existence of a measure on the entire algebra $\mathfrak{B}_R$, not with the much narrower question of whether or not the event $C \cup D$ has a probability. In any given experiment, C, C', D or D' may be observed, but, of course, never the complex event $C \cup D$. For computational purposes we may then in a natural way extend the probability measure to these events.

It is not possible here to study in any detail the relations between the orthocomplemented partial orderings and partial Boolean algebras, particularly the analysis of the similar but differing notions of simultaneously measurability, one of which is primitive in the theory of partial Boolean algebras, and the other of which is definable in a simple way in the theory of orthocomplemented partial orderings.

Either one of these theories certainly provides a more natural logic than that of classical Boolean algebra by itself as a structural foundation on which to erect the probabilistic superstructure of quantum mechanics because of the difficulties about improper or pseudo-probability distributions that arise in all too direct a fashion in the classical formulation.

Stanford University

NOTE

[1] I am indebted to Dana Scott for several helpful comments on this section.

PATRICK SUPPES

15. THE PROBABILISTIC ARGUMENT FOR A NONCLASSICAL LOGIC OF QUANTUM MECHANICS*

I. THE ARGUMENT

The aim of this paper is simple. I want to state as clearly as possible, without a long discursion into technical questions, what I consider to be the single most powerful argument for use of a nonclassical logic in quantum mechanics. There is a very large mathematical and philosophical literature on the logic of quantum mechanics, but almost without exception, this literature provides a very poor intuitive justification for considering a nonclassical logic in the first place. A classical example in the mathematical literature is the famous article by Birkhoff and von Neumann (1936). Although Birkhoff and von Neumann pursue in depth development of properties of lattices and projective geometries that are relevant to the logic of quantum mechanics, they devote less than a third of a page (p. 831) to the physical reasons for considering such lattices. Moreover, the few lines they do devote are far from clear. The philosophical literature is just as bad on this point. One of the better known philosophical discussions on these matters is that found in the last chapter of Reichenbach's book (1944) on the foundations of quantum mechanics. Reichenbach offers a three-valued truth-functional logic which seems to have little relevance to quantum-mechanical statements of either a theoretical or experimental nature. What Reichenbach particularly fails to show is how the three-valued logic he proposes has any functional role in the theoretical development of quantum mechanics. It is in fact fairly easy to show that the logic he proposes could not possibly be adequate for a systematic theoretical statement of the theory as it is ordinarily conceived. The reasons for this will become clear later on in the present paper.

The main premises of the argument I outline in this paper are few in number. I state them at this point without detailed justification in order to give the broad outline of the argument the simplest possible form.

* Reprinted from *Philosophy of Science* **33** (1966), 14–21.

PREMISE 1: *In physical or empirical contexts involving the application of probability theory as a mathematical discipline, the functional or working logic of importance is the logic of the events or propositions to which probability is assigned, not the logic of qualitative or intuitive statements to be made about the mathematically formulated theory.* (In the classical applications of probability theory, this logic of events is a Boolean algebra of sets; for technical reasons that are unimportant here this Boolean algebra is usually assumed to be countably additive, i.e., a σ-algebra.)

PREMISE 2: *The algebra of events should satisfy the requirement that a probability is assigned to every event or element of the algebra.*

PREMISE 3: *In the case of quantum mechanics probabilities may be assigned to events such as position in a certain region or momentum within given limits, but the probability of the conjunction of two such events does not necessarily exist.*

CONCLUSION: *The functional or working logic of quantum mechanics is not classical.*

From a scientific standpoint the conclusion from the premises is weak. All that is asserted is that the functional logic of quantum mechanics is not classical, which means that the algebra of events is not a Boolean algebra. Nothing is said about what the logic of quantum mechanics *is*. That question will be considered shortly. First I want to make certain that the support for the premises stated is clear, as well as the argument leading from the premises to the conclusion.

Concerning the first premise, the arguments in support of it are several. A source of considerable confusion in the discussion of the logic of quantum mechanics has been characterization of the class of statements whose logic is being discussed. On the one hand we are presented with the phenomenon that quantum mechanics is a branch of physics that uses highly developed mathematical tools, and on the other hand, discussions of logic deal with the foundations of mathematics itself. It is usually difficult to see the relation between characterization of the sentential connectives that seem appropriate for a new logic and the many mathematical concepts of an advanced character that must be available for actual work in quantum mechanics. The problem has often been posed as how can one consider changing the logic of quantum mechanics when the mathematics used in quantum mechanics depends in such a thorough fashion on classical logic. The point of this first premise is to narrow and sharpen the focus of the discussion of the logic of an empirical science. As in the case of quantum mechanics, we shall take it for granted that probability theory is involved in the mathematical statement of the theory. In every such case a logic of events is required as an underpinning for the probability theory. The structure of the algebra of events expresses in an exact way the logical structure of the theory itself.

Concerning the second premise the arguments for insisting that a probability may be assigned to every event in the algebra is already a part of classical probability theory. It is only for this reason that one considers an algebra, or σ-algebra, of sets as the basis for classical probability theory. If it were permitted to have events to which probabilities could not be attached, then we could always take as the appropriate algebra the set of all subsets of the basic sample space. The doctrine that the algebra of events must have the property asserted in the second premise is too deeply embedded in classical probability theory to need additional argument here. One may say that the whole point of making explicit the algebra of events is just to make explicit those sets to which probabilities may indeed be assigned. It would make no sense to have an algebra of events that was not the entire family of subsets of the given sample space and yet not be able to assign a probability to each event in the algebra.

Concerning the third premise it is straightforward to show that the algebra of events in quantum mechanics cannot be closed under conjunction or intersection of events. The event of a particle's being in a certain region of space is well defined in all treatments of classical quantum mechanics. The same is true of the event of the particle's momentum's being in a certain region as well. If the algebra of events were a Boolean algebra we could then ask at once for the probability of the event consisting of the conjunction of the first two, that is, the event of the particle's being in a certain region at a given time t and also having its momentum lying in a certain interval at the same time t. What may be shown is that the probability of such a joint event does not exist in the classical theory. The argument goes back to Wigner (1932), and I have tried to make it in as simple and direct a fashion as possible in Suppes (1961b).[1†] The detailed argument shall not be repeated here. Its main line of development is completely straightforward. In the standard formalism, we may compute the expectation of an operator when the quantum-mechanical system is in a given state. In the present case the operator we choose is the usual one for obtaining the characteristic function of a probability distribution of two variables. Having obtained the characteristic function we then invert it by the usual Fourier methods. Inversion should yield the density corresponding to the joint probability distribution of position and momentum. It turns out that for most states of any quantum-mechanical system the resulting density function is not the density function of any genuine joint probability distribution. We conclude that in general the joint distribution of two random variables like position and momentum does not exist in quantum mechanics and, consequently, we cannot talk about the conjunction of two events defined in terms of these two random variables. From the standpoint of the logic of science, the fundamental character of this result is at a much deeper level than the uncertainty principle itself, for there is nothing in the uncertainty principle as ordinarily formulated that runs counter to classical probability theory.

The inference from the three premises to the conclusion is straightforward enough hardly to need comment. From premise (1) we infer that the functional logic of events is the formal algebra of events on which a probability measure is defined. According to premise (2) every element, i.e., event, of the algebra must be assigned a probability. According to premise (3) the algebra of events in quantum mechanics cannot be closed under the conjunction of events and satisfy premise (2). Hence the algebra of events in quantum mechanics is not a Boolean algebra, because every Boolean algebra is closed under conjunction. Whence according to premise (1) the functional logic of quantum mechanics is not a Boolean algebra and thus is not classical.

II. THE LOGIC

Although the conclusion of the argument was just the negative statement that the logic of quantum mechanics is not classical, a great deal more can be said on the positive side about the sort of logic that does seem appropriate. To begin with it will be useful to record the familiar definition of an algebra, and σ-algebra, of sets.

DEFINITION 1: *Let X be a nonempty set. $\mathscr{F}$ is a classical algebra of sets on X if and only if $\mathscr{F}$ is a nonempty family of subsets of X and for every A and B in $\mathscr{F}$:*

1. $\sim A \in \mathscr{F}$.
2. $A \cup B \in \mathscr{F}$.

Moreover, if $\mathscr{F}$ *is closed under countable unions, that is, if for* $A_1, A_2, \ldots,$ $A_n, \ldots \in \mathscr{F}$,

$$\bigcup_{i=1}^{\infty} A_i \in \mathscr{F},$$

then $\mathscr{F}$ *is a classical* σ*-algebra on* X.

It is then standard to use the concepts of Definition 1 in defining the concept of a classical probability space. In this definition we assume that the set-theoretical structure of X, $\mathscr{F}$ and P is familiar; in particular, that X is a nonempty set, $\mathscr{F}$ a family of subsets of X and P a real-valued function defined on $\mathscr{F}$.

DEFINITION 2: *A structure* $\mathscr{X} = \langle X, \mathscr{F}, P \rangle$ *is a finitely additive classical probability space if and only if for every* A *and* B *in* $\mathscr{F}$:

P1. $\mathscr{F}$ *is a classical algebra of sets on* X;
P2. $P(A) \geqslant 0$;
P3. $P(X) = 1$;
P4. *If* $A \cap B = 0$, *then* $P(A \cup B) = P(A) + P(B)$.

Moreover, $\mathscr{X}$ *is a classical probability space (without restriction to finite additivity) if the following two axioms are also satisfied:*

P5. $\mathscr{F}$ *is a* σ*-algebra of sets on* $\mathscr{X}$;
P6. *If* $A_1, A_2, \ldots,$ *is a sequence of pairwise incompatible events in* $\mathscr{F}$, *i.e.,* $A_i \cap A_j = 0$ *for* $i \neq j$, *then*

$$P\left(\bigcup_{i=1}^{\infty} A_i\right) = \sum_{i=1}^{\infty} P(A_i).$$

In modifying the classical structures characterized in Definitions 1 and 2 to account for the truculent "facts" of quantum mechanics, there are a few relatively arbitrary choice points. One of them needs to be described in order to explain an aspect of the structures soon to be defined. I pointed out earlier that the joint probability of two events does not necessarily exist in quantum mechanics. A more particular question concerns the joint probability of two disjoint events. In this case there is no possibility of observing both of them, since the very structure of the algebra of events rules this out. On the other hand, it is theoretically convenient to include the union of two such events in the algebra of sets, or a denumerable sequence of pairwise disjoint events, in the case of a σ-algebra. This liberal attitude toward the concept of event has been adopted here, but it should be noted that it would be possible to take a stricter attitude without affecting the concept of an observable in any important way. (This stricter attitude is taken by Kochen and Specker, 1965, but they also deliberately exclude all probability questions in their consideration of the logic of quantum mechanics.)

So the logic of quantum mechanics developed here permits the union of disjoint events apart from any question of noncommuting random variables' being involved in their definition. A more detailed discussion of this point may be found in Suppes (1965b). Roughly speaking, the definitions that follow express the idea that the probability distribution of a single quantum-mechanical random variable is classical, and the deviations arise only when several random variables or different kinds of events are considered.

The approach embodied in Definition 3 follows Varadarajan (1962); it differs in that Varadarajan does not consider an algebra of sets, but only the abstract algebra.

DEFINITION 3: *Let X be a nonempty set. $\mathscr{F}$ is a quantum-mechanical algebra of sets on X if and only if $\mathscr{F}$ is a nonempty family of subsets of X and for every A and B in $\mathscr{F}$:*

1. $\sim A \in \mathscr{F}$;
2. *If* $A \cap B = 0$ *then* $A \cup B \in \mathscr{F}$.

Moreover, if $\mathscr{F}$ is closed under countable unions of pairwise disjoint sets, that is, if $A_1, A_2, \ldots$ is a sequence of elements of $\mathscr{F}$ such that for $i \neq j$, $A_i \cap A_j = 0$

$$\bigcup_{i=1}^{\infty} A_i \in \mathscr{F},$$

then $\mathscr{F}$ is a quantum-mechanical σ-algebra of sets.

The following elementary theorem is trivial.

THEOREM 1: *If $\mathscr{F}$ is a classical algebra (or σ-algebra) of sets on X then $\mathscr{F}$ is also a quantum-mechanical algebra (or σ-algebra) of sets on X.*

The significance of Theorem 1 is apparent. It shows that the concept of a quantum-mechanical algebra of sets is a strictly weaker concept than that of a classical algebra of sets. This is not surprising in view of the breakdown of joint probability distributions in quantum mechanics. We cannot expect to say as much, and the underlying logical structure of our probability spaces reflects this restriction.

It is hardly necessary to repeat the definition of probability spaces, because the only thing that changes is the condition on the algebra $\mathscr{F}$, but in the interest of completeness and explicitness it shall be given.

DEFINITION 4: *A structure* $\mathscr{X}=\langle X, \mathscr{F}, P\rangle$ *is a finitely additive quantum-mechanical probability space if and only if for every A and B in* $\mathscr{F}$:

P1. $\mathscr{F}$ *is a quantum-mechanical algebra of sets on X;*

P2. $P(A)\geqslant 0$;

P3. $P(X)=1$;

P4. *If* $A\cap B=0$, then $P(A\cup B)=P(A)+P(B)$.

Moreover, X is a quantum-mechanical probability space (without restriction to finite additivity) if the following two axioms are also satisfied:

P5. $\mathscr{F}$ *is a quantum-mechanical σ-algebra of sets on X;*

P6. *If* $A_1, A_2, \ldots,$ *is a sequence of pairwise incompatible events in* $\mathscr{F}$, i.e., $A_i\cap A_j=0$ *for* $i=j$, *then*

$$P\left(\bigcup_{i=1}^{\infty} A_i\right)=\sum_{i=1}^{\infty} P(A_i).$$

It is evident from the close similarity between Definitions 2 and 4 that we have as an immediate consequence of Theorem 1 the following result:

THEOREM 2: *Every classical probability space is also a quantum-mechanical probability space.*

It goes without saying that in the case of both of these theorems it is easy to give counterexamples to show that their converses do not hold.

Quantum-mechanical probability spaces can be used as the basis for an axiomatic development of classical quantum mechanics, but the restriction to algebras of sets in order to stress the analogy to classical probability spaces is too severe. The spaces defined are adequate for developing the theory of all observables that may be defined in terms of position and momentum, but not for the more general theory. The fundamental characteristic of the general theory is that not every quantum-mechanical algebra may be embedded in a Boolean algebra, and thus is not isomorphic to a quantum-mechanical algebra of sets, because every such algebra of sets is obviously embeddable in the Boolean algebra of the set of all subsets of X.

It is thus natural to consider the abstract analogue of Definition 3 and define the general concept of a quantum-mechanical algebra. (The axioms given here simplify those in Suppes, 1965b, which are in turn based on Varadarajan, 1962.) Let A be a non-empty set, corresponding to the family $\mathscr{F}$ of Definition 2, let $\leqslant$ be a binary relation on A – the relation $\leqslant$ is the abstract analogue of set inclusion, let $'$ be a unary operation on A – the operation $'$ is the abstract analogue of set complementation, and let 1 be an element of A – the element 1 is the abstract analogue of the sample space X. We then have:

DEFINITION 5: *A structure* $\mathfrak{A}=\langle A, \leqslant, ', 1\rangle$ *is a quantum-mechanical algebra if and only if the following axioms are satisfied for every a, b and c in A:*

1. $a\leqslant a$;
2. *If* $a\leqslant b$ *and* $b\leqslant a$ *then* $a=b$;
3. *If* $a\leqslant b$ *and* $b\leqslant c$ *then* $a\leqslant c$;
4. *If* $a\leqslant b$ *then* $b'\leqslant a'$;
5. $(a')'=a$;
6. $a\leqslant 1$;
7. *If* $a\leqslant b$ *and* $a'\leqslant b$ *then* $b=1$;
8. *If* $a\leqslant b'$ *then there is a c in A such that* $a\leqslant c$, $b\leqslant c$, *and for all d in A if* $a\leqslant d$ *and* $b\leqslant d$ *then* $c\leqslant d$;
9. *If* $a\leqslant b$ *then there is a c in A such that* $c\leqslant a'$, $c\leqslant b$ *and for every d in A if* $a\leqslant d$ *and* $c\leqslant d$ *then* $b\leqslant d$.

The only axioms of any complexity are the last three. If the operation of addition for disjoint elements were given the three axioms would be formulated as follows:

7! $a+a'=1$;
8! $a+b$ *is in* A;
9! *If* $a\leqslant b$ *then there is a c in A such that* $a+c=b$.

The difficulty with the operation of addition is that we do not want it to be defined except for disjoint elements, i.e., elements a and b of A such that $a\leq b'$.

It should also be apparent that we obtain a σ-algebra by adding to the axioms of Definition 5 the condition that for any sequence of pairwise disjoint elements $a_1, a_2, \ldots, a_n, \ldots$ of A there is a c in A such that for all n, $a_n \leqslant c$ and for every d in A, if for every n, $a_n \leqslant d$, then $c \leqslant d$.

Although it may be apparent, in the interest of explicitness, it is desirable to prove the following theorem.

THEOREM 3: *Every quantum-mechanical algebra of sets is a quantum-mechanical algebra in the sense of Definition* 5.

Proof: Let $\mathscr{F}$ be a quantum-mechanical algebra of sets on X. The relation $\leqslant$ of Definition 5 is interpreted as set inclusion $\subseteq$, and Axioms 1–3 immediately hold. The complementation is interpreted as set complementation with respect to X, and Axioms 4 and 5 hold in this interpretation. The Unit I is interpreted as the set X, and Axiom 6 holds because for any A in $\mathscr{F}$, $A \subseteq X$. In the case of Axiom 7 it is evident from elementary set theory that if $A \subseteq B$ and $\sim A \subseteq B$, then $A U \sim A \subseteq B$, whence $X \subseteq B$, but $B \subseteq X$, and so $B = X$. Regarding Axiom 8, if $A \subseteq \sim B$ then $A \cap B = 0$, so $A \cup B \in \mathscr{F}$ by virtue of the second axiom for algebras of sets, and we may take $C = A \cup B$ to satisfy the existential requirement of the axiom, because $A \subseteq A \cup B$, $B \subseteq A \cup B$, and if $A \subseteq D$ and $B \subseteq D$ then $A \cup B \subseteq D$. Finally, as to Axiom 9, if $A \subseteq B$ then we first want to show that $B \sim A \in \mathscr{F}$. By hypothesis $A, B \in \mathscr{F}$, whence $\sim B \in \mathscr{F}$, and since $A \subseteq B$, $A \cap \sim B = 0$ and thus $A \cup \sim B \in \mathscr{F}$, but then because $\mathscr{F}$ is closed under complementation, $\sim(A \cup \sim B) = \sim A \cap B = B \sim A \in \mathscr{F}$, as desired. It is easily checked, in order to verify Axiom 9 that because $A \subseteq B$, we have $B \sim A \subseteq \sim A$, $B \sim A \subseteq B$ and for every set D in $\mathscr{F}$, if $A \subseteq D$ and $B \sim A \subseteq D$ then $B \subseteq D$, since $A \cup (B \sim A) \subseteq D$ and $A \cup (B \sim A) = B$. Thus $B \sim A$ is the desired C, which completes the proof.

To obtain a sentential calculus for quantum-mechanical algebras, we define the notion of validity in the standard way. More particularly, in the calculus implication $\rightarrow$ corresponds to the relation $\leqslant$ and negation $\neg$ to the complementation operation $'$. We say that a sentential formula is quantum-mechanically valid if it is satisfied in all quantum-mechanical algebras, i.e., if under the expected interpretation the formula designates the element 1 of the algebra. The set of such valid sentential formulas characterizes the sentential logic of quantum mechanics. The axiomatic structure of this logic will be investigated in a subsequent paper.

I conclude with a brief remark about Reichenbach's three-valued logic.

It is easy to show that the quantum-mechanical logic defined here is not truth-functional in his three values (for more details see Suppes, 1965b). It seems clear to me that his three-valued logic has little if anything to do with the underlying logic required for quantum-mechanical probability spaces, and I have tried to show why the logic of quantum-mechanical probability is *the* logic of quantum mechanics. What I have not been able to do within the confines of this paper is to make clear precisely why the algebras characterized in Definition 5 are exactly appropriate to express the logic of quantum-mechanical probability. The argument in support of this choice is necessarily rather long and technical. A fairly good case is made out in detail in Varadarajan (1962).

However, apart from giving a mathematically complete argument for Definition 5, it may be seen that quantum-mechanical algebras have many intuitive properties in common with Boolean or classical algebras. The relation of implication or inclusion has most of its ordinary properties, the algebras are closed under negation, and the classical law of double negation holds. What is lacking are just the properties of closure under union and intersection – or disjunction and conjunction – that would cause difficulties for nonexistent joint probability distributions.

NOTE

[1†] Article 13 in this volume.

MARTIN STRAUSS

FOUNDATIONS OF QUANTUM MECHANICS*

1. Introduction and Summary

When [in 1905] the c-theory [Special Theory of Relativity] was born, both the mathematical formalism and its physical interpretation were established simultaneously; merely some questions of physical logic and axiomatics remained to be clarified. The story is quite different in the case of h-theory. To start with, two apparently quite different mathematical theories emerged, known as 'wave mechanics' (Schroedinger) and 'matrix mechanics' (Heisenberg-Born-Jordan), respectively. The underlying physical conceptions and, hence, the first physical interpretations were entirely different: Schroedinger believed he had reduced the quantum phenomena to a classical eigenvalue problem of the sort known from the theory of oscillations while Heisenberg-Born-Jordan understood their theory as a fundamental generalization of classical mechanics satisfying Bohr's principle of correspondence. The progress achieved in the following time consisted of three main steps.

First, after Schroedinger had shown how one of the two mathematical theories could be translated into the other, von Neumann showed that the two mathematical formalisms were *isomorphic*, to wit, different realizations [models$_2$] of the axiomatically defined [abstract] *Hilbert space*. To the Schroedinger function $\psi(q)$ corresponds a line or column U_a of an infinite quadratic unitary matrix [and vice versa] and to the [normalisation] condition

$$\int\int\int \psi^*\psi \, d^3q = 1$$

corresponds the equation

$$\sum_a U_a^* \, U_a = 1 \, .$$

Both $\psi(q)$ and U_a are *normalized vectors in Hilbert space*. At the same time it became clear that the difference between the two versions of the theory resides merely in the fact that different systems of coordinates in Hilbert space are used as preferential systems. The study of these questions led to the so-called *quantum mechanical transformation theory* (Jordan-Wigner, Dirac), which is an exact counterpart to the kinematical transformation theory of c-mechanics. While the kinematical transformations [so-called Lorentz transformations] of c-theory mean the transition from one space-time frame of reference to another one [within a so-called uniform motion equivalence], the unitary transformations of quantum theory represent the 'transition' between different measuring arrangements [or rather: different external conditions suitable] for measuring noncommensurable quantities represented by [noncommuting] Hermitean operators in Hilbert space. While it was clear from the beginning that the *eigenvalues* of such an operator represent the possible values of the quantity concerned, the physical meaning of these unitary transformations (rotations in Hilbert space) remained a question to be clarified.

A first answer [in the right direction] was the *statistical interpretation* given by M. Born. Its general formulation is as follows. Let

$$U_{ab}^{AB} = \int (\psi_a^A)^* \, \psi_b^B \, dq = \langle a \mid b \rangle \tag{1}$$

BORN SCHROEDINGER DIRAC

be the [elements of] the matrix of unitary transformation from the system of eigenvectors of A with eigenvalues a to the system of eigenvectors of B with eigenvalues b; then:

$$|U_{ab}^{AB}|^2 = |\langle a \mid b \rangle|^2 \tag{2}$$

is the [value of the] probability that the quantity [represented by] B has the value b, if [and when] [the quantity represented by] A has the value a. According to this interpretation quantum mechanics would be a *statistical theory in the classical meaning of this word*; its statements would refer in principle to statistical ensembles of like systems only [but not to single systems and elementary processes].

The third [and final] step consisted in replacing the statistical by the *probabilistic* [*stochastic*] *interpretation.* According to this interpretation, the words

'*has* the value b'

in the above formulation have to be replaced by

> '*takes* the value b when an interaction [of the given system with external conditions] comes into play that corresponds to a measurement of [the quantity represented by] B'.

Thereby, the quantum mechanical probabilities refer to *transitions* and *not* to statistical *distributions*.

It is now clear how the logical reconstruction of the theory has to proceed. Just as in c-theory the concept of [constant] *velocity* has to be analysed and axiomatised, so here the concept of [transition] *probability* has to be analysed and axiomatised. Just as the Lorentz transformation appeared as [true irreducible] representation [in x-y-z-t-space] of the [abstract] velocity group, the transformation theory of quantum mechanics will appear as a representation [or model$_2$] of the axiomatic [abstract] theory of [transition] probabilities. On this basis the quantum mechanical 'law of motion' is obtained in a way analogous to the way in which c-dynamics is obtained [on the basis of c-kinematics]: instead of Lorentz invariance we have to demand here invariance under the group of unitary transformations, and instead of the limit relation for $c \to \infty$ we have to demand here an analogous relation for $h \to 0$.

As all other analogies, that between the two transformation theories is also incomplete: while the [invariant] quantity c appears already in the transformation equations, the [invariant] quantity h only appears in the next step which introduces dynamical quantities. This is of fundamental importance for the logical structure of the theory and its proper understanding. It implies that the transition $h \to 0$ cannot be carried out for the underlying mathematical formalism (Hilbert space). [Physically,] this has to do with the probabilistic [stochastic] character of h-theory: In the limit $h \to 0$ all operators will commute (become c-numbers, in Dirac' terminology) and consequently all transition probabilities will cease to exist. Thus classical mechanics appears far more *conceptually degenerate* from the standpoint of h-mechanics than it does from the standpoint of c-mechanics. This state of affairs is most clearly expressed in terms of formal logic: h-mechanics is based not only on a different mathematical formalism but also on a different predicate logic (complementarity logic); only the latter, but not the mathematical formalism, is a proper generalisation of classical theory.

2. Statistical probability

The classical (statistical) concept of probability may be defined as that concept which occurs in statements of the following form:

> The probability that a subject with property E_1 also has the property E_2 is equal to p.

For this we write

$$\text{prob}_1 (E_1; E_2) = p. \tag{1}$$

If we take casting dice as an example, 'E_1' would be the predicate 'properly cast' and 'E_2' would stand for 'lying on the table with a six, say, on the top face'. Now predicates may be negated and [if of the same syntacto-semantic type] combined by 'and' ($\wedge$) and 'or' ($\vee$) [to give new predicates of the same type]. The rules of the classical calculus of predicates or classes apply. Thus, the arguments of the classical probability functor 'prob_1' are elements of a *Boolean algebra* (also called *distributive orthocomplemented lattice*). The axioms of classical prob theory are conditions for the function prob_1 [the rules of Boolean algebra being taken for granted]. These axioms are chosen such that they permit the *interpretation of* prob_1 *as relative frequency*; in this interpretation the axioms are tautologically satisfied or become mathematical identities.

The frequency [statistical] interpretation has the form

$$\text{prob}_1 (E_1; E_2) = \frac{f(E_1 \wedge E_2)}{f(E_1)}; \tag{2}$$

it reduces the two-place function prob_1 to the one-place function f. $f(E)$ is the number of objects with the property E. From this meaning of 'f' and the meaning of '$\vee$' it follows that

$$f(E_1 \vee E_2) = f(E_1) + f(E_2) - f(E_1 \wedge E_2). \tag{3}$$

Any function satisfying this functional equation is called an *additive function over a Boolean algebra*.

It is easy to prove that (2) together with (3) is equivalent to the axioms of the classical theory of probability as formulated, e.g., by H. Reichenbach in 1932. In this proof the meaning of f does not play any role. Thus, from the mathematical point of view the statistical theory of probability is identical with the *theory of additive functions over a Boolean algebra.*

This fundamental result is due to Kolmogoroff (1933). In his work a *set system* (consisting of all subsets of a given set) is used as model_2 of the Boolean algebra. In contrast to this we maintain the view that the arguments of prob_1 and f are predicates: only this view can be taken over, in a generalized form, into quantum mechanics.

3. REACTIVE PROPERTIES–COMPLEMENTARITY LOGIC

As shown above, the classical (statistical) theory of probability rests on the classical calculus of predicates. Hence it presupposes that the logical conjunction '$E_1 \wedge E_2$' is meaningful if 'E_1' and 'E_2' are meaningful. This presupposition need not however be satisfied for predicates representing reactive ['dispositional'] properties. Such predicates will be represented in the following by 'X', 'Y', 'Z'.

A typical predicate of this kind is 'soluable in water'. As this example shows, such predicates cannot be defined explicitly; they admit of only a partial definition of the form

$$E_1 \wedge X \equiv E_2 \tag{1}$$

(in the example: E_1 = is in water, E_2 = is dissolving in water).

If a second predicate of this kind (say: 'soluable in alcohol') with the definition

$$E_3 \wedge Y \equiv E_4 \tag{2}$$

is considered, the conjunction '$X \wedge Y$' is merely defined by

$$(E_1 \wedge E_3) \wedge (E \wedge Y) \equiv E_2 \wedge E_4 . \tag{3}$$

Now it may happen that the properties E_1 and E_3, and the properties E_2 and E_4, exclude one another so that

$$E_1 \wedge E_3 \equiv E_2 \wedge E_4 \equiv O \tag{4}$$

where 'O' stands for the contradiction. In this case (3) is identically satisfied, i.e., $X \wedge Y$ remains *completely undefined* [and hence possibly without meaning][1]. Thus, classical logic contains the inherent possibility of annulling itself.

It is precisely this possibility that has become a reality in quantum mechanics.

Following N. Bohr, predicates the logical conjunction of which is undefined and hence without meaning will be called *complementary* (to each other) (in the strict sense). The corresponding properties will also be called complementary (to each other) or *incommensurable*.

In view of the logical identity

$$X \vee Y \equiv \overline{\bar{X} \wedge \bar{Y}}$$

it follows that for complementary predicates the disjunction is likewise undefined.

From the definition given above it follows that complementarity is an irreflexive symmetrical relation between predicates while nothing follows concerning the question of its being transitive or intransitive.

Since expressions without meaning should not occur in scientific language, the formation of compound predicates out of complementary predicates is to be forbidden by a syntactical rule. It then follows: *complementary predicates are inconnectible*. Inconnectibility thus appears as the *syntactic formulation of complementarity*. The sentential and predicate calculus resulting from the admission of inconnectible predicates [or rather from *restricted* connectibility] is called *complementarity logic*.

In complementarity logic the universal connectibility of predicates and sentences is abolished. Hence its algebraic structure is no longer that of a Boolean algebra. However, since the relation of complementarity is not [by definition] transitive, complementarity logic in general admits connectible predicates. Hence its algebraic structure is called *partial Boolean*.

4. Theory of Probability on the Basis of Complementarity Logic

Logical connectibility is replaced in quantum mechanics by transition probability. If X and Y are complementary properties there always exists a number p such that

$$\text{prob}_2(X;\, Y) = p\,. \tag{1}$$

The question now arising is this: what calculus applies to the transition probability prob_2?

The answer is not difficult to find: we must have a prob calculus on the basis of complementarity logic, i.e., the arguments of prob_2 belong to a partial Boolean algebra. As far as these arguments are connectible the rules of the ordinary calculus of probability must apply. This leads to the result: *the rules of the classical theory of probability still hold provided the arguments (predicates) exist.*

The problem is thus reduced to that of finding a *mathematical model*$_2$ *of complementarity logic.* The solution found by quantum mechanics may be stated in generalized form as follows: *every predicate of complementarity logic is represented mathematically by a projection operator in a linear vector space.* The connectives for predicates and operators, respectively, are correlated as follows:

$$\begin{aligned} X &\leftrightarrow P_X && \\ X \dot{\wedge} Y &\leftrightarrow P_X P_Y && \text{(a)} \\ X \dot{\vee} Y &\leftrightarrow P_X + P_Y - P_X P_Y && \text{(b)} \\ \dot{\bar{X}} &\leftrightarrow I - P_X && \text{(c)} \end{aligned} \qquad (2)$$

With this scheme a *mathematical criterion* of complementarity is found: *two predicates are complementary iff the projection operators representing them do not commute.* Indeed, if two projection operators do not commute their product is no longer a projection operator and hence a corresponding [compound] predicate does not exist.

The remaining problem is that of finding the *general solution of the functional equations* for prob_2 in terms of projection operators. Obviously, the solution must again have the form

$$\text{prob}_2(X; Y) = \frac{S(P_X P_Y)}{S(P_X)} \qquad (3)$$

where S is a real-valued additive function:

$$S(P_X + P_Y) = S(P_X) + S(P_Y). \qquad (4)$$

The only function satisfying this condition is the *trace* (Tr), which may be defined as the sum of the eigenvalues of the argument operator.

Hence the complete solution of our problem is given in general form by the equation

$$\boxed{\text{prob}_2(X; Y) = \frac{Tr(P_x P_Y)}{Tr\, P_X}} \qquad (5)$$

This is the formula replacing Kolmogoroff's formula (2). At the same time it is the *fundamental formula of quantum mechanics*.

The space in which the projection operators are defined is the *representation space of complementarity logic*. As long as nothing else is known, number of dimensions and metric of the representation space are arbitrary, i.e., all vector spaces have equal rights to be considered. However, if (5) is taken into account it is easy to show that *unitary* (and not *real*) metric has to be chosen, i.e., the vectors of the representation space have to be *complex* quantities. Indeed, if the representation space were a *real* vector space, (5) would yield a real number for non-existing probabilities! The use of *complex*-valued functions (vectors in a Hilbert space with *unitary* metric) is thus not a mathematical trick that could in principle be avoided, as in [many parts of] classical physics, but an essential characteristic of the theory, to wit: a *necessary condition for obtaining meaningless answers to meaningless questions*.

Formula (5) covers two essentially different cases. If X and Y are *complementary* so that P_X and P_Y do not commute $\text{prob}_2(X; Y)$ is a *transition probability* which may take any *real* value between 0 and 1. If P_X and P_Y commute and hence X and Y are *commensurable* or *non-complementary* $\text{prob}_2(X; Y)$ is a *relative frequency* that takes only *rational* values between 0 and 1.

If these relative frequencies are written as fractions numerator and denominator are to be interpreted either as *statistical weights* of degenerated states or else as the frequency of a non-degenerate state in a certain *mixture* [non-uniform ensemble = ensemble of like systems in different states].

If only non-degenerate states ('pure cases') are considered, we have

$$Tr\, P_X = |(\psi_X, \psi_X)|^2\ [=1] \tag{6}$$

and

$$Tr\,(P_X P_Y) = |(\psi_Y, \psi_X)|^2 \tag{7}$$

where ψ_X, ψ_Y are the *state vectors*. If the latter are normalized, expression (6) takes the value 1 and (5) takes the familiar form

$$\text{prob}_2\,(X; Y) = |(\psi_Y, \psi_X)|^2. \tag{8}$$

The mathematical expressions on the right-hand sides of (5)–(8) are *invariants under the group of unitary transformations* and hence independent of the choice of the orthonormalized system of reference in Hilbert space. Different choices correspond to what are called different *representations* of the theory. The representation in which the position operator is diagonal, i.e., in which the eigenvectors of this operator are chosen as system of reference, is called the *Schroedinger representation*. In the p-representation often used by Dirac the momentum operator is diagonal, and in the *Heisenberg representation* the energy operator is diagonal. A further representation [used in perturbation theory] is the *interaction representation* in which the interaction energy is diagonal. The choice of a representation, just as the choice of a coordinate system in ordinary space, may be of practical importance for computations but has no fundamental import: by a unitary transformation any representation can be obtained from any other one.

5. The Quantum Mechanical Concept of 'Physical Quantity' and Its Relation to Quantum Mechanical Properties (Modes of Reaction)

Historically, the concept of a quantum mechanical quantity or 'observable' resulted from a reinterpretation of the classical concept in the spirit of the correspondence principle. In classical mechanics the state of a system with f degrees of freedom is fixed by the values of $2f$ variables, to wit: f general coordinates q_i and f canonically conjugated momenta $p_i (i = 1, \ldots, f)$. Hence any continuous function of the q_i, p_i is taken to represent a physical quantity. In quantum mechanics the variables q_i, p_i are replaced by operators Q_i and P_i satisfying the *commutation rules*

$$[P_i, Q_k] =_{\text{df}} P_i Q_k - Q_k P_i = \pm \sqrt{-1}\, \hbar\, \delta_{ik}. \qquad (1)$$

Hence only such operator-valued functions of the Q_i, P_i could be admitted as physical quantities that possess a real-valued eigenvalue spectrum. In the language of [meta]mathematics such quantities are called *hypermaximal Hermitean operators* (J. von Neumann). They are characterized by the fact that they admit of a representation in the form

$$A = \int_{-\infty}^{+\infty} a\, dE^A(a) \qquad (2)$$

where the integral is a Stieltjes integral and where the operator-valued function $E^A(a)$ is the so-called *resolution of unity* belonging to the operator A. Now these $E^A(a)$ are a *one-parameter family of commuting projection operators* with the properties

$$\begin{aligned} &E^A(-\infty) = O, \quad E^A(+\infty) = I \\ &E^A(a_1)\, E^A(a_2) = E^A(a_1) \quad \text{if} \quad a_1 \leqslant a_2 . \end{aligned} \tag{3}$$

(Where the function $E^A(a)$ is discontinuous we have discrete eigenvalues of A, where it is continuously increasing we have the continuous part of the eigenvalue spectrum of A.)

Thus, historically the projection operators entered quantum mechanics in a roundabout way *via* the concept of quantum mechanical quantity and its mathematical analysis.

In a rational reconstruction of the theory this way is to be reversed. We know already that properties of quantum mechanical systems are reactive properties, to be represented mathematically by projection operators. Hence, equation (2) is to be considered as the *definition of the quantum mechanical concept of physical quantity: any resolution of unity $E(a)$ defines, according to (2), a quantum mechanical quantity.* This definition is completely independent of all correspondence considerations and thus does not borrow anything from classical mechanics.

A correspondence to classical mechanics is only established by the following *first axiom of correspondence*: if, for a system with f degrees of freedom, we have $2f$ operators P_i, Q_i satisfying (1) these operators correspond to the general coordinates and momenta of classical mechanics.

A question that suggests itself in this connection is this: is there a classical quantity, i.e., a continuous function of the q_i, p_i, to every quantum mechanical quantity, i.e., to every resolution of unity? This question cannot be answered for sure at present. However, it can be shown mathematically and emerges most clearly from the quantum mechanical theory of diffraction that there is a *continuum* of quantum mechanical quantities between P and Q, all with a continuous eigenvalue spectrum.

The connection between quantum mechanical *quantities* and quantum mechanical *properties* is as follows. Let $E^A(a)$ be any resolution of unity defining a quantum mechanical quantity A according to (2). Then: the projection operator

$$E^A(a_1, a_2] \underset{\text{df}}{=} E^A(a_2) - E^A(a_1) \quad (a_1 < a_2)$$

represents the following property: *in case of a 'measurement of A' the quantity A takes a value from the interval* $(a_1, a_2]$. *A 'measurement of A'* is, by definition, any external action that forces the system considered to jump into one of the eigenstates of A.

6. The dynamical law of h-mechanics–Bohr's relation of indeterminacy

The dynamical law of h-mechanics determines the time dependence of the transition probabilities. It is a generalization of (3.5) and reads

$$\text{prob}_2(X; Y; t) = \frac{Tr\, P_X S(t)^{-1} P_Y S(t)}{Tr\, P_X}. \tag{1}$$

Here, $S(t)$ is the unitary transformation operator

$$S(t) = \exp\left(\frac{i}{h} Ht\right), \quad H = \text{Hamilton operator}. \tag{2}$$

Mathematically, (1) may be interpretated in two [different but equivalent] ways, known as *Heisenberg picture* and *Schroedinger picture*, respectively. The Heisenberg picture rests on the mathematical identity

$$P_{U^{-1}YU} = U^{-1} P_Y U \tag{3}$$

for any unitary operator U. In view of this we have

$$S(t)^{-1} P_Y S(t) = P_{Y(t)}$$

with

$$Y(t) = S(t)^{-1}\, Y\, S(t). \tag{4}$$

Thus, in the Heisenberg picture the time dependence of the transition probabilities is attributed to a time dependence of the 'observables' [or rather: the reactive properties] while the state vector, here represented by the projection operator P_X, is considered time independent.

In the Schroedinger picture it is just the other way round. Since the trace is invariant under a unitary transformation of its argument we can write (1) in the form

$$\mathrm{prob}_2(X;\,Y;\,t) = \frac{Tr\,S(t)\,P_X S(t)^{-1}\,P_Y}{Tr\,S(t)\,P_X S(t)^{-1}} \tag{1'}$$

The projection operator

$$P_X(t) =_{\mathrm{df}} S(t)\,P_X S(t)^{-1} \tag{5}$$

corresponds to the time-dependent state vector

$$\psi(t) = S(t)\,\psi(o) = \exp\left(\frac{i}{\hbar}Ht\right)\psi(o) \tag{6}$$

which is the formal solution of the SCHROEDINGER equation

$$\frac{i}{\hbar}H\psi = \frac{\partial}{\partial t}\psi\,. \tag{6'}$$

Equation (2) or (6′) may be called the *second correspondence axiom*. Together with (1) it says that the time dependence is uniquely determined by the Hamilton function, just as in classical mechanics. It should, however, be noted that (1) demands merely the existence of a time dependent unitary transformation and not the existence of a Hamiltonian. This has proved of great importance for the further development of quantum theory: The modern theory (S-matrix theory, dispersion theory) works only with a unitary S-operator which exists also in cases where a Hamiltonian H may not exist. It is certainly a further advantage of the present reconstruction of quantum theory that it gives logical priority to the S-operator over the Hamilton operator.

In the above equations the time variable t appears as a seemingly classical (non-quantized) quantity, in contrast to the coordinates [and all other quantities]. However, [the Schroedinger equation] (6′) can be considered as a formal solution or representation of the operator equation

$$H\,T - T\,H = -\,\hbar/i \tag{7}$$

for energy and time, [corresponding to (5.1)]. In line with this we have an indeterminacy relation

$$\Delta E\,\Delta t \geqslant \hbar \tag{8}$$

between energy and time. However, [as time is not a state variable in either classical or quantum mechanics] the physical interpretation of (8) is somewhat different from that of the Heisenberg relation

$$\Delta p \, \Delta q \geqslant \hbar$$

[which of course follows from the probabilistic interpretation of the general formalism]. According to Bohr, (8) means that within a time interval Δt the energy of a system can only be determined up to $\pm \Delta E$ [with a reasonable degree of certainty] and [more important] that in a state of mean lifetime Δt the energy of the system is only determined up to $\pm \Delta E$. This interpretation has proved correct; it gives, e.g., the empirically known relation between the mean lifetime of excited atoms and the coherence length of the light emitted by them [on the one hand and the natural line breadth on the other hand].[2]

NOTES

* Translated from 'Grundlagen der modernen Physik – Teil III: *h*-Theorie (Quantenmechanik)', in *Mikrokosmos-Makrokosmos*, Vol. **2** (ed. by H. Ley and R. Löther), Berlin 1967.

[1] [Viz., without meaning unless some meaning can be derived from the semantic axioms of the whole theory in an indirect way.]

[2] [This corrects a careless mistake in the original text. Coherence length Δl and mean lifetime Δt are of course classically related by $\Delta l = c \, \Delta t$. On the other hand, the natural line breadth in frequency measure is given by $\Delta \nu = \Delta E / h$. Together with (8) this yields $\Delta \nu \geqslant c / \Delta l$. Since this result can also be derived from classical Fourier analysis, an inconsistency would arise if (8) would not hold.]

JAMES C. T. POOL

BAER *-SEMIGROUPS AND THE LOGIC OF QUANTUM MECHANICS†

ABSTRACT. The theory of orthomodular ortholattices provides mathematical constructs utilized in the quantum logic approach to the mathematical foundations of quantum physics. There exists a remarkable connection between the mathematical theories of orthomodular ortholattices and Baer *-semigroups; therefore, the question arises whether there exists a phenomenologically interpretable role for Baer *-semigroups in the context of the quantum logic approach. Arguments, involving the quantum theory of measurements, yield the result that the theory of Baer *-semigroups provides the mathematical constructs for the discussion of "operations" and conditional probabilities.

0. INTRODUCTION

An affirmative answer to the following question would be extremely useful in the quantum logic approach to the foundations of quantum physics:

Question I. Does the collection of events pertaining to a physical system, which exhibits quantum effects, admit a phenomenologically interpretable orthomodular ortholattice structure?

If the word "ortholattice" is replaced by "orthoposet", then the answer is evidently affirmative. This aspect of Question I will be reviewed in Section I.

There exists a remarkable connection between orthomodular ortholattices and Baer *-semigroups. If $(S, \circ, *, ')$ is a Baer *-semigroup, then there exists an orthomodular ortholattice $(P'(S), \leqq, ')$ with $P'(S) \subset S$. If $(L, \leqq, ')$ is an orthomodular ortholattice, then there exists a Baer *-semigroup $(S(L), \circ, *, ')$ where $S(L)$ consists of a set of mappings from L into L and there exists an injective mapping $j: L \rightarrow S(L)$. Since orthomodular ortholattices evidently have a role in the quantum logic approach[1] and since orthomodular ortholattices and Baer *-semigroups are closely related mathematical objects, the following question arises:

Question II. Do Baer *-semigroups have a phenomenologically interpretable role in the quantum logic approach?

In Section II, this question will be answered positively provided one accepts a number of assertions of the conventional quantum theory of

C. A. Hooker (ed.), The Logico-Algebraic Approach to Quantum Mechanics, 365–394.

measurements[2]. Indeed, the theory of Baer *-semigroups will provide mathematical constructs for the discussion of operations[3] and conditional probabilities within the context of the quantum logic approach. A corollary to the affirmative answer of Question II will be the assertion that the orthoposet of events in Question I is an ortholattice. Furthermore, a new approach to the phenomenological interpretation of the lattice operations will be obtained.

Necessary definitions and theorems from the theories of orthomodular ortholattices and Baer *-semigroups are included in an Appendix.

I. EVENT-STATE STRUCTURES

The quantum logic approach to the mathematical foundations of quantum physics studies two distinguished sets, the set of events and the set of states, pertaining to a physical system. Some formulations of the quantum logic approach treat events as primitive entities and states as derived entities (see, for example, [3, 4, 12, 13, 23]). Other formulations treat the events and the states as equally primitive entities (see, for example, [7, 9, 15, 17, 18, 25, 28, 30]). Although the collection of axioms varies from one formulation to another, the following definition yields a mathematical structure which is widely utilized.

DEFINITION 1.1. An *event-state structure* is a triple $(\mathscr{E}, \mathscr{S}, P)$ where

(i) $\mathscr{E}$ is a set called the *logic* of the event-state structure and an element of $\mathscr{E}$ is called an *event*,

(ii) $\mathscr{S}$ is a set and an element of $\mathscr{S}$ is called a *state*,

(iii) P is a function $P:\mathscr{E}\times\mathscr{S}\to[0, 1]$ called the *probability function* and if $p\in\mathscr{E}$ and $\alpha\in\mathscr{S}$, then $P(p, \alpha)$ is called *the probability of occurrence of the event p in the state α*,

(iv) if $p\in\mathscr{E}$, then the subsets $\mathscr{S}_1(p)$ and $\mathscr{S}_0(p)$ of $\mathscr{S}$ are defined by

$$\mathscr{S}_1(p)=\{\alpha\in\mathscr{S}:P(p,\alpha)=1\}$$
$$\mathscr{S}_0(p)=\{\alpha\in\mathscr{S}:P(p,\alpha)=0\}$$

and if $\alpha\in\mathscr{S}_1(p)$ (respectively, $\alpha\in\mathscr{S}_0(p)$) then the event p is said *to occur* (respectively, *non-occur*) *with certainty in the state α, and*

(v) Axioms 1.1 through 1.7 are satisfied.

AXIOM 1.1. If $p, q \in \mathscr{E}$ and $\mathscr{S}_1(p) = \mathscr{S}_1(q)$, then $p = q$.

AXIOM 1.2. There exists an event $1 \in \mathscr{E}$ such that $\mathscr{S}_1(1) = \mathscr{S}$.

AXIOM 1.3. If $p, q \in \mathscr{E}$ and $\mathscr{S}_1(p) \subset \mathscr{S}_1(q)$, then $\mathscr{S}_0(q) \subset \mathscr{S}_0(p)$.

AXIOM 1.4. If $p \in \mathscr{E}$, then there exists an event $p' \in \mathscr{E}$ such that $\mathscr{S}_1(p') = \mathscr{S}_0(p)$ and $\mathscr{S}_0(p') = \mathscr{S}_1(p)$.

AXIOM 1.5. If
(i) $p_1, p_2, \ldots \in \mathscr{E}$ and
(ii) $\mathscr{S}_1(p_i) \subset \mathscr{S}_0(p_j)$ for $i \neq j$,
then there exists a $p \in \mathscr{E}$ such that
(a) $\mathscr{S}_1(p_i) \subset \mathscr{S}_1(p)$ for all i,
(b) if $q \in \mathscr{E}$ and $\mathscr{S}_1(p_i) \subset \mathscr{S}_1(q)$ for all i, then $\mathscr{S}_1(p) \subset \mathscr{S}_1(q)$, and
(c) if $\alpha \in \mathscr{S}$, then

$$P(p, \alpha) = \sum_i P(p_i, \alpha).$$

AXIOM 1.6. If $\alpha, \beta \in \mathscr{S}$ and $P(p, \alpha) = P(p, \beta)$ for all $p \in \mathscr{E}$, then $\alpha = \beta$.

AXIOM 1.7. If
(i) $\alpha_1, \alpha_2, \ldots \in \mathscr{S}$,
(ii) $t_1, t_2, \ldots \in [0, 1]$, and
(iii) $\sum_i t_i = 1$,
then there exists an $\alpha \in \mathscr{S}$ such that

$$P(p, \alpha) = \sum_i t_i P(p, \alpha_i)$$

for all $p \in \mathscr{E}$.

The phenomenological interpretation of the mathematical system, event-state structure, may be specified by selecting a collection of rules for the interpretation of the primitive entities: events, states, and probability function. The following collection is a possible (but obviously not the only) choice for these rules. An event-state structure $(\mathscr{E}, \mathscr{S}, P)$ is associated with the class of physical systems of a specified kind. A state may be identified with a "state-preparation procedure", that is, instructions for an apparatus which produces sample physical systems of the specified kind. An event may be identified with the "occurrence or non-occurrence" of a particular phenomenon pertaining to physical systems of the specified kind. More specifically, an event may be identified with an "observation procedure", that is, instructions for an apparatus which interacts with a sample physical system and indicates either yes or no corresponding to the occurrence or non-occurrence of the phenomenon [4]. The interpretation of $P(p, \alpha)$ for $p \in \mathscr{E}$ and $\alpha \in \mathscr{S}$ would then be the following. Prepare an ensemble of sample physical systems utilizing a state-preparation procedure corresponding to α. Determine the occurrence or non-occurrence of the event p utilizing an observation procedure for p with each sample of this ensemble. If the ensemble is sufficiently large, then the frequency of occurrence of P should be close to $P(p, \alpha)$.

The phenomenological interpretation of the general aspects of the quantum logic approach are discussed in [12], in particular, Chapters 5 and 6; however, brief comments on the specific axioms adopted above will be necessary.

Axiom 1.1 asserts that if p and q are events and if the set of states in which p occurs with certainty coincides with the set of states in which q occurs with certainty, then the events p and q are identical. This axiom is stronger than the corresponding axiom adopted, for example, in [18] and [25]. Its adoption is motivated by the phenomenological interpretation of the relation $\leqq$ introduced in the following definition.

Definition 1.2. If $(\mathscr{E}, \mathscr{S}, P)$ is an event-state structure, then the relation $\leqq$ on $\mathscr{E}$, called the *relation of implication*, is defined as follows: for $p, q \in \mathscr{E}$, $p \leqq q$ means $\mathscr{S}_1(p) \subset \mathscr{S}_1(q)$.

The relation $\leqq$ is evidently reflexive and transitive, since $\subset$ is a reflexive and transitive relation [5]. Axiom 1.1 and the antisymmetry of $\subset$ imply that $\leqq$ is antisymmetric; hence, the relation $\leqq$ is a partial ordering of $\mathscr{E}$. The phenomenological interpretation of the relation $\leqq$ may be briefly summarized: $p \leqq q$ means if p occurs with certainty, then q occurs with certainty. Indeed, if $\alpha \in \mathscr{S}$ is any state and if p occurs with certainty in the state α, then $\alpha \in \mathscr{S}_1(p) \subset \mathscr{S}_1(q)$ when $p \leqq q$ and q occurs with certainty in the state α. This interpretation of $\leqq$ evidently corresponds to the phenomenological concept of implication (see [12, 13, 23, 24, and 27]) more closely than the relation $\lesssim$ on $\mathscr{E}$ defined as follows (see [18, 19, 25, and 31]): for $p, q \in \mathscr{E}$, $p \lesssim q$ means $P(p, \alpha) \leqq P(q, \alpha)$ for all $\alpha \in \mathscr{S}$.

Axioms 1.1 and 1.2 assert the existence of a unique event $1 \in \mathscr{E}$ such that $\mathscr{S}_1(1) = \mathscr{S}$ (and, hence, $\mathscr{S}_0(1) = \emptyset$); moreover, 1 is the greatest element of $\mathscr{E}$ with respect to $\leqq$ since $p \leqq 1$ for all $p \in \mathscr{E}$. Axioms 1.1 and 1.4 assert if $p \in \mathscr{E}$, then there exists a unique $p' \in \mathscr{E}$ such that $\mathscr{S}_1(p') = \mathscr{S}_0(p)$ and $\mathscr{S}_0(p') = \mathscr{S}_1(p)$. Axiom 1.4 applied to the event $1 \in \mathscr{E}$ yields the unique event 0 in $\mathscr{E}$ such that $\mathscr{S}_1(0) = \emptyset$ and $\mathscr{S}_0(0) = \mathscr{S}$, namely, $1'$; moreover, 0 is the least element of $\mathscr{E}$ with respect to $\leqq$ since $0 \leqq p$ for all $p \in \mathscr{E}$. These remarks motivate introduction of the following terminology.

Definition 1.3. Let $(\mathscr{E}, \mathscr{S}, P)$ be an event-state structure.

(a) The unique event $1 \in \mathscr{E}$ such that $\mathscr{S}_1(1) = \mathscr{S}$ and $\mathscr{S}_0(1) = \emptyset$ is called the *certain* event.

(b) If $p \in \mathscr{E}$, then the unique event $p' \in \mathscr{E}$ such that $\mathscr{S}_1(p') = \mathscr{S}_0(p)$ and $\mathscr{S}_0(p') = \mathscr{S}_1(p)$ is called the *negation* of p.

(c) The unique event 0, namely, $1'$, of $\mathscr{E}$ such that $\mathscr{S}_1(0) = \emptyset$ and $\mathscr{S}_0(0) = \mathscr{S}$ is called the *impossible* event.

Axiom 1.3 asserts if $p, q \in \mathscr{E}$ and $\mathscr{S}_1(p) \subset \mathscr{S}_1(q)$ (that is, "if p occurs with certainty, then q occurs with certainty"), then $\mathscr{S}_0(q) \subset \mathscr{S}_0(p)$ (that is, "if q non-occurs with certainty, then p non-occurs with certainty"). Consequently, in terms of $\leqq$ and $'$, Axiom 1.3 asserts if $p, q \in \mathscr{E}$ and $p \leqq q$, then $q' \leqq p'$. From the defining property of p' and Axiom 1.1, it is also evident that $(p')' = p$ for all $p \in \mathscr{E}$. Since $\mathscr{S}_1(p) \cap \mathscr{S}_1(p') = \emptyset$, the greatest lower bound of p and p' with respect to $\leqq$ exists and equals the impossible event 0. Since $\mathscr{S}_0(p) \cap \mathscr{S}_0(p') = \emptyset$ and Axioms 1.2 and 1.3 are valid, it also follows that the least upper bound of p and p' with respect to $\leqq$ exists and equals the certain event 1. These remarks are summarized by the following theorem.

Theorem 1.1. If $(\mathscr{E}, \mathscr{S}, P)$ is an event-state structure, then

(a) $(\mathscr{E}, \leqq)$ is a poset,

(b) 0 and 1 are the least and greatest events, respectively, of the poset $(\mathscr{E}, \leqq)$,

(c) $p \to p'$ is an orthocomplementation of the poset $(\mathscr{E}, \leqq)$,

(d) if $p, q \in \mathscr{E}$, then the following are equivalent:

(i) $p \leqq q$,

(ii) $\mathscr{S}_1(p) \subset \mathscr{S}_1(q)$,

(iii) $\mathscr{S}_0(q) \subset \mathscr{S}_0(p)$,

(e) if $p, q \in \mathscr{E}$ then the following are equivalent:

(i) $p \perp q$ (for definition, see Appendix),

(ii) $\mathscr{S}_1(p) \subset \mathscr{S}_0(q)$,

(iii) $p \leqq q'$.

(f) and if $p \in \mathscr{E}$, then the following are equivalent:

(i) $p = 0$,

(ii) $\mathscr{S}_1(p) = \emptyset$,

(iii) $\mathscr{S}_0(p) = \mathscr{S}$.

Proof. Only assertions (e) and (f) remain to be proven. The relation $\perp$ on $\mathscr{E}$ is defined by $p \perp q$ means $p \leqq q'$. $p \leqq q'$ is equivalent to $\mathscr{S}_1(p) \subset \subset \mathscr{S}_1(q')$ and, hence, also to $\mathscr{S}_1(p) \subset \mathscr{S}_0(q)$, since $\mathscr{S}_0(q) = \mathscr{S}_1(q')$. Assertion (f) follows immediately from (d) by taking $q = 0$.

The following definitions are useful for the discussion of Axiom 1.5, 1.6, and 1.7.

DEFINITION 1.4. Let $(\mathscr{E}, \mathscr{S}, P)$ be an event-state structure.

(a) If $p, q \in \mathscr{E}$ and $p \perp q$, then p and q are *mutually exclusive* events.

(b) If $\alpha \in \mathscr{S}$, then the function $\mu_\alpha : \mathscr{E} \to [0, 1]$ is defined by

$$\mu_\alpha(p) = P(p, \alpha), \quad p \in \mathscr{E}.$$

(c) $\hat{\mathscr{S}}$ denotes the set defined by

$$\hat{\mathscr{S}} = \{\mu_\alpha : \alpha \in \mathscr{S}\}.$$

For $p, q \in \mathscr{E}$, p and q are mutually exclusive events if and only if "p occurs with certainty whenever q non-occurs with certainty". Consequently, $p \perp q$ is a generalization of the concept of mutually exclusive events of conventional probability theory. Axiom 1.5, therefore, asserts if $p_1, p_2, \ldots$ is a countable set of pairwise mutually exclusive events, then there exists a $p \in \mathscr{E}$ such that

(a) $p_i \leqq p$ for all i,

(b) if $q \in \mathscr{E}$ and $p_i \leqq q$ for all i, then $p \leqq q$, and

(c) if $\alpha \in \mathscr{S}$, then

$$P(p, \alpha) = \sum_i P(p_i, \alpha)$$

or

$$\mu_\alpha(p) = \sum_i \mu_\alpha(p_i).$$

(a) and (b) express the fact that p is the least upper bound, $\bigvee_i p_i$, of the set $\{p_1, p_2, \ldots\}$ of events. Consequently, Axiom 1.5 asserts the existence of the least upper bound of countable sets of pairwise mutually exclusive events and, furthermore, the law of additivity of probabilities for mutually exclusive events (see [14]).

THEOREM 1.2. If $(\mathscr{E}, \mathscr{S}, P)$ is an event-state structure, then

(a) $(\mathscr{E}, \leqq, ')$ is an orthomodular σ-orthoposet,

(b) $\mathscr{S}$ is a strongly-order-determining, σ-convex set of probability measures on $(\mathscr{E}, \leqq, ')$,

(c) $\alpha \to \mu_\alpha$ is a bijection of $\mathscr{S}$ onto $\hat{\mathscr{S}}$.

Proof. $(\mathscr{E}, \leqq, ')$ is a σ-orthoposet and μ_α is a probability measure on $(\mathscr{E}, \leqq, ')$ for each $\alpha \in \mathscr{S}$ because of Axiom 1.5. Axiom 1.6 and the definition of $\hat{\mathscr{S}}$ assert that $\alpha \to \mu_\alpha$ is a bijection. Axiom 1.7 asserts the σ-convexity of $\hat{\mathscr{S}}$. $\hat{\mathscr{S}}$ is strongly-order-determining because for $p, q \in \mathscr{E}$,

$$\text{if } \mathscr{S}_1(p) \subset \mathscr{S}_1(q), \quad \text{then} \quad p \leqq q$$

is equivalent to

$$\text{if } \{\mu \in \hat{\mathscr{S}} : \mu(p) = 1\} \subset \{\mu \in \hat{\mathscr{S}} : \mu(q) = 1\}, \quad \text{then} \quad p \leqq q.$$

$(\mathscr{E}, \leqq, ')$ is orthomodular, since any orthoposet possessing a separating set of probability measures is orthomodular.

The proof of the converse of Theorem 1.2 is straightforward and left to the reader.

THEOREM 1.3. *If*

(a) $(\mathscr{X}, \lesssim, \perp)$ is a σ-orthoposet,

(b) $\mathscr{M}$ is a σ-convex, strongly-order-determining set of probability measures on $\mathscr{X}$, and

(c) $P : \mathscr{X} \times \mathscr{M} \to [0, 1]$ is defined by

$$P(x, m) = m(x), \quad x \in \mathscr{X}, \quad m \in \mathscr{M},$$

then $(\mathscr{X}, \mathscr{M}, P)$ is an event-state structure; moreover,

(i) for $x, y \in \mathscr{X}$, $x \lesssim y$ if and only if $x \leqq y$,

(ii) for $x \in \mathscr{X}$, $x^\perp = x'$, and

(iii) $\mathscr{M} = \hat{\mathscr{M}}$.

(where $x \leqq y$, x' and $\hat{\mathcal{M}}$ are defined using the definitions relating to event-state structures).

Consequently, an event-state structure may be viewed either as a triple $(\mathscr{E}, \mathscr{S}, P)$ satisfying Axioms 1.1 through 1.7 or a pair $(\mathscr{E}, \hat{\mathscr{S}})$ where $\hat{\mathscr{S}}$ is a σ-convex, strongly-order-determining set of probability measures on an orthomodular σ-orthoposet. Both of these points of view will be employed in the following.

Example 1.1. If $\mathscr{P}(H)$ is the set of *all* orthogonal projections on a separable complex Hilbert space H of dimension greater than two, if $(\mathscr{P}(H), \mathscr{S}, P)$ is an event-state structure, if $\leqq$ coincides with the usual order of projections,

$$P \leqq Q \quad \text{if} \quad PQ = P,\ P, Q \in \mathscr{P}(H)$$

and if P' is the orthogonal complement of $P(P' = I - P)$ for $P \in \mathscr{P}(H)$, then there exists a bijection $\alpha \in \mathscr{S} \to D_\alpha$ of $\mathscr{S}$ onto the set $\mathscr{D}(H)$ (the set of density operators) of all positive, trace-class operators with trace equal to one such that

$$P(P, \alpha) = \mathrm{Tr}(D_\alpha P)$$

for all $P \in \mathscr{P}(H)$, $\alpha \in \mathscr{S}$. This, of course, is the event-state structure of von Neumann's Hilbert space model of quantum mechanics (see [29] and [18], pp. 71–81).

Example 1.2. The event-state structure $(\mathscr{E}, \hat{\mathscr{S}})$ where $\mathscr{E}$ is a σ-algebra of subsets of a set X and $\hat{\mathscr{S}}$ is a σ-convex, strongly-order-determining set of probability measures on $\mathscr{E}$ corresponds to the Kolmogorov model of probability theory (see [14] and [21]) with the additional feature that many probability measures are considered instead of one distinguished probability measure.

The formulations of the quantum logic approach to the foundations of quantum physics presented in [7, 18, 25, 28, 31] are evidently more general than the formulation adopted here. Indeed, these formulations replace the strongly-order-determining property of $\hat{\mathscr{S}}$ by at least one of the following consequences of this property: (1) $\hat{\mathscr{S}}$ is order-determining and (2) if $p \in \mathscr{E}$ and $p \neq 0$, then there exists an $\alpha \in \mathscr{S}$ such that $\mu_\alpha(p) = 1$. Discussions of the condition of strong-order-determining may be found in [8, 17, and 30].

II. EVENT-STATE-OPERATION STRUCTURES

Heuristic arguments have motivated the study of mathematical constructs corresponding to a number of physical concepts. For example, the notion of compatibility (or simultaneous observability) of events corresponds to a distinguished relation C on $\mathscr{E}$ (see [13, 18, and 25]). There exists at most one relation C on $\mathscr{E}$ with the following properties:

(a) if $p, q \in \mathscr{E}$ and $p \leqq q$, then $p \mathsf{C} q$;
(b) if $p, q \in \mathscr{E}$ and $p \mathsf{C} q$, then
  (i) $p \mathsf{C} q$,
  (ii) $q \mathsf{C} p$,
  (iii) $p \wedge q$ and $p \vee q$ exist in $\mathscr{E}$, and
(c) if $p_1, p_2, q \in \mathscr{E}$, $p_1 \mathsf{C} p_2$, $p_1 \mathsf{C} q$, and $p_2 \mathsf{C} q$, then
  (i) $p_1 \wedge p_2 \mathsf{C} q$,
  (ii) $(p_1 \wedge p_2) \vee q = (p_1 \vee q) \vee (p_2 \vee q)$.

Indeed, the relation C is determined by the following property: for $p, q \in \mathscr{E}$, $p \mathsf{C} q$ if and only if there exists a Boolean sublogic $\mathscr{B} \subset \mathscr{E}$ such that $p, q \in \mathscr{B}$. The existence of a relation C satisfying (a), (b), and (c) is not asserted; however, there always exists a relation C which satisfies properties (a), (b), and the following:

(c′) if $p_1, p_2, q \in \mathscr{E}$, $p_1 \mathsf{C} p_2$, $p_1 \mathsf{C} q$, $p_2 \mathsf{C} q$ and $(p_1 \vee q) \wedge (p_2 \vee q)$ exists in $\mathscr{E}$, then
  (i) $p_1 \wedge p_2 \mathsf{C} q$,
  (ii) $(p_1 \wedge p_2) \vee q = (p_1 \vee q) \wedge (p_2 \vee q)$.

This relation C may be defined as follows: for $p, q \in \mathscr{E}$, $p \mathsf{C} q$ means there exists $p_0, q_0, r \in \mathscr{E}$ such that
  (i) $p_0 \perp q_0$,
  (ii) $p_0 \perp r$ and $p = p_0 \vee r$,
  (iii) $q_0 \perp r$ and $q = q_0 \vee r$.

If $p, q \in \mathscr{E}$ and $p \mathsf{C} q$, then $p \wedge q$ exists in $\mathscr{E}$. The lattice property of $(\mathscr{E}, \leq, ')$ discussed in Question I, therefore, becomes the following: if $p, q \in \mathscr{E}$ and $p \mathsf{C} q$ does not hold, then does $p \wedge q$ exist in $\mathscr{E}$? The corresponding phenomenological question is evidently the following (see [12], pp. 74–78): if observation procedures for two incompatible (i.e., non-simultaneously observable) events are given, then how does one describe the observation procedure for the "and" (or conjunction) of these two events? Although answers to this question have been attempted in [1, 2, 12, 17,

23 and 24], no completely adequate answer is currently available. For example, in the context of an event-state structure, the arguments of [12, 23, and 24] reduce to the assertion of the universal validity of the hypothesis of the following theorem.

THEOREM 2.1. *Let* $(\mathscr{E}, \mathscr{S}, P)$ *be an event-state structure. If* $p_1, p_2 \in \mathscr{E}$ *and there exists an event* $p \in \mathscr{E}$ *such that*

$$\mathscr{S}_1(p) = \mathscr{S}_1(p_1) \cap \mathscr{S}_1(p_2),$$

then the greatest lower bound $p_1 \wedge p_2$ *of* p_1 *and* p_2 *with respect to* $\leqq$ *exists and equals* p.

Proof. Since p satisfies $\mathscr{S}_1(p) = \mathscr{S}_1(p_1) \cap \mathscr{S}_1(p_2)$ by hypothesis, $\mathscr{S}_1(p) \subset \mathscr{S}_1(p_1)$ and $\mathscr{S}_1(p) \subset \mathscr{S}_1(p_2)$; hence $p \leqq p_1$ and $p \leqq p_2$. Let $q \in \mathscr{E}$, $q \leqq p_1$ and $q \leqq p_2$. It follows that $\mathscr{S}_1(q) \subset \mathscr{S}_1(p_1)$ and $\mathscr{S}_1(q) \subset \mathscr{S}_1(p_2)$; hence $\mathscr{S}_1(q) \subset \mathscr{S}_1(p_1) \cap \mathscr{S}_1(p_2) = \mathscr{S}_1(p)$ and $q \leqq p$. Consequently, if $q \in \mathscr{E}$, $q \leqq p_1$ and $q \leqq p_2$, then $q \leqq p$. Therefore, the greatest lower bound of p_1 and p_2 exists and equals p. Q.E.D.

One result of this section will be to provide a new approach to Question I by introducing the theory of Baer *-semigroups into the context of the quantum logic approach.

The introduction of the concept of conditional probability in conventional probability theory greatly enhances the utility of the theory and deepens the mathematical structure of the theory (see [14 and 21]). The concept of conditional probability is expressed as a mathematical object defined constructively in terms of the primitive entities of the theory in an intuitively obvious fashion. In the case of a general event-state structure, there apparently exists no manifestly evident way of defining a mathematical construct corresponding to conditional probability in terms of the primitive objects of the theory in a constructive fashion. However, there exists a mathematical construct in von Neumann's Hilbert space model of quantum mechanics which is widely employed to represent the concept of conditional probability. These remarks provide the initial motivation for considering *event-state-operation structures*, event-state structures equipped with an additional primitive entity corresponding essentially to conditional probability. A role for Baer *-semigroups in the quantum logic approach will emerge from the study of these event-state-operation structures.

DEFINITION 2.1. Let $(\mathscr{E}, \mathscr{S}, P)$ be an event-state structure.

(a) Σ denotes the set of all maps $x:\mathscr{D}_x \to \mathscr{R}_x$ with domain $\mathscr{D}_x \subset \mathscr{S}$ and range $\mathscr{R}_x \subset \mathscr{S}$. If $x \in \Sigma$ and $\alpha \in \mathscr{D}_x$, then $x(\alpha)$ (or $x\alpha$, for brevity) denotes the image of α under x.

(b) If $x, y \in \Sigma$, then $x = y$ means

(i) $\mathscr{D}_x = \mathscr{D}_y$ and

(ii) $x\alpha = y\alpha$ for all $\alpha \in \mathscr{D}_x$.

(c) $0:\mathscr{D}_0 \to \mathscr{R}_0$ is defined by $\mathscr{D}_0 = \emptyset$.

(d) $1:\mathscr{D}_1 \to \mathscr{D}_1$ is defined by

(i) $\mathscr{D}_1 = \mathscr{S}$ and

(ii) $1\alpha = \alpha$ for all $\alpha \in \mathscr{D}_1$.

(e) If $x, y \in \Sigma$, then $x \circ y:\mathscr{D}_{x \circ y} \to \mathscr{R}_{x \circ y}$ is defined by

(i) $\mathscr{D}_{x \circ y} \{\alpha \in \mathscr{D}_y : y\alpha \in \mathscr{D}_x\}$ and

(ii) $(x \circ y)\,\alpha = x(y\alpha)$ for all $\alpha \in \mathscr{D}_{x \circ y}$.

In all manipulations with the elements of Σ care must be taken to examine the domains of definition (as, for example, domains of definitions must be checked for unbounded operators on Hilbert space). It is evident that $(\Sigma, \circ)$ is a semigroup with a unit element 1 and a zero element 0.

DEFINITION 2.2. An *event-state-operation structure* is a 4-tuple $(\mathscr{E}, \mathscr{S}, P, \Omega)$ where $(\mathscr{E}, \mathscr{S}, P)$ is an event-state structure and Ω is a mapping $\Omega:\mathscr{E} \to \Sigma (p \in \mathscr{E} \to \Omega_p \in \Sigma)$ which satisfies Axioms 2.1 through 2.7. If $p \in \mathscr{E}$, then Ω_p is called *the operation corresponding to the event p* (relative to Ω). If $p \in \mathscr{E}$ and $\alpha \in \mathscr{D}_{\Omega_p}$, then $\Omega_p\alpha$ is called *the state conditioned on the event p and the state* α (relative to Ω). If, moreover, $q \in \mathscr{E}$, then $P(q, \Omega_p\alpha)$ is called *the probability of q conditioned on the event p and the state* α (relative to Ω). S_Ω denotes the subset of Σ defined by

$$S = \{\Omega_{p_1} \circ \Omega_{p_2} \circ \cdots \circ \Omega_{p_n} : p_1, p_2, \ldots, p_n \in \mathscr{E}\}.$$

An element of S_Ω is called an *operation*.

AXIOM 2.1. If $p \in \mathscr{E}$, then the domain $\mathscr{D}_{\Omega_p}$ of Ω_p coincides with the set $\mathscr{D}_p$ defined by

$$\mathscr{D}_p = \{\alpha \in \mathscr{S} : P(p, \alpha) \neq 0\}.$$

AXIOM 2.2. If $p \in \mathscr{E}$, $\alpha \in \mathscr{D}_p$ and $P(p, \alpha)=1$, then

$$\Omega_p \alpha = \alpha .$$

AXIOM 2.3. If $p \in \mathscr{E}$ and $\alpha \in \mathscr{D}_p$, then $P(p, \Omega_p \alpha)=1$.

AXIOM 22.4. If $p_1, p_2, \ldots, p_n, q_1, q_2, \ldots, q_m \in \mathscr{E}$ and

$$\Omega_{p_1} \circ \Omega_{p_2} \circ \cdots \circ \Omega_{p_n} = \Omega_{q_1} \circ \Omega_{q_2} \circ \cdots \circ \Omega_{q_m},$$

then

$$\Omega_{p_n} \circ \Omega_{p_{n-1}} \circ \cdots \circ \Omega_{p_1} = \Omega_{q_m} \circ \Omega_{q_{m-1}} \circ \cdots \circ \Omega_{q_1} .$$

AXIOM 2.5. If $x \in \mathscr{S}_\Omega$, then there exists a $q_x \in \mathscr{E}$ such that

$$\mathscr{S}_1(q_x) = C\mathscr{D}_x = \{\alpha \in \mathscr{S} : \alpha \notin \mathscr{D}_x\} .$$

AXIOM 2.6. If $p, q \in \mathscr{E}$, $q \leqq p$ and $\alpha \in \mathscr{D}_p$, then

$$P(q, \Omega_p \alpha) = \frac{P(q, \alpha)}{P(p, \alpha)} .$$

AXIOM 2.7. If $p, q \in \mathscr{E}$, $p \mathbf{C} q$ and $\alpha \in \mathscr{D}_p$, then

$$P(q, \Omega_p, \alpha) = P(p \wedge q, \Omega_p \alpha) .$$

The rules of interpretation for an event-state structure must be augmented to include the concept of operation. The rule of interpretation adopted here depends upon the following phenomenological assertion. *If $p \in \mathscr{E}$, then an observation procedure for p can be selected to fulfill the following "gentleness" requirement: after utilizing this observation procedure with a sample physical system to determine the occurrence or non-occurrence of p, the resulting physical system is again a member of the class of physical systems corresponding to* $(\mathscr{E}, \mathscr{S}, P)$. A critical discussion of this assertion may be found in [12 and 19]. If $p \in \mathscr{E}$ and $\alpha \in \mathscr{S}$ with $P(p, \alpha) \neq 0$, then the following describes a state-preparation procedure:

Step A. Produce a sample physical system utilizing a state-preparation procedure corresponding to α.

Step B. Determine the occurrence or non-occurrence of the event p utilizing an observation procedure corresponding to p.

Step C. If the event p occurred, then accept the physical system resulting from this observation procedure as a sample physical system; if the event p did not occur, then do not accept the resultant physical system as a sample.

There should exist a state in $\mathscr{S}$ corresponding to this state-preparation procedure. The rule of interpretation for Ω_p adopted here is the assertion that this state is $\Omega_p\alpha$. The terminology *operation* is employed since this rule of interpretation corresponds essentially to a special case of the "operations" utilized by Haag and Kastler in the algebraic approach to quantum field theory [10].

Example 2.1. The event-state structure $(\mathscr{P}(H), \mathscr{S}, P)$ of von Neumann's Hilbert space model of quantum mechanics admits an operation map Ω. Indeed, for $P\in\mathscr{P}(H)$, Ω_P may be defined as follows: if $\alpha\in\mathscr{S}$ is the state with density operator $D_\alpha\in\mathscr{D}(H)$ and

$$P(P, \alpha)=\mathrm{Tr}(D_\alpha P)\neq 0,$$

then $\Omega_P\alpha$ is the state $\alpha'\in\mathscr{S}$ with the density operator $D_{\alpha'}\in\mathscr{D}(H)$ given by

$$D_{\alpha'}=\frac{PD_\alpha P}{\mathrm{Tr}(D_\alpha P)}.$$

This is the usual way of introducing "conditional probability" in quantum mechanics (see, for example, [20], p. 333 and [16]). The verification of all the axioms except Axiom 2.5 is straightforward. If $x\in\mathscr{S}_\Omega$ and $x=\Omega_{P_1}\circ\cdots\circ\Omega_{P_n}$, where $P_1, P_2,\ldots, P_n\in\mathscr{P}(H)$, then the projection Q on the null space of $P_1P_2\ldots P_n$ satisfies Axiom 2.5.

Example 2.2. The event-state structure $(\mathscr{E}, \hat{\mathscr{S}})$ of Example 1.2 also admits an operation map. For $p\in\mathscr{E}$, Ω_p is defined as follows: if $\mu\in\hat{\mathscr{S}}$ and $\mu(p)\neq 0$, then $\Omega_p\mu$ is the element of $\hat{\mathscr{S}}$ defined by

$$(\Omega_p\mu)(q)=\frac{\mu(p\wedge q)}{\mu(p)},\quad q\in\mathscr{E}.$$

This is the usual formulation of conditional probability from the Kolmogorov model of probability theory. The verification of the axioms is straightforward in this case.

The motivation for Axioms 2.1 through 2.7 will now be discussed utilizing the previously adopted rule of interpretation. If $p\in\mathscr{E}$ and $\alpha\in\mathscr{S}$, then the rule of interpretation yields a state in the case $P(p, \alpha) \neq 0$; however, the rule of interpretation does not yield a state when $P(p, \alpha) = 0$, since the samples will not satisfy the condition that the event p occurs. Consequently, the domain $\mathscr{D}_{\Omega_p}$ of Ω_p should, indeed, be the set $\mathscr{D}_p = \{\alpha \in \mathscr{S} : P(p, \alpha) \neq 0\}$. Axioms 2.2 and 2.3 evidently assert that the observation procedure corresponding to $p \in \mathscr{E}$ may be selected to be a measurement of the first kind in the sense of Pauli (see [12 and 22]). Axioms 2.6 and 2.7 are immediate consequences of the assertion: if $p, q \in \mathscr{E}$ and $p \mathsf{C} q$, then the observation of the event q should not disturb the results of the observation of the event p (since p and q are compatible events) and, hence, the arguments about frequencies of occurrence of conventional probability theory should be applicable.

Consequently, Axioms 2.1, 2.2, 2.3, 2.6, and 2.7 are explicitly part of the conventional quantum theory of measurements. Axioms 2.4 and 2.5 are implicitly part of the conventional quantum theory of measurements since Example 2.1 satisfies these axioms; however, the role of these axioms has evidently not been previously discussed.

If $x \in S_\Omega$, then there exist $p_1, p_2, \ldots, p_n \in \mathscr{E}$ such that

$$x = \Omega_{p_1} \circ \Omega_{p_2} \circ \cdots \circ \Omega_{p_n}.$$

The element x of S_Ω, therefore, represents the experimental procedure of first executing the operation Ω_{p_n}, then executing the operation $\Omega_{p_{n-1}}$, and so on until finally executing the operation Ω_{p_1}. The experimental procedure obtained by executing these operations in the reverse order yields an element of S_Ω also, namely, $\Omega_{p_n} \circ \Omega_{p_{n-1}} \circ \cdots \circ \Omega_{p_1}$. It, therefore, seems desirable to introduce a mapping * of S_Ω into S_Ω which corresponds to this reversal of the order of the execution of operations. Consequently, x^* would be the element $\Omega_{p_n} \circ \Omega_{p_{n-1}} \circ \cdots \circ \Omega_{p_1}$ of S_Ω. However $x \to x^*$ might not be a well-defined mapping. Indeed, there might also exist $q_1, q_2, \ldots, q_m \in \mathscr{E}$ such that $x = \Omega_{q_1} \circ \Omega_{q_2} \circ \cdots \circ \Omega_{q_m}$, ($x$ also represents the experimental procedure of first executing Ω_{q_m} then executing $\Omega_{q_{m-1}}$, and so on until finally executing Ω_{q_1}) but such that the "reversal" of this experimental procedure does not coincide with the "reversal" of the experimental procedure corresponding to the p_i's; that is,

$$\Omega_{p_1} \circ \Omega_{p_2} \circ \cdots \circ \Omega_{p_n} = \Omega_{q_1} \circ \Omega_{q_2} \circ \cdots \circ \Omega_{q_m}$$

but

$$\Omega_{p_n}\circ\Omega_{p_{n-1}}\circ\cdots\circ\Omega_{p_1}\neq\Omega_{q_{m-1}}\circ\cdots\circ\Omega_{q_1}.$$

Axiom 2.4 asserts that this does *not* happen; consequently, the following mapping $*:S_\Omega\to S_\Omega$ is well-defined.

DEFINITION 2.3. Let $(\mathscr{E}, \mathscr{S}, P, \Omega)$ be an event-state-operation structure. The mapping $*:S_\Omega\to S_\Omega$ is defined as follows: if $x\in S_\Omega$, then select $p_1, p_2,\ldots, p_n\in\mathscr{E}$ such that

$$x=\Omega_{p_1}\circ\Omega_{p_2}\circ\cdots\circ\Omega_{p_n}$$

and define x^* to be the element

$$x^*=\Omega_{p_n}\circ\Omega_{p_{n-1}}\circ\cdots\circ\Omega_{p_1}.$$

THEOREM 2.2. *If* $(\mathscr{E}, \mathscr{S}, P, \Omega)$ *is an event-state-operation structure, then* S_Ω *is a subsemigroup of* Σ*; moreover,*

(a) $\Omega_0=0$, $\Omega_1=1$,

(b) *if* $p\in\mathscr{E}$, *then*

$$\Omega_p\circ\Omega_p=\Omega_p$$

and the range of Ω_p *equals* $\mathscr{S}_1(p)$*;*

(c) $*:S_\Omega\to S_\Omega$ *is the unique mapping at* S_Ω *into* S_Ω *such that*

(i) $*$ *is an involution for the semigroup* $(S_\Omega, \circ)$,

(ii) $(\Omega_p)^*=\Omega_p$ *for all* $p\in\mathscr{E}$*;*

(d) if $p, q\in\mathscr{E}$, *then the following are equivalent properties:*

(i) $p\leqq q$,

(ii) $\mathscr{S}_1(p)\subset\mathscr{S}_1(q)$,

(iii) $\mathscr{S}_0(q)\subset\mathscr{S}_0(p)$,

(iv) $\Omega_q\circ\Omega_p=\Omega_p$,

(v) $\Omega_p\circ\Omega_q=\Omega_p$.

Proof. S_Ω is obviously a subsemigroup of Σ relative to the composition $\circ$ since

$$S_\Omega=\{\Omega_{p_1}\circ\cdots\circ\Omega_{p_n}:p_1,\ldots,p_n\in\mathscr{E}\}.$$

Since

$$\mathscr{D}_0=\{\alpha\in\mathscr{S}:P(0,\alpha)\neq0\}=\emptyset,$$

the domain of Ω_0 is $\emptyset$ and, hence, $\Omega_0=0$. Since

$$\mathscr{D}_1=\{\alpha\in\mathscr{S}:P(1,\alpha)\neq 0\}=\mathscr{S},$$

the domain of Ω_1 is $\mathscr{S}$. If $\alpha\in\mathscr{S}$, then $P(1,\alpha)=1$ by Axiom 1.3 and hence, $\Omega_1\alpha=\alpha$ by Axiom 2.2; consequently, $\Omega_1=1$.

Assertion (b) is a consequence of Axioms 2.1, 2.2, and 2.3. Let $p\in\mathscr{E}$. If $\alpha\in\mathscr{D}_p$, then $P(p,\Omega_p\alpha)=1$ by Axiom 2.3. Hence, $\Omega_p\alpha\in\mathscr{D}_p$ for $\alpha\in\mathscr{D}_p$ and since

$$\mathscr{D}_{\Omega_p\circ\Omega_p}=\{\alpha\in\mathscr{D}_p:\Omega_p\alpha\in\mathscr{D}_p\},$$

it follows that $\mathscr{D}_{\Omega_p\circ\Omega_p}=\mathscr{D}_{\Omega_p}=\mathscr{D}_p$. Since $P(p,\Omega_p\alpha)=1$ for $\alpha\in\mathscr{D}_p$,

$$(\Omega_p\circ\Omega_p)\,\alpha=\Omega_p(\Omega_p\alpha)=\Omega_p\alpha$$

by Axiom 2.2; hence, $\Omega_p\circ\Omega_p=\Omega_p$. Since $P(p,\Omega_p\alpha)=1$ for $\alpha\in\mathscr{D}_p$, the range of Ω_p is contained in $\mathscr{S}_1(p)$. If $\alpha\in\mathscr{S}_1(p)$, then $\Omega_p\alpha=\alpha$ by Axioms 2.1 and 2.2; hence, the range of Ω_p contains $\mathscr{S}_1(p)$.

It is evident that $x\to x^*$ is an involution such that $(\Omega_p)^*=\Omega_p$ for every $p\in\mathscr{E}$ and, moreover, it is the only such involution.

The equivalence of (i), (ii), and (iii) of assertion (d) is a general property of event-state structures. The equivalence of (iv) and (v) is an obvious consequence of assertion (c). Assume $\mathscr{S}_1(p)\subset\mathscr{S}_1(q)$. If $\alpha\in\mathscr{D}_p$, then $\Omega_p\alpha\in\mathscr{S}_1(p)$ by Axiom 2.3, hence, if $\alpha\in\mathscr{D}_p$, then $\Omega_p\alpha\in\mathscr{D}_q$ and

$$\mathscr{D}_{\Omega_q\circ\Omega_p}=\{\alpha\in\mathscr{D}_p:\Omega_p\alpha\in\mathscr{D}_q\}=\mathscr{D}_p.$$

If $\alpha\in\mathscr{D}_p$, then $\Omega_p\alpha\in\mathscr{S}_1(q)$ and by Axiom 2.2

$$(\Omega_q\circ\Omega_p)\,\alpha=\Omega_q(\Omega_q\alpha)=\Omega_p\alpha.$$

Thus, $\Omega_q\circ\Omega_p=\Omega_p$ if $\mathscr{S}_1(p)\subset\mathscr{S}_1(q)$. If $\Omega_p\circ\Omega_q=\Omega_p$, then

$$\mathscr{D}_p=\mathscr{D}_{\Omega_p\circ\Omega_q}=\{\alpha\in\mathscr{D}_q:\Omega_q\alpha\in\mathscr{D}_p\}\subset\mathscr{D}_q;$$

hence, $\mathscr{S}_0(q)\subset\mathscr{S}_0(p)$ since $\mathscr{D}_p=C\mathscr{S}_0(p)$ and $\mathscr{D}_q=C\mathscr{S}_0(q)$. Hence, (i) implies (iii). Q.E.D.

Consequently, Axiom 2.4 provides the semigroup $(S_\Omega,\circ)$ with an involution. The existence of this involution then yields a characterization of the partial order $\leqq$ of $(\mathscr{E},\leqq,')$ in terms of product $\circ$ of $(S_\Omega,\circ)$. In terms of the theory of involution semigroups (see Appendix), the theorem asserts: $(S_\Omega, \circ, {}^*)$ is an involution semigroup such that

(i) For each $p \in \mathscr{E}$, Ω_p is a *projection*, that is, Ω_p is an element of

$$P(S_\Omega) = \{e \in S_\Omega : e \circ e = e^* = e\}$$

(ii) $p \in \mathscr{E} \to \Omega_p \in P(S_\Omega)$ is an order preserving map of $(\mathscr{E}, \leqq)$ into $(P(S_\Omega), \leqq)$ where

$$e \leqq f \quad \text{means} \quad e \circ f = e$$

for $e, f \in P(S_\Omega)$.

If $x \in S_\Omega$, then there exist $p_1, p_2, \ldots, p_n \in \mathscr{E}$ such that

$$x = \Omega_{p_1} \circ \Omega_{p_2} \circ \cdots \circ \Omega_{p_n}.$$

Let $p_{n+1} = 1$. $\alpha \in \mathscr{S}$ is an element of the domain, $\mathscr{D}_x$, of x if and only if α is an element of the domain of $\Omega_{p_1} \circ \Omega_{p_2} \circ \cdots \circ \Omega_{p_n}$. Consequently, $\alpha \in \mathscr{D}_x$ if and only if

$$\Omega_{p_{j+1}} \circ \Omega_{p_{j+2}} \circ \cdots \circ \Omega_{p_{n+1}} \alpha \in \mathscr{D}_{p_j}$$

for $j = n, n-1, \ldots, 1$. Therefore, $\alpha \notin \mathscr{D}_x$ if and only if there exists an i, $n \geqq i \geqq 1$, such that

$$\Omega_{p_{j+1}} \circ \cdots \circ \Omega_{p_{n+1}} \alpha \in \mathscr{D}_{p_j}$$

for $j = i+1, \ldots, 1$ and $\Omega_{p_{i+1}} \circ \cdots \circ \Omega_{p_{n+1}} \alpha \notin \mathscr{D}_{p_i}$. Because of Axiom 2.1, this characterization of $C\mathscr{D}_x$ may be expressed as follows: $\alpha \in C\mathscr{D}_x$ if and only if there exists an i, $n \geqq i \geqq 1$, such that

$$P(p_j, \Omega_{p_{j+1}} \circ \cdots \circ \Omega_{p_{n+1}} \alpha) \neq 0$$

for $j = i+1, \ldots, 1$ and

$$P(p_i, \Omega_{p_{i+1}} \circ \cdots \circ \Omega_{p_{n+1}} \alpha) = 0.$$

This characterization of $C\mathscr{D}_x$ evidently provides an experimental procedure for determining whether a state belongs to $C\mathscr{D}_x$. Axiom 2.5 asserts the existence of an event q such that q occurs with certainty in the state α if and only if $\alpha \in C\mathscr{D}_x$, that is,

$$\mathscr{S}_1(q) = C\mathscr{D}_x.$$

If $q_1 \in \mathscr{E}$ and $\mathscr{S}_1(q_1) = C\mathscr{D}_x$ also, then $\mathscr{S}_1(q_1) = \mathscr{S}_1(q)$ and, hence $q_1 = q$.

DEFINITION 2.4. If $(\mathscr{E}, \mathscr{S}, P, \Omega)$ is an event-state-operation structure, then the mapping $': S_\Omega \to P(S_\Omega)$ is defined as follows: for $x \in S_\Omega$, x' is the element Ω_{q_x} of $P(S_\Omega)$, where $q_x \in \mathscr{E}$ is the unique element of $\mathscr{E}$ such that $\mathscr{S}_1(q_x) = C\mathscr{D}_x$.

The mapping $': S_\Omega \to P(S_\Omega)$, provided by Axiom 2.5, gives the involution semigroup $(S_\Omega, \circ, {}^*)$ the structure of a Baer *-semigroup (see Appendix).

THEOREM 2.3. *If $(\mathscr{E}, \mathscr{S}, P, \Omega)$ is an event-state-operation structure, then $(S_\Omega, \circ, {}^*, ')$ is a Baer *-semigroup; moreover, the mapping $p \in \mathscr{E} \to \Omega_p \in P(S_\Omega)$ is an isomorphism of the orthomodular orthoposet $(\mathscr{E}, \leqq, ')$ onto the orthomodular orthoposet $(P'(S_\Omega), \leqq, ')$ (see Appendix for a discussion of $P'(S_\Omega)$).*

Proof. Let $x \in S_\Omega$. $(S_\Omega, \circ, {}^*, ')$ is a Baer *-semigroup provided: if $y \in S_\Omega$, then $x \circ y = 0$ is equivalent to $x' \circ y = y$. $x \circ y = 0$ is equivalent to

$$\emptyset = \mathscr{D}_{x \circ y} = \{\alpha \in \mathscr{D}_y : y\alpha \in \mathscr{D}_x\}$$

or to the assertion: (A) if $\alpha \in \mathscr{D}_y$, then $y\alpha \in C\mathscr{D}_x$. Consequently, if $\alpha \in \mathscr{D}_y$, then $y\alpha \in \mathscr{D}_{x'}$, since $\mathscr{D}_{x'} = C\mathscr{D}_x$, and $\alpha \in \mathscr{D}_{x' \circ y}$. Since $\mathscr{D}_{x' \circ y} \subset \mathscr{D}_y$, it follows that $\mathscr{D}_{x' \circ y} = \mathscr{D}_y$ when assertion (A) holds. Since $C\mathscr{D}_x = \mathscr{S}_1(q_x)$, assertion (A) is equivalent to the following assertion by Axioms 2.2 and 2.3: (B) If $\alpha \in \mathscr{D}_y$, then $\Omega_{q_x}(y\alpha) = y\alpha$. Consequently, assertion (A) is equivalent to the assertion: (C) $\mathscr{D}_y = \mathscr{D}_{x' \circ y}$ and if $\alpha \in \mathscr{D}_y$, then $(\Omega_{q_x} \circ y)\,\alpha = y\alpha$. Since $\Omega_{q_x} = x'$, $x \circ y = 0$ is equivalent to $x' \circ y = y$.

If $p \in \mathscr{E}$, then $\Omega_p \in P(S_\Omega)$; moreover, $(\Omega_p)' = \Omega_{p'}$. Indeed, if $p \in \mathscr{E}$, then

$$\begin{aligned}\mathscr{S}_1(p') &= \mathscr{S}_0(p) = CC\mathscr{S}_0(p)\\ &= C\mathscr{D}_p = C\mathscr{D}_{\Omega_p}.\end{aligned}$$

and p' satisfies the criterion of Axiom 2.5 for the case $x = \Omega_p$; hence $(\Omega_p)' = \Omega_{p'}$. If $p \in \mathscr{E}$, then

$$(\Omega_p)'' = (\Omega_{p'})' = \Omega_{p''} = \Omega_p$$

and, hence, Ω_p is a closed projection, that is, $\Omega_p \in P'(S_\Omega)$. The mapping $p \in \mathscr{E} \to \Omega_p \in P'(S_\Omega)$ preserves order, since

$$p \leqq q \quad \text{if and only if} \quad \Omega_p \circ \Omega_q = \Omega_p,$$

and preserves orthocomplementation, since $(\Omega_p)' = \Omega_{p'}$. This mapping is injective and it is surjective, since $P'(S_\Omega)=\{x':x\in S_\Omega\}$ and $x'=\Omega_{q_x}$ for $x\in S_\Omega$. Consequently, $p\in\mathscr{E}\to\Omega_p\in P'(S_\Omega)$ is an isomorphism of the orthoposet $(\mathscr{E}, \leqq, ')$ onto the orthoposet $(P'(S_\Omega), \leqq, ')$. Q.E.D.

III. ON THE LATTICE STRUCTURE OF $(\mathscr{E}, \leqq, ')$

The event-state structure may be viewed as a *passive* picture for the description of physical systems since it considers only the probability of occurrence of events. The introduction of the concept of operation provides an *active* picture; indeed, the operations in S_Ω correspond to *filtering* experiments. The orthoposet $(\mathscr{E}, \leqq, ')$ of events is isomorphic to the orthoposet $(P'(S_\Omega), \leqq, ')$ under the mapping $p\in\mathscr{E}\to\Omega_p\in P'(S_\Omega)$. In $(P'(S_\Omega), \leqq, ')$, the order relation $\leqq$ is defined in terms of the composition $\circ$ of operations; indeed, for $p, q\in\mathscr{E}$,

$$p\leqq q \quad \text{if and only if} \quad \Omega_p\circ\Omega_q=\Omega_p.$$

The question, therefore, arises whether the greatest lower bound $p\wedge q$ of p and q in $\mathscr{E}$, an order theoretic construct in $(\mathscr{E}, \leqq, ')$, can be interpreted in terms of the composition $\circ$ of the Baer *-semigroup $(S_\Omega, \circ, *, ')$.

THEOREM 3.1. *If $(\mathscr{E}, \mathscr{S}, P, \Omega)$ is an event-state-operation structure, then $(\mathscr{E}, \leqq, ')$ is an ortholattice; moreover, if $p, q\in\mathscr{E}$, then*

$$\Omega_{p\wedge q}=(\Omega_{p'}\circ\Omega_q)'\circ\Omega_q.$$

Proof. $(P'(S_\Omega), \leqq, ')$ is an orthomodular ortholattice such that if e, $f\in P'(S_\Omega)$, then

$$e\wedge f=(e'\circ f)'\circ f$$

(see Appendix). The theorem follows immediately from the fact that $p\in\mathscr{E}\to\Omega_p\in P'(S_\Omega)$ is an isomorphism of $(\mathscr{E}, \leqq, ')$ onto $(P'(S_\Omega), \leqq, ')$.

Q.E.D.

Consequently, $(\mathscr{E}, \leqq, ')$ is an ortholattice for an event-state-operation structure; however, the greatest lower bound $p\wedge q$ in $\mathscr{E}$ is represented in $P'(S_\Omega)$ utilizing not only the composition $\circ$ of operations but also the mapping $':S_\Omega\to P'(S_\Omega)$.

Since the compatibility relation C discussed at the beginning of Section

II involves only the order and orthocomplementation of $(\mathscr{E}, \leqq, ')$, it must also be expressible in terms of the order and orthocomplementation of the isomorphic ortholattice $(P'(S_\Omega), \leqq, ')$.

THEOREM 3.2. *If* $(\mathscr{E}, \mathscr{S}, P, \Omega)$ *is an event-state-operation structure and* $p, q \in \mathscr{E}$*, then the following are equivalent:*

(a) $p \mathbf{C} q$,

(b) $\Omega_p \circ \Omega_q = \Omega_q \circ \Omega_p$;

moreover, if $p \mathbf{C} q$*, then*

$$\Omega_{p \wedge q} = \Omega_p \circ \Omega_q.$$

Proof. The relation $\check{\mathbf{C}}$ may be defined in the ortholattice $(P'(S_\Omega), \leqq, ')$ as follows: for $e, f \in P'(S_\Omega)$, $e \check{\mathbf{C}} f$ means there exists a triple $e_0, f_0, g \in P'(S_\Omega)$ such that

(i) $e_0 \perp f_0$,

(ii) $e_0 \perp g$ and $e = e_0 \vee g$,

(iii) $f_0 \perp g$ and $f = f_0 \vee g$.

It is a fact from the theory of Baer *-semigroup that for $e, f \in P'(S_\Omega)$, $e \check{\mathbf{C}} f$ is equivalent to $e \circ f = f \circ e$ and, moreover, if $e \check{\mathbf{C}} f$, then $e \wedge f = e \circ f$. The assertion of the theorem then follows from the fact that $(\mathscr{E}, \leqq, ')$ and $(P'(S_\Omega), \leqq, ')$ are isomorphic under $p \rightarrow \Omega_p$. Q.E.D.

Consequently, the compatibility of events corresponds to commutativity of the associated operations. Furthermore, in the case of compatibility, the greatest lower bound $p \wedge q$, of p and q (which is interpreted as the conjunction or "and" of p and q) corresponds to the composition of the associated operations Ω_p and Ω_q. This, of course, is an intuitively reasonable result.

IV. COMMENTS

Although Axioms 2.6 and 2.7 have not been utilized, they are included in the definition of an event-state-operation structure because of their equivalence to the conventional expression for conditional probabilities involving compatible events.

THEOREM 4.1. *If* $(\mathscr{E}, \mathscr{S}, P, \Omega)$ *is a 4-tuple which satisfies Axiom* 2.1*, then Axioms* 2.6 *and* 2.7 *are equivalent to the following: if* $p, q \in \mathscr{E}$, $p \mathbf{C} q$*, and*

$\alpha\in\mathscr{D}_p$, then

$$P(q,\,\Omega_p\alpha)=\frac{P(q\wedge p,\,\alpha)}{P(p,\,\alpha)}.$$

Proof. Assume Axioms 2.6 and 2.7 and let $p,\,q\in\mathscr{E}$, $p\,\mathbf{C}\,q$, and $\alpha\in\mathscr{D}_p$. Since $p\,\mathbf{C}\,q$,

$$P(q,\,\Omega_p\alpha)=P(q\wedge p,\,\Omega_p\alpha)$$

by Axiom 2.7. Since $q\wedge p\leqq p$,

$$P(q\wedge p,\,\Omega_p\alpha)=\frac{P(q\wedge p,\,\alpha)}{P(p,\,\alpha)}$$

by Axiom 2.6; hence,

$$P(a,\,\Omega_p\alpha)=\frac{P(q\wedge p,\,\alpha)}{P(p,\,\alpha)}.$$

Conversely, assume the validity of

$$P(q,\,\Omega_p\alpha)=\frac{P(q\wedge p,\,\alpha)}{P(p,\,\alpha)}$$

for $p,\,q\in\mathscr{E}$, $p\,\mathbf{C}\,q$ and $\alpha\in\mathscr{D}_p$. If $p,\,q\in\mathscr{E}$, $q\leqq p$ and $\alpha\in\mathscr{D}_p$, then

$$P(q,\,\Omega_p\alpha)=\frac{P(q\wedge p,\,\alpha)}{P(p,\,\alpha)}=\frac{P(q,\,\alpha)}{P(p,\,\alpha)};$$

hence, Axiom 2.6 is valid. If $p,\,q\in\mathscr{E}$, $p\,\mathbf{C}\,q$, and $\alpha\in\mathscr{D}_p$, then

$$\begin{aligned}P(q\wedge p,\,\Omega_p\alpha)&=\frac{P((q\wedge p)\wedge p,\,\alpha)}{P(p,\,\alpha)}\\&=\frac{P(q\wedge p,\,\alpha)}{P(p,\,\alpha)}\\&=P(q,\,\Omega_p\alpha);\end{aligned}$$

hence, Axiom 2.7 is valid. Q.E.D.

The relation of the operations in S_Ω to the operation discussed in [10] may be examined by considering Example 2.1. If $\mathscr{L}_p(H)$ is the set of finite products of projections in $\mathscr{P}(H)$,

$$\mathscr{L}_p(H)=\{P_1P_2\ldots P_n{:}\,P_1,\,P_2,\ldots,\,P_n\in\mathscr{P}(H)\},$$

then $(\mathscr{L}_p(H), \circ, *, ')$ is a Baer *-semigroup contained in the Baer *-semigroup $(\mathscr{L}_c(H), \circ, *, ')$ (see Appendix). Each $A \in \mathscr{L}_p(H)$ yields an element of S_Ω for the Example 2.1. If

$$A = P_1 P_2 \dots P_n,\ P_1, P_2, \dots, P_n \in \mathscr{P}(H),$$

then $x_A = \Omega_{p_1} \circ \cdots \circ \Omega_{p_n}$ is an element of S_Ω. A simple calculation proves: the domain of x_A is

$$\mathscr{D}_{x_A} = \{\alpha \in \mathscr{S} : \mathrm{Tr}(D_\alpha A^* A) \neq 0\}$$

and if $\alpha \in \mathscr{D}_{x_A}$, with density operator D_α, then $\alpha' = x_A \alpha$ has density operator $D_{\alpha'}$

$$D_{\alpha'} = \frac{A D_\alpha A^*}{\mathrm{Tr}(D_\alpha A^* A)}.$$

However, if $B \in \mathscr{L}_c(H)$ and $B = \lambda A$ where $\lambda \in \mathrm{C}$ (the field of complex numbers) and $\lambda \neq 0$, then

$$\frac{B D_\alpha B^*}{\mathrm{Tr}(D_\alpha B^* B)} = \frac{A D_\alpha A^*}{\mathrm{Tr}(D_\alpha A^* A)}.$$

Consequently, the Baer *-semigroup $(\mathscr{L}_c(H)/\equiv, \circ, *, ')$ is evidently the relevant semigroup in the approach adopted here instead of $(\mathscr{L}_c(H), \circ, *, ')$. $\equiv$ is the relation defined on $\mathscr{L}_c(H)$ as follows: for $A, B \in \mathscr{L}_c(H)$, $A \equiv B$ means there exists a $\lambda \in \mathrm{C}$, $\lambda \neq 0$, such that $A = \lambda B$. $\equiv$ is an equivalence relation which respects the Baer *-semigroup structure of $(\mathscr{L}_c(H), \circ, *, ')$ (see remark after Thm. A.2.); hence, $(\mathscr{L}_c(H)/\equiv, \circ, *, ')$ is also a Baer *-semigroup. However $\equiv$ does not respect the additive structure of $\mathscr{L}_c(H)$; indeed, if $A_1 \equiv B_1$ and $A_2 \equiv B_2$, then $A_1 + A_2 \not\equiv B_1 + B_2$. This remark indicates that *operations* and *observables* are evidently quite different kinds of entities. For example, there exists a phenomenological interpretation for the multiplication of operations but there exists a phenomenological interpretation for the addition of observables. It is evidently a property of examples like Example 2.1 that both operations and observables have simple descriptions in terms of the same mathemathical object, namely, an operator on a Hilbert space.

The connection between the mathematical theories of orthomodular ortholattices and Baer *-semigroups is explicit:

(a) If $(S, \circ, *, ')$ is a Baer *-semigroup, then there exists an orthomodular ortholattice $(P'(S), \leqq, ')$ with

$$P'(S) = \{x \in S : x \circ x = x^* = x'' = x\}.$$

(b) If $(L, \leqq, ')$ is an orthomodular ortholattice, then there exists a Baer *-semigroup $(S(L), \circ, *, ')$ where $S(L)$ consists of a set of mappings from L into L and there exists an injective mapping $j: L \to S(L)$.

The orthomodular orthoposet $(\mathscr{E}, \leqq, ')$ associated with an event-state structure $(\mathscr{E}, \mathscr{S}, P)$ is not necessarily an ortholattice. However, the introduction of operations to form an event-state-operation structure $(\mathscr{E}, \mathscr{S}, P, \Omega)$ makes $(\mathscr{E}, \leqq, ')$ into an orthomodular ortholattice and provides a Baer *-semigroup S_Ω which admits a phenomenological interpretation. S_Ω is a set of mappings of the space $\mathscr{S}$ into itself. Hence, S_Ω is not the Baer *-semigroup $S(\mathscr{E})$ mentioned in part b) of the connection between orthomodular ortholattices and Baer *-semigroups (when we take $(\mathscr{E}, \leqq, ')$ for the $(L, \leqq, ')$ of part b)). $S(\mathscr{E})$ is a collection of mappings of E into $\mathscr{E}$. The role of $S(\mathscr{E})$ will be discussed in [26].

Finally, the question arises whether the introduction of Baer *-semigroups yields any useful contributions to the quantum logic approach to the foundations of quantum physics. In general, a given mathematical construct in the theory of orthomodular ortholattices has a corresponding mathematical construct in the theory of Baer *-semigroups and vice versa. There exist a number of lattice-theoretic constructs which are extremely useful mathematical tools for the quantum logic approach but which do not possess a phenomenological interpretation. In several cases the associated construct in the theory of Baer *-semigroups, indeed, possesses an intuitively reasonable phenomenological interpretation. For example, the semimodularity of $(\mathscr{E}, \leqq, ')$ is a critical property in the proof of the "concrete representation" theorems in [17] and [23]; however, no phenomenological interpretation of this lattice-theoretic concept is available. In [26], it will be shown that the semimodularity of $(\mathscr{E}, \leqq, ')$ when $\mathscr{E}$ is atomic is equivalent to the following requirement: every $x \in S_\Omega$ is a *pure* operation [10], that is, if $\alpha \in \mathscr{D}_x$ and α is a pure state (an extreme point of the convex set $\mathscr{S}$), then $x\alpha$ is a pure state.

APPENDIX

The first part of this appendix is a review of concepts from the theory of orthomodular ortholattices while the remainder presents the necessary aspects of the theory of Baer *-semigroups.

DEFINITION A.1. A relation R on a set $\mathscr{X}$ is a subset R of the Cartesian product $\mathscr{X} \times \mathscr{X}$; notation: xRy means $(x, y) \in R$.

DEFINITION A.2. A relation R on a set $\mathscr{X}$ is said to be

(a) *symmetric:* if $x, y \in \mathscr{X}$ and $x R y$, then $y R x$,

(b) *anti-symmetric:* if $x, y \in \mathscr{X}$, $x R y$ and $y R x$, then $x = y$.

(c) *reflexive:* if $x \in \mathscr{X}$, then $x R x$.

(d) *transitive:* if $x, y, z \in \mathscr{X}$, $x R y$ and $y R z$, then $x R z$.

DEFINITION A.3. A *poset* is a pair $(\mathscr{X}, \leqq)$ where $\mathscr{X}$ is a set and $\leqq$ is an anti-symmetric, reflexive, transitive relation (a *partial ordering*) on $\mathscr{X}$.

DEFINITION A.4. Let $(\mathscr{X}, \leqq)$ be a poset and $\mathscr{Y} \subset \mathscr{X}$.

(a) $x \in \mathscr{X}$ is an *upper bound* for $\mathscr{Y}$ provided: if $y \in \mathscr{Y}$, then $y \leqq x$.

(b) $x \in \mathscr{X}$ is a *least upper bound* for $\mathscr{Y}$ provided:

(i) x is an upper bound for $\mathscr{Y}$,

(ii) if z is an upper bound for $\mathscr{Y}$, then $x \leqq z$.

(c) The least upper bound of $\mathscr{Y}$, if it exists, is denoted by $\bigvee \mathscr{Y}$; in case $\mathscr{Y} = \{y_1, y_2\}$, $\bigvee \mathscr{Y}$ is denoted by $y_1 \vee y_2$.

(d) *Lower bound, greatest lower bound,* $\bigwedge \mathscr{Y}$ and $y_1 \wedge y_2$ are defined dually.

(e) An element $0 \in \mathscr{X}$ (respectively $1 \in \mathscr{X}$) such that $0 \leqq x$ (respectively, $x \leqq 1$) for all $x \in \mathscr{X}$ is called a *least* (respectively, *greatest*) element of $\mathscr{X}$.

(f) $(\mathscr{X}, \leqq)$ is a lattice if $x_1, x_2 \in \mathscr{X}$ implies $x_1 \wedge x_2$ and $x_1 \vee x_2$ exist.

The set R of real numbers has a partial ordering, the usual ordering of real numbers. The collection 2^X of all subsets of a set X has a partial order, namely, the set-theoretic relation of inclusion. If H is a complex Hilbert space and $\mathscr{P}(H)$ is the set of all projection operators in H, then the relation $\leqq$ is a partial ordering where

$$P \leqq Q \quad \text{means} \quad PQ = P, \qquad P, Q \in \mathscr{P}(H).$$

Each of these examples is a lattice.

DEFINITION A.5. Let $(\mathscr{X}, \leqq)$ be a poset with 0 and 1.

(a) A mapping $':\mathscr{X}\to\mathscr{X}$ is an *orthocomplementation* provided:

(i) if $x\in\mathscr{X}$, then $(x')'=x$,

(ii) if $x, y\in\mathscr{X}$ and $x\leqq y$, then $y'\leqq x'$,

(iii) if $x\in\mathscr{X}$, then $x\wedge x'$ and $x\vee x'$ exist and equal 0 and 1, respectively.

(b) If $':\mathscr{X}\to\mathscr{X}$ is an orthocomplementation, the relation $\perp$, the relation of *orthogonality*, is defined as follows: for $x, y\in\mathscr{X}$, $x\perp y$ means $x\leqq y'$.

(c) An *orthoposet* $(\mathscr{X}, \leqq, ')$ is a poset $(\mathscr{X}, \leqq)$ together with an orthocomplementation of $(\mathscr{X}, \leqq)$ such that if $x, y\in\mathscr{X}$ and $x\perp y$, then $x\vee y$ exists.

(d) An orthoposet $(\mathscr{X}, \leqq, ')$ is a *σ-orthoposet* provided: if $x_1, x_2,\ldots\in\mathscr{X}$ and $x_i\perp x_j$ for $i\neq j$, $i, j=1, 2,\ldots$, then $\bigvee_i x_i$ exists.

(e) An orthoposet $(\mathscr{X}, \leqq, ')$ is *orthomodular* provided: if $x, y\in\mathscr{X}$ and $x\leqq y$, then $y=x\vee(x'\wedge y)$.

2^X admits an orthocomplementation, namely, the set-theoretic complementation. The mapping $P\in\mathscr{P}(H)\to P'=I-P\in\mathscr{P}(H)$ is an orthocomplementation of $\mathscr{P}(H)$.

DEFINITION A.6. Let $(\mathscr{X}, \leqq, ')$ be a σ-orthoposet.

(a) A probability measure μ on $\mathscr{X}$ is a function $\mu:\mathscr{X}\to[0, 1]$ such that

(i) $\mu(0)=0$, $\mu(1)=1$,

(ii) if $x_1, x_2,\ldots\in\mathscr{X}$ and $x_i\perp x_j$ for $i\neq j$, then

$$\mu\left(\bigvee_i x_i\right)=\sum_i \mu(x_i).$$

Let $\mathscr{M}$ be a set of probability measures on $\mathscr{X}$.

(b) $\mathscr{M}$ is *order-determining* provided: if $x, y\in\mathscr{X}$ and $\mu(x)\leqq\mu(y)$ for all $\mu\in\mathscr{M}$, then $x\leqq y$.

(c) $\mathscr{M}$ is *strongly-order-determining* provided: if $x, y\in\mathscr{X}$ and

$$\{\mu\in\mathscr{M}:\mu(x)=1\}\subset\{\mu\in\mathscr{M}:\mu(y)=1\},$$

then $x\leqq y$.

(d) $\mathscr{M}$ is *separating* provided: if $x, y\in\mathscr{X}$ and $\mu(x)=\mu(y)$ for all $\mu\in\mathscr{M}$, then $x=y$.

(e) $\mathscr{M}$ is σ-convex provided: if $\mu_1, \mu_2,\ldots\in\mathscr{M}$, $t_1, t_2,\ldots\in[0, 1]$ and $\sum_i t_i=1$, then there exists a $\mu\in\mathscr{M}$ such that

$$\mu(x)=\sum t_i\mu_i(x) \quad \text{for all} \quad x\in\mathscr{X}.$$

THEOREM A.1. *Let $\mathcal{M}$ be a set of probability measures on a σ-orthoposet* $(\mathcal{X}, \leqq, ')$. *If $\mathcal{M}$ is separating, then* $(\mathcal{X}, \leqq, ')$ *is orthomodular. If $\mathcal{M}$ is order-determining and* $(\mathcal{X}, \leqq, ')$ *is orthomodular, then $\mathcal{M}$ is separating. If $\mathcal{M}$ is strongly-order-determining and* $(\mathcal{X}, \leqq, ')$ *is orthomodular, then $\mathcal{M}$ is order-determining.*

Proof. See, for example, [25].

For additional material on posets and lattices, see [2].

DEFINITION A.7. (a) A *semigroup* $(S, \circ)$ is a set S with a mapping $\circ: S\times S\to S((x, y)\in S\times S\to x\circ y\in S)$ such that if $x, y, z\in S$, then

$$(x\circ y)\circ z = x\circ(y\circ z)$$

i.e., $\circ$ is *associative.*

(b) If $(S, \circ)$ is a semigroup, then an element $0\in S$ (respectively, $1\in S$) is a *zero* (respectively, *unit*) provided $0\circ x = x\circ 0 = 0$ (respectively $1\circ x = x\circ 1 = x$) for all $x\in S$.

(c) An *involution semigroup* $(S, \circ, *)$ is a semigroup $(S, \circ)$ together with a mapping called an *involution*, $*: S\to S(x\in S\to x^*\in S)$, such that

(i) if $x\in S$, then $(x^*)^* = x$,

(ii) if $x, y\in S$, then $(x\circ y)^* = y^*\circ x^*$.

(d) If $(S, \circ, *)$ is an involution semigroup, then an element of $P(S)$ is called a *projection* where

$$P(S) = \{e\in S: e\circ e = e^* = e\}.$$

(e) If $(S, \circ, *)$ is an involution semigroup, then the relation $\leqq$ on $P(S)$ is defined as follows: for $e, f\in P(S)$, $e\leqq f$ means $e\circ f = e$.

If H is a complex Hilbert space, then $(\mathcal{L}_c(H), \circ)$ is a semigroup where $\mathcal{L}_c(H)$ is the set of all continuous (i.e., bounded) linear operators on H and $\circ$ is operator multiplication, if $A, B\in\mathcal{L}_c(H)$, then $A\circ B = AB$. The usual operator adjoint, $A\to A^*$, is an involution for $(\mathcal{L}_c(H), \circ)$; moreover, in this case, $P(H)$, the set of projection operators in H, coincides with $P(\mathcal{L}_c(H))$. The relation $\leqq$ of (e) is just the conventional partial ordering of projection operators. This illustrates the following theorem.

THEOREM A.2. *If* $(S, \circ, *)$ *is an involution semigroup, then* $(P(S), \leqq)$ *is a poset; moreover, if S has a zero* 0 *(respectively, unit* 1*), then* 0 *(respectively,* 1*) is the least (respectively, greatest) element of* $P(S)$.

Define the relation $\equiv$ on $\mathscr{L}_c(H)$ as follows: for A, $B \in \mathscr{L}_c(H)$, $A \equiv B$ means there exists a $\lambda \in \mathbf{C}$ (the complex number field) such that $\lambda \neq 0$ and $A = \lambda B$. $\equiv$ is obviously an equivalence relation (i.e., $\equiv$ is reflexive, symmetric and transitive). If $A \in \mathscr{L}_c(H)$, let C_A denote the equivalence class containing A,

$$C_A = \{B \in \mathscr{L}_c(H) : B \equiv A\}$$

and let $\mathscr{L}_c(H)/\equiv$ denote the set of all these equivalence classes. If A, A_1, B, $B_1 \in \mathscr{L}_c(H)$, $A_1 \equiv A$ and $B_1 \equiv B$, then $A_1 \circ B_1 \equiv A \circ B$; hence $\circ$ induces a composition in $\mathscr{L}_c(H)/\equiv$ by

$$C_A \circ C_B = C_{A \circ B}, \ A, B \in \mathscr{L}_c(H).$$

Similarly, if A, $B \in \mathscr{L}_c(H)$ and $A \equiv B$, then $A^* \equiv B^*$; hence, * induces an involution in $\mathscr{L}_c(H)/\equiv$ by $(C_A)^* = C_{A^*}$, $A \in \mathscr{L}_c(H)$. $(\mathscr{L}_c(H)/\equiv, \circ, {}^*)$ is an involution semigroup such that $A \to C_A$ is a homomorphism. However, if A_1, A, B_1, $B \in \mathscr{L}_c(H)$, $A_1 \equiv A$, and $B_1 \equiv B$, then $A_1 + B_1 \not\equiv A + B$, in general; indeed, if $A_1 = \lambda A$ and $B_1 = \mu B$, λ, $\mu \in \mathbf{C}$, λ, $\mu \neq 0$, then, in general, there will exist no $\nu \in \mathbf{C}$ such that

$$A_1 + B_1 = \lambda A + \mu B = \nu(A + B).$$

DEFINITION A.8. (a) A *Baer *-semigroup* $(S, \circ, {}^*, {}')$ is an involution semigroup $(S, \circ, {}^*)$ with a zero 0 and a mapping $' : S \to P(S)$ such that if $x \in S$, then

$$\{y \in S : x \circ y = 0\} = \{z \in S : z = x' \circ z\}.$$

(b) If $(S, \circ, {}^*, {}')$ is a Baer *-semigroup, then an element of

$$P'(S) = \{e \in P(S) : (e')' = e\}$$

is called a *closed* projection.

If $A \in \mathscr{L}_c(H)$, the *null space* of A is denoted by $\mathscr{N}_A$,

$$\mathscr{N}_A = \{\psi \in H : A\psi = 0\}$$

and the projection with range $\mathscr{N}_A$ is denoted by A'. The mapping $A \to A'$ makes $(\mathscr{L}_c(H), \circ, {}^*)$ into a Baer *-semigroup. Furthermore, if A, $B \in \mathscr{L}_c(H)$ and $A \equiv B$, then $A' \equiv B'$; consequently, both $(\mathscr{L}_c(H), \circ, {}^*, {}')$ and $(\mathscr{L}_c(H)/\equiv, \circ, {}^*, {}')$ are Baer *-semigroups where $(C_A)' = C_{A'}$ for $A \in \mathscr{L}_c(H)$.

THEOREM A.3. *Let* $(S, \circ, *, ')$ *be a Baer *-semigroup.*

(a) $P'(S) = \{x' : x \in S\}$.

(b) *If* $e \in P'(S)$, *then* $e' \in P'(S)$.

(c) $(P'(S), \leqq, ')$ *is an orthomodular ortholattice where* $\leqq$ *is the relation* $\leqq$ *on* $P(S)$ *restricted to* $P'(S)$ *and* $'$ *is the restriction of* $' : S \to P(S)$ *to* $P'(S)$; *moreover, if* $e, f \in P'(S)$, *then* $e \wedge f = (e' \circ f)' \circ f$.

(d) *If* $e, f \in P'(S)$, *then the following are equivalent:*

(i) *there exist* $e_0, f_0, g \in P'(S)$ *such that*

$$e_0 \perp f_0,\ e_0 \perp g,\ f_0 \perp g,\ e = e_0 \vee g \quad \text{and} \quad f = f_0 \vee g,$$

(ii) $e \circ f = f \circ e$;

moreover, if $e \circ f = f \circ e$, *then* $e \wedge f = e \circ f$.

The proofs of these theorems together with further details of the theory of Baer *-semigroups may be found in [6].

ACKNOWLEDGEMENTS

The author gratefully acknowledges numerous informative discussions with H. Ekstein and A. B. Ramsay and expresses his gratitude to R. Haag both for many stimulating discussions and for the hospitality of the II. Institut für Theoretische Physik der Universität Hamburg.

II. Institut für Theoretische Physik der Universität Hamburg
Argonne National Laboratory,
Argonne, Ill.

NOTES

† Supported in part by the United States Atomic Energy Commission.

1 For example, the set of events has the structure of an orthomodular ortholattice in von Neumann's Hilbert space model of quantum mechanics.

2 For a review of the quantum theory of measurements, see [11] and [12].

3 The concept of operation was introduced in the algebraic approach to quantum field theory by Haag and Kastler [10].

4 For comments on state-preparation procedures and observation procedures (in the context of the algebraic approach to quantum physics), see [5].

5 See the Appendix for definitions of terminology from the theory of orthomodular ortholattices.

BIBLIOGRAPHY

[1] Birkhoff, G., 'Lattices in Applied Mathematics', *Proceedings of Symposia in Pure*

Mathematics, Vol. 2, *Lattice Theory*, pp. 155–184, American Mathematical Society Providence, R.I., 1961.
[2] Birkhoff, G., *Lattice Theory, Colloquium Publications*, Vol. 25, 3rd ed., American Mathematical Society Providence, R.I., 1967.
[3] Birkhoff, G., and Neumann, J. von, 'The Logic of Quantum Mechanics', *Ann. Math.* **37** (1936), 823–843.
[4] Bodiou, G., *Theorie dialectique des probabilites*, Gauthier-Villars, Paris, 1964.
[5] Ekstein, H., 'Presymmetry', *Phys. Rev.* **153** (1967), 1397–1402.
[6] Foulis, D. J., 'Baer *-Semigroups', *Proc. Am. Math. Soc.* **11** (1960), 648–654.
[7] Gudder, S., 'Spectral Methods for a Generalized Probability Theory', *Trans. Am. Math. Soc.* **119** (1965), 428–442.
[8] Gudder, S., 'Uniqueness and Existence Properties of Bounded Observables', *Pacific J. Math.* **19** (1966), 81–93.
[9] Gunson, J., 'On the Algebraic Structure of Quantum Mechanics', *Commun. Math. Phys.* **6** (1957), 262–285.
[10] Haag, R. and Kastler, D., 'An Algebraic Approach to Quantum Field Theory', *J. Math. Phys.* **5** (1964), 848–861.
[11] Jammer, M., 'The Conceptual Development of Quantum Mechanics', McGraw-Hill, New York, 1966.
[12] Jauch, J. M., *Foundations of Quantum Mechanics*, Addison-Wesley, Reading, Mass., 1968.
[13] Kamber, F., 'Die Struktur des Aussagenkalkuls in einer physikalischen Theorie', *Nachr. Akad. Wiss. Göttingen Math.-Phys. Kl. II* (1964), 103–124.
[14] Kolmogorov, A. N., *Foundations of Probability Theory*, Chelsea Publishing Co., New York, 1950.
[15] Ludwig, G., 'Versuch einer axiomatischen Grundlegung der Quantenmechanik und allgemeinerer physikalischer Theorien', *Z. Physik* **181** (1964), 233–260.
[16] Lüders, G., 'Über die Zustandsänderung durch den Meßprozeß', *Ann. Physik* **8** (1951), 322–328.
[17] MacLaren, M. D., 'Notes on Axioms for Quantum Mechanics', Argonne National Laboratory Report, ANL-7065 (1965).
[18] Mackey, G. W., *Mathematical Foundations of Quantum Mechanics*, W. A. Benjamin, Inc., New York, 1963.
[19] Margenau, H.: 'Philosophical Problems Concerning the Meaning of Measurement in Physics', *Philos. Sci.* **25**, (1958), 23–33.
[20] Messiah, A., *Quantum Mechanics*, Vol. 1, North-Holland Publishing Company, Amsterdam, 1961.
[21] Nevue, J., *Mathematical Foundations of the Calculus of Probability*, Holden-Day, San Francisco, 1965.
[22] Pauli, W., 'Die allgemeinen Prinzipien der Wellenmechanik', *Handbuch der Physik*, Vol. 1, Part 1, 1–168. Berlin-Göttingen-Heidelberg 1958.
[23] Piron, C., 'Axiomatique quantique', *Helv. Phys. Acta* **37** (1964), 439–468.
[24] Piron, C., 'De l'interpretation des treillis complets faiblement modulaires', Preprint, Institut de Physique Théorique de l'Université de Genève.
[25] Pool, J. C. T., 'Simultaneous Observability and the Logic of Quantum Mechanics', Thesis, State University of Iowa, Department of Physics, Report SUI-63-17 (1963).
[26] Pool, J. C. T., 'Semimodularity and the Logic of Quantum Mechanics', To appear in *Commun. Math. Phys.*
[27] Randall, C. H., 'A Mathematical Foundation for Empirical Science with Special

Reference to Quantum Theory', Ph.D. Thesis, Rensselaer Polytechnic Institute (1966).
[28] Varadarajan, V. S., 'Probability in Physics and a Theorem on Simultaneous Observability', *Commun. Pure Appl. Math.* **15** (1962), 189–217.
[29] Neumann, J. von, *Mathematical Foundations of Quantum Mechanics* (transl. by R. T. Beyer), Princeton Univ. Press, Princeton, 1955.
[30] Zierler, N., 'Axioms for Non-Relativistic Quantum Mechanics', *Pacific J. Math.* **11** (1961), 1151–1169.
[31] Zierler, N. and Schlessinger, M., 'Boolean Embeddings of Orthomodular Sets and Quantum Logic', *Duke Math. J.* **32** (1965), 251–262.

JAMES C. T. POOL

SEMIMODULARITY AND THE LOGIC OF QUANTUM MECHANICS*

ABSTRACT. If $(\mathscr{E}, \mathscr{S}, P, \Omega)$ is an event-state-operation structure, then the events form an orthomodular ortholattice $(\mathscr{E}, \leqq, ')$ and the operations, mappings from the set of states $\mathscr{S}$ into $\mathscr{S}$, form a Baer *-semigroup $(S_\Omega, \circ, *, ')$. Additional axioms are adopted which yield the existence of a homomorphism θ from $(S_\Omega, \circ, *, ')$ into the Baer *-semigroup $(S(\mathscr{E}), \circ, *, ')$ of residuated mappings of $(\mathscr{E}, \leqq, ')$ such that $x \in S_\Omega$ maps states while $\theta_x \in S(\mathscr{E})$ maps supports of states. If $(\mathscr{E}, \leqq, ')$ is atomic and there exists a correspondence between atoms and pure states, then the existence of θ provides the result: $(\mathscr{E}, \leqq, ')$ is semimodular if and only if every operation $x \in S_\Omega$ is a pure operation (maps pure states into pure states).

0. INTRODUCTION

The theory of orthomodular ortholattices provides the mathematical constructs for the quantum logic approach to the foundations of quantum physics. A role for the theory of Baer *-semigroup, a mathematical theory closely related to the theory of orthomodular ortholattices, was exhibited in [15]. The definitions and terminology introduced in [15] will be utilized in this paper without further explanation. If $(\mathscr{E}, \mathscr{S}, P, \Omega)$ is an event-state-operation structure, then $(\mathscr{E}, \leqq, ')$ is an orthomodular ortholattice and $(S_\Omega, \circ, *, ')$ is a Baer *-semigroup such that $p \in \mathscr{E} \to \Omega_p \in P'(S_\Omega)$ is an isomorphism of $(\mathscr{E}, \leqq, ')$ onto the orthomodular ortholattice $(P'(S_\Omega), \leqq, ')$ of closed projections in S_Ω. Each $x \in S_\Omega$ is a mapping, $x: \mathscr{D}_x \to \mathscr{R}_x$, with domain $\mathscr{D}_x$ and range $\mathscr{R}_x$ contained in $\mathscr{S}$.

The connection between the theories of orthomodular ortholattices and Baer *-semigroups includes the following: if $(L, \leqq, ')$ is any orthomodular ortholattice, then there exists a Baer *-semigroup $(S(L), \circ, *, ')$ where $S(L)$ is a set of mappings of L into L and there exists an injective mapping $j: L \to S(L)$. Section I is devoted to a discussion of $S(\mathscr{E})$ for the orthomodular ortholattice $(\mathscr{E}, \leqq, ')$. In particular, the relation of $(S(\mathscr{E}), \circ, *, ')$ to the Baer *-semigroup $(S_\Omega, \circ, *, ')$ of operations will be exhibited.

One goal of the quantum logic approach to the foundations of quantum physics is to augment the axioms of an event-state structure to obtain an axiomatic characterization of von Neumann's Hilbert space model for quantum mechanics: an axiomatic characterization where each

C. A. Hooker (ed.), The Logico-Algebraic Approach to Quantum Mechanics, 395–414.

axiom has a plausible physical interpretation. The currently available "concrete representation theorems" (the identification of $(\mathscr{E}, \leqq, ')$ with a lattice of subspaces of a vector space) require the following hypothesis: $(\mathscr{E}, \leqq, ')$ is semimodular. The correspondence between $(S_\Omega, \circ, *, ')$ and $(S(\mathscr{E}), \circ, *, ')$ will be utilized to obtain a direct phenomenological interpretation of the semimodularity of $(\mathscr{E}, \leqq, ')$ in Section III. The setting for the investigation of semimodularity will be developed in Section II.

Definitions and theorems relating to orthomodular ortholattices and Baer *-semigroups were presented in the Appendix of [15]. Additional definitions and theorems concerning residuated mappings, atomicity, and semimodularity are included in an Appendix to this paper.

I. A ROLE FOR RESIDUATED MAPPINGS OF $(\mathscr{E}, \leqq, ')$

The set $S(L)$ of residuated mappings of any orthomodular ortholattice $(L, \leqq, ')$ admits the structure of a Baer *-semigroup $(S(L), \circ, *, ')$; moreover, the orthomodular ortholattice $(P'(S(L)), \leqq, ')$ of closed projections in $S(L)$ is isomorphic to $(L, \leqq, ')$. Therefore, if $(\mathscr{E}, \mathscr{S}, P, \Omega)$ is an event-state-operation structure, then the residuated mappings of the orthomodular ortholattice $(\mathscr{E}, \leqq, ')$ form a Baer *-semigroup $(S(\mathscr{E}), \circ, *, ')$. The question arises whether this Baer *-semigroup is related to the Baer *-semigroup $(S_\Omega, \circ, *, ')$ of operations in a phenomenologically interpretable way. The answer to this question requires a review of the concept of support of a state.

DEFINITION 1.1. Let $(\mathscr{E}, \mathscr{S}, P)$ be an event-state structure.

(a) If $\alpha \in \mathscr{S}$, then $\mathscr{E}_0(\alpha)$ and $\mathscr{E}_1(\alpha)$ are the subsets of $\mathscr{E}$ defined by

$$\mathscr{E}_0(\alpha) = \{p \in \mathscr{E} : P(p, \alpha) = 0\}$$
$$\mathscr{E}_1(\alpha) = \{p \in \mathscr{E} : P(p, \alpha) = 1\}.$$

(b) If $\alpha \in \mathscr{S}$, then $p \in \mathscr{E}$ is a *support* of α provided: for $q \in \mathscr{E}$, $P(q, \alpha) = 0$ if and only if $q \perp p$.

The validity of the following assertions is evident.

THEOREM 1.1. *Let $(\mathscr{E}, \mathscr{S}, P)$ be an event-state structure.*

(a) *If $\alpha \in \mathscr{S}$ and $p \in \mathscr{E}$, then the following are equivalent:*

(i) *p is a support of α,*

(ii) $\mathscr{E}_0(\alpha)=\{q\in\mathscr{E}:q\perp p\}$,
(iii) $\mathscr{E}_1(\alpha)=\{q\in\mathscr{E}:p\leqq q\}$,
(iv) *p is the least element of the subset $\mathscr{E}_1(\alpha)$ of $\mathscr{E}$.*

(b) *If $\alpha\in\mathscr{S}$, then there exists at most one $p\in\mathscr{E}$ such that p is a support of α.*

DEFINITION 1.2. If $(\mathscr{E}, \mathscr{S}, P)$ is an event-state structure and $\alpha\in\mathscr{S}$, then the support of α, provided it exists, is denoted by p_α.

The investigation of the role of $S(\mathscr{E})$ will involve the adoption of another axiom to supplement the seven axioms for event-state structures presented in [15].

AXIOM 1.8. (a) If $\alpha\in\mathscr{S}$, then the support p_α of α exists.

(b) If $p\in\mathscr{E}$ and $p\neq 0$, then there exists an $\alpha\in\mathscr{S}$ such that p is the support of α.

Axiom 1.8 is valid for von Neumann's Hilbert space model of quantum mechanics; indeed, this axiom is valid for a wide class of event-state structures [21].

Example 1.1. (See Example 1.1 of [15]). Let $(\mathscr{P}(H), \mathscr{S}, P)$ be the event-state structure for von Neumann's Hilbert space model of quantum mechanics. If $\alpha\in\mathscr{S}$, then the support of α is the operator-theoretic support projection of the density operator D_α corresponding to α, since Tr is a faithful normal trace on the von Neumann algebra $\mathscr{L}_c(H)$ of all continuous linear operators on H (see, for example, [4], [5] and [16]). The operator-theoretic support projection of D_α is the projection on the orthogonal complement of the null space of D_α; hence, in terms of the Baer *-semigroup $(\mathscr{L}_c(H), \circ, *, ')$ (see the Appendix of [15]), the support of α is $(D_\alpha)''$.

THEOREM 1.2. *Let $(\mathscr{E}, \mathscr{S}, P, \Omega)$ be an event-state-operation structure satisfying Axiom 1.8. If $p\in\mathscr{E}$, $\alpha\in\mathscr{S}$, $P(p, \alpha)\neq 0$ and $\beta=\Omega_p\alpha$, then the support p_β of β and the support p_α of α satisfy*

$$p_\beta\leqq(p_\alpha\vee p')\wedge p.$$

Proof. p_α is the support of α and $p'_\alpha\perp p_\alpha$; hence, $P(p'_\alpha, \alpha)=0$ by the defining property of support. Since $p'_\alpha\wedge p\leqq p_\alpha$, it follows that

$$0\leqq P(p'_\alpha\wedge p, \alpha)\leqq P(p'_\alpha, \alpha)=0$$

and, since $p'_\alpha \wedge p \leqq p$, Axiom 2.6 of [15] implies

$$P(p'_\alpha \wedge p, \Omega_p\alpha) = \frac{P(p'_\alpha \wedge p, \alpha)}{P(p, \alpha)};$$

hence, $P(p'_\alpha \wedge p, \Omega_p\alpha) = 0$. Since $\beta = \Omega_p\alpha$, one has

$$P(p_\alpha \vee p', \beta) = 1 - P(p'_\alpha \wedge p, \beta) = 1.$$

By assertion (a) of Theorem 1.1, $P(p_\alpha \vee p', \beta) = 1$ implies $p_\beta \leqq p_\alpha \vee p'$. Axiom 2.2 of [15] asserts $P(p, \Omega_p\alpha) = 1$; hence, $P(p, \beta) = 1$, since $\beta = \Omega_p\alpha$, and $p_\beta \leqq p$ again by Theorem 1.1. Therefore, $p_\beta \leqq p_\alpha \vee p'$ and $p_\beta \leqq p$; hence, $p_\beta \leqq (p_\alpha \vee p') \wedge p$. Q.E.D.

The assertion of Theorem 1.2 may be expressed in terms of residuated mappings as follows: if $p \in \mathscr{E}$, $\alpha \in \mathscr{D}_p$ and $\beta = \Omega_p\alpha$, then $p_\beta \leqq \phi_p(p_\alpha)$, where ϕ_p is the following residuated mapping of $\mathscr{E}$ into $\mathscr{E}$:

$$\phi_p(q) = (q \vee p') \wedge p, \quad q \in \mathscr{E}.$$

The example of von Neumann's Hilbert space model of quantum mechanics and the case of a compatible logic provide stronger results than Theorem 1.2.

Example 1.2. For the event-state-operation structure $(\mathscr{P}(H), \mathscr{S}, P, \Omega)$ (see Example 2.1 of [15]), Ω is defined as follows: if $P \in \mathscr{P}(H)$, $\alpha \in \mathscr{S}$, $P(P, \alpha) \neq 0$, and D_α is the density operator for α, then $\beta = \Omega_p\alpha$ is the state with density operator D_β,

$$D_\beta = \frac{PD_\alpha P}{Tr(D_\alpha P)}.$$

The supports of α and β are $P_\alpha = (D_\alpha)''$ and $P_\beta = (D_\beta)''$, respectively, in terms of the Baer *-semigroup $(\mathscr{L}_c(H), \circ, *, ')$. The support $(D_\beta)''$ coincides with $(PD_\alpha P)''$, since the positive number $(T_r(D_\alpha P))^{-1}$ is immaterial for supports. D_α is a positive operator on H; hence, there exists a $T_\alpha \in \mathscr{L}_c(H)$ such that $D_\alpha = T_\alpha^* T_\alpha$. Consequently, Theorem A.5 of the Appendix asserts

$$(D_\beta)'' = (PD_\alpha P)'' = ((D_\alpha)'' \vee P') \wedge P.$$

Therefore, the inequality of the conclusion of Theorem 1.2 is replaced by an equality

$$P_\beta = (P_\alpha \vee P') \wedge P$$

for the special case of von Neumann's Hilbert space model for quantum physics.

THEOREM 1.3. *Let* $(\mathscr{E}, \mathscr{S}, P, \Omega)$ *be an event-state-operation structure such that*

(i) *Axiom* 1.8 *is satisfied, and*
(ii) *if* $p, q \in \mathscr{E}$, *then* $p \mathsf{C} q$.

If $p \in \mathscr{E}$, $\alpha \in \mathscr{S}$, $P(p, \alpha) \neq 0$ *and* $\beta = \Omega_p \alpha$, *then*

$$p_\beta = p_\alpha \wedge p = (p_\alpha \vee p') \wedge p.$$

Proof. Because of hypothesis (ii), the relation of orthogonality may be characterized as follows (see, for example, [14]): for $p, q \in \mathscr{E}$, $p \perp q$ if and only if $p \wedge q = 0$. If $q \in \mathscr{E}$, then $p \mathsf{C} q$ and

$$P(q, \beta) = P(q, \Omega_p \alpha) = \frac{P(q \wedge p, \alpha)}{P(p, \alpha)}$$

by Theorem 4.1 of [15]; hence, for $q \in \mathscr{E}$, $P(q, \beta) = 0$ if and only if $P(q \wedge p, \alpha) = 0$. $P(q \wedge p, \alpha) = 0$ if and only if $q \wedge p \perp p_\alpha$ by the definition of the support of α. $q \wedge p \perp p_\alpha$ if and only if $q \perp p \wedge p_\alpha$, since $(q \wedge p) \wedge p_\alpha = q \wedge (p \wedge p_\alpha)$ and $\perp$ has the above characterization. Consequently, for $q \in \mathscr{E}$, $P(q, \beta) = 0$ if and only if $q \perp p \wedge p_\alpha$. This, however, is the defining property for the support p_β of β; hence, $p_\beta = p_\alpha \wedge p$. Since $p \mathsf{C} p_\alpha$, the characteristic properties of the relation C (see Section III of [15]) imply $(p_\alpha \vee p') \wedge p = p_\alpha \wedge p$. Q.E.D.

The following axiom for event-state-operation structures is, therefore, motivated by the general result of Theorem 1.2 and the results of the special cases of Example 1.2 and Theorem 1.3.

AXIOM 2.8. If $p \in \mathscr{E}$, $\alpha \in \mathscr{S}$, $P(p, \alpha) \neq 0$, $\beta = \Omega_p \alpha$, and the supports p_α and p_β of α and β exist, then

$$p_\beta = \phi_p(p_\alpha) = (p_\alpha \vee p') \wedge p.$$

THEOREM 1.4. *Let* $(\mathscr{E}, \mathscr{S}, P, \Omega)$ *be an event-state-operation structure satisfying Axioms* 1.8 *and* 2.8 *with* $x \in S_\Omega$, $p_1, p_2, \ldots, p_n \in \mathscr{E}$, *and* $x = \Omega_{p_1} \circ \Omega_{p_2} \circ \cdots \circ \Omega_{p_n}$.

(a) *If* $\alpha \in \mathscr{D}_x$, *then*

$$p_{x\alpha} = \phi_{p_1} \circ \phi_{p_2} \circ \cdots \circ \phi_{p_n}(p_\alpha).$$

(b) $\alpha \in C\mathscr{D}_x$ *if and only if*

$$\phi_{p_1} \circ \phi_{p_2} \circ \cdots \circ \phi_{p_n}(p_\alpha) = 0.$$

(c) $\alpha \in C\mathscr{D}_x$ *if and only if*

$$p_\alpha \leqq (\phi_{p_1} \circ \phi_{p_2} \circ \cdots \circ \phi_{p_n})^* (1)'.$$

(d) $q_x = (\phi_{p_1} \circ \phi_{p_2} \circ \cdots \circ \phi_{p_n})^* (1)'$.

Proof. Assertion (a) follows from Axiom 2.8 by induction on n. Let $p_{n+1} = 1$. The following characterization of $C\mathscr{D}_a$ was obtained in Section II of [15]: $\alpha \in C\mathscr{D}_x$ if and only if there exists an i, $1 \leqq i \leqq n$, such that

$$P(p_j, \Omega_{p_{j+1}} \circ \cdots \circ \Omega_{p_n} \circ \Omega_{p_{n+1}} \alpha) \neq 0$$

for $i \leqq j \leqq n$ and

$$P(p_i, \Omega_{p_{i+1}} \circ \cdots \circ \Omega_{p_n} \circ \Omega_{p_{n+1}} \alpha) = 0.$$

Since the supports of the states involved in these two expressions may be determined by utilizing (a), this characterization of $C\mathscr{D}_x$ yields the following: $\alpha \in C\mathscr{D}_x$ if and only if there exists an i, $1 \leqq i \leqq n$, such that

$$\phi_{p_{j+1}} \circ \cdots \circ \phi_{p_n} \circ \phi_{p_{n+1}}(p_\alpha) \perp p_j$$

for $i < j \leqq n$ and

$$\phi_{p_{i+1}} \circ \cdots \circ \phi_{p_n} \circ \phi_{p_{n+1}}(p_\alpha) \not\perp p_i.$$

For any $p, q \in \mathscr{E}$, $\phi_p(q) = 0$ if and only if $p \perp q$. Hence, the characterization of $C\mathscr{D}_x$ may be expressed as follows: $\alpha \in C\mathscr{D}_x$ if and only if $\phi_{p_1} \circ \phi_{p_2} \circ \cdots \circ \phi_{p_n}(p_\alpha) = 0$. Thus, assertion (b) is established. For any $q \in \mathscr{E}$ and $\phi \in S(\mathscr{E})$, $\phi(q) = 0$ if and only if $q \leqq \phi^*(1)'$; hence, assertion (c) follows immediately from (b). Because of the properties of p_α, (c) asserts

$$C\mathscr{D}_x = \{\alpha \in \mathscr{S} : P((\phi_{p_1} \circ \phi_{p_2} \circ \cdots \circ \phi_{p_n})^* (1)', \alpha) = 1\}.$$

Since q_x is the unique element of $\mathscr{E}$ such that

$$C\mathscr{D}_x = \mathscr{S}_1(q_x) = \{\alpha \in \mathscr{S} : P(q_x, \alpha) = 1\}$$

(see Definition 2.3 of [15]), assertion (d) is valid. Q.E.D.

THEOREM 1.5. *Let* $(\mathscr{E}, \mathscr{S}, P, \Omega)$ *be an event-state-operation structure satisfying Axioms* 1.8 *and* 2.8.

(a) *The mapping* $\theta: S_\Omega \to S(\mathscr{E})$ $(x \in S_\Omega \to \theta_x \in S(\mathscr{E}))$ *defined as follows is well-defined: if* $x \in S_\Omega$, *then select* $p_1, p_2, \dots, p_n \in \mathscr{E}$ *such that* $x = \Omega_{p_1} \circ \Omega_{p_2} \circ \dots \circ \Omega_{p_n}$ *and define* θ_x *by*

$$\theta_x = \phi_{p_1} \circ \phi_{p_2} \circ \dots \circ \phi_{p_n}.$$

(b) θ *is a homomorphism of the Baer* *-*semigroup* $(S_\Omega, \circ, *, ')$ *of operations into the Baer* *-*semigroup* $(S(\mathscr{E}), \circ, *, ')$ *of residuated mappings of* $(\mathscr{E}, \leqq, ')$.

(c) *If* $x \in S_\Omega$ *and* $\alpha \in \mathscr{D}_x$, *then*

$$p_{x\alpha} = \theta_x(p_\alpha).$$

(d) *If* $x \in S_\Omega$ *and* $\alpha \in \mathscr{S}$, *then the following are equivalent:*

(i) $\alpha \notin \mathscr{D}_x$,

(ii) $\theta_x(p_\alpha) = 0$,

(iii) $p_\alpha \leqq \theta_x^*(1)'$.

Proof. $\theta: S_\Omega \to S(\mathscr{E})$ is well-defined provided: if $x \in S_\Omega$, $p_1, p_2, \dots, p_n$, $q_1, q_2, \dots, q_m \in \mathscr{E}$, and

$$x = \Omega_{p_1} \circ \Omega_{p_2} \circ \dots \circ \Omega_{p_n} = \Omega_{q_1} \circ \Omega_{q_2} \circ \dots \circ \Omega_{q_m},$$

then

$$\phi_{p_1} \circ \phi_{p_2} \circ \dots \circ \phi_{p_n} = \phi_{q_1} \circ \phi_{q_2} \circ \dots \circ \phi_{q_m},$$

that is, if $p \in \mathscr{E}$, then

$$\text{(I)} \qquad \phi_{p_1} \circ \phi_{p_2} \circ \dots \circ \phi_{p_n}(p) = \phi_{q_1} \circ \phi_{q_2} \circ \dots \circ \phi_{q_m}(p).$$

Both sides of (I) are equal to 0, if $p = 0$. If $p \in \mathscr{E}$ and $p \neq 0$, then there exists an $\alpha \in \mathscr{S}$ such that p is the support p_α of α, $p = p_\alpha$, by Axiom 1.8. If $\alpha \in \mathscr{D}_x$, then both sides of (I) are equal to the support of $x\alpha$ by assertion (a) of Theorem 1.4. If $\alpha \notin \mathscr{D}_x$, then both sides of (I) are equal to 0 by assertion (b) of Theorem 1.4. Consequently, θ is well-defined.

θ obviously preserves $\circ$ and *,

$$\theta_{x \circ y} = \theta_x \circ \theta_y, \quad \theta_{x^*} = (\theta_x)^*$$

for $x, y \in S_\Omega$. If $x \in S_\Omega$, $p_1, p_2, \dots, p_n \in \mathscr{E}$ and $x = \Omega_{p_1} \circ \Omega_{p_2} \circ \dots \circ \Omega_{p_n}$, then $x = \Omega_{q_x}$ (see Definition 2.3 of [15]). By the definition of $'$ for $(S(\mathscr{E}), \circ, *, ')$,

$(\theta_x)' = \phi_\alpha$ where $q \in \mathscr{E}$ is given by

$$q = \theta_x^*(1)' = (\phi_{p_1} \circ \phi_{p_2} \circ \cdots \circ \phi_{p_n})^* (1)'.$$

(d) of Theorem 1.4 asserts that $q = q_x$; hence, $(\theta_x)' = \phi_{q_x}$. Since $x' = \Omega_{q_x}$, $(\theta_x)' = \theta_{x'}$ and θ preserves $'$.

Assertion (c) is an immediate consequence of assertion (a) of Theorem 1.4 while assertion (d) is an immediate consequence of (b) and (c) of Theorem 1.4. Q.E.D.

Consequently, if $(\mathscr{E}, \mathscr{S}, P, \Omega)$ is an event-state-operation structure satisfying Axioms 1.8 and 2.8, then, for each element x of the Baer *-semigroup $(S_\Omega, \circ, *, ')$ of operations, there is an element θ_x of the Baer *-semigroup $(S(\mathscr{E}), \circ, *, ')$ of residuated mappings of $(\mathscr{E}, \leq, ')$. x maps states while θ_x maps supports of states; specifically, if $\alpha \in \mathscr{D}_x$ then the support p_α of the state α is mapped into the support $p_{x\alpha}$ of the state $x\alpha$ by θ_x,

$$p_{x\alpha} = \theta_x(p_\alpha).$$

II. ATOMS AND PURE STATES

The purpose of this section is to augment the axioms of an event-state structure to provide a setting for investigating the role of semimodularity.

DEFINITION 2.1. Let $(\mathscr{E}, \mathscr{S}, P)$ be an event-state structure.

(a) If $\alpha_1, \alpha_2, \ldots, \in \mathscr{S}$, $t_1, t_2, \ldots, \in [0, 1]$, and $\sum_i t_i = 1$, then the unique $\alpha \in \mathscr{S}$ such that

$$P(p, \alpha) = \sum_i t_i P(p, \alpha_i),$$

for all $p \in \mathscr{E}$ (see Axioms 1.6 and 1.7 of [15]) is denoted by

$$\alpha = \sum_i t_i \alpha_i$$

and called the *mixture* of $\alpha_1, \alpha_2, \ldots$ with respective weights $t_1, t_2, \ldots$.

(b) A state $\alpha \in \mathscr{S}$ is *pure* provided: if $t \in (0, 1)$, $\alpha_1, \alpha_2 \in \mathscr{S}$, and

$$\alpha = t\alpha_1 + (1 - t)\,\alpha_2,$$

then $\alpha_1 = \alpha_2$; otherwise, α is *mixed*. The set of all pure states is denoted by $\tilde{S}$.

(c) The set of all atoms in $\mathscr{E}$ is denoted by $\tilde{\mathscr{E}}$.

For a general event-state structure, $\mathscr{E}$ is not atomic; indeed there may be no atoms in $\mathscr{E}$ (for the definitions of atom and atomic, see the Appendix). Furthermore, there may exist no pure states in $\mathscr{S}$. An additional axiom is necessary [12].

AXIOM 1.9. (a) If $p \in \mathscr{E}$ and $p \neq 0$, then there exists a pure state $\alpha \in \mathscr{S}$ such that $P(p, \alpha) = 1$.

(b) $\alpha \in \mathscr{S}$ is a pure state if and only of there exists a $p \in \mathscr{E}$ such that, for $\beta \in \mathscr{S}$, $P(p, \beta) = 1$ is equivalent to $\beta = \alpha$.

If $p \in \mathscr{E}$ and $p \neq 0$, then there exists an $\alpha \in \mathscr{S}$ such that $P(p, \alpha) = 1$ by assertion (f) of Theorem 1.1 of [15]. Part (a) of Axiom 1.9 asserts that this state may be selected to be a pure state. Part (b) of Axiom 1.9 asserts that a state is pure if and only if it may be prepared (see Theorem 3.1) and identified by observing a single event.

THEOREM 2.1. *If* $(\mathscr{E}, \mathscr{S}, P)$ *is an event-state structure satisfying Axiom* 1.8, *then the following are equivalent statements:*

(a) $(\mathscr{E}, \mathscr{S}, P)$ *satisfies Axiom* 1.9,

(b) $(\mathscr{E}, \mathscr{S}, P)$ *satisfies the following:*

(i) *if* $p \in \mathscr{E}$ *and* $p \neq 0$, *then there exists a pure state* $\alpha \in \mathscr{S}$ *such that* $P(p, \alpha) = 1$, *and*

(ii) $\alpha \in \mathscr{S}$ *is a pure state if and only if* p_α *is an atom and* α *is the unique state in* $\mathscr{S}$ *with support equal to* p_α; *and*

(c) $(\mathscr{E}, \mathscr{S}, P)$ *satisfies the following:*

(i) $(\mathscr{E}, \leqq, ')$ *is atomic and*

(ii) *there exists a mapping* $p \to \alpha_p$ *of the set of atoms,* $\tilde{\mathscr{E}}$, *onto the set of pure states,* $\tilde{\mathscr{S}}$, *such that, for* $p \in \tilde{\mathscr{E}}$, α_p *is the unique state* $\alpha \in \mathscr{S}$ *with* $P(p, \alpha) = 1$.

Moreover, if Axiom 1.9 *is satisfied and* $p \in \tilde{\mathscr{E}}$, *then* p *is the support of* α_p.

Proof. First, consider part (b) of Axiom 1.9. Suppose $\alpha \in \mathscr{S}$ and $p \in \mathscr{E}$ has the following property: for $\beta \in \mathscr{S}$, $P(p, \beta) = 1$ if and only if $\beta = \alpha$. Since $P(p, \alpha) = 1$, one has $p_\alpha \leqq 1$; consequently for $\beta \in \mathscr{S}$, $P(p_\alpha, \beta) = 1$ implies $P(p, \beta) = 1$ and, hence, $\beta = \alpha$. Conversely, if $\beta \in \mathscr{S}$ and $\beta = \alpha$, then $P(p_\alpha, \beta) = P(p_\alpha, 1) = 1$. Consequently, part (b) of Axiom 1.9 is equivalent to the following: $\alpha \in \mathscr{S}$ is pure if and only if, for $\beta \in \mathscr{S}$, $P(p_\alpha, \beta) = 1$ is equivalent to $\beta = \alpha$.

Suppose $\alpha \in \mathscr{S}$ and p_α satisfies the property: (I) for $\beta \in \mathscr{S}$, $P(p_\alpha, \beta) = 1$

is equivalent to $\beta = \alpha$. It is asserted first that p_α is an atom. Indeed, let $q \in \mathscr{E}$, $q \neq 0$, and $q \leqq p_\alpha$. By Axiom 1.8, there exists a state $\beta \in \mathscr{S}$ such that the support p_β of β equals q; hence, $P(p_\alpha, \beta) = 1$, since $p_\beta = q \leqq p_\alpha$. Therefore, $\beta = \alpha$ and $q_\beta = p = p_\alpha$; consequently, p_α is, indeed, an atom. Let $\beta \in \mathscr{S}$ and suppose the support of β equals p_α. Then $P(p_\alpha, \beta) = P(p_\beta, \beta) = 1$ and $\beta = \alpha$. Therefore, if p_α satisfies (I), then p_α satisfies: (II) p_α is an atom and α is the unique state with support equal to p_α. Suppose now that p_α satisfies (II). If $\beta \in \mathscr{S}$ and $P(p_\alpha, \beta) = 1$, then $p_\beta \leqq p_\alpha$ and, hence, $p_\beta = p_\alpha$, since p_α is an atom; consequently, $\beta = \alpha$ since α is the unique state with support equal to p_α. Therefore, if p_α satisfies (II), then p_α satisfies (I).

Consequently, (a) and (b) are equivalent. Assume now the validity of (b) for $(\mathscr{E}, \mathscr{S}, P)$. $\mathscr{E}$ is atomic provided: if $p \in \mathscr{E}$ and $p \neq 0$, then there exists an atom $q \leqq p$. If $p \in \mathscr{E}$ and $p \neq 0$, then there exists a pure state $\alpha \in \mathscr{S}$ such that $P(p, \alpha) = 1$; consequently, p_α is an atom such that $p\alpha \leqq p$. Therefore, $\mathscr{E}$ is atomic. If $p \in \tilde{\mathscr{E}}$, then select a pure state $\alpha \in \mathscr{S}$ such that $P(p, \alpha) = 1$. Since p is an atom and $p_\alpha \leqq p$, p coincides with the support of α; consequently, there exists exactly one $\alpha \in \mathscr{S}$ such that $P(p, \alpha) = 1$. Denote this α by α_p. The mapping $p \to \alpha_p$ from $\tilde{\mathscr{E}}$ into $\tilde{\mathscr{S}}$ is surjective; indeed, if $\alpha \in \mathscr{S}$ is pure, then $\alpha = \alpha_p$ where p is the support of α. Consequently, (b) implies (c).

Assume the validity of (c). If $p \in \tilde{\mathscr{E}}$, then p is the support of α_p. Indeed, $P(p, \alpha_p) = 1$ implies $p_\alpha \leqq p$ and, hence, $p_\alpha = p$, since p is an atom. Suppose $p \in \mathscr{E}$ and $p \neq 0$. Since $\mathscr{E}$ is atomic, there exists an atom $q \in \tilde{\mathscr{E}}$ such that $q \leqq p$. The pure state α_g satisfies $P(p, \alpha_g) = 1$ since q is the support of α_g and $q \leqq p$. Therefore, (c) implies (b). Q.E.D.

Consequently, for an event-state structure $(\mathscr{E}, \mathscr{S}, P)$ satisfying Axiom 1.8, Theorem 2.1 asserts that Axiom 1.9 is satisfied if and only if $\mathscr{E}$ is atomic and the restriction of the mapping $\alpha \to p_\alpha$ from $\mathscr{S}$ onto $\mathscr{E}$ to the set of pure states, $\tilde{\mathscr{S}}$, is a one-to-one mapping of $\tilde{\mathscr{S}}$ onto the set of atoms, $\tilde{\mathscr{E}}$, such that if $\alpha \in \tilde{\mathscr{S}}$ and $\beta \in \mathscr{S}$ with $p_\alpha = p_\beta$, then $\beta = \alpha$.

The event-state structure $(\mathscr{P}(H), \mathscr{S}, P)$ of von Neumann's Hilbert space model of quantum mechanics satisfies Axiom 1.9.

Example 2.1. The atoms of $\mathscr{P}(H)$ are the projections with one-dimensional range. A state $\alpha \in \mathscr{S}$ is pure if and only if the density operator D_α is a projection with one-dimensional range [10]; therefore, the support of a pure state α is the atom D_α.

III. SEMIMODULARITY AND PURE OPERATIONS

The recent interest in the quantum logic approach to the foundations of quantum physics, at least partially, stems from the recognition of the desirability of an axiomatic characterization of von Neumann's Hilbert space model for quantum mechanics. Ideally, a criterion for the adoption of each axiom of this characterization would be the existence of a phenomenological interpretation of the axiom. The currently available "concrete representation theorems," the identification of $(\mathscr{E}, \leqq, ')$ with an appropriate lattice of subspaces of a vector space (although not necessarily a Hilbert space), depend upon hypothesizing the atomicity and semimodularity of $(\mathscr{E}, \leqq, ')$ (see [10], [11], [12], and [13]). The purpose of this section is to provide a direct phenomenological interpretation of the property of semimodularity when Axioms 1.8 and 1.9 are satisfied.

Although von Neumann's early papers on quantum mechanics involved the quantum logic approach (see [19] and [20], pp. 247–254), the first formalization of the quantum logic approach was contained in the classic work by Birkhoff and von Neumann [2]. The hypothesis of modularity was imposed on the logic, $(\mathscr{E}, \leqq, ')$, of quantum mechanics in [2] despite the fact that $(\mathscr{P}(H), \leqq, ')$ is not modular for a separable infinite dimensional complex Hilbert space H. More recently, it has been recognized that the orthomodularity of $(\mathscr{E}, \leqq, ')$ is adequate to replace modularity for many purposes; indeed, the orthomodularity of $(\mathscr{E}, \leqq, ')$ is a consequence of the axioms for an event-state structure and possesses a phenomenological interpretation even when only the set of events is considered [13]. Definition A.4 and Theorems A.7 and A.8 of the Appendix indicate the interdependence of modularity, semimodularity, orthomodularity, and distributivity.

The semimodularity of $(\mathscr{E}, \leqq, ')$ will be discussed in terms of pure operations [9].

DEFINITION 3.1. If $(\mathscr{E}, \mathscr{S}, P, \Omega)$ is an event-state-operation structure, then $x \in S_\Omega$ is a *pure* operation provided: if $\alpha \in \mathscr{D}_x$ and α is a pure state, then $x\alpha$ is a pure state.

THEOREM 3.1. *Let $(\mathscr{E}, \mathscr{S}, P, \Omega)$ be an event-state-operation structure satis-*

fying Axioms 1.8 *and* 1.9. *If* $p \in \mathscr{E}$ *is an atom,* $\alpha \in \mathscr{S}$ *and* $P(p, \alpha) \neq 0$, *then* $\Omega_p \alpha = \alpha_p$; *hence,* Ω_p *is a pure operation.*

Proof. By Axiom 2.3 of [15], $P(p, \Omega_p \alpha) = 1$; since p is an atom, $\Omega_p \alpha = \alpha_p$ by assertion (c) of Theorem 2.1. α_p is a pure state; hence, Ω_p is trivially a pure operation. Q.E.D.

A first indication of a connection between semimodularity and pure operations is contained in the following theorem.

THEOREM 3.2. *Let* $(\mathscr{E}. \mathscr{S}, P, \Omega)$ *be an event-state-operation structure satisfying Axioms* 1.8 *and* 1.9. *If* $(\mathscr{E}, \leqq, ')$ *is semimodular, then* Ω_p *is a pure operation for every* $p \in \mathscr{E}$; *hence, every* $x \in S_\Omega$ *is a pure operation.*

Proof. If $p \in \mathscr{E}$ and $\alpha \in \mathscr{D}_{\Omega_p}$, then $P(p, \alpha) \neq 0$. Consequently, by the definition of support, $p_\alpha \not\perp p$. The support p_β of $\beta = \Omega_p \alpha$ satisfies

$$p_\beta \leqq (p_\alpha \vee p') \wedge p$$

because of Theorem 1.2. If, moreover, α is a pure state, then p_α is an atom since Axiom 1.9 holds. Since $(E, \leqq, ')$ is atomic, orthomodular and semimodular, $(p_\alpha \vee p') \wedge p$ is an atom (see Theorem A.8 of the appendix). p_β is the support of a state; hence, $p_\beta \neq 0$. Therefore, the support p_β of $\beta = \Omega_p \alpha$ is an atom,

$$p_\beta = (p_\alpha \vee p') \wedge p.$$

Consequently, β is a pure state and Ω_p is a pure operation. If $x \in S_\Omega$, then there exist $p_1, p_2, \ldots, p_n \in \mathscr{E}$ such that

$$x\alpha = \Omega_{p_1}(\Omega_{p_2}(\ldots \Omega_{p_n}\alpha \ldots))$$

for all $\alpha \in \mathscr{D}_x$; therefore, if $\alpha \in \mathscr{D}_x$ is pure, then $x\alpha$ is pure. Q.E.D.

The crucial step in the proof of Theorem 3.2 was noticing the fact that the semimodularity of the atomic, orthomodular ortholattice $(\mathscr{E}, \leqq, ')$ yields the following property of the residuated mapping ϕ_p: if $q \in \mathscr{E}$, $q \not\perp p$, and q is an atom, then

$$\phi_p(q) = (q \vee p') \wedge p$$

is an atom. Indeed, this is a characterization of the semimodularity of $(\mathscr{E}, \leqq, ')$ when $(\mathscr{E}, \leqq, ')$ is atomic. Consequently, if the connection between operations and residuated mappings is utilized, that is, if Axiom 2.8 is assumed, then a characterization of semimodularity is obtained.

THEOREM 3.3. *If* $(\mathscr{E}, \mathscr{S}, P, \Omega)$ *is an event-state-operation structure satisfying Axioms* 1.8, 1.9 *and* 2.8, *then the following are equivalent:*

(a) $(\mathscr{E}, \leqq, ')$ *is semimodular,*

(b) Ω_p *is a pure operation for every* $p \in \mathscr{E}$; *and*

(c) *every* $x \in S_\Omega$ *is a pure operation.*

Proof. (b) and (c) are equivalent obviously and (a) implies (b) by Theorem 3.2 (without the assumption of Axiom 2.8). Assume Ω_p is a pure operation for every $p \in \mathscr{E}$. To prove $(\mathscr{E}, \leqq, ')$ is semimodular it suffices to prove (see Theorem A.8 of the Appendix): if $p \in \mathscr{E}$, $q \in \mathscr{E}$ and q is an atom with $q \not\perp p$, then

$$\phi_p(q) = (q \vee p') \wedge p$$

is an atom. Consider the state α_q corresponding to the atom q. Since $p \not\perp q$ and q is the support of α_q, $P(p, \alpha_q) \neq 0$ and α_q is in the domain of Ω_p; moreover, α_q is a pure state. Since Ω_p is a pure operation, $\beta = \Omega_p(\alpha_q)$ is a pure state and the support p_β is an atom. By Axiom 2.8,

$$p_\beta = \phi_p(p_{\alpha_q}) = \phi p(q) = (q \vee p') \wedge p$$

and $\phi_p(q)$ is an atom. Q.E.D.

The rule for the phenomenological interpretation of operations presented in [15] provides a heuristic reason for asserting that Ω_p should be a pure operation for each $p \in \mathscr{E}$. The phenomenological characterization of the fact that a state $\alpha \in \mathscr{S}$ is pure is the following *indecomposability* of the corresponding ensemble. The ensemble of physical systems prepared by a state-preparation procedure corresponding to α can not be decomposed into two subensembles prepared by state-preparation procedures corresponding to two distinct states α_1 and α_2 with the following property: a system of the ensemble corresponding to α may be attributed to the ensemble corresponding to α_1 with probability t and to the ensemble corresponding to α_2 with probability $1-t$, where $0<t<1$. The ensemble corresponding to $\Omega_p\alpha$ is constructed by selecting the systems of the ensemble corresponding to α for which the observation procedure for p indicates that the event p occurs. It is, therefore, plausible that this ensemble is indecomposable provided the ensemble corresponding to α is indecomposable.

IV. SUMMARY

If $(\mathscr{E}, \mathscr{S}, P, \Omega)$ is an event-state-operation structure, then the set S_Ω of operations admits the structure of a Baer *-semigroup, $(S_\Omega, \circ, *, ')$. Since $(\mathscr{E}, \leqq, ')$ is an orthomodular ortholattice, the set $S(\mathscr{E})$ of residuated mappings of $(\mathscr{E}, \leqq, ')$ also admits the structure of a Baer *-semigroup, $(S(\mathscr{E}), \circ, *, ')$. Theorem 1.1 indicated a connection between these two Baer *-semigroups: if $p \in \beta$ and $\alpha \in \mathscr{S}$ with $P(p, \alpha) \neq 0$, then the supports p_β and p_α of $\beta = \Omega_p \alpha$ and α, respectively, satisfy

$$p_\beta \leqq \phi_p(p_\alpha),$$

where ϕ_p is the residuated mapping

$$\phi_p(q) = (q \vee p') \wedge p, \quad q \in \mathscr{E}.$$

Axiom 2.8 replaces the above inequality

$$p_\beta \leqq \phi_p(p_\alpha)$$

by an equality

$$p_\beta = \phi_p(p_\alpha).$$

Axiom 2.8 is satisfied, for example, by a compatible logic (that is, $p \mathsf{C} q$ for every pair $p, q \in \mathscr{E}$) and by von Neumann's Hilbert space model for quantum mechanics. If Axiom 2.8 is satisfied, the there exists a homomorphism $\theta: S_\Omega \to S(\mathscr{E})$ of the Baer *-semigroup $(S_\Omega, \circ, *, ')$ of operations into the Baer *-semigroup $(S(\mathscr{E}), \circ, *, ')$ of residuated mappings. The operation $x \in S_\Omega$ maps states while the residuated mapping θ_a maps supports of states, specifically, if $\alpha \in \mathscr{D}_x$, then

$$\theta_x(p_\alpha) = p_{x\alpha}$$

where p_α and $p_{x\alpha}$ are the supports of α and $x\alpha$, respectively.

When the logic $(\mathscr{E}, \leq, ')$ is atomic and there exists a correspondence between pure states and atoms, Theorems 3.2 and 3.3 provide a connection between semimodularity of $(\mathscr{E}, \leq, ')$ and pure operations (operations which map pure states into pure states). Indeed, if $(\mathscr{E}, \leq, ')$ is semimodular, then every $x \in S_\Omega$ is a pure operation. When Axiom 2.8 is imposed, $(\mathscr{E}, \leq, ')$ is semimodular if and only if every $x \in S_\Omega$ is a pure operation.

APPENDIX

This Appendix is devoted to the exposition of facts about the residuated mappings of an orthomodular ortholattice and about the semimodularity of lattices, specifically, atomic orthomodular ortholattices, and of the proof of one theorem for general Baer *-semigroups.

A.I. *Residuated Mappings*

Some definitions and theorems from the theory of posets will be needed [1, 3].

DEFINITION A.1. Let $(X, \leqq)$ be a poset.

(a) If Y is a subset of X, then $x \in Y$ is a *least* (respectively, *greatest*) element for Y provided: $x \leqq y$ (respectively, $y \leqq x$) for every $y \in Y$. (There exists at most one least (respectively, greatest) element of Y.)

(b) A mapping $\phi: X \to X$ is *isotone* provided: if $x, y \in X$ and $x \leqq y$, then $\phi(x) \leqq \phi(y)$.

(c) A mapping $\phi: X \to X$ is *residuated* provided:

(i) ϕ is isotone, and

(ii) if $x \in X$, then the subset

$$\{y \in X : \phi(y) \leqq x\}$$

is nonempty and possesses a greatest element.

(d) $S(X)$ denotes the set of all residuated mappings of X.

(e) If $\phi: X \to X$ and $\psi: X \to X$, then the mapping $\phi \circ \psi: X \to X$ is defined by

$$(\phi \circ \psi)(x) = \phi(\psi(x))$$

for all $x \in X$.

(f) If $\phi \in S(X)$, then the mapping $\phi^+: X \to X$ is defined as follows: for $x \in X$, $\phi^+(x)$ is the greatest element of the subset

$$\{y \in X : \phi(y) \leqq x\}$$

of X.

THEOREM A.1. *Let* $(X, \leqq)$ *be a poset.*

(a) *An isotone mapping* $\phi: X \to X$ *is residuated if and only if there exists an isotone mapping* $\psi: X \to X$ *such that*

$$(\psi \circ \phi)(x) \geqq x$$

and

$$(\phi \circ \psi)(x) \leqq x$$

for all $x \in X$*; moreover, if* ϕ *is residuated, then* ψ *is uniquely determined,* $\psi = \phi^+$.

(b) *If* $\phi, \psi \in S(X)$*, then* $\phi \circ \psi \in S(X)$ *and*

$$(\phi \circ \psi)^+ = \psi^+ \circ \phi^+.$$

(c) *Let* 0 *be a least element for* $(X, \leqq)$, $\phi \in S(X)$ *and* $x \in X$. $\phi(x) = 0$ *if and only if* $x \leqq \phi^+(0)$.

(d) *If* $(X, \leqq)$ *has least and greatest elements,* 0 *and* 1*, respectively, then the mappings* $0: X \to X$ *and* $1: X \to X$ *are residuated where*

$$0(x) = 0, \quad x \in X$$
$$1(x) = x, \quad x \in X;$$

moreover, $(S(X), \circ)$ *is a semigroup with zero,* 0*, and unit,* 1.

If $(X, \leqq)$ is not only a poset but also has an orthocomplementation, then $(S(X), \circ)$ admits an involution [6].

Definition A.2. If $': X \to X$ is an orthocomplementation of a poset $(X, \leqq)$ and $\phi \in S(X)$, then the mapping $\phi^*: X \to X$ is defined by

$$\phi^*(x) = \phi^+(x')', \quad x \in X.$$

Theorem A.2. *Let* $': X \to X$ *be an orthocomplementation of the poset* $(X, \leqq)$.

(a) *If* $\phi \in S(X)$ *and* $x \in X$*, then*

$$\phi^+(x) = \phi^*(x')'$$

and $\phi^*(x)$ *is the least element of the subset*

$$\{y' : \phi(y) \leqq x'\}$$

of X.

(b) *If* $\phi \in S(X)$*, then* $\phi^* \in S(X)$*; moreover,* ϕ^* *is the unique isotone map-*

ping $\psi: X \to X$ *such that*

$$\psi(\phi(x')') \leqq x$$

and

$$\phi(\psi(x')') \leqq x$$

for all $x \in X$.

(c) *Let* $(X, \leqq)$ *have least and greatest elements* 0 *and* 1, $\phi \in S(X)$, *and* $x \in X$. $\phi(x) = 0$ *if and only if* $x \perp \phi^*(1)$.

(d) $*$ *is an involution for the semigroup* $(S(X), \circ)$.

If $(X, \leqq, ')$ is not only an orthoposet but also an orthomodular ortholattice, then $S(X)$ admits the structure of a Baer *-semigroup [6].

DEFINITION A.3. Let $(L, \leqq, ')$ be an ortholattice.

(a) If $p \in L$, then the mapping $\phi_p: L \to L$ is defined by

$$\phi_p(q) = (q \vee p') \wedge p, \quad q \in L.$$

(b) If $\phi \in S(L)$, then the mapping $\phi': L \to L$ is defined by $\phi' = \phi_p$ where $p = \phi^*(1)'$.

THEOREM A.3. *Let* $(L, \leqq, ')$ *be an orthomodular ortholattice.*

(a) *If* $p \in L$, *then* $\phi_p \in S(L)$; *moreover,*

$$\phi_p^* = \phi_p \circ \phi_p = (\phi_p)'' = \phi_p.$$

(b) *For* $p, q \in L$, $\phi_p(q) = 0$ *if and only if* $p \perp q$.

(c) $(S(L), \circ, *, ')$ *is a Baer *-semigroup such that* $p \to \phi_p$ *is an isomorphism of the orthomodular ortholattice* $(L, \leqq, ')$ *onto the orthomodular ortholattice* $(P'(S(L)), \leqq, ')$ *of closed projections in* $S(L)$.

A.II. *Two Theorems on Baer *-semigroups*

The following theorem is from the general theory of Baer *-semigroups [6, 7, 8].

THEOREM A.4. *If* $(S, \circ, *, ')$ *is a Baer *-semigroup, then*

(a) *for* $x \in S$, $(x^* \circ x)'' = x''$,

(b) *for* $x, y \in S$, $(x \circ y)'' = (x'' \circ y)''$,

(c) *for* $e, f \in P'(S)$, $(e \circ f)'' = (e \vee f') \wedge f$.

The computation of the operator-theoretic support projection of *PDP*, where D is a positive operator and P is a projection on a Hilbert space H, is an application of the following theorem to the Baer *-semigroup $(\mathscr{L}_c(H), \circ, *, ')$ of continuous linear operators on H.

THEOREM A.5. *If* $(S, \circ, *, ')$ *is a Baer *-semigroup,* $y \in S$, $e \in P'(X)$ *and* $x = y^* \circ y$, *then* $(e \circ x \circ e)'' = (x'' \vee e') \wedge e$.

Proof. First, one notes

$$e \circ x \circ e = (y \circ e)^* \circ (y \circ e)$$

and, by (a) of Theorem A.4,

$$(e \circ x \circ e)'' = (y \circ e)''.$$

Utilizing (b) of Theorem A.4, one has

$$(y \circ e)'' = (y'' \circ e)'';$$

hence

$$(e \circ x \circ e)'' = (y'' \circ e)''.$$

Since $y'' \in P'(S)$ and $e \in P'(S)$, (c) of Theorem A.4 asserts

$$(y'' \circ e)'' = (y'' \vee e') \wedge e.$$

Since $x = y^* \circ y$, another application of (a) of Theorem A.4 yields

$$x'' = (y^* \circ y)'' = y'';$$

consequently,

$$(e \circ x \circ e)'' = (x'' \vee e') \wedge e,$$

the assertion of the theorem. Q.E.D.

A.III. *Semimodularity and Atomicity*

The following definition and theorem indicate the interdependence of distributivity, orthomodularity, semimodularity, and modularity [1, 11].

DEFINITION A.4. Let $(L, \leqq)$ be a lattice.

(a) If $p, q, r \in L$, then (p, q, r) is a *distributive triple*, written $(p, q, r)\,D$, provided:

$$(p \vee q) \wedge r = (p \wedge r) \vee (q \wedge r).$$

$(L, \leqq)$ is a *distributive* lattice provided: if $p, q, r \in L$, then $(p, q, r)\,D$.

(b) If $q, r \in L$, then (q, r) is a *modular pair*, written $(q, r)\, M$, provided: if $p \in L$ and $p \leqq r$, then $(p, q, r)\, D$. $(L, \leqq)$ is a *modular lattice* provided: if $p, q \in L$, then $(p, q)\, M$.

(c) $(L, \leqq)$ is *semimodular* provided: if $p, q \in L$ and $(p, q)\, M$, then $(q, p)\, M$.

THEOREM A.6. *An ortholattice $(L, \leqq, ')$ is orthomodular if and only if every orthogonal pair is a modular pair, i.e., $p, q \in L$ and $p \perp q$ implies $(p, q)\, M$.*

Consequently, every distributive lattice is modular and every modular ortholattice is orthomodular. It should be noted that the projection lattice of every von Neumann algebra is not only orthomodular but also semimodular [18]. Semimodularity admits a useful characterization in an atomic orthomodular ortholattice [17].

DEFINITION A.5. Let $(X, \leqq)$ be a poset with a least element 0.

(a) An element $x \in X$ is an *atom* provided:

(i) $x \neq 0$ and

(ii) if $y \in X$ and $y \leqq x$, then either $y = 0$ or $y = x$.

(b) $(X, \leqq)$ is *atomic* provided: if $x \in X$ and $x \neq 0$, then there exists an atom $y \in X$ such that $y \leqq x$.

THEOREM A.7. *A necessary and sufficient condition for an atomic orthomodular ortholattice $(L, \leqq, ')$ to be semimodular is the following: if $p \in L$, $q \in L$ is an atom and q is not orthogonal to p, $q \not\perp p$, then $(q \vee p') \wedge p$ is an atom.*

The author expresses his gratitude to A. B. Ramsay for a continuing dialogue on the mathematical foundations of quantum physics, to J. M. Jauch, C. Piron, and J. Sopka for numerous informative discussions, and to J. M. Jauch for the hospitality of the Institut de Physique Théorique de l'Université de Genève.

Institut de Physique Théorique, Université de Genève
and
Argonne National Laboratory

NOTE

* Supported in part by the United States Atomic Energy Commission and in part by the Fonds National Suisse.

BIBLIOGRAPHY

[1] Birkhoff, G., *Lattice theory, American Mathematical Society Colloquium Publications*, Vol. 25, 3rd. edn., Providence, R.I., 1967.
[2] Birkhoff, G. and Neumann, J. von, 'The Logic of Quantum Mechanics', *Ann. Math.* **37** (1936), 823–843.
[3] Derderian, J. C., 'Residuated Mappings', *Pacific J. Math.* **20** (1967), 35–43.
[4] Dixmier, J., *Les algebres d'operateurs dans l'espace hilbertian*, Gauthier-Villars, Paris, 1957.
[5] Dixmier, J., *Les C*-algebres et leurs representations*, Gauthier-Villars, Paris, 1964.
[6] Foulis, D. J., 'Baer *-semigroups', *Proc. Am. Math. Soc.* **11** (1960), 648–654.
[7] Foulis, D. J., 'A Note on Orthomodular Lattices', *Portugal. Math.* **21** (1962), 65–72.
[8] Foulis, D. J., 'Relative Inverses in Baer *-semigroups', *Michigan Math. J.* **10** (1963), 65–84.
[9] Haag, R. and Kastler, D., 'An Algebraic Approach to Quantum Field Theory', *J. Math. Phys.* **5** (1964), 848–861.
[10] Jauch, J. M., *Foundations of Quantum Mechanics*, Addison-Wesley, Reading, Mass., 1968.
[11] MacLaren, M. D., 'Atomic Orthocomplemented Lattices', *Pacific J. Math.* **14** (1964), 597–612.
[12] MacLaren, M. D., 'Notes on Axioms for Quantum Mechanics', Argonne National Laboratory Report, ANL-7065 (1965).
[13] Piron, C., 'Axiomatique quantique', *Helv. Phys. Acta* **37** (1964), 439–468.
[14] Pool, J. C. T., 'Simultaneous Observability and the Logic of Quantum Mechanics', Thesis, State University of Iowa, Department of Physics Report SUI-63-17 (1963).
[15] Pool, J. C. T., 'Baer *-Semigroups and the Logic of Quantum Mechanics', *Commun. Math. Phys.* **9** (1968), 118–141.
[16] Schatten, R., *Norm Ideals of Completely Continuous Operators*, Springer, Berlin-Göttingen-Heidelberg, 1960.
[17] Schreiner, E. A., 'Modular Pairs in Orthomodular Lattices', *Pacific J. Math.* **19** (1966), 519–528.
[18] Topping, D. M., 'Asymptoticity and Semimodularity in Projection Lattices', *Pacific J. Math.* **20** (1967), 317–325.
[19] Neumann, J. von, 'Wahrscheinlichkeitstheoretischer Aufbau der Quantenmechanik', *Nachr. Akad. Wiss. Göttingen Math.-Phys. Kl. II* (1927, 245–272.
[20] Neumann, J. von, *Mathematical Foundations of Quantum Mechanics*, (transl. by R. T. Beyer), Princeton Univ. Press, Princeton, 1955.
[21] Zierler, N., 'Axioms for Non-Relativistic Quantum Mechanics', *Pacific J. Math.* **11** (1961), 1151–1169.

P. D. FINCH

ON THE STRUCTURE OF QUANTUM LOGIC

0. INTRODUCTION

In the axiomatic development of the logic of nonrelativistic quantum mechanics it is not difficult to set down certain plausible axioms which ensure that the quantum logic of propositions has the structure of an orthomodular poset. This can be done in a number of ways, for example, as in Gunson [2], Mackey [4], Piron [5], Varadarajan [7] and Zierler [8], and we summarise one of these ways in Section II below. It is customary to impose further axioms which ensure that this logic is a complete atomic orthomodular lattice so that easy access is obtained to a representation of the logic as the lattice of closed subspaces of Hilbert space, for example, Jauch [3], Piron [5] and Zierler [8]. Not much is known about the structure of orthomodular posets and in Section I we show how they arise naturally in the study of certain sets of Boolean logics in which one can define common operations of implication and negation. In Section III we show that every completely orthomodular poset arises in this way.

Part of the interest of this result comes about because of the following remarks. When quantum logic is represented as the lattice of subspaces of Hilbert space the physical quantities, or observables, are represented by certain selfadjoint linear operators. It is these operators which play the role of the "random variables" of quantum theory in contrast to those of classical probability theory where the random variables are just real valued functions on a space of possible outcomes which are measurable with respect to a Boolean σ-algebra of subsets which can, in an obvious way, be thought of as a Boolean σ-logic of propositions. This fact has been one of the conceptual difficulties of quantum theory because its "random variables" do not seem to be quantities of the same nature as those of classical theory. However if the quantum logic of subspaces of Hilbert space is represented as a logical σ-structure, as defined below, certain equivalence classes of the set of random variables on the com-

C. A. Hooker (ed.), The Logico-Algebraic Approach to Quantum Mechanics, 415–425.

ponent Boolean σ-algebras determine, in a natural way, selfadjoint linear operators on the Hilbert space. Thus the physical quantities of quantum theory are, in a sense, just equivalence classes of the physical quantities of classical theory. A detailed account of this fact will be presented elsewhere.

Another consequence of the results obtained below is that "quantum logic" can be developed in an abstract way which owes little to the usual physical motivation. Thus logical structures, as defined below, would seem to deserve further study in their own right as interesting logical systems.

I. LOGICAL STRUCTURES

By a Boolean logic L we mean a Boolean algebra of propositions in which the Boolean lattice operations of join, meet and orthocomplementation correspond to the logical operations of disjunction, conjunction, and negation respectively. As is well known the ordering in L may be interpreted as a logical relation of implication between the propositions of L. We use 1 and 0 to denote the greatest and least elements of L.

In what follows we consider an indexed set $\mathscr{L}=\{L_\gamma:\gamma\in\Gamma\}$ of Boolean logics L_γ which are so related that one can define negation and implication in $\bigcup\{L_\gamma:\gamma\in\Gamma\}$. In order to specify the structure of $\mathscr{L}$ we need to refer to the logical operations in each L_γ; to avoid ambiguity we denote order in L_γ by ω_γ and the orthocomplementation in L_γ by N_γ. Thus for x and y in L_γ we mean by "$x\omega_\gamma y$" that the proposition x implies the proposition y in the logic L_γ and, to avoid unnecessary repetition, we adopt the convention that whenever an expression such as "$x\omega_\gamma y$" occurs it is to be understood that x and y are both in L_γ. Similarly for x in L_γ "$N_\gamma x$" denotes the negation in the logic L_γ of the proposition x, here again it is to be understood that when an expression "$N_\gamma x$" occurs the element x is in the logic L_γ. By $J_\gamma X$ we mean the lub in the logic L_γ of the set X, it being understood that in this expression X is a subset of L_γ and that the lub in question does exist in L_γ.

DEFINITION 1.1. *A logical structure is an indexed set $\mathscr{L}=\{L_\gamma:\gamma\in\Gamma\}$ of Boolean logics with the following properties,*

(i) *each L_γ has the same least element* 0;

(ii) *if x and y belong to $L_\alpha\cap L_\beta$ then $x\omega_\alpha y$ if and only if $x\omega_\beta y$;*

(iii) *if* $x\omega_\alpha y$ *and* $y\omega_\beta z$ *there is* γ *in* Γ *such that* $x\omega_\gamma z$;

(iv) *if* x *belongs to* $L_\alpha \cap L_\beta$ *then* $N_\alpha x = N_\beta y$;

(v) *if* x *and* y *belong to* $L_\alpha \cap L_\beta$ *then* $J_\alpha\{x, y\} = J_\beta\{x, y\}$;

(vi) *Suppose that* $y\omega_\alpha N_\alpha x$ *for some* x *and* y *in* L_α, *if* $x\omega_\beta z$ *and* $y\omega_\gamma z$ *there is* L_δ *which contains* x, y *and* z.

Let $\mathscr{L}$ be a logical structure and write

$$L = \bigcup \{L_\gamma : \gamma \in \Gamma\},$$

we call L the logic associated with the logical structure $\mathscr{L}$. One can introduce a partial ordering into L by decreeing that for x, y in L one has $x \leqslant y$ if and only if there is a γ in Γ such that $x\omega_\gamma y$. The verification that this decree does define a partial order in L uses only the properties (ii) and (iii) above. Because of (i) the common 0 of the L_γ is the least element of L. By (iv) one can define a map $N: L \to L$ by asserting that for each x in L one has $Nx = N_\gamma x$ for any L_γ which contains x. In particular the common $1 = NO$ of the L_γ is the greatest element in L. The map N has the following properties

(a) $NNx = x$, *for each* x *in* L,

(b) $x \leqslant y$ *implies* $Ny \leqslant Nx$,

(c) 1 *is the* lub *in* L *of* x *and* Nx.

Of these properties (a) and (b) are obvious consequences of the corresponding facts about each N_γ in its parent Boolean logic L_γ. To prove (c) let y be any upper bound in L to x and Nx. There are α and β in Γ such that $x\omega_\alpha y$ and $(Nx)\,\omega_\beta y$. But $(Nx)\,\omega_\beta y$ implies $(Ny)\,\omega_\beta x$ and, by (iii),

$$(Ny)\,\omega_\beta x \;\&\; x\omega_\alpha y \Rightarrow \exists\gamma, (Ny)\,\omega_\gamma y.$$

Since $Ny = N_\gamma y$ and L_γ is a Boolean algebra we have $N_\gamma y = 0$, that is $y = 1$ and this is the desired result.

In the terminology of lattice theory the three properties (a), (b) and (c) mean that the map N is an orthocomplementation of the poset L, we shall adopt the usual terminology, Birkhoff [1], and write $x^\perp$ instead of Nx. When $x \leqslant y$ we say that the proposition x implies the proposition y in the logic L and we call $x^\perp$ the negation of x in L. We have not yet used the properties (v) and (vi) of the definition of a logical structure, we do so to prove

THEOREM 1.1. *The logic L associated with a logical structure is an orthomodular poset, that is*

(a) $x \vee y$ *exists whenever* $y \leqslant x^{\perp}$, *and*

(b) *if* $y \leqslant x^{\perp}$ *and* $x \vee y = 1$ *then* $y = x^{\perp}$.

Proof. Let x, y be in L and suppose that $y \leqslant x^{\perp}$. There is L_{α} such that $y \omega_{\alpha} x^{\perp}$. Since $x^{\perp} = N_{\alpha} x$ is in L_{α} so is $x = N_{\alpha} N_{\alpha} x$ and the Boolean logic L_{α} contains x and y. Using (vi) we see that if z is any upper bound in L to x and y then there is L_{δ} which contains x, y and z. From $x \omega_{\delta} z$ and $y \omega_{\delta} z$ we deduce that $J_{\delta}\{x, y\} \, \omega_{\delta} z$, that is $J_{\delta}\{x, y\} \leqslant z$. But by (v) $J\{x, y\} = J_{\delta}\{x, y\}$ is defined independently of the particular L_{δ} which contains x and y, thus $\mathscr{E}\{x, y\} \leqslant z$. But $J\{x, y\}$ is an upper bound in L to x and y and so it is the desired least upper bound.

This proves (a), to prove (b) we need only observe that if $x \vee y = 1$ then $J\{x, y\} = J_{\delta}\{x, y\} = 1$ and since $y \omega_{\delta} x^{\perp}$ and L_{δ} is a Boolean algebra we have $y = x^{\perp}$.

COROLLARY. *If* x, y *are in L and* $x \geqslant y^{\perp}$ *then* $x \wedge y$ *exists in L.*

Proof. $x \wedge y = (x^{\perp} \vee y^{\perp})^{\perp}$ and $x^{\perp} \vee y^{\perp}$ exists in L since $y^{\perp} \leqslant x = (x^{\perp})^{\perp}$.

Remark. It is known (for example Zierler [8]) that in the definition of orthomodularity one may replace the condition (b) of Theorem 1.1 by any one of the following three implications, each of which, in the presence of condition (a), is equivalent to (b).

(1.1) if $x \geqslant y^{\perp}$ and $x \wedge y = 0$ then $x = y^{\perp}$,

(1.2) if $x \leqslant y^{\perp}$ then $(x \vee y) \wedge y^{\perp} = x$,

(1.3) if $x \leqslant y$ then $x \vee (y \wedge x^{\perp}) = y$.

In an orthocomplemented poset S (with orthocomplementation $x \rightarrow x^{\perp}$) one says that x is orthogonal to y, and then writes $x \perp y$, when $x \leqslant y^{\perp}$. Note that $x \perp y$ implies $y \perp x$ and $0 \perp x$ since $0 = 1^{\perp}$. A subset X of S is orthogonal when $x \perp y$ for any $x \neq y$ in X, in particular the empty set $\square$ and the one element subsets of S are orthogonal sets. Since the set union of a nondecreasing chain of orthogonal sets is itself orthogonal there exist maximal orthogonal subsets of S. We note

LEMMA 1.1. *Let X be a maximal orthogonal subset of the orthocomplemented poset S, then* $VX = 1$.

Proof. Let X be a nonempty orthogonal subset of S. If 1 is not the lub of X there is $x < 1$ in S which is an upper bound of X. We cannot have $x^\perp$ in X for this would imply $x^\perp \leqslant x$, that is $x=1$. On the other hand if $x^\perp$ is not in X then $X \cup \{x^\perp\}$ is an orthogonal set and so X is not maximal.

DEFINITION 1.2. *An orthocomplemented poset is separable when its orthogonal subsets are at most countable.*

DEFINITION 1.3. *A logical σ-structure is a logical structure in which each of its component Boolean logics is a Boolean σ-algebra and which has, in addition to the properties* (i) *to* (vi) *above, the properties*

(vii) *if $X \subseteq L_\alpha \cap L_\beta$ is at most countable then $J_\alpha X = J_\beta X$,*

(viii) *if X is a finite or countably infinite orthogonal subset of L and z is an upper bound to X in L then there is a γ in Γ such that $X \cup \{z\} \subseteq L_\gamma$.*

Note that in (vii) $J_\alpha X$ exists in L_α since this is a Boolean σ-algebra. If X is a countable orthogonal subset of L_α then by (vii) $JX = J_\alpha X$ is defined independently of the particular L_α which contains X. Taking $z=1$ in (viii) shows that each countable orthogonal subset of L is a subset of some L_α and an argument similar to that used in the proof of Theorem 1.1 shows that $VX = JX$ exists in L for any countable orthogonal subset of L.

If L is separable then each component Boolean algebra L_γ is separable. However, it is not true that L is always separable when $\mathscr{L}$ is a σ-logic each of whose component Boolean σ-algebras is separable. I am indebted to the referee for this comment. For example, let H be a Hilbert space of denumerably infinite dimension and let $\mathscr{L}$ be the set of all Boolean algebras of closed subspaces of H generated by countable (or finite) sets of pairwise orthogonal closed subspaces. Then $\mathscr{L}$ is a σ-structure but L is precisely the lattice $\mathscr{L}$ and it is not separable. However, when L is separable VX exists in L for any orthogonal subset X of L since these are finite or countably infinite, that is L is a completely orthomodular poset. Combining this remark with Theorem 1.1 we have

THEOREM 1.2. *If $\mathscr{L}$ is a logical σ-structure for which L is separable, then L is a separable completely orthomodular poset.*

II. THE DERIVATION OF QUANTUM LOGIC FROM ITS SET OF STATES

In the axiomatic derivation of the logic of quantum mechanics one arrives at a logic of propositions which has the structure of an orthomodular poset from considerations rather different from those which motivated the discussion in the last section. One starts with a nonempty set Q whose elements are propositions and a set $\mathcal{M}$ of possible states of Q. The elements of $\mathcal{M}$ are nonnegative real-valued functions on Q with interpretation that for x in Q and μ in $\mathcal{M}$ the quantity $\mu(x)$ is the probability attaching to the proposition x in the state specified by μ. One supposes that there are sufficient states to separate the elements of Q, that is one imposes the "quantum axiom".

QA 1. *For any* $x \neq y$ *in* Q *there is at least one* μ *in* $\mathcal{M}$ *such that* $\mu(x) \neq \mu(y)$.

Because of QA 1 one can partially order the set Q by decreeing that $x \leqslant y$ in Q means that $\mu(x) \leqslant \mu(y)$ for all μ in $\mathcal{M}$. This ordering is interpreted as a relation of implication between the propositions of Q. One requires that the poset Q has a least element 0 and with this end in mind one imposes the axiom

QA 2. *There is* 0 *in* Q *such that* $\mu(0) = 0$ *for all* μ *in* $\mathcal{M}$.

Note that here "0" is denoting both the least element of Q and the real number zero. A third axiom is now introduced so that one may have a negation operation in Q,

QA 3. *For each* x *in* Q *there is* $x^{\perp}$ *in* Q *such that* (a) *if* $x \neq 0$ *then* $x \nleqslant x^{\perp}$ *and* (b) *for all* μ *in* $\mathcal{M}$ *one has* $\mu(x) + \mu(x^{\perp}) = 1$.

It is an immediate consequence of QA 3(b) and QA 2 that (i) $0 \leqslant \mu(x) \leqslant 1$ for each x in Q and each μ in $\mathcal{M}$, (ii) $\mu(0^{\perp}) = 1$ for each μ in $\mathcal{M}$. From QA 1 we deduce that Q has a greatest element $1 = 0^{\perp}$. We prove

LEMMA 2.1. *Under* QA 1, 2 *and* 3 *the map* $x \to x^{\perp}$ *is an orthocomplementation of the poset* Q.

Proof. We have to verify the three properties (a), (b) and (c) listed after Definition 1.1. By QA 3(b) we have

$$\mu(x) + \mu(x^{\perp}) = 1 = \mu(x^{\perp}) + \mu(x^{\perp\perp})$$

and so $\mu(x) = \mu(x^{\perp\perp})$ for all μ in $\mathcal{M}$. By QA 1 we deduce that $x^{\perp\perp} = x$.

Next, using QA 1 and QA 3(b), we have

$$\begin{aligned} x \leqslant y &\Rightarrow \mu(x) \leqslant \mu(y), \quad \forall \mu \in \mathcal{M} \\ &\Rightarrow \mu(y^{\perp}) \leqslant \mu(x^{\perp}), \quad \forall \mu \in \mathcal{M} \\ &\Rightarrow y^{\perp} \leqslant x^{\perp}. \end{aligned}$$

Finally we show that 1 is the lub of x and $x^{\perp}$. To do so let t be any upper bound in Q to x and $x^{\perp}$. Then $x^{\perp} \leqslant t$ and so, by the result just proved, $t^{\perp} \leqslant x$. But $x \leqslant t$ and so $t^{\perp} \leqslant t$. By QA 3(a) we have $t^{\perp} = 0$, that is $t = t^{\perp\perp} = 0^{\perp} = 1$. This is the desired result.

Remark. The orthocomplementation $x \to x^{\perp}$ is usually interpreted as a negation operation on the propositions in Q.

The following axiom ensures that the orthocomplemented poset Q is orthomodular.

QA 4. *If X is a finite orthogonal subset of Q there is an element $U(X)$ in Q such that*

$$\mu\{U(X)\} = \sum_{x \in X} \mu(x)$$

for any μ in $\mathcal{M}$.

LEMMA 2.2. *Under* QA 1 to QA 4 *one has $U(X) = VX$ for any finite orthogonal subset X of Q.*

Proof. Let $x \perp y$ be in Q and let t be an upper bound to x and y, then $\{x, y, t^{\perp}\}$ is an orthogonal set and so

$$1 \geqslant \mu\{U(x, y, t^{\perp})\} = \mu\{U(x, y)\} + \mu(t^{\perp}).$$

It follows that

$$\mu\{U(x, y)\} \leqslant \mu(t), \quad \forall \mu \in \mathcal{M}.$$

Thus $t \geqslant U(x, y)$ and since $U(x, y)$ is clearly an upper bound to x and y we deduce that $U(x, y) = x \vee y$ the lub of x and y in Q. The general result now follows by induction.

COROLLARY. *Q is orthomodular.*

Proof. If $y \leqslant x^{\perp}$ and $x \vee y = 1$ then

$$\mu(x) + \mu(x^{\perp}) = 1 = \mu(x) + \mu(y)$$

and so $\mu(y) = \mu(x^{\perp})$ for all μ in $\mathcal{M}$, thus $y = x^{\perp}$ and Q is orthomodular.

In the axiomatic development of quantum mechanics one usually assumes a stronger version of QA 4 in which X is any countable orthogonal set. One is then assured that a countable orthogonal subset of Q has a lub. This assumption is usually supplemented by the requirement that any orthogonal subset of the logic Q is at most countably infinite. In this way one arrives at a quantum logic Q which has the structure of a separable completely orthomodular poset.

The derivation of quantum logic in this way is a fairly standard procedure, see for example Zierler [8], Piron [5] and Gunson [2] where further details are given, and it is a natural one when the motivation comes from a consideration of physical systems which, in practice, are equipped with sets of possible states. However, the results of the previous section show that logics with a similar structure arise naturally from sets of Boolean logics in which one can define common operations of implication and negation. It is natural to ask, therefore, if every orthomodular poset arises as the logic associated with a logical structure. In the next section we answer this question affirmatively for completely orthomodular posets.

III. COMPLETELY ORTHOMODULAR POSETS

We establish some preliminary lemmas.

LEMMA 3.1. *Let S be a completely orthomodular poset. If M is a maximal orthogonal subset of S and $X \subseteq M$ then $\vee(M \setminus X) = x^{\perp}$ where $x = \vee X$.*

Proof. $M \setminus X$ is an orthogonal subset of S. Since S is completely orthomodular $y = \vee(M \setminus X)$ exists in S. But $y \leqslant x^{\perp}$ and by Lemma 1.1 $x \vee y = 1$. Using orthomodularity in S we deduce that $y = x^{\perp}$.

By a frame of a completely orthomodular poset S we mean a set $M \setminus 0$ where M is a maximal orthomodular subset of S. We prove

LEMMA 3.2. *Let $P(F)$ be the set of all subsets of the frame F of a completely orthomodular poset S. Define a map $h: P(F) \to S$ by writing $Xh = \vee X$ for any subset X of F, then*

(3.1) $(F \setminus X)\, h = (Xh)^{\perp}$

and, for any X, Y in $P(F)$

(3.2) $X \subseteq Y \Leftrightarrow Xh \leqslant Yh$.

Proof. Note that $\square h = V\square = 0$. Equation (3.1) follows from Lemma 3.1. The forward implication in (3.2) is trivial, to establish the backward implication we suppose that X and Y are in $P(F)$, that $Xh \leqslant Yh$ and show that this implies $X \setminus Y = \square$. To do so suppose that x is in X but not in Y, then $x \leqslant VX \leqslant VY$. But x is in F and Y is a subset of the orthogonal set F and so $VY \leqslant x^{\perp}$. It follows that $x \leqslant x^{\perp}$ and so $x=0$, but 0 is not in F and this contradiction establishes the desired result.

COROLLARY. *$P(F)\,h$ is a sublattice of S which is an isomorphic copy of the Boolean algebra of all subsets of F.*

Proof. Because of Equation (3.2) the map h is an injection. Let X and Y be subsets of F and let $x = Xh$, $y = Yh$. If t in S is an upper bound to x and y then $t \geqslant x \geqslant a$ for each a in X and $t \geqslant y \geqslant b$ for each b in Y. Thus t is an upper bound to $X \cup Y$ and consequently $t \geqslant V(X \cup Y) = (X \cup Y)\,h$. But by Equation (3.2) $(X \cup Y)\,h$ is an upper bound to x and y and so $(X \cup Y)\,h = Xh \vee Yh$ is the least upper bound in S to Xh and Yh. By orthocomplementation (Equation (3.1)) we deduce that $(X \cap Y)\,h = Xh \wedge Yh$. Thus $P(F)\,h$ is a sublattice of S which is an isomorphic copy of the Boolean algebra $P(F)$. We may note that $P(F)\,h$ is the Boolean subalgebra of S generated by the frame F.

Remark. An instance of this result (namely the finite dimensional case) was given by Ramsay (Ramsay [6, Lemma 2]) but the statement of his lemma fails to recognize that the map h is an isomorphism and not just a homomorphism. We are now in the position to prove

THEOREM 3.1. *Let S be a completely orthomodular poset. Then S is the logic associated with a logical structure. If S is separable then the logical structure is a σ-structure.*

Proof. Let $\{F_\gamma : \gamma \in \Gamma\}$ be the set of frames of the completely orthomodular poset S. Let $S_\gamma = P(F_\gamma)\,h$, since each x in S with $0 < x < 1$ belongs to at least one frame we have

$$S = \bigcup \{S_\gamma : \gamma \in \Gamma\}.$$

As before let ω_γ, N_γ and J_γ denote the order, orthocomplementation and lattice join respectively in the complete Boolean algebra S_γ. To establish the first part of the proposition we have only to verify that we have the six properties listed in Definition 1.1. Since $\square h = 0$ each of the S_γ has the same least element and so property (i) holds. To establish (ii) we note that if x, y are in S_α then $x\omega_\alpha y$ if and only if $x \leqslant y$ in S. We prove (iii) by observing that if $x\omega_\alpha y$ and $y\omega_\beta z$ then $x \leqslant y \leqslant z$ and so $x \leqslant z$. But then, by orthomodularity (Equation (1.3)), $z = x \vee (z \wedge x^\perp)$ and it follows that any maximal orthogonal set which contains the orthogonal set $\{x, z \wedge x^\perp\}$ leads to a frame F_γ such that S_γ contains x and z, since $x \leqslant z$ we have $x\omega_\gamma z$. Next (iv) is established by the observation $N_\alpha x = x^\perp$. Property (v) is a consequence of the fact that each S_α is a sublattice of S so that if x and y are in S_α so is $x \vee y$ and $J_\alpha\{x, y\} = x \vee y$. Finally to prove (vi) we note that if $x\omega_\beta z$ and $y\omega_\gamma z$ then $x \leqslant z$ and $y \leqslant z$. If further x and y are in S_α and $y\omega_\alpha N_\alpha x$ then $x \vee y$ exists in S and so $x \vee y \leqslant z$. By orthomodularity in S (Equation (1.3))

$$x \vee y \vee \{z \wedge (x \vee y)^\perp\} = z,$$

thus any maximal orthogonal set which contains the orthogonal set $\{x, y, z \wedge (x \vee y)^\perp\}$ leads to a frame F_δ such that S_δ contains x, y and z. This is the desired result and concludes the proof of the fact that S is the logic associated with a logical structure.

When S, in addition to being completely orthomodular, is separable each S_α is separable and one has properties (vii) vnd (viii) of Definition 1.3. For if F is a frame of S it is countable and for any countable subset $X = \{x_k\}_{k=1}^\infty$ of $P(F)\,h$ we have

$$\bigvee_{k=1}^{\infty} x_k = \left\{ \bigcup_{k=1}^{\infty} X_k \right\} h,$$

where $X_k = x_k h^{-1} \subseteq F$. Thus the set X has a lub in S. It follows that if $X \subseteq S_\alpha$ is at most countable then $J_\alpha X = VX$ exists in S and so we have (vii), in fact since each S_α is complete this result holds for an arbitrary subset X of S_α. To establish (viii) we need only observe that if X is a countable orthogonal subset of S and z is an upper bound in S to X then $z \geqslant x = VX$. Thus $X \cup \{z \wedge x^\perp\}$ is an orthogonal subset of S and any maximal orthogonal set which contains it leads to a frame F_γ such that S_γ contains X, $z \wedge x^\perp$ and hence also $z = x \vee (z \wedge x^\perp)$. This concludes the proof of the theorem.

Monash University

BIBLIOGRAPHY

[1] Birkhoff, G., *Lattice Theory*, 3rd ed., *American Mathematical Society Colloquium Publications*, Vol. 25, Amer. Math. Soc., Providence, R.I., 1967.
[2] Gunson, J., 'On the Algebraic Structure of Quantum Mechanics', *Communications in Mathematical Physics*, **6** (1967), 262–285.
[3] Jauch, J. M., *Foundations of Quantum Mechanics*, Addison-Wesley, Reading, Mass., 1968.
[4] Mackey, G. W., *Mathematical Foundations of Quantum Mechanics*, W. A. Benjamin, New York, 1963.
[5] Piron, C., 'Axiomatique Quantique', *Helvetica Physica Acta* **36** (1964), 439–468.
[6] Ramsay, A., 'A Theorem on Two Commuting Observables', *Journal of Mathematics and Mechanics* **15** (1966), 227–234.
[7] Varadarajan, V. S., 'Probability in Physics and a Theorem on Simultaneous Observability', *Communications in Pure and Applied Mathematics* **15** (1962), 189–217.
[8] Zierler, N., 'Axioms for Non-Relativistic Quantum Mechanics', *Pacific Journal of Mathematics* **2** (1961), 1151–1169.

J. M. JAUCH AND C. PIRON

ON THE STRUCTURE OF QUANTAL PROPOSITION SYSTEMS*

ABSTRACT. It is shown that the axiom of atomicity and the covering law can be justified on the basis of a new and more satisfactory notion of state and the existence of ideal measurements of the first kind. These two axioms are thereby given a satisfactory justification in terms of empirical facts known about micro-systems. Furthermore the new notion of state introduced here does not involve any probability statements and there is therefore no difficulty attributing it to individual systems, which was not possible with the notion heretoforth used in quantum mechanics.

I. INTRODUCTION

One of the central problems in the foundation of quantum mechanics concerns the question to what extent the theory, as we know it today, is determined by the empirical facts that we observe in microsystems. Such a question does not have a precise answer, since it is clear that empirical facts alone do not determine a theory. Indeed the theory can only be constructed from the raw material of the facts by a process of induction which proceeds from a finite number of observations to an axiomatically formulated mathematical structure supplemented by the rules of interpretation. The best that one can hope to do then is to rule out certain of these structures on the basis of empirical evidence. One can never really *verify* a theory, one can only *falsify* it.

The axiomatic construction of the theory has the great advantage in that the theoretical structure and its rules of interpretation are introduced explicitly and the empirical foundation of the theory is thereby much easier to identify. If the theory is essentially determined by the axioms and if such a theory is falsified by a test of one of its consequences then (provided the mathematical deductions are free from error) one or several of the axioms must be modified.

Recent attempts [1] to reconstruct conventional quantum mechanics by such an axiomatic approach have shown, that quantum mechanics in Hilbert space can only be deduced if, in addition to empirically well supported axioms, certain additional axioms are introduced, which

C. A. Hooker (ed.), The Logico-Algebraic Approach to Quantum Mechanics, 427–436.

heretofore have not had a good support with empirical facts. We mean the two (related) axioms of *atomicity* and the *covering law.*

By a more careful analysis of the concept of the *state* of a physical system it has been possible to improve on this aspect and to give a better justification of these two axioms. At the same time it has become possible to clarify the notion of *state* and that of a *physical property.* The latter notion is closely related to that of *element of physical reality* introduced by Einstein, Rosen and Podolsky in the discussion of their paradox which bears their names [2].

It is significant that these three authors came to the conclusion that the notion of state as used in quantum mechanics cannot meaningfully be attributed to an individual system and that it is a statistical concept, applicable only to a suitably chosen assembly of systems. This criticism is justified for the usual definition of state in terms of a state vector (or Schrödinger function) in Hilbert space. However we shall show in this paper that a modified definition of state can be meaningfully applied to an individual system which represents all the properties (or elements of reality) provided the propositions of that system are an atomic lattice.

It is perhaps interesting to point out that this new notion of state, although fully quantum mechanical in its connotations, resembles the classical notion of that concept. In both areas of physics, classical and quantal, the state can only be determined by a statistical procedure, as in all physical measurements. Nevertheless in both areas it is possible and useful to define a notion of state which would correspond to an idealized set of measurements of infinite precision.

This possibility counters effectively some of the criticisms which have been formulated by several physicists and philosophers in various ways concerning the conventional notion of state and its implication for that of 'physical reality'.

Equally important for the reconstruction of conventional quantum mechanics is the axiom which we have called the *covering law.* It is shown that this axiom is closely related to the possibility of an *ideal measurement*, where 'ideal' will have to be properly defined. There is no doubt that such measurements are possible in many situations, or more precisely, that such measurements can be often simulated by actual (and therefore not ideal) measurements to any desired degree of accuracy. The covering law, which formerly had to be postulated ad hoc, obtains thereby a high degree of plausibility.

II. YES-NO EXPERIMENTS

The properties of a physical system are determined by measurements. A certain class of measurements play a particularly important rôle in the establishment of the physical properties of a system. It is the experiment with only two possible results which may be denoted by 1 or 0 (yes or no). We denote such experiments by Greek letters α, β, $\gamma, \ldots$ and shall refer to them as yes-no experiments.

If α is a yes-no experiment then there exists another one, denoted by $\alpha^{\vee}$, obtained from α by inverting the results yes and no. Thus if the result of α is 'yes' that of $\alpha^{\vee}$ is 'no' and vice versa. It is clear that $\alpha^{\vee}$ can be measured with the same physical equipment as that used for the measurement of α and that $(\alpha^{\vee})^{\vee}=\alpha$.

If $\alpha_i (i \in I$, same index set) is any family of yes-no experiments then one can define another such experiment, denoted by $\prod \alpha_i$ by the following procedure: One chooses at random one of the $\alpha_i (i \in I)$ and measures it. The result is the value of $\prod \alpha_i$. It follows that

$$(\prod \alpha_i)^{\vee} = \prod \alpha_i^{\vee}.$$

There exists a trivial and an absurd yes-no experiment denoted by I and ϕ, respectively. The first consists of the 'experiment' which verifies that the system exists and the second is $\phi = I^{\vee}$.

III. PROPERTIES OF A SYSTEM

We shall say that the yes-no experiment α is 'true' if a measurement of α will give the result yes with certainty. For the time being we are not concerned with the question how we can produce systems for which a given yes-no experiment is known to be 'true' nor how we obtain this knowledge.

It is an empirical fact that certain pairs α, β of yes-no experiments have the property

$$\alpha \text{ true} \Rightarrow \beta \text{ true}.$$

If this is the case, we write $\alpha < \beta$. This relation is a partial preorder relation, that is it satisfies the properties

$$(1)\ \alpha < \alpha, \qquad (2)\ \alpha < \beta \quad \text{and} \quad \beta < \gamma \Rightarrow \alpha < \gamma.$$

If two yes-no experiments α_1 and α_2 satisfy the relation $\alpha_1 < \alpha_2$ and $\alpha_2 < \alpha_1$ we shall call them *equivalent* and we denote it by $\alpha_1 \sim \alpha_2$. This relation is an equivalence relation, that is, it satisfies

$$(1)\ \alpha \sim \alpha, \qquad (2)\ \alpha \sim \beta \Rightarrow \beta \sim \alpha,$$
$$(3)\ \alpha \sim \beta \ \text{ and } \ \beta \sim \gamma \Rightarrow \alpha \sim \gamma.$$

Let α be any yes-no experiment. We denote by $a \equiv \{\alpha\}$ the class of all such experiments which are equivalent to α and we call it a *proposition*. Thus

$$a = \{\alpha_i \mid \alpha_i \sim \alpha\} = \{\alpha\}.$$

If α is true, then every $\alpha_i \sim \alpha$ is true too. Hence we see the proposition a is true if and only if any (and therefore all) of the $\alpha \in a$ are true. If the proposition a is true we shall call it a *property* of the system. We write $a \subset b$ if $\alpha \in a$, $\beta \in b$ and $\alpha < \beta$.

If α_i is a family of yes-no experiments all of which are true then $\prod \alpha_i$ is true too. We denote by $\bigcap_i a_i$ the equivalence class $\{\prod \alpha_i\}$ which contains the yes-no experiment $\prod \alpha_i$. It follows from the definition that $\bigcap_i a_i$ depends only on the equivalence classes $\{\alpha_i\}$ and not on the representatives of these classes. Hence the notation is justified. Thus if a_i are properties of a system then $\bigcap a_i$ is a property too.

If $b \subset \cap a_i$ then it follows from the definition that this is equivalent to $b \subset a_i$:

$$b \subset \cap a_i \Leftrightarrow b \subset a_i \ \forall i \in I.$$

Thus $\bigcap a_i$ is the *greatest lower bound* of the propositions a_i.

Similarly we can define the *least upper bound* by setting $\bigcup a_i \equiv \bigcap_{a_i \subset x} x$ and verify that it satisfies

$$\bigcup a_i \subset b \Leftrightarrow a_i \subset b \ \forall i \in I.$$

If L denotes the set of all propositions we have $\phi = \bigcap_{x \in L} x$ and $I = \bigcup_{x \in L} x$.

We have thus proved the

THEOREM. *The set of all propositions is a complete lattice.*

IV. THE COMPLEMENTS

Two propositions a and b are said to be *complements* of one another if they satisfy

$$a \cap b = \phi \quad \text{and} \quad a \cup b = I.$$

For a lattice which satisfies the distributive laws

$$a \cap (b \cup c) = (a \cap b) \cup (a \cap c),$$
$$a \cup (b \cap c) = (a \cup b) \cap (a \cup c).$$

the complement, if it exists, is unique.

The lattices which are encountered in quantal systems do not satisfy the distributive law and there exist usually many complements. Among these different complements we can still distinguish one, called the *compatible complement*, by the following:

DEFINITION. *The complement b is said to be a* compatible complement *of a if there exists a yes-no experiment $\alpha \in a$ such that $\alpha^{\vee} \in b$. We denote a compatible complement by a'.*

All the known physical systems have the property that every proposition has a compatible complement. We therefore formulate the

AXIOM C. *For every proposition $a \in L$ there exists at least one compatible complement a'.*

The lattices which satisfy the axiom C are still too general for quantal systems. The essential physically motivated axiom [1] which limits this generality is

AXIOM P. *If $a \subset b$ then the sublattice generated by (a, b, a', b') is Boolean.*

It follows from this axiom that $a \subset b \Leftrightarrow b' \subset a'$ so that the mapping $a \mapsto a'$ is an orthocomplementation. Furthermore the lattice L is weakly modular, that is we have

$$a \subset b \Rightarrow a \cup (a' \cap b) = b.$$

It is now possible to introduce the fundamental notion of *compatibility* by the following

DEFINITION. *Two propositions $a, b \in L$ are said to be* compatible $(a \leftrightarrow b)$ *if the sublattice generated by (a, b, a', b') is Boolean.*

In classical systems any pair of propositions is compatible. The greater richness of quantal systems appears through the presence of proposition pairs which are not compatible.

V. THE STATES

The classical notion of state is so familiar that it has influenced much of our thinking about quantal systems. A classical system is described by a number of real variables which define the *phase-space* of the system and a state is determined by a *point* in this space.

The propositions of a classical system can be identified with the subsets of the phase space with inclusion as the ordering relation. For any given state (identified with a point P in phase space) there exists then a class of propositions which are true in the sense defined before. They are in fact all subsets which contain the point P.

For quantal systems the phase space does not exist, but the property of states expressed in the last paragraph still persists and can be used as the defining property of states.

Guided by this analogy we are led to the

DEFINITION. *A state of a system is the set S of all true propositions of the system:*

$$S=\{x \mid x\in L,\ x \text{ true}\}.$$

The following remarks should clarify the meaning of this definition.

The definition is meant to imply that the state is a property of an individual system and not of a statistical ensemble of such systems. This was not possible in previous definitions of the state which involved probabilities (or probability amplitudes). Indeed, a probability is meaningful only with reference to a statistical ensemble. The definition we have given above refers only to true propositions, that is to what we have called properties of the system, and there is no objection in attributing these properties to an individual system.

We shall in fact assume that every individual system, be it an isolated system or a member of a statistical ensemble, is in a definite state as defined above.

It is important to distinguish the state of a system from the amount of information available about the system. This distinction is already important for classical systems and it appears again here for quantal systems. We attribute to every system a state in the sense defined above quite independently whether this state has been measured. We may think of the state as containing the maximal amount of information that is possible concerning an individual system. Thus we shall postulate that two states S_1 and S_2 cannot be subsets of one another.

The states defined here correspond to the so-called 'pure' states of quantum mechanics. In the view that we adopt here every individual system is in a pure state. Mixtures are only properties of statistical ensembles.

The following properties are elementary consequences of the definitions given earlier

(1) If $x \in S$ and $x \subset y$ then $y \in S$.
(2) If $x, y \in S$ then $x \cap y \in S$.
(2′) If $x_i \in S (i \in I)$ then $\bigcap_{i \in I} x_i \in S$.
(3) $\phi \notin S$, $I \in S$ for every state S.
(4) For any $x \in L$, $x \neq \phi$ there exists at least one state S such that $x \in S$.

The meaning of the last property is that a proposition x is different from ϕ if there exists at least one procedure which gives to the system the property x.

From the above it follows that for every state S, $e \equiv \bigcap_{x \in S} x$ is also contained in S and that it is an atom. Indeed if $y \subset e$ and $y \neq \phi$ then there exists a state S_0 such that $y \in S_0$. It follows then that $S \subset S_0$ so that S could not be a state. This contradicts the hypothesis. We have thus proved the

THEOREM. *For every state S, $e = \bigcap_{x \in S} x$ is an atom and $e \in S$.*

From this we obtain the

COROLLARY. *Every $a \neq \phi$ contains at least one atom e. In order to verify this it suffices to consider a state S such that $a \in S$. The proposition $e = \bigcap_{x \in S} x$ is then an atom and $e \subset a$.*

A lattice with this property is said to be atomic. Thus we have motivated the

AXIOM A_1. *The lattice of propositions is atomic.*

The preceding considerations show that every state may be represented by an atom e. The set of all the atoms is identical with the set of all the states. The state S associated with the atom e is the set

$$S=\{x \mid e\subset x\}.$$

In the analogy to the classical systems and the phase space, the atoms of L may be considered as the 'phase space' of the quantal system.

It is seen that the analogy to the phase space suggested here brings this new definition of states of quantal system much closer to the classical notion of states. In fact one of the essentially non-classical aspects of the states of quantal systems appears now if we consider the evolution of states in time. Classically the evolution of states is given by a transformation of phase space which maps every point of that space into another one.

This type of evolution may also occur in quantum mechanics and it is that evolution which is described in the Hilbert space formalism by a Schrödinger equation. We shall call it Schrödinger-type evolution of states. In the lattice-theoretic formulation a Schrödinger-type evolution is generated by a continuous automorphism of the lattice.

However in quantum mechanics one encounters other types of evolutions which play an equally important rôle. They are in fact at the root of most of the paradoxes in quantum mechanics. A state of a quantal system may also evolve according to a stochastic process. As we know from the examples studied in connection with the measuring process this always occurs if such a system is part of another quantal system with which it interacts. The unavoidable occurrence of probabilities in quantum mechanics is entirely due to this stochastic evolution of systems in interaction.

VI. IDEAL MEASUREMENTS

A measurement a is said to be *ideal* if every true proposition compatible with a is also true after the measurement.

A measurement of a is called of the *first kind* if the answer yes implies a true immediately after the measurement.

We shall suppose that for every proposition a there shall exist ideal measurements of the first kind.

Consider now a system in the state S defined by the atom e and let a be any proposition. We consider an ideal measurement of the first kind of a and ask what the state is going to be after such a measurement.

We consider the proposition $y=e\cup a'$. Since $e\subset y$ we have $y\in S$. Furthermore $y\leftrightarrow a$. Since $y\leftrightarrow a'$ (Axiom P) it follows $y\leftrightarrow a$ by the definition of compatibility.

The set S_a of propositions which are true immediately after such ideal measurement, in the case of answer yes, is therefore the set that is implied both by $y=e\cup a'$ and by a. Thus we are led to the conclusion that

$$S_a=\{x\mid (e\cup a')\cap a\subset x\}.$$

Now two possibilities may a priori arise. The first one is $(e\cup a')\cap a=\phi$. But this contradicts our hypotheses because this implies $e\subset a'$, which means that the proposition a' is true and then that the answer yes, as result of the measurement a, is impossible. The second possibility is $(e\cup a')\cap a\neq\phi$. The state after the measurement of a is then this set S_a which is maximal if and only if

$$e_a=(e\cup a')\cap a$$

is an atom. Thus we are led to the following conclusion:

For every $a\in L$ and every atom $e\in L$ the proposition $(e\cup a')\cap a$ is either ϕ or an atom.

It is now easy to show that this result is equivalent with the covering law. Consider an element $b\in L$ and an atom $e\in L$. Let x be such that

$$b\subset x\subset e\cup b.$$

It follows from this that

$$\phi\subset x\cap b'\subset(e\cup b)\cap b'.$$

Since x is compatible with b, as well as b', $\phi=x\cap b'$ implies $x\subset b$ thus $x=b$. Hence if $x\neq b$ then $x\cap b'\neq\phi$. Since $(e\cup b)\cap b'$ is an atom we have then $x\cap b'=(e\cup b)\cap b'$ from which follows $e\cup b=((e\cup b)\cap b')\cup b=(x\cap b')\cup b=x$.

Thus we have established

AXIOM A_2 (covering law):
For every proposition $b\in L$ and any atom $e\in L$, $b\subset x\subset e\cup b$ implies either $x=b$ or $x=e\cup b$.

Institut de Physique Théorique, University of Geneva, (Switzerland)

NOTE

* This is a revised version of a widely distributed preprint by C. Piron, *Sur l'interprétation des treillis orthocomplémentés.*

BIBLIOGRAPHY

[1] Jauch, J. M., *Foundations of Quantum Mechanics*, (Addison-Wesley, 1968), chap. 5, where further references are listed.
[2] Einstein, A., Podolsky, P., and Rosen, N., *Phys. Rev.* **47** (1935), 777.

SAMUEL S. HOLLAND, JR.

THE CURRENT INTEREST IN ORTHOMODULAR LATTICES*

I. INTRODUCTION

All published work on orthomodular lattice theory has appeared within the last fifteen years; no more than thirty people have ever worked on it; no more than fifty papers dealing explicitly with it have ever appeared: Orthomodular lattice theory is therefore a newly uncovered very small corner of mathematics.

Because of this fact, writing a survey article about the topic is both easy and difficult. It is easy because the field is still small enough to be surveyed completely. We can still read all the published work on the subject and can still keep within reach of understanding all its major results. It is easy also because of the extra motivation supplied by the excitement of fresh discoveries in this new field. It is difficult because we have very few significant results to describe. There being only a few major theorems, the greater part of our exposition must be concerned with conjectured or historical aspects of the theoretical development and these are more difficult to present in a systematic way.

The historical aspect is especially important and cannot be slighted. Orthomodular lattice theory grew out of the theory of von Neumann algebras; it can therefore trace its origin to the appearance of the "Rings of Operators" papers of F. J. Murray and J. von Neumann. There were four "Rings of Operators" papers, the first one appearing in the 1936 issue of the *Annals of Mathematics* [40]. To appreciate the reasons for studying orthomodular lattices, the significance of the kinds of questions asked, and the importance of the theorems proved one must know how the theory of orthomodular lattices evolved out of this work of Murray and von Neumann, the beginnings of which date back to 1936.

It is really too bad to have to go into the theory of von Neumann algebras, because it complicates what might otherwise be a very clean exposition. By themselves, the axioms defining an abstract orthomodular lattice are simple: an orthomodular lattice L is

C. A. Hooker (ed.), The Logico-Algebraic Approach to Quantum Mechanics, 437–496.

(1) *a lattice with* 0 *and* 1 (that is, a partially ordered set with least element 0 and greatest element 1 in which every pair a, b of elements has a least upper bound, denoted $a \vee b$, and a greatest lower bound, symbolized $a \wedge b$);

(2) *with an orthocomplementation* (a mapping $a \to a^{\perp}$ of L onto itself satisfying

(i) $a \vee a^{\perp} = 1$, $a \wedge a^{\perp} = 0$,

(ii) $a \leqslant b \Rightarrow a^{\perp} \geqslant b^{\perp}$, and

(iii) $a^{\perp\perp} = a$)

(3) *satisfying the orthomodular identity*, $a \leqslant b \Rightarrow b = a \vee (b \wedge a^{\perp})$.

Simple enough. More importantly, there is a common example that everyone has been familiar with since high school days – the lattice of all subspaces of three dimensional real Euclidean space. In this example L consists of these elements: the origin "0" $= (0, 0, 0)$, all lines through the origin (we mean "lines" that extend to infinity in both directions), all planes through the origin, and the whole of 3-space itself. The partial order relation $\leqslant$ is simply set-theoretic inclusion. The least upper bound of two lines is (if they are different) the plane they span; the least upper bound of a plane and a line not in that plane is the whole of 3-space (which is the "1" of L), the greatest lower bound of a plane and a line not in that plane is the origin, 0, and so on. The orthocomplement of a line a is the plane $a^{\perp}$ (through the origin) orthogonal to that line, and the orthocomplement of a plane is the line through the origin orthogonal to that plane. The reader can easily check mentally that this mapping $a \to a^{\perp}$ (incidentally, $0^{\perp} = 1$) satisfies the three conditions listed above. The only non-trivial case of the orthomodular identity: $a \leqslant b \Rightarrow b = a \vee (b \wedge a^{\perp})$ is this: b is a plane, a is a line, and a lies in b ($a \leqslant b$). Then $a^{\perp}$ is a plane, orthogonal to the line a and therefore orthogonal also to the plane b; the intersection of the plane b and the plane $a^{\perp}$ is a line, $b \wedge a^{\perp}$, lying in b, orthogonal to a. The equation $b = a \vee (b \wedge a^{\perp})$ has then this interpretation: The lines a and $b \wedge a^{\perp}$ both lie in the plane b, are orthogonal, and therefore span b. The orthomodular identity is therefore certainly true in this example.

So we have a compact set of axioms and a familiar example, and we could therefore continue by listing some more definitions, proving theorems, and developing, in nice logical fashion, the theory of the abstract orthomodular lattice as far as we know it. As appealing as this approach is, it would be quite misleading. For it would obscure the source of the theory's questions that have shaped its development, and it would rob the reader of the opportunity to place the theory in its proper mathematical and historical perspective. While orthomodular lattice theory is properly a branch of algebra, it has its roots in functional analysis and serves as another connecting link between algebra and analysis; this perspective is perhaps as important as the theory itself and must be properly stressed.

So we start the exposition with Section II: *The origin of orthomodular lattice theory*, including in this section a self-contained from-scratch exposition of the basic features of the theory of von Neumann algebras. Also included in this section is an exposition of the theory of continuous geometries, this theory being a precursor of orthomodular lattice theory, and a description of the 1955 papers of L. H. Loomis [29] and S. Maeda [38] which mark the beginning of orthomodular lattice theory proper. We try to describe the historical sequence, von Neumann algebras – continuous geometries – orthomodular lattices, in such a way that the reader can appreciate the motivations behind each successive piece of work and can trace the continuous thread of a common mathematical idea. The theory of continuous geometry (which was also invented by von Neumann) is perhaps the most important item in the sequence, because this theory showed the basic lattice-theoretic character of the theory of von Neumann algebras. It probably was the main inspiration behind the 1955 papers of Loomis and S. Maeda. One can think of the interrelations among these theories in this way: von Neumann algebra theory is the Mother Theory, continuous geometry is the firstborn son, and orthomodular lattice theory the second son. Our Section II studies this family tree.

Section III contains the basic elementary theory of the orthomodular lattice with emphasis on the day-to-day calculational and computational principles required in working with these lattices, and a description of some recently discovered strange examples of orthomodular lattices.

In Section IV, I go out on a limb and give my selection of the seven major results of the theory (so far). This is bound to provoke controversy and get me into trouble, but will also hopefully generate some interest, so it is worth the risk.

Section V contains a few brief remarks on a subject probably deserving a book of its own: the logic of quantum mechanics.

In summary, what I try to do in this paper is this: to fit orthomodular lattice theory into historical perspective to show how and why it was created and how it fits into mathematics; and, by implication, to give some idea of the major outstanding problems and the directions in which the theory is heading.

The presentation is as elementary as I could make it. It has been my observation that survey articles don't always successfully reach a group much larger than the inner circle of experts who don't need to read them anyway. I have tried to avoid this problem by keeping the technical details to a minimum and the explanations elementary and fully detailed, even though I felt at times a little foolish doing so. This paper is primarily written for an audience composed of mathematicians who know little or no lattice theory (and are interested in learning some) and graduate students who know some linear algebra and Hilbert space theory but little or no lattice theory. For the first group, I hope the paper will serve as a useful source of information on an unfamiliar area of mathematics; for the second group I hope that it may serve to excite the interest of some of these young researchers.

I should like to acknowledge here with thanks the money granted to me by the National Science Foundation under research grants GP-464 and GP-4242 which provided partial support for my researches and for the writing of this report.

II. THE ORIGIN OF ORTHOMODULAR LATTICE THEORY

In this part of the exposition we shall try to trace the thread of orthomodular lattice theory's historical development from its origin: in a sense to mark out the theory's family tree. We begin by describing the mother theory, von Neumann algebras. Next comes the firstborn son, Continuous Geometry, and finally we take up the (separate and independent) papers of Loomis and S. Maeda that mark the birth of orthomodular lattice theory itself (the "second son").

This is a bulky section. Including von Neumann Algebras and Continuous Geometry in our exposition has expanded it considerably, neither subject being easily explained. Indeed, books have been written about both topics [9, 36, 44]. To describe them fully we have had to give a rather lengthy discourse especially since we were writing for readers with a minimal mathematical background. The expert should accordingly do some judicious skipping. On the other hand, enough details have been included so that the non-expert should be able to get along quite well with just some elementary Hilbert space theory, for example the material in Chapter ten of Simmon's book [48].

The symbol H will always stand for a complex Hilbert space, and $\mathscr{B}(H)$ will denote the algebra of all bounded, everywhere defined, linear operators mapping H into itself. We first take up the study of this very important algebra and certain special operators on it.

A. *The projection lattice of $\mathscr{B}(H)$ and the orthomodular identity*

A *projection* is an operator E in $\mathscr{B}(H)$ that is self-adjoint and idempotent: $E = E^* = E^2$. There is a one-to-one correspondence between the set of such operators and all closed subspaces of H. Going from projections to subspaces, we get the closed subspace M corresponding to a projection E as simply E's image, $M = E(H)$. (Equivalently, M is the set of those $x \in H$ satisfying $Ex = x$.) In the other direction, given a closed subspace M, we can write an arbitrary $z \in H$ uniquely as a sum, $z = x + y$, where $x \in M$ and y is orthogonal to M; the projection corresponding to M is then that operator E determined by $Ez = x$. Each of these transformations between projections and closed subspaces is inverse to the other, and together they constitute the one-to-one correspondence between projections and subspaces cited above. If E and M correspond as described, then we say that E is the projection on M.

The closed subspaces of H form a lattice under set-theoretic inclusion, the greatest lower bound of two closed subspaces M and N being simply their set-theoretic intersection, and the least upper bound the closure of their sum. We shall use $M \wedge N$ for the greatest lower bound and $M \vee N$ for the least upper bound. One might guess that the sum-subspace $M + N$, the set of all $m + n$, $m \in M$, $n \in N$, would be always closed and thus serve as the least upper bound. This is not true – a highly significant fact. If H is infinite dimensional, then we can always find closed subspaces M, N whose sum $M + N$ is not closed; for a proof see Stone [49, p. 21] or Halmos [20, p. 28]. This fact is connected with the non-modularity of the lattice of all closed subspaces of infinite dimensional H; we shall be coming to this later. We stress: the least upper bound of two closed subspaces is given by: $M \vee N = \text{closure}\,(M + N)$, and, if H is infinite dimensional, then there are subspaces such that $M \vee N \neq M + N$. (Of course if H is finite dimensional, then all subspaces are closed and always $M \vee N = M + N$.)

Owing to the one-to-one correspondence between projections and closed subspaces, this lattice structure can be transplanted to the projections of $\mathscr{B}(H)$. We shall use the same symbols for the lattice operations: If E is the projection on the closed subspace M, and F the projection on N, then $E \vee F$ will denote the projection on $M \vee N$ and $E \wedge F$ that on $M \wedge N$. We shall write $E \leqslant F$ to mean that M is contained in N.

There are some interesting connections between these lattice operations and those algebraic operations that are available to the projections as elements of the algebra $\mathscr{B}(H)$. Indeed, the ordering itself can be characterized purely algebraically: $E \leqslant F \Leftrightarrow E = EF$. Also if $EF = FE$, then both the glb and lub have simple algebraic expressions: $E \wedge F = EF$ and $E \vee F = E + F - EF$. However, this is about as far as it goes; nothing as simple is available if E and F do not commute. Keep in mind that the lattice operations are operations in the set of projections and never lead out of this set; however, the algebraic operations are not, neither the sum nor the product of projections being in general a projection.

There is a largest projection, the identity operator I, and a smallest projection, the zero operator 0. If M is a closed subspace, then $M^{\perp}$, the set of vectors in H orthogonal to M, is also a closed subspace. It is routine to verify that the map $M \to M^{\perp}$ is a one-to-one map of the set of closed subspace onto itself satisfying the three conditions for an orthocomplementation:

(1) $M \vee M^{\perp} = I$, $M \wedge M^{\perp} = 0$;
(2) $M \leqslant N \Rightarrow M^{\perp} \geqslant N^{\perp}$; and
(3) $M = M^{\perp\perp}$.

If E is the projection on M, we use $E^{\perp}$ for the projection on $M^{\perp}$, and call $E^{\perp}$ the *orthocomplement* of E. The orthocomplement has a nice algebraic expression: $E^{\perp} = I - E$. If $E \leqslant F$ we say that E and F are *orthogonal* and write $E \perp F$. Since $E \leqslant F^{\perp}$ is characterized algebraically by $E = EF^{\perp} = E(I - F) = E - EF$, (this is the algebraic characterization of order we gave above) $E \perp F$ is characterized by $EF = 0$ (simply subtract E from both sides). Some other simple facts: If $E \perp F$ then $E \vee F = E + F$ (because $E \perp F \Rightarrow EF = FE = 0 \Rightarrow E \vee F = E + F - EF = E + F$); if $E \leqslant F$, then $F \wedge E^{\perp} = F - E$. Using these last two simple facts we are able to deduce easily the validity in the lattice of projections of $\mathscr{B}(H)$ the *orthomodular identity*:

$$E \leqslant F \Rightarrow F = E \vee (F \wedge E^{\perp}).$$

Proof. The projections E and $F \wedge E^{\perp}$ are orthogonal and therefore their lattice span is simply their sum, $E \vee (F \wedge E^{\perp}) = E + (F \wedge E^{\perp})$. Since $E \leqslant F$, the projection $F \wedge E^{\perp}$ is just $F - E$; putting this together with the equation just derived we get

$$E \vee (F \vee E^{\perp}) = E + (F - E) = F.$$

End of proof.

This algebraic proof can be generalized to arbitrary $*$-rings. A $*$-ring is a ring R together with a mapping $a \rightarrow a^*$ of R onto itself satisfying $(a+b)^* = a^* + b^*$, $(ab)^* = b^* a^*$, and $a^{**} = a$. For example $\mathscr{B}(H)$ is a $*$-ring, the mapping $a \rightarrow a^*$ of $\mathscr{B}(H)$ being the ordinary operator-adjoint operation. Define the element e of the $*$-ring R to be a *projection* if $e^2 = e^* = e$: a projection is therefore a ring element that is idempotent and self-adjoint. In $\mathscr{B}(H)$ these abstract projections are the usual ones, so that the new terminology is consistent with the old. Easy ring-theoretic computations show that the definition: $e \leqslant f \Leftrightarrow e = ef$ yields a partial ordering on the set of projections of an arbitrary $*$-ring R, and, if this ring has an identity 1, then setting $e^{\perp} = 1 - e$ we define an orthocomplementation on the set of projections. (This is a little tricky since these projections may not constitute a lattice. The assertion is this: e and $e^{\perp}$ always have a lub and a glb and $e \vee e^{\perp} = 1$, $e \wedge e^{\perp} = 0$. The important point is this: *If the projections in a $*$-ring with identity form a lattice at all, then this lattice is orthomodular, that is, fulfills the identity*

$$e \leqslant f \Rightarrow f = e \vee (f \wedge e^{\perp}).$$

The proof is virtually the same as it was for the ring $\mathscr{B}(H)$. Hence the orthomodular identity can be given a purely algebraic setting.

On the other hand if we use subspaces instead of projections, then we can interpret the orthomodular identity as a basic fact about the geometry of Hilbert space; namely, that if the closed subspace M is contained in the closed subspace N, then N is the orthogonal direct sum of M and $N - M$ (the orthocomplement of M in N). In symbols: $M \leqslant N \Rightarrow N = M \oplus (N - M)$. See Stone [49, p. 21], Halmos [20, p. 28] or Simmons [48, p. 251].

Araki and Amemiya [2] have proved a converse to this result. Suppose E is an inner product space (sometimes "pre-Hilbert" space). The subspaces M of E that satisfy $M = M^{\perp\perp}$ constitute an orthocomplemented lattice. Araki and Amemiya proved that if this lattice is orthomodular, then E is a Hilbert space. See [37] for a proof.

The simplest example of an orthocomplemented non-orthomodular lattice is the "benzene ring": The map $x \to x^{\perp}$ satisfies the requirements

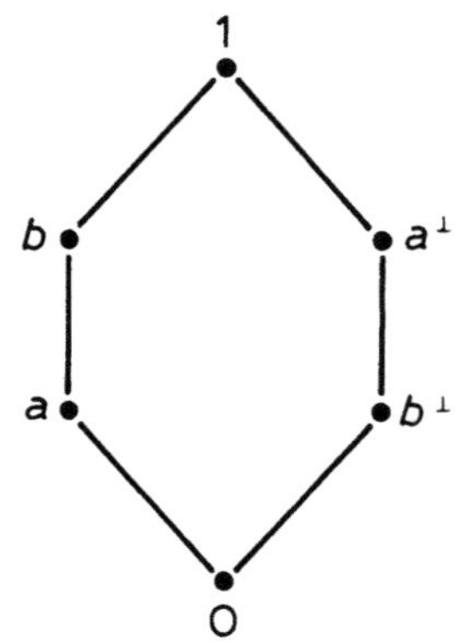

we have set down for it (simply assume that $x^{\perp\perp} = x$ and set $0^{\perp} = 1$), but $a \vee (b \wedge a^{\perp}) = a \vee 0 = a \neq b$ even though $a \leqslant b$. Hence the orthomodular identity does not hold in this particular orthocomplemented lattice. In [39] there is a thorough discussion of postulates equivalent to the orthomodular identity and of weaker postulates that one can impose on an orthocomplemented lattice.

So that is the orthomodular identity, the key axiom. It is a lattice-theoretic version of a geometric property of Hilbert space, or, equivalently, an algebraic property of $\mathscr{B}(H)$. And it can be given a quite general algebraic setting as a statement about the lattice of projections in a $*$-ring with identity.

Now, in a sense, this entire paper is an explanation of this axiom – what body of theory it can support, to what extent it characterizes projection lattices of von Neumann algebras, and so on. Comments concerning it will occur again and again throughout this paper. But I shall nonetheless make a few general remarks here to set the stage. (For the definition of a von Neumann algebra see the following sections.)

First, the orthomodular axiom seems to have had no serious rivals so far. No other axiom has appeared with its combination of virtues: basic, easily verified, and general enough to include all projection lattices of von Neumann algebras. So it is – in this sense – "natural."

Second, the theory has developed as a blend of two easily distinguished points of view. Point of view number one: to treat the abstract orthomodular lattice as an algebraic entity deserving of study on its own merits in the same manner as other kinds of lattices are studied. Point of view number two: to put the emphasis on projection lattices, using the abstract orthomodular lattice as a general organizing vehicle in which to set the developing theory. The history of the subject records a fruitful blend of these two equally valid and equally useful points of view. But the second point of view, centering attention as it does on the projection lattice of von Neumann algebras, tends to induce a more critical attitude toward the orthomodular identity, and keeps its proponents alert to the possibility that some, yet undiscovered, stronger axiom may yet prove more suitable for an abstract theory of projection lattices. We'll leave it to the reader to estimate for himself the chances of such a discovery.

One last point to close out this section: the projection lattice of $\mathscr{B}(H)$ has this additional important property, it is *complete*. Not only do pairs of projections have least upper bounds and greatest lower bounds, but *any* non-empty family of projections (of arbitrary number) has a least upper bound and greatest lower bound. Nothing more is needed to prove this fact than the simple observation that the intersection of any family of closed subspaces is a closed subspace. This intersection gives the greatest lower bound of the family of closed subspaces (and, using the one-to-one correspondence between closed subspaces and projections, the greatest lower bound of the corresponding family of projections). The least upper bound of an arbitrary family of projections is the greatest lower bound of the set of projections having this property: each majorizes every projection in the original family. We can give a different and often more convenient construction for this least upper bound. If $\mathscr{M}$ is a set of closed subspaces then the set of all finite sums $m_1+m_2+\cdots+m_k$, $m_i \in M_i \in \mathscr{M}$, constitute a subspace of H. The closure of this subspace is the desired least upper bound. We shall write $\bigvee(E;\ E\in\mathscr{E})$ for the least upper bound of the set $\mathscr{E}$ of projections, and $\bigwedge(E;\ E\in\mathscr{E})$ for the greatest lower bound.

We close this section with a summary.

There is a one-to-one correspondence between the closed subspaces of the complex Hilbert space H and the projection operators in the algebra $\mathscr{B}(H)$. This correspondence transfers the lattice structure of the closed subspaces to the projections, the set of projections constituting then a complete orthomodular lattice.

We shall prove later on that this lattice is modular if and only if H is finite dimensional.

B. *Von Neumann Algebras*

A *von Neumann algebra* $\mathscr{A}$ is a non-empty subset of $\mathscr{B}(H)$ that satisfies the following four requirements:

(1) $\mathscr{A}$ is a subalgebra of $\mathscr{B}(H)$; that is, if S, $T \in \mathscr{A}$, and α is any complex number, then $\alpha S \in \mathscr{A}$, $S + T \in \mathscr{A}$, and $ST \in \mathscr{A}$.

(2) $\mathscr{A}$ is $*$-closed: $T \in \mathscr{A} \Rightarrow T^* \in \mathscr{A}$.

(3) $\mathscr{A}$ contains the identity operator I.

(4) $\mathscr{A}$ is its own double centralizer: $\mathscr{A} = \mathscr{A}''$. (Where $\mathscr{A}'$ symbolizes the set of all operators in $\mathscr{B}(H)$ that commute with each operator in $\mathscr{A}$; symbolically, $\mathscr{A}' = (T \in \mathscr{B}(H);\ TS = ST$ for every $S \in \mathscr{A}$).)

Properties (1), (2), and (3) describe simply a $*$-subalgebra of $\mathscr{B}(H)$ containing I. Item (4), the "double centralizer property," is the crucial property distinguishing von Neumann algebras among the $*$-subalgebras of $\mathscr{B}(H)$.

Probably the most important single property possessed by a von Neumann algebra is this: it is generated by its projection operators. (We shall discuss this fact in detail shortly making precise the term "generated.") This qualitative feature is a distinguishing characteristic of von Neumann algebras. One finds it described in the literature by such phrases as "a lavish supply of projections," "a rich projection lattice," and the like.

Even more is true, as the last quote indicates: not only are there "enough" projections in a von Neumann algebra, but this set of projections is closed under the lattice operations (that we discussed in the previous section). The projections in a von Neumann algebra constitute a complete orthomodular lattice in their own right, a sub-lattice of the lattice of all projections of $\mathscr{B}(H)$. Murray and von Neumann's original analysis of these algebras was based mainly on a detailed study of these projection lattices. Not only was this fact basic for their original study, but it really *is* the point we are trying to make in this section: that the theory of orthomodular lattices originated with the theory of von Neumann algebras. The central importance of this result necessitates our giving a proof.

First some preliminary facts:

Fact 1. *$\mathscr{B}(H)$ itself and (λI), the set of complex multiples of the identity operator I, are both von Neumann algebras.* Clearly in both cases we have a $*$-subalgebra of $\mathscr{B}(H)$ containing I. The only non-obvious point is the double centralizer property. That property is an easy consequence of this fact: $\mathscr{B}(H)'=(\lambda I)$. That is, the only operators commuting with *every* bounded linear transformation on H are the scalar multiples of the identity. We outline the proof of this fact, which is an elementary application of Hilbert space theory and leave it to the reader to fill in the details. If $S\in\mathscr{B}(H)$ commutes with every $T\in\mathscr{B}(H)$, then certainly S commutes with every one-dimensional projection in $\mathscr{B}(H)$. It follows that S maps every one-dimensional subspace into itself [48, p. 275], hence that $Sx=\lambda_x x$ for every non-zero $x\in H$, where λ_x is a complex number possibly depending on x. Last step: show that λ_x does not depend on x. Having $\mathscr{B}(H)'=(\lambda I)$ then, since obviously $(\lambda I)'=\mathscr{B}(H)$, we have $(\mathscr{B}(H))''=((\mathscr{B}(H))')'=(\lambda I)'=\mathscr{B}(H)$ and $(\lambda I)''=((\lambda I)')'=\mathscr{B}(H)'=(\lambda I)$ so that both $\mathscr{B}(H)$ and (λI) have the double centralized property and are thus von Neumann.

Fact 2. *If $\mathscr{M}$ is an arbitrary non-empty subset of $\mathscr{B}(H)$, there is a unique smallest von Neumann algebra containing $\mathscr{M}$, called the algebra generated by $\mathscr{M}$.* There are two ways to get this algebra:

(1) as the intersection of all von Neumann algebras containing $\mathscr{M}$, or

(2) as $(\mathscr{M}\cup\mathscr{M}^*)''$, where $\mathscr{M}^*$ denotes the set of all T^*, $T\in\mathscr{M}$. As to (1): we have already shown that $\mathscr{B}(H)$ itself is a von Neumann algebra so that every $\mathscr{M}$ is contained in at least one von Neumann algebra. The intersection of an arbitrary family of von Neumann algebras is again a von Neumann algebra, as one can easily show by a little manipulating with the centralizer operation [9, p. 2, Prop. 1], and hence the procedure indicated in (1) does in fact give the unique smallest von Neumann algebra containing $\mathscr{M}$. As to (2): observe first that if $\mathscr{N}$ is any non-empty $*$-closed subset of $\mathscr{B}(H)$, then $\mathscr{N}'$ is a von Neumann algebra. A moment's thought will convince you that $\mathscr{N}'$ is a subalgebra. Its $*$-closure follows from that of $\mathscr{N}$, for if $T\in\mathscr{N}'$ then T commutes with every $S\in\mathscr{N}$ and,

since $\mathcal{N}$ is $*$-closed, with every S^*, $S \in \mathcal{N}$. Thus $TS^* = S^*T$ for every $S \in \mathcal{N}$, and, taking adjoints, we get $ST^* = T^*S$ for every $S \in \mathcal{N}$, that is, $T^* \in \mathcal{N}'$. It is an easy formal consequence of the definition of the centralizer operation that $\mathcal{N}' = \mathcal{N}'''$ for any non-empty subset $\mathcal{N}$, and we conclude: $(\mathcal{M} \cup \mathcal{M}^*)''$ is a von Neumann algebra obviously containing $\mathcal{M}$. If $\mathcal{A}$ is another von Neumann algebra containing $\mathcal{M}$, then $\mathcal{M} \cup \mathcal{M}^* \subseteq \mathcal{A}$ (since $\mathcal{A}$ is $*$-closed) and then $(\mathcal{M} \cup \mathcal{M}^*)'' \subseteq \mathcal{A}'' = \mathcal{A}$ since the double centralizer operation preserves inclusion. Thus the von Neumann algebra $(\mathcal{M} \cup \mathcal{M}^*)''$ is contained in any other von Neumann algebra containing $\mathcal{M}$.

The following notation will be convenient. For an arbitrary non-empty subset $\mathcal{M}$ of $\mathcal{B}(H)$, let $\mathcal{P}(\mathcal{M})$ denote the projection operators in $\mathcal{M}$. Thus $T \in \mathcal{P}(\mathcal{M}) \Leftrightarrow T \in \mathcal{M}$ and $T = T^* = T^2$. The set $\mathcal{P}(\mathcal{B}(H))$ is the lattice of all projection operators on H that we discussed in section A.

THEOREM. *If $\mathcal{A}$ is a von Neumann algebra the subset $\mathcal{P}(\mathcal{A})$ generates $\mathcal{A}$. Moreover $\mathcal{P}(\mathcal{A})$ is itself a complete orthomodular lattice (with the previously defined lattice operations of $\mathcal{P}(\mathcal{B}(H))$.*

Proof. $\mathcal{P}(\mathcal{A})$ is never empty containing at least the projections 0, I. As a set of projections, $\mathcal{P}(\mathcal{A})$ is $*$-closed. Hence $\mathcal{P}(\mathcal{A})''$ is the von Neumann algebra generated by $\mathcal{P}(\mathcal{A})$. Since $\mathcal{P}(\mathcal{A}) \subseteq \mathcal{A}$, it follows that $\mathcal{P}(\mathcal{A})' \supseteq \mathcal{A}'$, the centralizer operation being inclusion reversing. We shall prove that the reverse inclusion also holds, $\mathcal{P}(\mathcal{A})' \subseteq \mathcal{A}'$, which will lead to $\mathcal{P}(\mathcal{A})' = \mathcal{A}'$, hence $\mathcal{P}(\mathcal{A})'' = \mathcal{A}'' = \mathcal{A}$, the desired result. So the proof of the first statement of the theorem boils down to verifying this inclusion: $\mathcal{P}(\mathcal{A})' \subseteq \mathcal{A}'$. At this point we must refer to the spectral theorem for bounded self-adjoint transformations on a Hilbert space. The formulation given to this theorem by Halmos [20, pp. 68–69] is particularly convenient for us. Putting together his Theorem 43.1 with his Theorem 41.2 we obtain this: if $T \in \mathcal{B}(H)$ is self-adjoint with spectral family $(E(\lambda), -\infty < \lambda < \infty)$ and if S commutes with T, then each $E(\lambda)$, $-\infty < \lambda < \infty$, commutes with S. Briefly, $(E(\lambda);\ -\infty < \lambda < \infty) \subseteq \{T\}''$. If $\mathcal{A}$ is a von Neumann algebra, and the self-adjoint T belongs to $\mathcal{A}$, then $\{T\}'' \subseteq \mathcal{A}'' = \mathcal{A}$. We conclude that if $\mathcal{A}$ contains a self-adjoint operator T, then $\mathcal{P}(\mathcal{A})$ contains all the spectral projections of T. Hence if $S \in \mathcal{P}(\mathcal{A})'$, then S commutes with every spectral projection of every self-adjoint T in $\mathcal{A}$ and hence with every self-adjoint $T \in \mathcal{A}$. Since an arbitrary $T \in \mathcal{A}$ can be written as a sum $A + iB$ with A and B self-adjoint operators in $\mathcal{A}$.

S commutes with every $T \in \mathscr{A}$. This means $S \in \mathscr{A}'$ and we have concluded our proof of $\mathscr{P}(\mathscr{A})' \subseteq \mathscr{A}'$.

To complete the proof of the theorem we need to show that $\mathscr{P}(\mathscr{A})$ is closed under the lattice operations inherited from $\mathscr{P}(\mathscr{B}(H))$. Knowing $E^{\perp} = I - E$, we conclude immediately that if $E \in \mathscr{A}$, then $E^{\perp} \in \mathscr{A}$, $\mathscr{A}$ being a subalgebra containing I. It remains to show that $\mathscr{A}$ is closed under the taking of lub's and glb's. Actually, one half of this verification is enough because by using the easily established identity $\bigwedge(E; E \in \mathscr{E}) = [\bigvee(E^{\perp}; E \in \mathscr{E})]^{\perp}$ together with the just proved closure under $\perp$ we can deduce glb-closure from lub-closure. Hence we are left with this: if $\mathscr{E} \subseteq \mathscr{P}(\mathscr{A})$ then $\bigvee(E; E \in \mathscr{E}) \in \mathscr{P}(\mathscr{A})$. Let F be the lub in question, $F = \bigvee(E; E \in \mathscr{E})$. We have to show $F \in \mathscr{P}(\mathscr{A})$ or, what is the same thing, since we already know F to be a projection, $F \in \mathscr{A}$. Pick $T \in \mathscr{A}'$. If $E \in \mathscr{E}$, then $TE = ET$. Let us multiply the operator $FT - TF$ on the right by E. Using $E = EF$ (since $E \leqslant F$)

$$(FT - TF)E = FTE - TFE = FET - TE = ET - TE = 0.$$

We conclude: The closed subspace on which E projects is contained in the null space of the operator $FT - TF$. This null space is closed; let K be the projection on it. We have shown $K \geqslant E$ for every $E \in \mathscr{E}$ and since F is the *least* of the upper bounds of $\mathscr{E}$, then $K \geqslant F$ also. Hence $(FT - TF)F = 0$ or $FTF = TF$. But $T \in \mathscr{A}'$ was arbitrary; if we repeat the same argument with T^* and take the adjoint of the resulting equation then we get $FTF = TF = FT$. Thus F commutes with every $T \in \mathscr{A}'$ so that $F \in \mathscr{A}'' = \mathscr{A}$, proving closure. The orthomodular identity is inherited from $\mathscr{P}(B(H))$ and is therefore automatically valid in $\mathscr{A}$. All parts of the theorem are proved.

This theorem is of key importance for us.

First, it provides a rich source of supply of orthomodular lattices by reason of its assertion that each von Neumann algebra contains an orthomodular lattice as its full set of projection operators. As we shall soon see, there are many such algebras and – accordingly – many accompanying examples of lattices. A plentiful and varied supply of lattices arises in this way; some atomic, some non-atomic; some modular, some non-modular; some reducible, some irreducible. (This terminology will be explained shortly.) In fact, the only *obvious* lattice-theoretic property shared by all these examples is orthomodularity. And thus the focus of attention on this particular lattice-theoretic property. And thus indeed a paper with the title, "The Current Interest in Orthomodular Lattices."

Second, by reason of its assertion that the projection lattice generates the algebra, the theorem suggests that the mathematics of the lattice and that of its containing algebra are closely intertwined and that much is to be gained by exploring this connection. This turns out to be the case: indeed right at the outset Murray and von Neumann classified the algebras by means of a careful and detailed analysis of their projection lattices. After that, von Neumann's invention of continuous geometry showed that, for a restricted class of operator algebras, their "dimension theory" was a special instance of a purely lattice dimension theory. So in these early examples lattice theory provided functional-analytic information. Lately, the flow has been reversed, the algebra being used to deduce some purely lattice-theoretic facts about projection lattices. We have here a fascinating interplay between functional analysis and lattice theory.

We shall devote the remainder of this section to the first instance of this interaction – the now famous Murray-von Neumann classification of algebras into types I, II, and III. This is one of the principal results of their classic 1936 paper [40]. This will not only illustrate how lattice theory can have an impact on analysis, but will also provide, as we mentioned earlier, many interesting examples of orthomodular lattices.

The cardinal definition in the Murray-von Neumann analysis is this: If E and F are projections in the von Neumann algebra $\mathscr{A}$, then E and F are said to be *equivalent relative to* $\mathscr{A}$, (or to have the *same dimension relative to* $\mathscr{A}$) if there is an operator W in $\mathscr{A}$ such that $W^*W=E$, $WW^*=F$. This idea is the basis for the entire "dimension theory" of von Neumann algebras. It is an inspiration of genius, notwithstanding the fact that it is easy enough now in retrospect to argue for its "naturalness." The definition has an equivalent geometrical formulation: if E projects on the closed subspace M, and F on N, then there is an operator W in $\mathscr{A}$ having $M^{\perp}$ for null space, N for image; and whose restriction to M is a Hilbert space isomorphism (that is, preserves the inner product) of M on N. Note two features of this definition:

(1) the requirement that W belong to $\mathscr{A}$. The phrase "relative to $\mathscr{A}$" stresses this fact.

(2) the heavy use of operator theory.

Even though it refers to the projection lattice, the definition of equivalence uses other operators from the algebra. Question: Does it have a purely lattice-theoretic formulation? We shall not attempt to touch on this question now – it is a refrain that will occur again and again throughout this paper.

We shall use the notation $E \sim F$ to signify that the projections E and F are equivalent. If it is necessary to distinguish the particular algebra relative to which this equivalence is taken, we shall write $E \sim F(\mathscr{A})$ (for equivalence relative to $\mathscr{A}$) but for us this distinction will rarely be necessary, a fixed von Neumann algebra $\mathscr{A}$ being generally understood.

Murray and von Neumann established then the following facts about this relation:

(1) It is an equivalence relation on $\mathscr{P}(\mathscr{A})$.

(2) It is completely additive. By this we mean: if $(E_\alpha;\ \alpha \in A)$, $(F_\alpha;\ \alpha \in A)$ are two families of projections from $\mathscr{P}(\mathscr{A})$ indexed by the same set A, each family being orthogonal, $E_\alpha \perp E_\beta$ if $\alpha \neq \beta$ and $F_\alpha \perp F_\beta$ if $\alpha \neq \beta$, and the E_α's and F_α's pairwise equivalent, $E_\alpha \sim F_\alpha$ for all $\alpha \in A$, then $\bigvee (E_\alpha;\ \alpha \in A) \sim \bigvee (F_\alpha;\ \alpha \in A)$.

(3) It is completely divisible. This perhaps inappropriate terminology means this: If $E, F \in \mathscr{P}(\mathscr{A})$, $E \sim F$, and if $E = \bigvee (E_\alpha;\ \alpha \in A)$ where the family $(E_\alpha;\ \alpha \in A)$ is orthogonal, then F can also be written as the span of an orthogonal family $F = \bigvee (F_\alpha;\ \alpha \in A)$ where, especially, $F_\alpha \sim E_\alpha$ for all $\alpha \in A$.

(4) It obeys the "parallelogram law," $E \vee F - E \sim F - E \wedge F$ for any E, $F \in \mathscr{P}(\mathscr{A})$.

(5) It fulfills the Cantor-Bernstein property: $E \sim F_1 \leqslant F$, $F \sim E_1 \leqslant E$ where $E, F, E_1, F_1 \in \mathscr{P}(\mathscr{A})$, then $E \sim F$.

They concentrated their attention on the equivalence classes of $\mathscr{P}(\mathscr{A})$ consisting of the pairwise disjoint subsets of $\mathscr{P}(\mathscr{A})$ of mutually equivalent projections, and observed that by (5) the relation

$$(*) \qquad [E] \leqslant [F] \Leftrightarrow [E \sim F_1 \leqslant F \quad \text{for some} \quad F_1 \in \mathscr{P}(\mathscr{A})]$$

(where $[E]$ denotes the equivalence class of projections equivalent to $E \in \mathscr{P}(\mathscr{A})$) effectively defined a partial order on the set of equivalence classes.

An immense technical simplification takes place if we make the following restrictive assumption about $\mathscr{A}$: that its center consist only of the complex multiples of the identity operator. The center of $\mathscr{A}$ is just $\mathscr{A} \cap \mathscr{A}'$, so the assumption is this: $\mathscr{A} \cap \mathscr{A}' = (\lambda I)$. Von Neumann algebras satisfying this additional assumption are called *factors*. This assumption was made by Murray and von Neumann and used to establish the following key result: *If* $\mathscr{A}$ *is a factor, then the set of equivalence classes is totally ordered by* (*). We shall use the notation $[\mathscr{P}(\mathscr{A})]$ to denote this totally ordered set, and shall use the symbol $[E]$ to denote a particular element of $[\mathscr{P}(\mathscr{A})]$, the equivalence class containing the projection E.

Since the ordering in $\mathscr{P}(\mathscr{A})$ is characterized algebraically ($E \leqslant F \Leftrightarrow E = EF$), a $*$-isomorphism between von Neumann algebras will preserve order and therefore the lattice structure. The equivalence relation is also defined algebraically so that a $*$-isomorphism between factors will yield an isomorphism of the totally ordered sets of equivalence classes. We conclude: *the order type of the totally ordered set* $[\mathscr{P}(\mathscr{A})]$ *is an algebraic invariant of the factor* $\mathscr{A}$. Two factors $\mathscr{A}$ and $\mathscr{B}$ cannot be $*$-isomorphic unless their totally ordered sets $[\mathscr{P}(\mathscr{A})]$, $[\mathscr{P}(\mathscr{B})]$ have the same order type. (That is, having the same order type for $[\mathscr{P}(\mathscr{A})]$, $[\mathscr{P}(\mathscr{B})]$ is a necessary condition for the $*$-isomorphism of the factors $\mathscr{A}$ and $\mathscr{B}$. It is not sufficient.) (Interesting recent result: L. T. Gardner has shown that isomorphic von Neumann algebras are necessarily $*$-isomorphic [16].)

Murray and von Neumann made an analysis of this totally ordered set $[\mathscr{P}(\mathscr{A})]$ more detailed than a simple determination of its order type. The major additional new concept involved is that of *finiteness*, a projection $E \in \mathscr{A}$ being finite if it is not equivalent to a proper subprojection of itself. That is, $E \sim F \leqslant E$ for $F \in \mathscr{P}(\mathscr{A})$ implies $F = E$. The key theorem is this: *for every factor* $\mathscr{A}$ *on a separable Hilbert space there is a mapping* $D: \mathscr{P}(\mathscr{A}) \to \{x;\ 0 \leqslant x \leqslant \infty\}$ *that has the following properties* (E, F *representing projections in the factor* $\mathscr{A}$):

(1) $D(E) = 0 \Leftrightarrow E = 0$;
(2) *If* $E \perp F$, $D(E+F) = D(E) + D(F)$;
(3) E *is finite* $\Leftrightarrow D(E) < \infty$;
(4) $E \sim F \Leftrightarrow D(E) = D(F)$.

Moreover, there is the following uniqueness result for such a D: *If* D' *satisfies* (1) *through* (4) *then* $D' = \beta D$ *for some* $\beta > 0$.

This function D, unique up to multiplication by a positive real number, they termed the *dimension function*. Since it is characterized uniquely, up to constant positive factor, by the properties (1)–(4) that involve only $*$-isomorphic invariant concepts, D itself is a $*$-isomorphic invariant.

Hence its range (again neglecting multiplication by a positive real constant) is the same for $*$-isomorphic $\mathscr{A}$'s. It follows from property (4) that D is constant on the equivalence classes of $\mathscr{P}(\mathscr{A})$ and can therefore be considered as a function on the equivalence classes. Moreover it is an easy consequence of (1)–(4) that D is an order isomorphism of $[\mathscr{P}(\mathscr{A})]$ into $\{x;\ 0 \leqslant x \leqslant \infty\}$. Hence we have associated with every factor $\mathscr{A}$ a subset of $\{x;\ 0 \leqslant x \leqslant \infty\}$ of the same order type of $[\mathscr{P}(\mathscr{A})]$ and having this strong invariance property: a necessary condition for the $*$-isomorphism of two factors is that the subset of $\{x;\ 0 \leqslant x \leqslant \infty\}$ associated with one factor be obtainable from that associated with the other by multiplication by a positive number. (This is a stronger condition than saying merely that the $[\mathscr{P}(\mathscr{A})]$'s have the same type.)

This phase of Murray and von Neumann's work culminated in the determination of all possibilities for the ranges of the dimension functions of factors. It turned out that there were only five principal "types." We list these possibilities in Table I along with some other information.

TABLE I

Murray-von Neumann classification of factors on a separable Hilbert space. ($\mathscr{A}$ is the factor, $\mathscr{P}(\mathscr{A})$ its projection lattice.)

The range of $\mathscr{A}$'s dimension function is precisely one of the following sets (neglecting multiplication by an $\alpha > 0$).	Type of $\mathscr{A}$	Example of such an $\mathscr{A}$	$\mathscr{P}(\mathscr{A})$ is complete, orthomodular,
$\{0, 1, 2, \ldots, n\}$	I_n	$\mathscr{B}(H)$, $\dim(H) = n$	modular, atomic, satisfies the chain conditions, but is non-distributive if $n \geqslant 2$.
$\{0, 1, 2, \ldots, \infty\}$	I_∞	$\mathscr{B}(H)$, $\dim(H) = \aleph_0$	non-modular, and atomic
$\{x;\ 0 \leqslant x \leqslant 1\}$	II_1	new, non-classical	modular, non-distributive, and has no atoms
$\{x;\ 0 \leqslant x \leqslant \infty\}$	II_∞	–	non-modular, and has no atoms
$\{0, \infty\}$	III	–	non-modular, and has no atoms.

This classification is a remarkable accomplishment made even more remarkable by the fact that it took four years to complete the classification by showing that factors of "type III" did actually exist.

Since each von Neumann algebra contains a complete lattice of projections, each of the examples in Table I provides us with an example of a complete orthomodular lattice. We shall be referring back to the information contained in Table I.

We summarize our discussion:

THEOREM. *Let $\mathscr{A}$ be a factor on a separable Hilbert space. There exists a mapping $D:\mathscr{P}(\mathscr{A})\to\{x;\ 0\leqslant x\leqslant\infty\}$ satisfying the following criteria:*

(1) $D(E)=0\Leftrightarrow E=0$;

(2) *If* $E\perp F$, $D(E+F)=D(E)+D(F)$;

(3) E *is finite* $\Leftrightarrow D(E)<\infty$;

(4) $E\sim F\Leftrightarrow D(E)=D(F)$.

This function, called the dimension function, is uniquely determined, up to a constant positive multiple, by conditions (1)–(4). *The range of D, neglecting changes resulting by multiplication by a real number* >0, *is an isomorphism invariant of $\mathscr{A}$ and must be one of the 5 types as listed in Table I. A factor of each type does actually occur.*

C. *The Various Kinds of Factors and Their Projection Lattices*

We refer to Table I. We have already verified that $\mathscr{B}(H)$ is a von Neumann algebra. It is a factor because $\mathscr{B}(H)'=(\lambda I)$ so that $\mathscr{B}(H)\cap\mathscr{B}(H)'=\mathscr{B}(H)\cap(\lambda I)=(\lambda I)$. Similarly, (λI) is a factor. The projection lattice $\mathscr{P}(\mathscr{B}(H))$ is, of course, the lattice of *all* projections in $\mathscr{B}(H)$ that we discussed earlier. The dimension function in this case is quite familiar: it is nothing more than the function that assigns to each closed subspace of H (or to the corresponding projection operator) its ordinary Hilbert space dimension. The equivalence relation in $\mathscr{P}(\mathscr{B}(H))$ is just that of having the same Hilbert space dimension, and the totally ordered set $[\mathscr{P}(\mathscr{B}(H))]$ has the same order type as the set of non-negative integers, together with a last element ∞ if $\dim(H)=\aleph_0$. The situation for (λI) is totally trivial since here we have only the projections 0 and I.

We have referred in Table I to the fact that $\mathscr{P}(\mathscr{B}(H))$ is "atomic." An *atom* in a lattice L with smallest element 0 is a non-zero element a such that $x\in L$, $x\leqslant a$ together imply $x=0$ or $x=a$. In $\mathscr{P}(\mathscr{B}(H))$, the one-dimensional projections are atoms, clearly, and the statement that $\mathscr{P}(\mathscr{B}(H))$ is *atomic* refers to the fact that every non-zero projection is the span of the atoms (one-dimensional projections) it majorizes. The meaning of the term "chain condition" is the standard one (see [4]), and the reader should have no difficulty in seeing that $\mathscr{P}(\mathscr{B}(H))$ satisfies the chain conditions if and only if H is finite dimensional.

The factors of types II and III have totally different qualitative characteristics, and their discovery was an extraordinary achievement. We cannot go into their construction but shall discuss instead their properties, and especially the properties of their associated lattices. The lattices $\mathscr{P}(\mathscr{A})$ for $\mathscr{A}$ of type II or III differ fundamentally from the type I case we have just discussed.

Consider the type II case. Here the range of the dimension function is continuous, a fact that has the following consequence: $\mathscr{P}(\mathscr{A})$ has no atoms. For if $E \in \mathscr{P}(\mathscr{A})$, $E \neq 0$, then $D(E) > 0$ and we simply pick a projection F_1 with $D(F_1) = \frac{1}{2}D(E)$ (if $D(E) = \infty$ simply pick F_1 with $D(F_1)$ finite). Then F_1 $F < E$ as can be easily verified using the properties of D, so we have found $F < E$, $F \neq 0$. It follows that $\mathscr{P}(\mathscr{A})$ *has no atoms in the type II case.* (The same is true in type III but a little proof is needed.) (Hence the type I and type II–type III $\mathscr{P}(\mathscr{A})$'s exhibit a difference similar to that distinguishing atomic and continuous Boolean algebras.) It follows that a type II or type III factor contains no projections at all that have ordinary finite Hilbert space dimension, that is, projections whose subspaces are spanned by finitely many linearly independent vectors. It is quite clear then that these $\mathscr{P}(\mathscr{A})$'s satisfy no chain conditions.

The projection lattice of type II_1 factor has this additional property as listed in Table I: *it is modular.* We now prove this important fact.

We refer to Professor Birkhoff's article (page 1 and following) for an authoritative discussion of the historical origin of the modular law and the part it has played in lattice theory. The modular law goes back to Dedekind. After the distributive lattices, modular lattices are the most thoroughly investigated. In fact they seem to be the only non-distributive lattices that have a reasonably complete theory.

The modular law is this: $E \leqslant F \Rightarrow (E \vee X) \wedge F = E \vee (X \wedge F)$. (for arbitrary X in $\mathscr{P}(\mathscr{A})$). Let $P = (E \vee X) \wedge F$, $Q = E \vee (X \wedge F)$. Since $E \leqslant E \vee X$ and $E \leqslant F$ (by assumption), we have $E \leqslant P$, P being the *greatest* of the lower bounds of $E \vee X$ and F. Obviously $X \vee F \leqslant P$ so $Q \leqslant P$. The problem is to show $Q=P$. The trick is this: the projection $Z=(X-X\wedge Q)\vee(X\vee Q)^{\perp}$ is at the same time a complement of P and a complement of Q (see, for example, [22]). Then by the parallelogram law,

$$\begin{aligned}P&=I-P^{\perp}=Z^{\perp}\vee P^{\perp}-P^{\perp}\sim Z^{\perp}-Z^{\perp}\wedge P^{\perp}\\&=Z^{\perp}=Z^{\perp}-Z^{\perp}\wedge Q^{\perp}\sim Z^{\perp}\vee Q^{\perp}-Q^{\perp}\\&=I-Q^{\perp}=Q.\end{aligned}$$

Hence $P\sim Q$ and so $D(P)=D(Q)$. But then

$$D(Q)=D(P+(Q-P)=D(P)+D(Q-P)$$

whence $D(Q-P)=0$ and so $Q=P$ (we have used the properties of the dimension function as listed in the previous section, leaving it to the reader to check where each particular property was used).

The modularity of the projection lattice of a type II_1 factor is unexpected since the projection lattice of $\mathscr{B}(H)$ is non-modular (for H infinite dimensional). As we remarked earlier, this fact is a consequence of the existence of non-closed sums of closed subspaces. We prove the non-modularity of $\mathscr{P}(B(H))$ now. Simply select two closed subspaces M and N whose sum $M+N$ is not closed. Select a vector $b\in M\vee N=$closure $(M+N)$ not in $M+N$, and let $Z=M+(b)$ where (b) symbolizes the one-dimensional subspace spanned by b. It is easy to see that Z is closed, that $M\leqslant Z$, and that $N\wedge Z=N\wedge M$. Clearly $N\wedge Z\geqslant N\wedge M$. If $x\in N\wedge Z$ then $x\in N$ and $x=m+\lambda b$, λ a complex number $\Rightarrow\lambda b=x-m\in M+N\Rightarrow\lambda=0$ or $b\in M+N$. We know $b\notin M+N$, so $\lambda=0$, from which fact we conclude easily that $x\in N\wedge M$. Thus the reverse inequality $N\wedge Z\leqslant N\wedge M$ also holds, and so equality.) If $\mathscr{B}(H)$ were modular, then since $M\leqslant Z$, we would have

$$(M\vee N)\wedge Z=M\vee(N\wedge Z)$$

as we had above in type II_1 factor. However, by the equation $N\wedge Z=N\wedge M$ just proved, the subspace on the right, $M\vee(N\wedge Z)$, is just M and therefore does not contain the vector b. However b clearly belongs to the subspace $(M\vee N)\wedge Z$. The modular law fails.

Note where the argument that was used to prove the modularity in type II_1 case breaks down here. The parts of this proof up to and including the existence of a common complement for $(M\vee N)\wedge Z$ and $M\vee(N\wedge Z)$ are still valid in $\mathscr{P}(\mathscr{B}(H))$. However the last step of the proof, where we used the existence of a *finite valued* dimension function, is not valid because we have no such function in $\mathscr{P}(\mathscr{B}(H))$, its (unique) dimension function assuming the value $+\infty$.

There is a way of looking at the concepts of distributivity, modularity, and orthomodularity that helps one to understand the meaning of these concepts and the relationships between them. Suppose L is an (abstract) orthocomplemented lattice. In this discussion it will be more convenient for us to use the distributive identities L6′, L6″ contained in Professor Birkhoff's book [4, p. 11] rather than the postulate L6 listed in his paper. Substituting in these distributive identities an arbitrary triple a, b, c of elements of L, and permuting a, b, c in all possible ways, we obtain six propositions, which may or may not be true, concerning the distributivity of the triple a, b, c. These are:

(1) $(a \vee b) \wedge c = (a \wedge c) \vee (b \wedge c)$
(2) $(b \vee c) \wedge a = (b \wedge a) \vee (c \wedge a)$
(3) $(c \vee a) \wedge b = (c \wedge b) \vee (a \wedge b)$

and propositions (1′), (2′), (3′) obtained from these by interchanging $\vee$ and $\wedge$. For example, proposition (2′) is $(b \wedge c) \vee a = (b \vee a) \wedge (c \vee a)$.

If all of the propositions are true let us say that (a, b, c) is a *distributive triple.*

L is distributive (by definition) *if and only if every triple is distributive.*

L is modular if and only if $a \leqslant b$ *implies that* (a, b, x) *is a distributive triple for every* $x \in L$. To prove this simply note that in the presence of the assumption $a \leqslant b$ the six laws listed above reduce to $a \vee (x \wedge b) = (a \vee x) \wedge b$ which is Professor Birkhoff's L5. Hence the modular law is a weakening of the distributivity requirement in that we require that only certain selected triples be distributive.

L is orthomodular if and only if $a \leqslant b$ *implies that* $(a, b, a^{\perp})$ *is a distributive triple.* This is again easily checked by noting that the six distributive laws reduce under the stated assumptions to the orthomodular identity. Hence the orthomodular lattice is seen to be a step further removed from distributivity than the modular lattice for we now require not that every triple (a, b, x) be distributive if $a \leqslant b$ but only the particular one $(a, b, a^{\perp})$.

The implications:

$$\text{distributivity} \Rightarrow \text{modularity} \Rightarrow \text{orthomodularity}$$

obviously follow from the statements given above. Neither arrow can be reversed, as our examples have shown. The historical process seems to have followed this chain of implications, distributive lattices being studied first, modular lattices next, and orthomodular lattices last.

D. *Continuous Geometries*

The previous sections summarize those portions of the "Mother Theory" (the theory of von Neumann algebras) that bear on the theory of orthomodular lattices. Those sections constitute the first stage of our exposition. Now comes the second stage: continuous geometries.

By a *continuous geometry* von Neumann meant a complete modular lattice L that satisfies these two additional conditions:

(1) It is *complemented*, which means this: for any $a \in L$ there is an $x \in L$ such that $a \vee x = 1$, $a \wedge x = 0$. Such an x is called a *complement* of a.

(2) It is *continuous*, which means this: if $\mathcal{M}$ is any totally ordered subset of L, then for any $a \in L$, we have

$$a \wedge [\bigvee (m; m \in \mathcal{M})] = \bigvee (a \wedge m; m \in \mathcal{M})$$

and

$$a \vee [\bigwedge (m; m \in \mathcal{M})] = \bigwedge (a \vee m; m \in \mathcal{M})$$

Let's discuss these axioms.

The terms "complete" and "modular" have already come up, and there is nothing to add here.

The complementation axiom (1) is reminiscent of the orthocomplementation that has been a constant part of our discussion up to now. But "orthocomplemented" is stronger because an orthocomplemented lattice not only needs to satisfy (1) but also has to have a one-to-one order inverting involutory map $a \rightarrow a^{\perp}$ onto itself such that for every a, $a^{\perp}$ is a complement of a.

The continuity axioms (2) are brand new; we have not run into anything like them before. We shall be discussing them in some detail later on when we take up Kaplansky's theorem, and so for the present content ourselves with borrowing a few items from that discussion. Kaplansky's theorem states that if axiom (1) is strengthened to read "orthocomplemented" instead of just "complemented" then the axioms (2) are automatically satisfied. An equivalent form: in a complete orthomodular lattice, modularity implies continuity. Note that the continuity axioms are trivial for finite chains $\mathcal{M}$. Hence any lattice in which all chains are finite is trivially continuous.

There are many examples of continuous geometries. Any complete Boolean lattice is a continuous geometry. The modularity is, of course, a special case of the much stronger distributivity, and the continuity is also a special case of "infinite distributivity" – this too will be explained later on when we discuss Kaplansky's theorem. Any modular geometric lattice (see Section 7 of Professor Birkhoff's article) being a modular lattice of finite length is both complete and continuous and hence a continuous geometry. Likewise, the classical projective geometries are continuous geometries.

The projection lattice of a type I_n factor is a continuous geometry since it is essentially the complex $(n-1)$-dimensional projective geometry (the lattice of all subspaces of complex n-space). The projection lattice of a type II_1 factor is another example. We have already proved that it is modular, and the continuity then follows by Kaplansky's theorem (since the projection lattice of any von Neumann algebra is complete and orthomodular). Here, however, infinite chains *do* exist and the validity of (2) is far from trivial. Professor Birkhoff cites another interesting example (his Section 8) namely, the continuous geometry $CG(D)$ over an arbitrary division ring D.

So much for the axioms and examples; now for the methods and results.

Von Neumann based his analysis on three main concepts: (1) independence, (2) perspectivity, and (3) distributivity.

(1) *Independence.* The family $(a_\sigma; \sigma \in I)$ of elements of the continuous geometry L is *independent* [44, p. 8] if for any pair J, K of disjoint subsets of the indexing set I we have

$$[\bigvee(a_\sigma; \sigma \in J)] \wedge [\bigvee(a_\sigma; \sigma \in K)] = 0.$$

(2) *Perspectivity.* The elements a, b of L are called *perspective* if they have a common complement. Namely, if there is an $x \in L$ such that $a \vee x = b \vee x = 1$, $a \wedge x = b \wedge x = 0$. [44, p. 16.] Write $a \sim b$ if a and b are perspective.

(3) *Distributivity.* Write $(a, b, c)D$ if the distributive law

$$(a \vee b) \wedge c = (a \wedge c) \vee (b \wedge c)$$

is valid for the particular ordered triple (a, b, c) of elements from L, and write $(a, b, c)\,D$ if the dual equation got by exchanging $\vee$ and $\wedge$ is valid. The notation $(a, b)\,D$ means $(a, b, x)\,D$ for every x in L, and $(a)\,D$ means $(a, x, y)\,D$ for every x, y in L [44, p. 32].

Using these concepts, von Neumann proved that in a continuous geometry L:

The relation $(a, b, c)\,D$ *is self-dual and is independent of the order of* a, b, c [44, Part I, Theorem 5.1].

In terminology we have already used, if $(a, b, c)\,D$, then (a, b, c) is a distributive triple.

The following conditions on an element $a \in L$ *are equivalent:*

(i) $(a)\,D$;

(ii) *a has only one complement;*

(iii) *a has a complement b such that* $(a, b)\,D$.

(The set of such elements he called the *center* of the continuous geometry.) He proved further: *The center Z contains the* 0 *and* 1 *of L, contains along with any subset* $\mathcal{M}$ *the least upper bound* $\bigvee(m;\ m \in \mathcal{M})$ *and greatest lower bound* $\bigwedge(m;\ m \in \mathcal{M})$ *as computed in L (and thus is a complete lattice in its own right with respect to the operations inherited from L), is distributive with respect to these operations, and contains along with every element its unique complement; in brief, the center of L is a complete Boolean sublattice of L. Moreover the center of the "interval"* $L(0, a) = (x \in L;\ 0 \leqslant x \leqslant a)$ (easily seen to be itself a continuous geometry with the inherited operations) *consists of all* $z \wedge a$, *z running through the center of L.* [44, Part I, Theorem 5.3; Part III, Theorems 1.2 and 1.6.]

If L is a projection lattice of a von Neumann algebra $\mathcal{A}$, then this lattice theoretic center is precisely the same as the set of projections in $\mathcal{A} \cap \mathcal{A}'$; that is, consists precisely of those projections in $\mathcal{A}$ commuting with all operators in $\mathcal{A}$. So here we have the first instance of a lattice-theoretic version of an operator-theoretic concept: the *center*. Corresponding to a "factor" we have then an *irreducible continuous geometry*: one whose center consists solely of the elements 0 and 1. The assumption of irreducibility for continuous geometries allows similar technical simplifications as in the case of operator algebras.

Perspectivity is transitive. [44; Part III, Theorem 2.2] This is one of the hardest theorems to prove. It follows by means of a series of intricate arguments based on all the above concepts. No simple proof is known even today. Perspectivity is the lattice-theoretic version of the equivalence of projections. Recall that two projections E and F in a von Neumann algebra $\mathscr{A}$ are *equivalent relative to* $\mathscr{A}$ if there is an operator W in $\mathscr{A}$ such that $W^*W=E$, $WW^*=F$. We cannot transplant this definition to a lattice since it involves operators from the algebra; perspectivity is its replacement. While the transitivity of equivalence is trivial, the transitivity of perspectivity lies very deep, and von Neumann's proof of this fact using only the lattice structure was an impressive achievement.

(The main result) *If L is irreducible, then there is a unique mapping $D:L\to[0,1]$ satisfying:*

(i) $D(a)=0\Leftrightarrow a=0,\ D(1)=1$;

(ii) $a\wedge b=0\Rightarrow D(a\vee b)=D(a)+D(b)$;

(iii) *a and b are perspective* $\Leftrightarrow D(a)=D(b)$.

This D is the dimension function. It generalizes the projection lattice dimension function in type I_n and II_1 factors. And it comes from only lattice theory; there isn't a Hilbert space in sight!

In the face of this extraordinary achievement one's optimism is naturally stimulated and one asks whether it might be possible to generalize von Neumann's theory of continuous geometries to include *all* the projection lattices, not only those in factors of types I_n and II_1. Kaplansky called for such a "more general lattice project" in the preface to his notes on Rings of Operators [28]. And in [35] F. Maeda wrote: "...we may conjecture that with respect to dimensionality there is a lattice theory which contains both the continuous geometry and the operator rings." (This was in 1941.) This project, more-or-less vaguely formulated, has been the main impetus behind the rapid growth of orthomodular lattice theory. It is far from being successfully completed, and remains one of the major undone tasks of orthomodular lattice theory.

Progress has been made however. In separate work Loomis [29] and S. Maeda [38], assuming the existence of an equivalence relation on a complete orthomodular lattice and assuming basic postulates for this equivalence relation, proceeded to then derive the major part of the dimension theory of operator algebras in this purely lattice-theoretic setting. Mac Laren [33] and Ramsay [45] showed how to get such an equivalence relation from the lattice structure in certain cases, including some non-modular ones.

Other progress has not been concerned so specifically with this dimension-type equivalence relation. In [23] there is a discussion of some simple lattice properties shared by all projection lattices but not possessed by the general abstract orthomodular lattice. This points up the fact that while complete orthomodular lattices are the natural vehicle for generalizing projection lattices of von Neumann algebras, they are by themselves too wide a class. In [11], Peter Fillmore proved the remarkable result that in the projection lattice of any von Neumann algebra, perspectivity and unitary equivalence are the same, from which he obtained many interesting lattice-theoretic results – one notable result being the transitivity of perspectivity in any projection lattice. Even more recent is Topping's proof [50] that the projection lattice of a von Neumann algebra is semi-modular (or M-symmetric). These are two more properties of projection lattices not shared by general orthomodular lattices. Work along these lines is continuing, with a view to understanding more fully just what is "extra special" about the projection lattice of a von Neumann algebra as compared with the general complete orthomodular lattice.

I shall be discussing all of these pieces of work more fully later on – in case the reader has become somewhat anxious at the sudden spate of undefined terms. But I did want to scan briefly ahead at this point to give the reader some perspective. For this is a crucial point in the exposition. As I announced in the introduction, one of my purposes in writing this paper is to show the historical motivation behind the theory of orthomodular lattices. At this point the reader should have a "feel" for how the subject arose from the interaction of the theory of operator algebras and the theory of continuous geometries. If the reader at this stage does feel a natural curiosity as to whether von Neumann's theory of continuous geometries *can* in fact be generalized so as to include all projection lattices, and does feel a bit of hope that it can, and does feel a sense of anticipation for the value and success of such a study, then I shall have done my job well.

In back of all this stands von Neumann's wonderful achievement – the invention of continuous geometry. Of course continuous geometry is much more than a generalization of projection lattices, even though the stress that we have put on that aspect might lead the reader to believe otherwise. For a more balanced account of the other aspects, one can consult Professor Birkhoff's article and [42].

One final note on notation. The current terminology for a complemented modular lattice satisfying axioms (1) and (2) is; *von Neumann lattice*. We have used the older term "continuous geometry" solely to maintain historical continuity in this survey article.

E. *The Work of Loomis and S. Maeda*

In 1955, about twenty years after the first appearance of the work on von Neumann algebras and continuous geometry, Loomis and S. Maeda independently published papers on extending the Murray-von Neumann dimension theory to non-modular lattices. Loomis worked with a complete orthomodular lattice, while S. Maeda employed a somewhat more general type of lattice that included the complete orthomodular lattice as a special case. Since we are concerned with orthomodular lattices, we shall follow Loomis' formulation.

Loomis' paper was apparently the first published work studying the orthomodular identity as an explicit lattice-theoretic axiom and containing the explicit definition of an orthomodular lattice. His first paper also derived some elementary theory of these lattices, although dimension theory rather than lattice theory was the main theme of this paper. (Kaplansky had also isolated the orthomodular axiom and had done some unpublished work on these lattices. The term "orthomodular" is due to Kaplansky.)

Loomis started with a complete orthomodular lattice L. He then postulated at the outset an equivalence relation on L satisfying certain properties possessed by the dimensional equivalence relation in a von Neumann algebra. A structure of this type – complete orthomodular lattice plus an equivalence relation – he called a dimension lattice. He then developed, on the basis of his axioms, the dimension theory of von Neumann algebras, without any further reference to Hilbert spaces or operator algebras. S. Maeda did much the same thing under somewhat more general hypotheses. The conclusion: modulo this mysterious *ad hoc* equivalence relation, the dimension theory of projection lattice can be made purely lattice-theoretic. The Loomis-S. Maeda generalization included all projection lattices, not just the modular ones. But their accomplishment was not directly comparable to von Neumann's invention of continuous geometry, for they *assumed* the existence of an equivalence relation, whereas he *proved* the existence of one (perspectivity).

Their work was nevertheless vitally important for many reasons. For one thing, it opened up the theory of orthomodular lattices, because their work contained the explicit definition of these lattices, some fragments of their elementary theory, and the implicit conclusion that these lattices were the natural vehicle for the Murray-von Neumann dimension theory. It therefore focussed attention on the class of abstract orthomodular lattices. It also raised the questions: Why was it necessary to *assume* an equivalence relation? What *about* abstract orthomodular lattices? Do they have a general theory? (I asked this question of Kaplansky. He was optimistic: "I'm sure there *is* a theory.")

In the section "Major Results" we shall discuss in greater detail the work of Loomis and S. Maeda and shall also discuss the recent work of Mac Laren and Ramsay bearing on the question: is it necessary to *assume* an equivalence relation? We shall accordingly go no further into these matters here.

But I should like to add a few historical comments about the beginning of the general theory of orthomodular lattices. I became interested in this question while writing my dissertation under Loomis on some aspects of Segal's paper [47] on "non-commutative integration." Segal's integral was defined on a subset of a von Neumann algebra having therefore operators rather than functions for its domain. It turned out that a great deal of his material could be carried over to dimension lattices. While working on this transfer of concepts and theorems into the framework of abstract dimension lattices, I became interested in the study of orthomodular lattices for their own sake. After completing my thesis in 1960, I devoted much of the next year to the study of these lattices.

D. J. Foulis took up the study of the general theory of orthomodular lattices at about the same time. I shall have more to say about his work and the work of his students in the next section. Foulis and I were apparently the first mathematicians to seriously take up the project of devising a general theory of orthomodular lattices. Foulis also got into the subject through the door opened by Loomis and S. Maeda.

III. SOME BASIC ORTHOMODULAR LATTICE THEORY

In Section II we have traced the lineage of orthomodular lattice theory from its beginnings in the "Rings of Operators" papers of Murray and von Neumann up through the appearance of the 1955 papers of Loomis and S. Maeda. In this Section we shall take up some of the post 1955 work on the general theory of the abstract orthomodular lattice.

We shall gradually abandon the historical approach now as we get into the "clean" theory of these lattices. Even though most of the results we discuss can be traced directly to theorems about projection lattices or continuous geometries, motivating every result would simply take too much time. Moreover the reader will, if he is interested enough, hopefully be able to supply much of this motivation himself on the basis of the background material we have already given.

The sample of results that we have selected for presentation here constitutes part of the basic elementary theory of orthomodular lattices. The coverage is necessarily sketchy, and readers wishing to go more deeply into these matters can consult the lecture notes of Foulis [14] or Janowitz [25] or Professor Birkhoff's book [4].

A. *Commutativity*

Let us say that the element a of the orthocomplemented lattice L commutes with an element b if $a=(a \wedge b) \vee (a \wedge b^{\perp})$. This definition first appeared in print apparently in the 1936 paper of Birkhoff and von Neumann [5, p. 833]. It also occurs in F. Maeda's book [36, p. 277] on continuous geometry. We shall abbreviate this relation aCb. It makes sense in any orthocomplemented lattice and Nakamura [41] discovered this interesting fact: *The orthocomplemented lattice L is orthomodular if and only if the relation C is symmetric in L.* It is an interesting exercise to verify that this lattice-theoretic commutativity and common algebraic commutativity ($ab=ba$) agree if L is the projection lattice of a von Neumann algebra (indeed they are the same if L is just the projection lattice of a $*$-ring with identity).

Commutativity is a very useful concept in the study of orthomodular lattices. For example, we can characterize the *center* of an orthomodular lattice L as the set of elements that commute with every other element of L, maintaining a perfect parallel with the common algebraic notion of the center. This characterization of the center agrees with von Neumann's characterizations: either as the set of "neutral" elements z (z is neutral if $\{z, a, b\}$ is a distributive triple for every $a, b \in L$) or as the set of elements with unique complements.

Foulis and I independently discovered the following theorem which turned out to be a very useful principle in the day-to-day calculations of the theory and in the systematic presentation of the results. *If one of the elements a, b, c commutes with the other two, then (a, b, c) is a distributive triple* (meaning that any distributive law you care to write down for the elements a, b, c holds). This result makes a systematic theory out of a mysterious bag of tricks, and makes it possible to give beginning students a swift well-motivated classroom introduction to the subject. In the use of this result the following additional (easily proved) facts are required: If $a \leqslant b$ or $a \perp b$, then aCb, if aCb then $a^{\perp}Cb$, and if $a_1 Cb$ and $a_2 Cb$ then $(a_1 \vee a_2)Cb$ and $(a_1 \wedge a_2)Cb$.

In the presence of commutativity the infinite distributive laws hold: if $a_\alpha Cb$ for all α, then $b \wedge (\bigvee a_\alpha) = \bigvee (b \wedge a_\alpha)$ and

$$b \vee (\bigvee a_\alpha) = \bigwedge (b \vee a_\alpha).$$

(Here we need to assume L complete.)

If a non-empty subset M of the orthomodular lattice L is closed under three operations $\wedge$, $\vee$, and $\perp$ let us say with Foulis that M is a *sub-orthomodular lattice* of L. If S is any non-empty subset of L, let S' symbolize the set of all x of L that commute with every element of S. Facts: S' is always a sub-orthomodular lattice of L, and if L is complete so is S'. There are many other properties of this prime operation all analogies of the centralizer operation in von Neumann algebras. Note that the center of L is just L'. For proofs and additional material see Foulis' papers [12, 13] and also [22].

B. *The "Independence" Theorem*

A family $\{a_\alpha; \alpha \in A\}$ of elements of an orthomodular lattice is called *orthogonal* when $a_\alpha \perp a_\beta$ if $\alpha \neq \beta$. The result we refer to is this:

THEOREM. *Let $\{a_\alpha; \alpha \in A\}$ be an orthogonal family of elements in the complete orthomodular lattice L. Suppose that for each $\alpha \in A$ we have a family of elements $\{a_{\alpha\beta}; \beta \in B\}$ satisfying $a_{\alpha\beta} \leqslant a_\alpha$ for all $\beta \in B$, where B is a fixed non-empty indexing set. Then*

$$\bigvee_\alpha \bigwedge_\beta a_{\alpha\beta} = \bigwedge_\beta \bigvee_\alpha a_{\alpha\beta}$$

We have modelled the statement of this theorem after the statement of the corresponding result in [1]. Here Amemiya and Halperin prove the corresponding theorem for complete complemented modular lattices with "strongly independent family" substituted for "orthogonal family." Both theorems have their origin in von Neumann's theory of independent families in a continuous geometry.

The proof of Amemiya and Halperin applies equally well to the orthomodular case and is so short we can give it in full here. Let $x=\bigwedge_\beta \bigvee_\alpha a_{\alpha\beta}$. It is clear that $x \geqslant \bigvee_\alpha \bigwedge_\beta a_{\alpha\beta}$, so that we need only prove $x \leqslant \bigvee_\alpha \bigwedge_\beta a_{\alpha\beta}$. But

$$x \leqslant \bigvee_\alpha a_{\alpha\beta} = a_{\gamma\beta} \vee \left(\bigvee_{\alpha \neq \gamma} a_{\alpha\beta}\right) \leqslant a_{\gamma\beta} \vee \left(\bigvee_{\alpha\neq\gamma} a_\alpha\right) = a_{\gamma\beta} \vee (a \wedge a_\gamma^\perp)$$

where $a = \bigvee a_\alpha$. Since this is true for every β,

$$x \leqslant \bigwedge_\beta [a_{\gamma\beta} \vee (a \wedge a_\gamma^\perp)] = \left(\bigwedge_\beta a_{\gamma\beta}\right) \vee (a \wedge a_\gamma^\perp)$$

the infinite distributive law discussed in the previous section being available to us because each $a_{\gamma\beta}$ commutes with each $a \wedge a_\gamma^\perp$. Let $c_\gamma = \bigwedge_\beta a_{\gamma\beta}$ so that $c_\gamma \leqslant a_\gamma$. To make the final calculation of the proof clearer we use the following suggestive notation: if $a \perp b$ we write $a \oplus b$ for $a \vee b$, and if $a \leqslant b$, $b-a$ for $b \wedge a^\perp$. It is easy to show that the "orthomodular identity" is in this notation simply $a \leqslant b \Rightarrow b = a \oplus (b-a)$ and its dual is

$$a \leqslant b \Rightarrow a = b-(b-a).$$

Then, since $x \leqslant c_\gamma \vee (a-a_\gamma)$ for all $^\gamma \in A$,

$$\begin{aligned} x &\leqslant \bigwedge_\gamma (c_\gamma \oplus (a-a_\gamma)) = \bigwedge_\gamma (a-(a_\gamma - c_\gamma)) = \bigwedge_\gamma [a \wedge (a_\gamma - c_\gamma)^\perp] \\ &= a \wedge \left[\bigvee_\gamma (a_\gamma - c_\gamma)\right]^\perp = a \wedge [a - \bigvee c_\gamma]^\perp \\ &= a-(a-\bigvee c_\gamma) = \bigvee c_\gamma \end{aligned}$$

as was to be proved. Another proof of this result from a different point of view is given in [45]; compare the remark following Amemiya and Halperin's proof.

By specializing to an indexing set B with two elements we obtain the following corollary:

COROLLARY 1. *If* $\{a_\alpha \vee b_\alpha;\ \alpha \in A\}$ *constitute an orthogonal family in* L *(i.e., if* $(a_\alpha \vee b_\alpha) \perp (a_\beta \vee b_\beta)$ *when* $\alpha \neq \beta$*), then*

$$\bigvee (a_\alpha \wedge b_\alpha) = (\bigvee a_\alpha) \wedge (\bigvee b_\alpha).$$

In this form the result was first obtained independently by Fillmore [11] and me [23]. The most useful consequence so far is this (let us say that "a and b have a common complement in c" if there is an $x \in L$ such that $a \vee x = b \vee x = c$, $a \wedge x = b \wedge x = 0$):

COROLLARY 2. *If* a_α *and* b_α *have a common complement in* c_α *for each* $\alpha \in A$, *and if* $\{c_\alpha;\ \alpha \in A\}$ *is an orthogonal family, then* $\bigvee a_\alpha$ *and* $\bigvee b_\alpha$ *have a common complement in* $\bigvee c_\alpha$.

We can describe the result of this corollary as the "restricted orthogonal additivity of perspectivity"; it is likewise due independently to Fillmore and me. For applications see [11, 23, 30].

C. *Distributivity and Perspectivity*

The center of an orthomodular lattice L consists of those a in L that make every triple (a, x, y) (for any choice of x and y) distributive. It is natural to inquire about *pairs* of elements a, b that make every triple (a, b, x) distributive (for any choice of x). It is often more convenient to study pairs that satisfy also $a \wedge b = 0$. What we are asking then, of the pair a, b is whether $a \wedge b = 0$ and all six identities

(1) $(a \vee b) \wedge x = (a \wedge x) \vee (b \wedge x)$
(2) $(b \vee x) \wedge a = x \wedge a$
(3) $(x \vee a) \wedge b = x \wedge b$
(1′) $x = (x \vee a) \wedge (x \vee b)$
(2′) $(b \wedge x) \vee a = (b \vee a) \wedge (x \vee a)$
(3′) $(x \wedge a) \vee b = (x \vee b) \wedge (a \vee b)$

hold for any choice of $x \in L$.

Well, it turns out that in a projection lattice (whether modular or not), the validity (for all $x \in L$) of any one of these six identities implies the validity of five others. But in a general orthomodular lattice this is no longer true. We can derive all six identities from any one of the last five, but cannot infer any of them from (1) above. We make therefore the following two definitions:

$a \nabla b$ if $a \wedge b = 0$ and (a, b, x) is a distributive triple for all $x \in L$;

$a S b$ if $a \wedge b = 0$ and $(a \vee b) \wedge x = (a \wedge x) \vee (b \wedge x)$ for all $x \in L$.

(Actually the relation $a \nabla b$ was introduced by F. Maeda to mean (3) only.) Always $\nabla \Rightarrow S$. The reverse implication is valid in any projection lattice and any *modular* orthomodular lattice, but not in the general orthomodular lattice.

These notions are connected with perspectivity. Recall that a and b are *perspective* when there is an x such that $a \vee x = b \vee x = 1$, $a \wedge x = b \wedge x = 0$. In his work on continuous geometry, von Neumann makes constant use of the fact that if a and b are perspective, then an $x \in L$ can be found so that $a \vee x = b \vee x = a \vee b$, $a \wedge x = b \wedge x = 0$. That is to say, if a and b are perspective, then they are perspective in their own span. This is another result that fails to be true in a general orthomodular lattice, but in this instance we can pin-point the reason why the result does not obtain: the failure of the modular law. If there is an $x \in L$ such that $a \vee x = b \vee x = a \vee b$, $a \wedge x = b \wedge x = 0$, let us say that a and b are *strongly perspective*. Then, in an orthomodular lattice L, *perspectivity implies strong perspectivity if and only if L is modular* [23].

Following notation introduced by Loomis, let us say that a and b are *related* if there exist elements a_1, $b_1 0 \neq a_1 \leqslant a$, $0 \neq b_1 \leqslant b$ with a_1 and b_1 perspective, and *strongly related* if we can find a_1, $b_1 0 \neq a_1 \leqslant a$, $0 \neq b_1 \leqslant b$ with a_1 and b_1 strongly perspective. We have the following interesting connections between the distributive type relations ∇ and S introduced above, and these concepts:

(1) $a \nabla b \Leftrightarrow$ *a and b are not related* (this is due to M. F. Janowitz);

(2) $a S b \Leftrightarrow$ *a and b are not strongly related.*

In a projection lattice of a von Neumann algebra, "strongly related" and "related" are the same.

These results were all suggested by theorems from continuous geometry, and are the fruits of efforts to extend the ideas of continuous geometry to orthomodular lattices. It turns out, as illustrated above, that many theorems can be extended to projection lattices but not to general orthomodular lattices, such theorems being therefore not dependent on the modular law for their validity (since they hold in non-modular projection lattices) but instead on some more subtle lattice-properties of projection lattices. Stated in another way, the failure of these theorems in general orthomodular lattices reflects the extent to which abstract orthomodular lattices over-generalize projection lattices.

Some other aspects of this same phenomenon: In a projection lattice L, as in a continuous geometry, the center of any interval $L(0, a)$, $a \in L$ consists of all $z \wedge a$, $z \in$ center (L). It is suggested to refer to this fact as the "relative center property." The relative center property does *not* hold in general orthomodular lattices. However, we do have this positive result (due independently to Fillmore and Janowitz): *If* $S \Rightarrow \nabla$ *then the relative center property holds.* (They are in fact equivalent, but the reverse implication is easy.) Janowitz has also shown that if *strong perspectivity and perspectivity agree on orthogonal elements* (always true in a projection lattice) *then the relative center property holds.* Surprisingly the reverse implication does not obtain here.

We don't understand this whole situation very well. The projection lattices have a limited usefulness in suggesting results, since distributivity and perspectivity are much more well behaved in the projection lattice case than in the general case. Although some understanding about the general case is beginning to emerge as results accumulate (mainly from the active pen of M. F. Janowitz), it would seem that we are far from understanding the role of distributivity and perspectivity in orthomodular lattices.

D. *Greechie's Examples* [18]

R. J. Greechie has discovered a bewildering array of examples of orthomodular lattices that lie at the opposite end of the spectrum from projection lattices. Projection lattices are "nice," almost every desirable theorem holding in them. Greechie's examples are "wild" and constitute a seemingly exhaustive family of counterexamples (that is, if a theorem holds in all Greechie's examples, then it is true).

Except for projection lattices, examples of orthomodular nonmodular lattices are hard to come by. Greechie's method furnishes us with a rich and varied supply of such examples, and it is the only such general method available now. That is the reason for its importance and its inclusion in this paper. This presentation is based mainly on Theorem 3.8 of [18].

The basic method is a construction procedure for "pasting" together a family of partially ordered sets. Suppose that $(L_\alpha;\, \alpha \in A)$ is a family of two or more distinct partially ordered sets, with a partially ordered subset K_α singled out in each one. (Actually, any non-empty subset of L_α is a partially ordered subset in its inherited ordering and thus can serve as a K_α.) Assume that all subsets K_α are order-isomorphic to a fixed partially ordered set K.

We now form a new partially ordered set L by pasting together the L_α's along the K_α's. Imagine the fixed partially ordered set K standing vertically amidst the L_α's, and imagine that each $z \in K$ is connected to its isomorphism-correspondent in K_α by a string. Then each $z \in K$ will have card (A) strings radiating from it, each α-string going to the element $z_\alpha \in K_\alpha$ that corresponds to z under the isomorphism between K_α and K. Now pull all the strings taut. Each K_α will merge into K. The L_α's will cluster around K, with K as the "maypole," each element in a set-theoretic difference $L_\alpha - K_\alpha$ maintaining its individual identity in L but each element in a K_α becoming identified with its partner in K.

Let L denote the resulting set. The set L then consists of two kinds of elements: elements that result from identifications and elements that do not. The elements that result from identifications are in one-to-one correspondence with K, and we shall regard K as a subset of L. For an x in L that is also in K, we shall use a subscripted symbol x_α to denote the old element of K_α corresponding to x under the isomorphism between K_α and K. We shall also occasionally think of an element x in K as the set of all x_α's, $\alpha \in A$. The non-identified elements of L consist simply of all the elements in the set-theoretic differences $L_\alpha - K_\alpha$. In particular, if A is a finite set of $n \geqslant 2$ elements, if each L_α has l_α elements (finitely many in each case), and if each K_α (and hence K) has m elements, then the set L will consist of a total of

$$m + \sum (l_\alpha - m) = m + \sum l_\alpha - nm = N - m(n-1)$$

elements, where $N = \sum (l_\alpha;\ \alpha \in A)$ is the total number of distinct elements we began with.

L inherits a more-or-less natural partial ordering from its constituents. Recalling that for each $x \in L$ exactly one of the following mutually exclusive possibilities must hold: either $x \in K$ or $x \in L_\alpha - K_\alpha$ for some unique α, we can define this partial ordering case-by-case as follows. For $x, y \in L$, $x \neq y$, we declare $x < y$ in these and only these cases:

(1) $x \in K$, $y \in K$ and $x < y$ in K;
(2) $x \in K$, $y \in L_\alpha - K_\alpha$ and $x_\alpha < y$ in L_α;
(3) $x \in L_\alpha - K_\alpha$, $y \in K$ and $x < y_\alpha$ in L_α;
(4) $x \in L_\alpha - K_\alpha$, $y \in L_\alpha - K_\alpha$ and $x < y$ in L_α;
(5) $x \in L_\alpha - K_\alpha$, $y \in L_\beta - K_\beta$, where $\alpha \neq \beta$ and there exists $z \in K$ such that $x < z$ (using 3) and $z < y$ (using 2).

This prescription effectively defines a relation on L. The verification that this relation is a partial ordering on L presents no particular difficulties – the transitivity being built in by (5) – and we leave the details to the reader. L is the partially ordered set obtained by "pasting together" the L_α's along the K_α's.

Question: When is this partial ordering a lattice ordering? Not always, not even in the simple case of two finite Boolean lattices pasted together along a sublattice. Greechie gives a set of sufficient conditions that seems to apply to most cases that arise in practice. Of course, it is always possible in a particular construction to simply verify after the fact that L is a lattice.

According to Greechie the partial ordering on L will be a lattice ordering (and in fact complete) if these four conditions are satisfied:

(1) Each L_α is a complete lattice.

(2) Each K_α is closed under the taking of arbitrary least upper bounds and greatest lower bounds in L_α. In particular we are postulating closure under the taking of lub's and glb's of *finite* subsets of K_α so that K_α is a sublattice of L_α. Requiring closure under the taking of lub's and glb's of *arbitrary* subsets of K_α (computing these lub's and glb's in L_α) secures not only the completeness of the lattice K_α but guarantees in addition that (arbitrary) lub's and glb's are the same whether computed in K_α or in L_α. Moreover the "maypole" K will be a complete lattice since it is isomorphic to each K_α.

(3) Each K_α is the disjoint set-theoretic union of a proper order ideal I_α and a proper dual order ideal J_α. (The non-empty subset I of a lattice L is an *order ideal* if I contains along with any element a all $x \in L$ satisfying $x \leqslant a$. I is *proper* if $I \neq L$.) Note that we do *not* require I to be a lattice ideal, although in some useful special cases it will be. For example we shall discuss some instances where $I_\alpha = [0, a_\alpha]$, $J_\alpha = [b_\alpha, I_\alpha]$, so that K_α is the disjoint set-theoretic union of a principal ideal and a principal dual ideal. On the other hand, we shall also discuss cases where $I_\alpha =$

$[0_\alpha, a_\alpha] \cup [0_\alpha, b_\alpha]$ where $a_\alpha \wedge b_\alpha = 0$ (and correspondingly for J_α). In this case I_α is always an order ideal but need not be a lattice ideal.

(4) If ϕ_α is the isomorphism of K_α onto K, and ϕ_β that of K_β onto K, then the map $\phi_\beta^{-1}\phi_\alpha$ (which is obviously an isomorphism of K_α onto K_β) takes the order ideal I_α onto the order ideal I_β, and takes the dual order ideal J_α onto J_β. This requirement ensures that in the matching of the K_α's with the common lattice K, things don't get turned over.

If these four conditions are satisfied, then the ordering of the set L, which we get by pasting the L_α's together along the K_α's, will be a lattice ordering. These conditions, although complicated in appearance, are not difficult to verify in practice; we shall provide some concrete illustrations shortly. The proof that fulfillment of these conditions guarantees a lattice ordering is rather tedious and we refer the reader to [18] for the details.

We are primarily interested in the orthomodular case. The constructed lattice L will be orthomodular if in addition to (1) through (4) we also require that: each L_α be orthomodular, the dual order ideal J_α consist precisely of the orthocomplements of the elements of the order ideal I_α, the lattice K be orthomodular and the isomorphisms ϕ_α preserve orthocomplements. Again, we refer the reader to [18] for the details.

Let's look at some examples obtained by Greechie's construction.

(1) *Horizontal Sums.* This special case of Greechie's construction was discovered earlier by many people; the terminology is due to Mac Laren [33]. Here the L_α's are arbitrary orthomodular lattices, each I_α is just the zero element $\{0_\alpha\}$ of L_α, and each J_α is just its largest element, $\{1_\alpha\}$. Thus for each α, $K_\alpha = \{0_\alpha, 1_\alpha\}$ and the common lattice K is simply the two-element orthomodular lattice $K = \{0, 1\}$, $0 < 1$, $0^\perp = 1$. In this simple case we can describe the construction quite vividly: We place all the L_α's side-by-side, tie together all the 1_α's and tie together all the 0_α's. In between the lattices are untouched. The resulting configuration L has the obvious component-by-component order and orthocomplementation. By Greechie's general result L is a complete orthomodular lattice. In this special instance more can be said. If each L_α has more than two elements, then L is irreducible (its center consists of just 0 and 1) and is never modular unless each L_α has length 2. (That is, L is modular if and only if in each L_α $x_\alpha < y_\alpha$ implies $x_\alpha = 0_\alpha$ or $y_\alpha = 1_\alpha$.) In Figure 1 we show the horizontal sum of the Boolean lattice of four elements and that of eight.

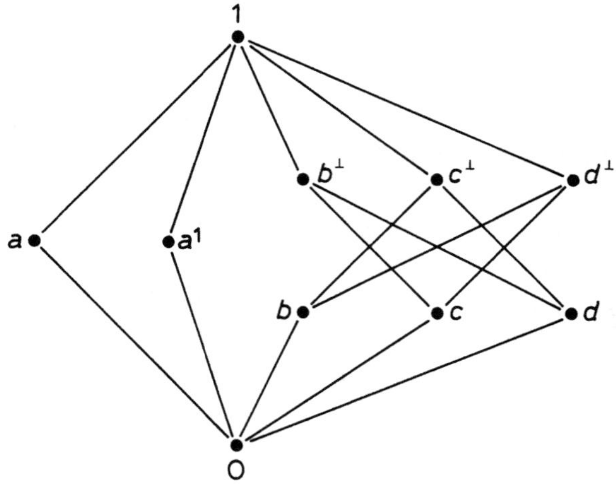

Fig. 1. Horizontal sum of 2^2 and 2^3.

It is the smallest non-modular irreducible orthomodular lattice.

(2) *Dilworth's lattice.* This lattice is obtained by pasting together two copies of the eight element Boolean lattice 2^3 and then pasting another copy of 2^3 to the result. The construction as schematized in Figure 2 is self-explanatory. The result is a sixteen element non-modular irreducible orthomodular lattice which is not the horizontal sum of Boolean lattices.

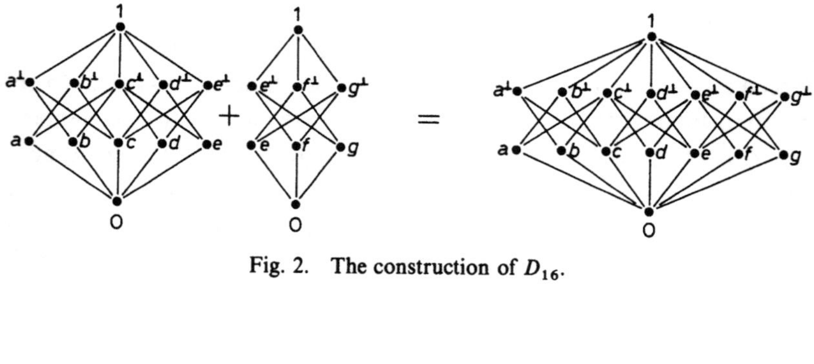

Fig. 2. The construction of D_{16}.

It was first discovered by Dilworth [7]. For some properties of this lattice see [23].

(3) *Greechie's lattice, "G_{32}."* The following schematic conventions introduced by Greechie are very convenient in describing this lattice. He uses a "top view" rather than the "side view" schematizing 2^3 thus: ·—·—·. The twelve element lattice of the first line of Figure 2 obtained by pasting together two copies of 2^3 he schematizes thus:

the straight line break locating the place at which the pasting occurs. Thus the top view of Dilworth's lattice is:

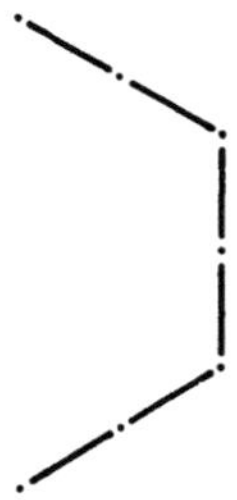

"G_{32}" is obtained by pasting together two copies of Dilworth's lattice along $K = \{[0, a] \cup [0, g]\} \cup \{[a^{\perp}, 1] \cup [g^{\perp}, 1]\}$ (refer to Figure 2). We get:

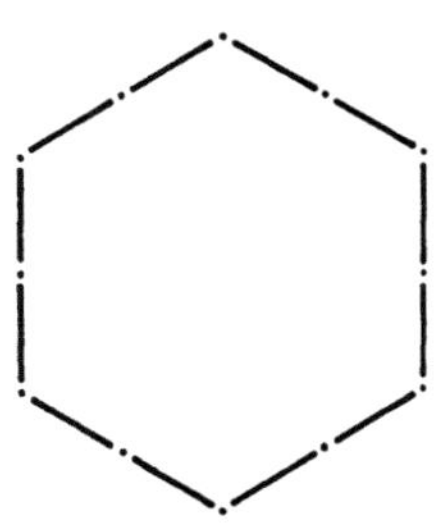

Next we paste together in succession four copies of 2^3 to get:

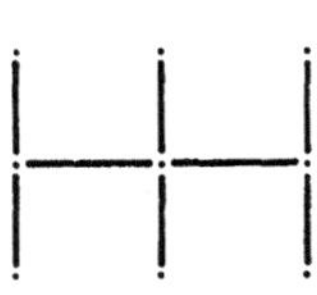

Twist this whole configuration (no pastings) and insert it in the ring pasting it in at six places. This last pasting is not covered by Greechie's four conditions, but we can easily verify after the fact that the resulting partially ordered set, shown in Figure 3, is a lattice. It is in fact a 32

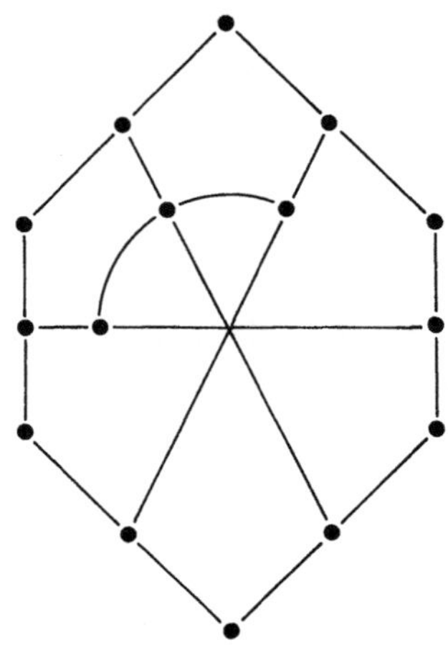

Fig. 3. G_{32} (top view – see text).

element non-modular irreducible, orthomodular lattice in which all elements $\neq 0, 1$ are perspective.

IV. THE PRINCIPAL RESULTS

In this section I shall list seven results (or classes of results) that I regard as having special significance for orthomodular lattice theory. I have taken the risk of going on record with a subjective judgment like this because I do believe that (if my information is fairly complete and my judgment substantially correct) such a listing of major results has considerable value. My seven candidates are:

A. *The Results of Loomis* [29] *and S. Maeda* [38]

I have already discussed these papers in II, E. Orthomodular lattice theory essentially begins with the appearance of these papers; hence their listing here as the first item. I have been more or less vague about the results of these papers, and it is perhaps worthwhile to give some specific details. Loomis makes the following definition:

A dimension lattice *is a complete orthomodular lattice together with an equivalence relation* $\sim$ *satisfying these four axioms:*

(A) $a \sim 0 \Rightarrow a = 0$;

(B) *If* $a_1 \perp a_2$ *and* $b \sim a_1 \vee a_2$, *then* $b = b_1 \vee b_2$ *with* $b_1 \perp b_2$ *and* $b_1 \sim a_1$, $b_2 \sim a_2$;

(C) *If* $(a_\alpha;\ \alpha \in A)$ *is a family of pairwise orthogonal elements* ($a_\alpha \perp a_\beta$ *if* $\alpha \neq \beta$) *and if* $(b_\alpha;\ \alpha \in A)$ *is a second such family with* $a_\alpha \sim b_\alpha$ *for all* $\alpha \in A$, *then* $\bigvee a_\alpha \sim \bigvee b_\alpha$;

(D′) *If* a *and* b *have a common complement, then* $a \sim b$. He gives the dimension-theoretic definitions:

$a \in L$ is *finite* if $a \sim b \leqslant a$ implies $b = a$; otherwise a is *infinite*;

a and b are *related* if there exists a_1, b_1, $0 \neq a_1 \leqslant a$,

$0 \neq b_1 \leqslant b$, $a_1 \sim b_1$;

a is *simple* if x and $a \wedge x^\perp$ are not related for every $x \leqslant a$,

and

e is *invariant* if e and $e^\perp$ are not related,

and then establishes these basic facts:

If a is finite and $b \sim a_1 \leqslant a$, then b is finite.

If a is simple, then a is finite.

If a is simple and $b \sim a_1 \leqslant a$, then b is simple.

If $a \sim b_1 \leqslant b$ and $b \sim a_1 \leqslant a$, then $a \sim b$ (Schroeder-Bernstein) and especially: Given a, $b \in L$, there exists $a_0 \leqslant a$, $b_0 \leqslant b$ such that the elements $a \wedge a_0^\perp$ and $b \wedge b_0^\perp$ are not related, and $a_0 \sim b_0$.

Loomis first discusses the irreducible case which he defines this way: any two nonzero elements are related. In a few pages he derives the existence of the dimension function (that we discussed earlier). He then goes on to discuss in the same spirit the technically more complicated reducible case. S. Maeda does not discuss the irreducible case separately, but carries through the general analysis from the beginning.

The dimension theory of projection lattices is a special case of the Loomis-S. Maeda Theory (once you have verified that the intrinsically defined equivalence relation in projection lattices satisfies the Loomis-S. Maeda axioms). It is worth repeating the conclusion that we have already drawn from this: the Murray-von Neumann dimension theory is basically lattice-theoretic, the complex Hilbert space and operator algebra structure being unnecessary for its existence.

B. *The Work of Foulis and his Students on the Abstract Orthomodular Lattice*

The orthomodular lattice, as an algebraic structure, can and should be studied as such an entity, in much the same way as groups and rings are studied as abstract algebraic structures. In fact, if we were not so close to the beginning of the subject, this would be the way to approach it: as the structure theory of an axiomatic-algebraic object.

Thanks to the pioneering work of G. Birkhoff and others, there is a general lattice theory in which the appropriate lattice-concepts have been isolated and studied. This general theory serves as the framework to guide the study of the abstract orthomodular lattice and, since 1960, D. J. Foulis and his students have actively worked on such a study. Whatever general theory we have today is due primarily to them. The best sources for the accumulated results are two unpublished sets of lecture notes, Foulis' University of Florida lecture notes of 1963 [14], and Janowitz's note of 1965 [25] from the University of New Mexico.

Foulis and his people have worked on such topics as: homomorphisms of various types including the specially important "residuated mappings" (Derderian), distributivity and the center (Janowitz), irreducibility and various strengthened forms of irreducibility (Catlin), modular pairs (Schreiner), examples (Greechie), cartesian products, ideal theory, quantifiers (Janowitz), and so on. All of this material is given in full detail in the notes referred to above. We have given some brief exerpts in Section III. Since it is probably not particularly valuable for a survey article of this kind to list many technical details, we shall not give any further description of the material here but refer the reader to the available notes. However one novel idea due to Foulis deserves mention, and that is his coordinatization of orthomodular lattice by Baer $*$-semigroups [12].

Let S be an involution semigroup. That is, a set with an associative binary operation and a mapping $x \to x^*$ of S into itself satisfying $(ab)^* = b^*a^*$, $(a^*)^* = a$. A *projection* e is defined as usual: $e = e^2 = e^*$. We assume also that the semigroup has a zero element 0, (characterized uniquely by $x0 = 0 = 0x$ for all $x \in S$), and identity 1, (characterized uniquely by $1x = x1 = x$ for all $x \in S$). Foulis calls S a *Baer $*$-semigroup* if for every $x \in S$, the set of all z in S satisfying $xz = 0$ is precisely the set eS for some projection e. In other words, the right annihilator of an arbitrary $x \in S$

is a principal right ideal generated by a projection e. The projection e is uniquely determined by x and Foulis denotes it x'. Call a projection f *closed* if $f = f''$. Then we have: *the set of closed projections of a Baer $*$-semigroup is an orthomodular lattice under the ordering $e \leqslant f \Leftrightarrow e = ef$ and the orthocomplementation $e^{\perp} = e'$.*

Foulis showed, more importantly, that *every* orthomodular lattice arises in this way! That is, given an orthomodular lattice L, there is a Baer $*$-semigroup S such that L is isomorphic (as an orthocomplemented lattice) to the lattice of closed projections in S. A great deal of attention has been focussed on this lattice-semigroup connection, and the interplay between these structures has helped formulate conjectures and prove theorems about orthomodular lattices.

Janowitz has generalized Foulis' idea to arbitrary lattices [24]. He (Janowitz) defines a Baer semigroup as a semigroup S with 0 such that the left (resp. right) annihilator of every element is a principal left (resp. right) ideal generated by a (not necessarily unique) idempotent. Janowitz shows that the set of left annihilators of elements of S forms a lattice under set-theoretic inclusion, and that every lattice can be obtained in this way (is isomorphic to such a lattice of left annihilators of elements). Janowitz's theorem shows that lattice theory is part of semigroup theory. Ideas of J. C. Derderian are vital in this work.

Foulis gives an encompassing description of these and related results in an unpublished survey paper [15]. We have taken much of this material from his paper. He raises the general problem: in this correspondence between lattices and semigroups, how do the various lattice properties and semigroup properties match up? (This problem is complicated by the fact that the semigroup is not uniquely determined by the lattice.) Foulis cites some known theorems. (Terminology: if the lattice L is isomorphic to the left annihilator lattice of elements of the semigroup S, we say that S *coordinatizes* L.) *L is complemented modular $\Leftrightarrow$ it can be coordinatized by a left regular Baer semigroup* (Janowitz). (S is *left regular* if for every $x \in S$, Sx is the left annihilator of some $y \in S$.) *L admits an orthocomplementation making it into an orthomodular lattice $\Leftrightarrow$ L can be coordinatized by a Baer $*$-semigroup* (Foulis).

C. *Kaplansky's Theorem*

The result we refer to is this:

THEOREM [Kaplansky, 27]. *Let L be a complete orthomodular lattice. Then if L is modular it is necessarily continuous.*

Hence the continuity axioms that von Neumann needed to "face the infinities" are, in a complete orthomodular lattice, an automatic consequence of the modular law.

Continuity you will recall means this: if $\mathcal{M}$ is any totally ordered subset of L (a "chain") and a is an arbitrary element of L then both

$$(1) \qquad a \wedge [\bigvee(m; m \in \mathcal{M})] = \bigvee (a \wedge m; m \in \mathcal{M})$$

and

$$(2) \qquad a \vee [\bigwedge(m; m \in \mathcal{M})] = \bigwedge (a \vee m; m \in \mathcal{M}).$$

In these formulas $\mathcal{M}$ is a totally ordered subset of L – this is a crucial point. With $\mathcal{M}$ unrestricted, (1) and (2) are the much stronger infinite distributive laws [4, p. 118]. We have already seen instances of infinite distributivity, namely the result cited in Section III, A, that if an element a of a complete orthomodular lattice L commutes with every element m of an (arbitrary) subset $\mathcal{M}$ of L, then (1) and (2) both hold. Since a complete Boolean lattice is just a complete orthomodular lattice in which all pairs commute, every complete Boolean lattice is infinitely distributive and therefore continuous. But continuity is much weaker than infinite distributivity – since it depends on the validity of (1) and (2) for only *chains* $\mathcal{M}$. It can hold in the complete absence of any distributivity, either finite or infinite.

Continuity does not obtain in the general orthomodular lattice. It fails to hold for example in the lattice $\mathcal{P}(\mathcal{B}(H))$ of all closed subspaces of infinite dimensional Hilbert space. Suppose H separable with orthonormal basis $\{e_1, e_2, \ldots\}$. Let M_n be the finite dimensional closed subspace spanned by the n vectors $\{e_1, e_2, \ldots, e_n\}$, and let A be the one-dimensional subspace spanned by the single vector $a = \sum_{i=1}^{\infty} (1/i)\, e_i$. The M_n constitute a totally ordered subset of $\mathcal{P}(\mathcal{B}(H))$ and $\bigvee M_n = H$ so that $A \wedge (\bigvee M_n) = A \wedge H = A$. But $A \wedge M_n = 0$ for every n (as an easy check shows) so that $\bigvee (A \wedge M_n) = 0$. Hence $\bigvee (A \wedge M_n) \lneqq A \wedge (\bigvee M_n)$ and the lattice is not continuous. Of course neither is it modular so we have no contradiction to Kaplansky's theorem.

Some technical points concerning the laws (1) and (2): It is easy to see that (1) is trivially true if $\mathcal{M}$ has a largest element and (2) is trivially true if $\mathscr{M}$ has a smallest element. The simplest nontrivial instance of (1) is when $\mathscr{M}$ is a sequence $m_1 < m_2 < \cdots$ with no largest element (and dually for (2)). The totally ordered subsets $\mathscr{M}$ can be replaced by well-ordered or directed subsets without affecting the definition; for a proof see the appendix to [36]. Also, in an orthocomplemented lattice we can get either one of the continuity laws from the other by taking orthocomplements; hence in the orthocomplemented case we need prove only one of these laws.

Amemiya and Halperin [1] have given an "elementary" (but technically complicated) proof of Kaplansky's theorem. We are surely indebted to these authors for this self-contained proof, because Kaplansky's original proof employed some formidable technical machinery indeed, using in particular von Neumann's coordinatization theorem that a complete complemented modular lattice with four or more independent perspective elements of lub 1 is isomorphic to the lattice of principal right ideals of a regular ring. Complete and detailed proofs of von Neumann's coordinatization theorem generally fill about half a good-size book [36, 44].

Actually, Amemiya and Halperin make a thorough and detailed study of complemented modular lattices, going considerably beyond a proof of Kaplansky's theorem. They do however, in the appendix to their paper, supply a self-contained, four-page proof that a complete orthocomplemented modular lattice is "finite," that is, cannot possesses an infinite independent family of non-zero pairwise perspective elements. One can then get continuity from finiteness as in the proof of their Theorem 9.4, or as follows:

If L is a complete orthocomplemented modular lattice, then it is easy to check that $L(a, b)$, the set of all $x \in L$ satisfying $a \leqslant x \leqslant b$, is also a complete orthocomplemented modular lattice with the lattice operations inherited from L and the orthocomplementation $x^* = (x^\perp \vee a) \wedge b$. Now, given arbitrary elements a and x in L, check that the elements $a_1 = a - [(a \wedge x) \vee (a \wedge x^\perp)]$ and $b_1 = a^\perp - [(a^\perp \wedge x) \vee (a^\perp \wedge x^\perp)]$ are orthogonal and perspective (*via* x). (For that fact we need only orthomodularity.) By using a Zorn's lemma argument (or transfinite induction) obtain a maximal family of pairs of non-zero elements $\{a_\alpha, b_\alpha\}$ with these properties: $a_\alpha \perp b_\alpha$ and $a_\alpha \sim b_\alpha$ all α, and $(a_\alpha \vee b_\alpha) \perp (a_\beta \vee b_\beta)$ if $\alpha \neq \beta$. Let $a = \bigvee a_\alpha$, $b = \bigvee b_\alpha$ and $c = a \vee b$ and observe that $a \sim b$ by the additivity of perspectivity discussed in Section III, B (the "independence" theorem). Note also that $L(0, c^{\perp})$ is Boolean and therefore certainly continuous) because by the maximality of the family $\{a_\alpha, b_\alpha\}$ it is not possible to find in $L(0, c^{\perp})$ orthogonal non-zero elements a_1, b_1 with $a_1 \sim b_1$ and therefore for every $a, x \in L(0, c^{\perp})$, $a=(a \wedge x) \vee (a \wedge x^{\perp})$ which means that all pairs of elements of $L(0, c^{\perp})$ commute. Now by Amemiya and Halperin's Theorem 4.3 we can reduce the problem to proving that $L(0, a)$ is continuous. But $L(0, a)$ can be "doubled" in the terminology of Amemiya and Halperin. And now the methods of the Corollary to their Theorem 3.4 can be easily adapted to show that if $L(0, a)$ is not continuous it has an independent sequence of non-zero pairwise perspective elements. This contradicts finiteness.

However Kaplansky's original proof, like most original proofs, contains the germ of the original discovery and for this reason maintains a permanent independent value in spite of its heavy prerequisites. For this is not an easy theorem to guess – the implication modularity $\Rightarrow$ continuity is not something one would naturally expect – and it is valuable to know why Kaplansky suspected it was true. Kaplansky says (in a letter) that the idea stemmed directly from the fact that modularity is the same as the sum of closed subspaces being closed; and he refers in this connection to Mackey's paper [31].

Mackey was the first to study in detail the lattice properties, especially modularity, of the closed subspaces of topological and metric vector spaces. In [31] one finds a thorough study of the lattice of "closed" subspaces of "linear systems," the theorem that the closed subspaces M, N form a dual modular pair if and only if they have a closed sum, the proof that dual modularity is a symmetric relation in the lattice of closed subspaces, theorems relating modularity to "tangency," and the like. The lattice of closed subspaces of real normed linear spaces is also studied by Mackey; he proves for example that the lattice of closed subspaces of a real normed linear space is modular if and only if the space is finite dimensional, and that in the lattice of closed subspaces of a Banach space, modularity and dual modularity are equivalent. Mackey's paper [31] has provided many valuable ideas and probably has still many more of offer.

Kaplansky applied these methods to an abstract orthocomplemented modular lattice by first using von Neumann's coordinatization theorem to represent the lattice as the set of projections in a *-regular ring, and then exploiting the analogy between these rings and rings of operators on a Hilbert space.

One final historical note: Kaplansky once asked von Neumann whether the idea that orthocomplemented complete modular $\Rightarrow$ continuity had crossed his mind. Von Neumann's reply: he never looked at whether algebraic assumptions could lead to continuity.

We shall have occasion later on to refer to the following corollary of Kaplansky's theorem (L continues to symbolize a complete orthomodular lattice):

If L is modular, then perspectivity is transitive in L.

D. *The Results of Mac Laren* [33] *and Ramsay* [45]

Mac Laren and Ramsay published closely related (though separate) papers dealing with the question: for what complete orthomodular lattices can we derive from the lattice structure a dimensional equivalence relation satisfying Loomis' axioms? Not for all, because there are complete orthomodular lattices that do not admit any dimensional equivalence relation satisfying Loomis' axioms. On the other hand, if we impose on L the very strong assumption of modularity, then by the previous theorem of Kaplansky and by von Neumann's work on continuous geometry we know that perspectivity itself will do. Can we find some middle ground?

To describe the results of these authors we need some definitions from their papers. A dimension lattice L is *locally finite* if every element in L is the lub of finite elements. A complete orthomodular lattice L is *locally modular* if it contains a subset I that satisfies the following three conditions:

(1) I is a modular sublattice of L;
(2) I is an ideal, and
(3) every element of L is the lub of elements from I.

A pair $\{a, b\}$ is a *modular pair* if for every x in the lattice L satisfying $x \leqslant b$ we have $(x \vee a) \wedge b = x \vee (a \wedge b)$. L is semi-modular (or M-symmetric) if the relation of being a modular pair is symmetric. A complete orthomodular lattice L is *nearly modular* if it satisfies the following conditions:

(1) L is semi-modular, and
(2) if we denote by S the set of those elements a of L satisfying the two conditions:

(i) $L(0, a)$ is a modular lattice, and
(ii) (x, a) is a modular pair for every $x \in L$, then every element of L is the lub of elements from S.

With these definitions we can summarize a main result of these authors as follows:

THEOREM. *The following conditions on a complete orthomodular lattice L are equivalent:*
(1) *L is a locally finite dimension lattice;*
(2) *L is locally modular;*
(3) *L is nearly modular.*

We refer the reader to the original papers for the details of the proof and other related results.

Mac Laren gets his dimensional equivalence relation this way: Denote by $e(a)$ the *central cover* of the element $a \in L$, that is, the glb of all central elements majorizing a. (This glb exists since the center of a complete orthomodular lattice is a complete sublattice, see III.A.) Then, given two elements a and b of L, Mac Laren calls them equivalent (or having the same relative dimension) if and only if $e(a) = e(b)$ *and* for every $z \in$ center (L) such that $z \wedge a \in S$, $z \wedge a$ and $z \wedge b$ are perspective. If L has a trivial center this boils down to: elements of S are equivalent if and only if they are perspective; all elements not in S are mutually equivalent.

The principal tools of proof used in both papers are Kaplansky's theorem (discussed in the preceding section) and von Neumann's theory of continuous geometries.

Yet another approach to this problem of making an arbitrary complete orthomodular lattice into a dimension lattice has been developed by M. F. Janowitz in an interesting paper [26].

E. *Fillmore's Theorem: Unitary Equivalence and Perspectivity Are the Same*

While the projection lattice of a von Neumann algebra is always complete and orthomodular, the converse is false, as we have seen. But the failure of the converse forces upon us this very natural question: under what additional conditions is a complete orthomodular lattice ortho-isomorphic to the projection lattice of a von Neumann algebra? Clearly, this will be a very difficult question to answer. While some progress on it has been made, namely, a result of Zierler [52, 53] (see also [34]) giving conditions that guarantee an orthoisomorphism onto the full projection lattice of $\mathscr{B}(H)$, we are still apparently a long way from characterizing the projection lattices among the general orthomodular lattices. Much more striking progress has been made on the less ambitious task of investigating lattice-properties of projection lattices.

A very nice recent result along these lines is Topping's theorem [50]: *The projection lattice of a von Neumann algebra is semi-modular.* (See p. 64 for the definition of semi-modular.)

However, the result we wish particularly to stress here is the following:

THEOREM. [P. A. Fillmore, 11.] *If L is the projection lattice of a von Neumann algebra $\mathscr{A}$, then two elements are perspective in L if and only if they are unitarily equivalent in $\mathscr{A}$.*

(Two projections E and F of a von Neumann algebra $\mathscr{A}$ are unitarily equivalent in $\mathscr{A}$ if there is a unitary operator U in $\mathscr{A}$ such that $U^*EU=F$. An operator $U\in\mathscr{B}(H)$ is unitary if $U^*U=UU^*=1$.)

As it stands this result belongs to operator algebra theory rather than lattice theory. However it has a number of extremely interesting lattice-theoretic consequences. In all that follows we shall use L to symbolize a complete orthomodular lattice.

COROLLARY 1. *If L is the projection lattice of a von Neumann algebra, then perspectivity is transitive in L.*

The proof uses only the fact that the product of two unitary operators is unitary. For if E is perspective to F and F to G, then by Fillmore's theorem there are unitary operations U and V in the algebra such that $U^*EU=F$ and $V^*FV=G$. The operator $W=UV$ is also in the algebra and is unitary (as a trivial manipulation shows) and $W^*EW=G$. Hence E and G are unitarily equivalent and therefore (by Fillmore's theorem again) perspective.

While the corollary to Kaplansky's theorem tells us that the modularity of L guarantees the transitivity of perspectivity, Corollary 1 reveals that modularity is by no means essential for this transitivity since transitivity obtains in the projection lattice of any von Neumann algebra modular or not. In his expository paper [21], Halperin (referring to continuous geometry) states: "It would be interesting to track down just which parts of the axioms are essential to prove transitivity of perspectivity." Corollary 1 seems to say that in the orthomodular case the transitivity of perspectivity has really nothing to do with either modularity or continuity.

While the transitivity of perspectivity is itself lattice-theoretic, and could be adopted as an axiom, one would hope that it could be traced back to a simpler, more basic lattice property.

It would indeed be interesting to attempt Halperin's program and try to isolate such a property.

COROLLARY 2. *Suppose L is the projection lattice of a von Neumann algebra. If the elements E, F of L are perspective, then there is an ortho-automorphism θ of L such that $\theta(E)=F$.*

Roughly speaking, this corollary states that every perspectivity in L is the restriction of an ortho-automorphism of L. (An ortho-automorphism θ of the orthomodular lattice L is a one-to-one mapping of L onto itself such that $E \leqslant F \Leftrightarrow \theta(E) \leqslant \theta(F)$ and $\theta(E^{\perp})=\theta(E)^{\perp}$.) Corollary 2 is, like Corollary 1, a simple consequence of Fillmore's theorem. For if E and F are perspective in L, then there is a unitary U such that $U^{*}EU=F$. Then a routine check shows that the map $\theta(X)=U^{*}XU$ is an ortho-automorphism of L such that $\theta(E)=F$.

Corollary 2 ties the internal structure of L to its automorphism group and exhibits a purely lattice-theoretic regularity of projection lattices not shared by the general run of orthomodular lattices. We refer the reader to [11] for some other interesting consequences of the theorem.

Since the proof of Fillmore's theorem is rather technical, I shall not attempt to summarize it. However, there is a special case which is really the heart of the proof and I shall say a few words about this. Consider a separable infinite dimensional Hilbert space H with orthonormal basis $\{\ldots, e_{-2}, e_{-1}, e_0, e_1, e_2, \ldots\}$. Let E be the closed subspace spanned by the vectors $\{e_0, e_1, e_2, \ldots\}$ and F that spanned by $\{e_1, e_2, \ldots\}$. Then $F \leqslant E$ and $E-F$ is one dimensional being spanned by the vector e_0. Moreover E and F are unitarily equivalent in H by the "shift" operator. Hence, if Fillmore's theorem is really true, then E and F must be perspective. That is, there must exist a closed subspace X of H such that $E \vee X = F \vee X = H$, $E \wedge X = F \wedge X = 0$. Puzzle for the reader: Find such a subspace (explicitly in terms of the e_i).

F. *Gleason's Theorem*

As Professor Birkhoff notes in his paper (Section 10), in modern measure theory (espoused particularly by Caratheodory) one considers measures defined on an abstract Boolean lattice, without requiring in advance any specific representation of this lattice as a field of sets. The defining conditions are the same as usual (let B denote the Boolean lattice in question):

(1) $0 \leqslant \mu(x) \leqslant +\infty$ for all $x \in B$;

(2) $\mu(0)=0$;

(3) if $\{a_\alpha; \alpha \in A\}$ is an orthogonal family, then $\mu(\bigvee a_\alpha)=\sum \mu(a_\alpha)$. By specifying the cardinality of the set A in (3) (and adding appropriate assumptions about the completeness of B) we can obtain completely additive measures (card (A) arbitrary), countably additive measures (card $(A)=\aleph_0$), and finitely additive measures (card $(A)=2$). The other qualifying adjectives can also be added: μ is *finite if* $\mu(1)<\infty$, and so on.

The point we wish to stress about this definition is that it makes no reference at all to the *distributivity* of the underlying lattice. The orthogonality is all that is needed. Hence the definition applies as it stands to the arbitrary orthomodular lattice. (Orthomodular rather than just simply orthocomplemented because an orthocomplemented nonorthomodular lattice can carry a faithful finite measure μ which has the undesirable property that for some a, b we have $a<b$ but yet $\mu(a)=\mu(b)$.)

Such measures on general orthomodular lattices have been considered by Bodiou [6], Gudder [19], Holland [22], Ludwig [30], Mackey [32], Randall [46], Varadarajan [51], and probably others. (In fact some of these authors, notably Mackey, consider measures on general "orthomodular partially ordered sets.") Also the dimension function of Murray and von Neumann [40] is a special kind of measure on the projection lattice of a factor, the specialization consisting essentially in the requirement that the measure take the same value on equivalent projections. Hence, all in all, there has already been a great deal of attention devoted to these general measures, and, as a consequence, we have a large number of theorems about them. But the one result above all others that points to the existence of a respectable theory of measures on non-distributive lattices, even though it concerns a very particular lattice, is the following result due to Gleason:

THEOREM [Gleason, 17]. *Let H be a real or complex separable Hilbert space of dimension* $\geqslant 3$. *Then every finite countably additive measure* μ *on the orthomodular projection lattice* $\mathscr{P}(\mathscr{B}(H))$ *has the form*

$$\mu(E) = \operatorname{trace}(TE) \text{ for all } E \in \mathscr{P}(\mathscr{B}(H))$$

where T is an operator of trace class (a fixed T for each μ*).*

This is a Radon-Nikodym theorem, the operator T being the Radon-Nikodym derivative of the measure μ with respect to the canonical measure "trace." The remarkable feature about Gleason's result, and the feature that allows us to quote this as part of lattice theory rather than operator algebra theory, is its generality in assuming that μ is only a *measure* on the projection lattice as contrasted with the stronger assumption that μ is a *linear functional* on the algebra $\mathscr{B}(H)$. In fact, if we know that μ were the restriction of a positive linear functional, then Gleason's theorem would follow from known results even without the restriction $\dim(H) \geqslant 3$ (see for example Dixmier [9, p. 54, Theorem 1]).

This brings us to one of the major differences between "distributive" and "non-distributive" measure theory. Given the measure μ of Gleason's theorem we can easily extend it to $\mathscr{B}(H)$ by first extending to the self-adjoint operators using the spectral theorem,

$$\mu(S) = \int_{-\infty}^{+\infty} \lambda \, d\mu(E_\lambda),$$

where (E_λ) is the spectral family of the arbitrary self-adjoint operator S, and then extending to non-self-adjoint S by additivity. This is essentially the same procedure as is used in extending an ordinary measure on sets to an integral on functions. The difference is this: in the "distributive" case it is an easy matter to prove that the integral is additive, but in the non-distributive case it is extremely difficult. In fact if we could prove the additivity of the extension just described, we would have a proof of Gleason's theorem (since μ would then be the restriction of a positive linear functional).

It was exactly this same difficulty that plagued Murray and von Neumann in their extension of a dimension function to a trace. However, here they had the advantage that the measure to be extended (namely the dimension function) was invariant under equivalence. Even so the proof was (and still is) extremely difficult. Gleason, of course, assumed no invariance.

After Gleason's theorem the next natural question is: what is the corresponding result for measures on the projection lattices of other factors? As far as I know, nothing is known. If the answers were known, one would then hope to extend the projection lattice results to measures on abstract orthomodular lattices (see [22]). However, such a program should probably wait until we get the answers in the projection lattice case and until we learn more about the structure of orthomodular lattices.

We close this section with a historical note. The question that Gleason answered by determining all the measures on $\mathscr{P}(\mathscr{B}(H))$ arose in connection with investigations of the foundations of quantum mechanics and was posed originally by Mackey.

G. *Dye's Theorem*

The theorem we refer to is this:

THEOREM [Dye, 10]. *Let M be a von Neumann algebra with no direct summands of type I_2 in the large. Then any orthoisomorphism between the projection lattice of M and that of a von Neumann algebra N is implemented by the direct sum of a $*$-isomorphism and a $*$-anti-isomorphism.*

It would take us too far into the theory of von Neumann algebras to clarify completely the statement and meaning of Dye's theorem. We can for our purposes put across its main idea with the loose paraphrase. *If the lattices are isomorphic, then the algebras are isomorphic.*

We have already mentioned this special case of von Neumann's coordinatization theorem: A complete orthocomplemented modular lattice with four or more independent perspective elements [of lub 1] is isomorphic to the lattice of projections of a complete $*$-regular ring. One would like to prove a similar theorem for complete orthomodular lattices, or for a subclass of the complete orthomodular lattices large enough to contain all the projection lattices. (Of course, then regular rings are out and a suitable generalization has to be found.) One valuable feature of Dye's theorem is that it gives us in advance a prototype uniqueness theorem for such a representation.

As we have said we cannot go too deeply into this theorem without becoming enmeshed in the technicalities of the theory of von Neumann algebras. But we recommend [10] to the reader for a closer study of this important result.

V. SOME BRIEF REMARKS ON QUANTUM MECHANICS

Orthomodular lattice theory is closely connected with the study of "quantum logic," more general "logic of physical systems" and indeed, although this latter connection has yet to be explored in any depth, with "pure" mathematical logic. The first item – the lattice-theoretic aspect of the foundations of quantum mechanics – is now under especially intense scrutiny, both by mathematicians and physicists alike. In fact the extraordinary effort being applied to this problem accounts for a substantial part of the current interest being shown in orthomodular lattices. I therefore regret that, because of lack of space (and of time), I was not able to include a complete discussion of this topic. However, a fine survey report by Mac Laren [34] is available for those who wish to pursue it further. Mac Laren has a representative 24-item bibliography, and the recent report by Randall [46] contains a rich up-to-date 130-item bibliography. We have included in part three of our bibliography a few items not listed by either Mac Laren or Randall. Taken all together, these listings constitute as complete and current a bibliography as is probably possible to give for such a rapidly developing subject. With these materials the interested reader should have no trouble working his way into the subject. The basic references that are referred to most often are the 1936 paper of Birkhoff and von Neumann [5], von Neumann's book [43], and Mackey's book [32]. Incidentally, Randall's impressive and scholarly report [46], the beginning of an ambitious attempt to provide a "universal mathematical foundation for all empirical sciences," makes very interesting reading.

Perhaps I should make a few general comments here to whet the curiosity of the reader who might be wondering what in the world lattice theory has got to do with quantum mechanics. Roughly, the connection is this: The yes-or-no questions, or propositions, about a physical system (example of such a proposition: is the energy greater than six ergs?) are partially ordered by "implication" and have the natural orthocomplementation wherein the orthocomplement of a proposition a is its negation, *not-a*. In classical theory, these propositions constitute a

Boolean lattice (Boolean algebra). But in quantum theory this is no longer true. The distributivity appears to be inconsistent with the essential new features of quantum mechanics, notably the Heisenberg uncertainty principle. There seems to be general agreement among all investigators that these propositions constitute an "orthomodular partially ordered set"; but there the agreement ends. There seems to be no clear-cut reason why they should form a lattice, and indeed one may argue the question both ways (see [34, p. 10] and [3]). A standard form of quantum mechanics can be obtained by identifying the set of propositions with the lattice of all projections on a separable infinite dimensional complex Hilbert space – our old friend $\mathscr{P}(\mathscr{B}(H))$. As we now well know $\mathscr{P}(\mathscr{B}(H))$ is rather special even among the orthomodular lattices, let alone among the orthomodular partially ordered sets, and a great deal of current work is directed toward answering the questions: Is such a specialization necessary? Can it be justified on physical (or philosophical) grounds? Are there other alternatives? Perhaps the reader of these lines will now be sufficiently intrigued to sally forth and tackle these vexing problems for himself.

University of Massachusetts, Amherst

Note added January 1969

Since this expository and survey article contains much historical material, and since very recent advances in the theory are taken up in the Proceedings of the Conference on Orthomodular Lattices and Empirical Logic (held at the University of Massachusetts in the summer of 1968), I have decided, in spite of the long delay in publication, to make no revision of it, and am publishing it here exactly as it was written in the summer of 1966.

NOTE

* Reprinted from *Trends in Lattice Theory* (ed. by J. C. Abbott), Van Nostrand Reinhold Math. Studies #31, Van Nostrand Reinhold, New York, 1970.

BIBLIOGRAPHY

This bibliography is in three parts. The first part contains works referred to in the text.

The second part contains papers not referenced in the text but nonetheless relevant to the subjects treated there. The papers on the logic of quantum mechanics I have treated as a separate category. For them there are already the excellent bibliographies in the reports of Mac Laren [34] and Randall [46]. I have accordingly limited myself here to listing recent quantum mechanics papers not referenced by them; this material is in part three. These three parts, together with Mac Laren's and Randall's bibliographies, constitute (I hope) a reasonably complete up-to-date bibliography for orthomodular lattice theory and related topics.

PART 1. WORKS REFERENCED IN THE TEXT

[1] Amemiya, I. and Halperin, I., 'Complemented Modular Lattices', *Canad. J. Math.* **11** (1950), 481–520.

[2] Amemiya, I. and Araki, H., 'A Remark on Piron's Paper', *Publ. Res. Inst. Math. Sci. Ser. A*, **12** (1966/67), 423–427.

[3] Birkhoff, G., 'Lattices in Applied Mathematics', from *Lattice Theory, Proc. Symp. Pure Math.* Vol. 11, Amer. Math. Soc., Providence 1961.

[4] Birkhoff, G., *Lattice Theory*, 3rd ed., Amer. Math. Soc. Colloq. Publ. Vol. 25 (1966).

[5] Birkhoff, G. and Neumann, J. von, 'The Logic of Quantum Mechanics', *Ann. of Math.* **37** (1936), 823–842.

[6] Bodiou, G., *Théorie dialectique des probabilités*, Gauthier-Villars, Paris, 1964.

[7] Dilworth, R. P., 'On Complemented Lattices', *Tohoku Math. J.* **47** (1940), 18–23.

[8] Dixmier, J., 'Position relative de deux variétés linéaires fermées dans un espace de Hilbert', *La Revue Scientifique*, Fasc. 7, **86** (1948), 387–399.

[9] Dixmier, J., *Les algèbres d'opérateurs dans l'espace hilbertien*, Gauthier-Villars, Paris, 1957.

[10] Dye, H. A., 'On the Geometry of Projections in Certain Operator Algebras', *Ann. of Math.* **61** (1955), 73–89.

[11] Fillmore, P. A., 'Perspectivity in Projection lattices', *Proc. Amer. Math. Soc.* **16** (1965), 383–387.

[12] Foulis, D. J., 'Baer *-Semigroups', *Proc. Amer. Math. Soc.* **11** (1960), 648–654.

[13] Foulis, D. J., 'A Note on Orthomodular Lattices', *Portugal. Math.* **21** (1962), 65–72.

[14] Foulis, D. J., 'Notes on Orthomodular Lattices', lecture notes, Univ. of Florida, 1963 (unpublished).

[15] Foulis, D. J., 'Lattices and Semigroups', mimeographed, unpublished.

[16] Gardner, L. T., 'On Isomorphisms of C*-Algebras', *Amer. J. Math.* **87** (1965), 384–396.

[17] Gleason, A. M., 'Measures on the Closed Subspaces of a Hilbert Space', *J. Math. Mech.* **6** (1957), 885–894.

[18] Greechie, R. J., 'On the Structure of Orthomodular Lattices Satisfying the Chain Condition', (to appear).

[19] Gudder, S. P., 'Spectral Methods for a Generalized Probability Theory', *Trans. Amer. Math. Soc.* **119** (1965), 428–442.

[20] Halmos, P. R., *Introduction to Hilbert Space and the Theory of Spectral Multiplicity*, Chelsea, New York, 1957.

[21] Halperin, I., 'Complemented Modular Lattices', from *Lattice Theory, Proc. Symp. Pure Math.* Vol. II, Amer. Math. Soc., Providence, 1961.

[22] Holland, Jr., S. S., 'A Radon-Nikodym Theorem in Dimension Lattices', *Trans. Amer. Math. Soc.* **108** (1963), 66–87.

[23] Holland, Jr., S. S., 'Distributivity and Perspectivity in Orthomodular Lattices', *Trans. Amer. Math. Soc.* **112** (1964), 330–343.

[24] Janowitz, M. F., 'A Semigroup Approach to Lattices', *Canad. J. Math.* **18** (1966), 1212–1223.
[25] Janowitz, M. F., 'Notes on Orthomodular Lattices', Univ. of New Mexico, 1965 (unpublished).
[26] Janowitz, M. F., 'On Conditionally Continuous Lattices', Univ. of New Mexico Tech. Report no. 107 (1966).
[27] Kaplansky, I., 'Any Orthocomplemented Complete Modular Lattice is a Continuous Geometry', *Ann. of Math.* **61** (1955), 524–541.
[28] Kaplansky, I., 'Rings of Operators', University of Chicago Mimeographed Notes, 1955.
[29] Loomis, L. H., 'The Lattice Theoretic Background of the Dimension Theory of Operator Algebras', *Mem. Amer. Math. Soc.*, **18** (1955), 36 pp.
[30] Ludwig, G., 'Versuch einer axiomatischen Grundlegung der Quantenmechanik und allgemeiner physikalischer Theorien', *Z. Physik* **181** (1964), 233–260 (mimeographed English translation available).
[31] Mackey, G. W., 'On Infinite-Dimensional Linear Spaces', *Transl. Amer. Math., Soc.* **57** (1945), 155–207.
[32] Mackey, G. W., *Mathematical Foundations of Quantum Mechanics*, Benjamin, New York, 1963.
[33] Mac Laren, M. D., 'Nearly Modular Orthocomplemented Lattices', *Trans. Amer. Math. Soc.* **114** (1965), 401–416.
[34] Mac Laren, M. D., 'Notes on Axioms for Quantum Mechanics', Argonne National Lab. Rept. ANL-7065, July 1965 (available from the Clearinghouse for Federal Scientific and Technical Information, National Bureau of Standards, U.S. Dept. of Commerce, Springfield, Virginia, $ 1.00).
[35] Maeda, F., 'Relative Dimensionality in Operator Rings', *J. Sci. Hiroshima Univ. Ser. A* 11 (1941), 1–6.
[36] Maeda, F., *Kontinuierliche geometrien*, Springer, Berlin, 1958.
[37] Maeda, F. and Maeda, S., *Theory of Symmetric Lattices*, book in press.
[38] Maeda, S., 'Dimension Functions on Certain General Lattices', *J. Sci. Hiroshima Univ. Ser. A-*1 *Math.* **19** (1955), 211–237.
[39] Maeda, S., 'On Conditions for the Orthomodularity', *Proc. Japan Acad.* **42** (1966), 247–251.
[40] Murray, F. J. and Neumann, J. von, 'On Rings of Operators', *Ann. of Math.* **37** (1936), 116–229.
[41] Nakamura, M., 'The Permutability in a Certain Orthocomplemented Lattice', *Kodai Math. Series. Rep.* **9** (1957), 158–160.
[42] Neumann, J. von, 'Continuous Geometry', *Proc. Natl. Acad. Sci. U.S.A.* **22** (1936), 92–100.
[43] Neumann, J. von, *Mathematical Foundations of Quantum Mechanics*, Princeton Univ. Press, Princeton, 1955.
[44] Neumann, J. von, *Continuous Geometry*, Princeton Univ. Press, Princeton 1960.
[45] Ramsay, A., 'Dimension Theory in Complete Orthocomplemented Weakly Modular Lattices', *Trans. Amer. Math. Soc.* **116** (1965), 9–31.
[46] Randall, C. H., 'A Mathematical Foundation for Empirical Science – with Special Reference to Quantum Theory. Part 1, A Calculus of Experimental Propositions', Knolls Atomic Power Lab. Report. KAPL-3147. June 1966. Available from the Clearinghouse for Federal Scientific and Technical Information. National Bureau of Standards, U.S. Department of Commerce. Springfield. Virginia, $ 6.00.

[47] Segal, I. E., 'A Non-Commutative Extension of Abstract Integration', *Ann. of Math.* (2) **57** (1953), 401–457.

[48] Simmons, G. F., *Topology and Modern Analysis*, McGraw-Hill, 1963.

[49] Stone, M. H., 'Linear Transformations in Hilbert Space', *Amer. Math. Soc., Colloq. Publ.* Vol. 15, Amer. Math. Soc., Providence, 1932.

[50] Topping, D. M., 'Asymptoticity and Semimodularity in Projection Lattices', to appear.

[51] Varadarajan, V. S., 'Probability in Physics and a Theorem on Simultaneous Observability', *Comm. Pure and Appl. Math.* **15**, (1962), 189–217.

[52] Zierler, N. 'Axioms for Non-Relativistic Quantum Mechanics', *Pacific J. Math.* **11** (1961), 1151–1169.

[53] Zierler, N., 'On the Lattice of Closed Subspaces of Hilbert Space', *Pacific J. Math.* **19** (1966), 583–586.

PART 2. WORKS NOT REFERENCED IN THE TEXT

Berberian, S. K., 'On the Projection Geometry of a Finite AW*-Algebra', *Trans. Amer. Math. Soc.* **83** (1956), 493–509.

Bodiou, G., 'Probabilité sur une trellis non modulair', *Publ. Inst. Statist.*, Univ. Paris **6** (1957), 11–25.

Bodiou, G., *Recherches sur les fondements du calcul quantique des probabilité dans les cas purs*, Masson, Paris, 1950.

Croisot, R., 'Applications Residuées', *Ann. Sci. Ecole Norm. Sup.* **73** (1956), 453–474.

Dilworth, R. P., 'The Structure of Relatively Complemented Lattices', *Ann. of Math.* **51** (1950), 348–359.

Davis, C., 'Separation of Two Linear Subspaces', *Acta Sci. Math.* (Szeged) **19** (1958), 172–187.

Dye, H. A., 'The Radon-Nikodym Theorem for Finite Rings of Operators', *Trans. Amer. Math. Soc.* **72** (1952), 243–280.

Foulis, D. J., 'Conditions for the Modularity of an Orthomodular Lattice', *Pacific J. Math.* **11** (1961), 889–895.

Foulis, D. J., 'Relative Inverses in Baer *-Semigroups', *Mich. Math. J.* **10** (1963), 65–84.

Foulis, D. J., 'Semigroups Coordinatizing Orthomodular Geometries', *Canad. J. Math.* **17** (1960), 40–51.

Gratzer, G. and Schmidt, E. T., 'Ideals and Congruence Relations in Lattices', *Acta. Math. Acad. Sci. Hungar* **9** (1958), 137–175.

Gratzer, G. and Schmidt, E. T., 'Standard Ideals in Lattices', *Acta. Math. Acad. Sci. Hungar* **12** (1961), 17–86.

Halmos, P. R., *Algebraic Logic*, Chelsea, New York, 1962.

Halperin, I., 'Introduction to von Neumann Algebras and Continuous Geometry', *Canad. Math. Bull.* **3** (1960), 273–288.

Iqbalunnisa, 'Neutrality in Weakly Modular Lattices', *Acta. Math. Acad. Sci. Hunger.* **26** (1965), 325.

Iqbalunnisa, 'On Neutral Elements in a Lattice', *J. Indian Math. Soc.* (New series) **28** (1964), 25–31.

Janowitz, M. F., 'Quasi-Orthomodular Lattices', Univ. of New Mexico Tech. Report No. 42 (1963).

Janowitz, M. F., 'AC-Lattices', Univ. of New Mexico Tech. Report No. 45 (1963).

Janowitz, M. F., 'Quantifiers and Orthomodular Lattices', *Pacific J. Math.* **13** (1963), 1241–1249.

Janowitz, M. F., 'On the Antitone Mappings of a Poset', *Proc. Amer. Math. Soc.* **15** (1964), 529–533.
Janowitz, M. F., 'Projective Ideals and Congruence Relations', Univ. of New Mexico Tech. Report No. 51 (1964).
Janowitz, M. F., 'Projective Ideals and Congruence Relations II', Univ. of New Mexico Tech. Report No. 63 (1964).
Janowitz, M. F., 'Residuated Closure Operators', Univ. of New Mexico Tech. Report No. 79 (1965).
Janowitz, M. F., 'A Note on Normal Ideals', Univ. of New Mexico Tech. Report No. 95 (1965).
Janowitz, M. F., 'A Characterization of Standard Ideals', *Acta. Math. Acad. Sci. Hungar* **26** (1965), 289–301.
Janowitz, M. F., 'Quantifier Theory on Quasi-Orthomodular Lattices', *Illinois J. Math.* **9** (1965), 660–676.
Janowitz, M. F., 'Baer-Semigroups', *Duke Math. J.* **32** (1965), 85–96.
Janowitz, M. F., 'Independent Complements in Lattices', Univ. of New Mexico Tech. Report No. 87 (1965).
Janowitz, M. F., 'The Center of a Complete Relatively Complemented Lattice is a Complete Sublattice', Univ. of New Mexico Tech. Report No. 105 (1966).
Kaplansky, I., 'Projections in Banach Algebras', *Ann. of Math.* **53** (1951), 235–249.
Mac Laren, M. D., 'Atomic Orthocomplemented Lattices', *Pacific J. Math.* **14** (1964), 597–612.
Maeda, F., 'Direct Sums and Normal Ideals of Lattices', *J. Sci. Hiroshima Univ. Ser. A-I Math.* **14** (1949), 85–92.
Maeda, F., 'Representations of Orthocomplemented Lattices', *J. Sci. Hiroshima Univ. Ser. A* **14** (1950), 1–4.
Maeda, F., 'Decomposition of General Lattices into direct Summands of Types I, II and III', *J. Sci. Hiroshima Univ. Ser. A* **23** (1959), 151–170.
Maeda, S., 'On the Lattices of Projections of a Baer *-Ring', *J. Sci. Hiroshima Univ. Ser. A* **22** (1958), 76–88.
Maeda, S., 'On Relatively Semi-Orthocomplemented Lattices', *J. Sci. Hiroshima Univ. Ser. A* **24** (1960), 155–161.
Maeda, S., 'On a Ring whose Principal Right Ideals Generated by Idempotents Form a Lattice', *J. Sci. Hiroshima Univ. Ser. A* **24** (1960), 510–525.
Maeda, S., 'Dimension Theory on Relatively Semi-Orthocomplemented Complete Lattices', *J. Sci. Hiroshima Univ. Ser. A* **25** (1961), 369–404.
Maeda, S., 'On the Symmetry of the Modular Relation in Atomic Lattices', *J. Sci. Hiroshima Univ. Ser. A-I* **29** (1965), 165–170.
Morgado, J., 'On the Automorphisms of the Lattice of Closure Operators of a Complete Lattice', *Rev. Un. Mat. Argentina* **20** (1960), 188–193.
Morgado, J., 'Some Results on Closure Operators of Partially Ordered Sets', *Portugal Math.* **19** (1960), 101–139.
Morgado, J., 'Note on the Automorphisms of the Lattice of Closure Operators of a Complete Lattice', *Nederl. Akad. Wetensch. Proc. Ser. A* **64** (1961), 211–218.
Morgado, J., 'Some Remarks on Quasi-Isomorphisms between Finite Lattices', *Portugal Math.* **20** (1961), 137–145.
Morgado, J., 'Quasi-Isomorphisms between Complete Lattices', *Portugal Math.* **20** (1961), 17–31.
Nakamura, N., 'Center of Closure Operators and a Decomposition of a Lattice', *Math. Japan* **4** (1964), 49–52.

Neumann, J. von, *Collected Works*, 6 vols., Pergamon, 1961.
Sachs, D., 'Partition and Modulated Lattices', *Pacific J. Math.* **11** (1961), 325–345.
Sasaki, U., 'On an Axiom of Continuous Geometry', *J. Sci. Hiroshima Univ. Ser. A* (1950), 100–101.
Sasaki, U., 'On Orthocomplemented Lattices Satisfying the Exchange Axiom', *J. Sci. Hiroshima Univ. Ser. A* **17** (1954), 293–302.
Sasaki, U., 'Lattices of Projections in AW∗-Algebra', *J. Sci. Hiroshima Univ. Ser. A* **19** (1955), 1–30.
Schmidt, E. T., 'Remark on a Paper of M. F. Janowitz', *Acta. Math. Acad. Sci. Hungar.* **16** (1965), 435.
Schreiner, E. A., 'Modular Pairs in Orthomodular Lattices', *Pacific J. Math.* **19** (1966), 519–528.
Topping, D. M., 'Jordan Algebra of Self-Adjoint Operators', *Mem. Amer. Math. Soc.* **53** (1965), 48 pp.
Wright, F. B., 'A Reduction for Algebras of Finite Type', *Ann. of Math.* **60** (1954), 560–570.
Wright, F. B., 'Some Remarks on Boolean Duality', *Portugal Math.* **16** (1957), 109–117.
Zierler, N. and Schlessinger, M., 'Boolean Embeddings of Orthomodular Sets and Quantum Logic', *Duke Math. J.* **32** (1965), 251–262.

PART 3. SOME RECENT PAPERS ON QUANTUM MECHANICS NOT LISTED IN [34] OR [46]

Jordan, P., 'Quantenlogik und das kommutative Gesatz', from *The Axiomatic Method* (ed. by L. Henkin, P. Suppes, and A. Tarski, Amsterdam 1959.
Ludwig, G., 'An Axiomatic Foundation of Quantum Mechanics on a Non-Subjective Basis', Marburg/L. Mimeographed, unpublished.
Ludwig, G., 'Attempt of an Axiomatic Foundation of Quantum Mechanics and More General Physical Theories II', mimeographed, unpublished.
Rose, G., 'Zur Orthomodularität von Wahrscheinlichkeitsfeldern', *Z. Physik* **181** (1964), 331–332.
Weizsacker, C. G. von, 'Komplementarität und Logik I', *Naturwissenschaften* **42** (1955), 521.
Weizsacker, C. G. von, II, *Ž. Naturforsch.* **13a** (1958), 245.
Weizsacker, C. G. von, III, *Z. Naturforsch.* **13a** (1958), 705.

N. S. KRONFLI

INTEGRATION THEORY OF OBSERVABLES

ABSTRACT. Representations of abstract observables on a generalised logic are given in terms of bounded vector-valued Borel measures on the real line whose ranges are in the dual space X^* of the Banach space of states X. Each bounded observable is furthermore represented by an element u^* of X^* such that for any proper state $p \in X$, $u^*(p)$ is the expectation value of u when the system is in the state p.

I. GENERALISED QUANTUM THEORY

By *generalised quantum* theory is meant in this paper the list $(\mathscr{L}, \mathscr{S}, \mathscr{O})$ where $\mathscr{L}$ is the proposition system assumed to form an orthocomplemented weakly modular σ-lattice which we call *generalised logic*, $\mathscr{S}$ is the set of (proper) states consisting of all the probability measures on $\mathscr{L}$ and $\mathscr{O}$ is the set of observables consisting of all the σ-homomorphisms on the Borel σ-algebra $\mathscr{B}$ of the real line R into $\mathscr{L}$. Another ingredient to consider is the group $\mathrm{Aut}(\mathscr{S})$ consisting of all the convex automorphisms of $\mathscr{S}$. This contains the symmetry operations of the system under consideration. For definitions see Varadarajan (1968, pp. 105–130).

With such weak conditions on $\mathscr{L}$, the theory is very general and includes both quantum and classical mechanics as special cases. Extra conditions must be imposed on the logic so as to reproduce the conventional quantum formalism in terms of a separable Hilbert space. These, however, are not very clear physically. Restricting ourselves to the generalised case $(\mathscr{L}, \mathscr{S}, \mathscr{O})$, it is necessary to carry out further mathematical developments in order to investigate rigorously questions connected with particles, localisability, dynamics, symmetries and scattering – some of the principal considerations of any physical theory. The main technical problem is that of obtaining useful representations of $\mathscr{S}$, $\mathscr{L}$, $\mathrm{Aut}(\mathscr{S})$ and $\mathscr{O}$ in terms of entities connected with some topological vector space. Our aim in this series is to obtain such representations which, although not as sharp as in the quantum case, are strong enough so as to bring the subject into grips with modern analysis. In the rest of this section are outlined the

C. A. Hooker (ed.), The Logico-Algebraic Approach to Quantum Mechanics, 497–502.

results obtained showing that the generalised theory is quite manageable mathematically.

In a previous paper (Kronfli, 1970) representations of $\mathscr{S}$ and Aut$(\mathscr{S})$ were obtained. Firstly we chose the most important topology, from the physical point of view, on $\mathscr{S}$, namely that defined by the *natural metric*

$$\rho(p, q)=\sup\{|p(a)-q(a)|: a\in\mathscr{L}\} \quad (p, q\in\mathscr{S})$$

with $\mathscr{S}$ being assumed separating. Let X_1 be the set of all signed measures on $\mathscr{L}$ with finite variations. Defining

$$\|p\|=|p|(I) \quad (p\in X_1)$$

where $a\to|p|(a)$ is the total variation of p at $a\in\mathscr{L}$ and I is the identity on $\mathscr{L}$ the following were proved

THEOREM 1.1. *$(X_1, \|\cdot\|)$ is a real Banach space containing $\mathscr{S}$ as a closed convex subset with the norm $\|\cdot\|$ inducing the natural metric.*

THEOREM 1.2. *Each convex automorphism of $\mathscr{S}$ is represented uniquely by a unit-normed linear one-one operator on $(X, \|\cdot\|)$ onto itself, where X is the closure of the linear span of $\mathscr{S}$ in $(X_1, \|\cdot\|)$.*

These two theorems bear some resemblance to the corresponding ones in quantum theory and for this reason we call X the *Banach space of states*. Problems connected with symmetries, for example, are now easy to handle using operator theory. For an application to abstract scattering see Kronfli (1969).

To complete the picture we consider in this paper the representation theory of the observables $\mathscr{O}$ on the generalised logic $\mathscr{L}$. For quantum logic the observables are represented by projection-valued (spectral) measures on the Borel subsets of R which is equivalent to representing them by the corresponding self-adjoint operators on the Hilbert space of (quantum) states $\mathscr{H}$. It is now the practice to regard each self-adjoint operator on $\mathscr{H}$ as a quantum observable although whether it can actually be observed or not is still a question of debate. Here we shall obtain results rather similar to the quantum case. Each observable u on the generalised logic $\mathscr{L}$ is represented (one-one) by an X^*-valued weakly countably additive Borel measure on R satisfying special boundedness properties similar to those of spectral measures in the quantum case. The set of all such measures is denoted by $\mathscr{M}$. Here X^* is the (Banach) dual of X. The question of regarding each element of $\mathscr{M}$ as an observable is investigated. In fact we give conditions on $\mathscr{O}$ such that each measure in $\mathscr{M}$ represents an observable. When $\mathscr{O}$ satisfies these conditions $\mathscr{M}$ and $\mathscr{O}$ can be identified.

Finally we consider the bounded observables $\mathscr{O}_0$ on $\mathscr{L}$. We shall prove that each $u \in \mathscr{O}_0$ is represented simply by a continuous linear functional u^* on X, i.e. by an element $u^* \in X^*$, with the desirable property that for any state $p \in \mathscr{S}$, $u^*p = \langle p, u^* \rangle$ is the expectation value of u when the system is in the state p. This gives an important physical role to X^* and we can call it the *Banach space of bounded observables.*

II. INTEGRATION THEORY OF OBSERVABLES

The first result is an injection of the logic $\mathscr{L}$ into the dual X^* of X. We shall adopt the following notation: for any linear functional f on X we write

$$\langle p, f \rangle \quad \text{for} \quad f(p) \quad (p \in X)$$

PROPOSITION 2.1. *There exists a natural injection $T: \mathscr{L} \to X^*$ such that*

(i) $T(\emptyset) = 0$,

(ii) $\|T(a)\| \leqslant 1 \quad (a \in \mathscr{L}), \quad \|T(I)\| = 1$,

(iii) *T is weakly countably additive on $\mathscr{L}$.*

Proof: For each $a \in \mathscr{L}$ consider the map $T(a): p \to p(a)$ on $X \to R$. This is clearly a single-valued real linear functional on X such that $T(\emptyset) = 0$. Furthermore, for any $p \in X$,

$$|\langle p, T(a) \rangle| = |p(a)| \leqslant |p|(a) \leqslant |p|(I) = \|p\|$$

implying that $T(a) \in X^*$ and $\|T(a)\| \leqslant 1$. Also for $p \in \mathscr{S}$, $\langle p, T(I) \rangle = p(I) = 1$ which means that $\|T(I)\| = 1$.

The map $a \to T(a)$ is an injection because $T(a) = T(b)$ implies that $p(a) = p(b)$ for all $p \in X$ and since $\mathscr{S}$ is separating and is contained in X this in turn implies that $a = b$.

Thus T is one-one.

It remains to prove (iii). Let (a_n) be a disjoint sequence in $\mathscr{L}$. Then for any $p \in X$

$$\langle p, T(\bigvee_n a_n) \rangle = p(\bigvee_n a_n) = \sum_n p(a_n) = \sum_n \langle p, T(a_n) \rangle.$$

This completes the proof. ■

COROLLARY 2.2. *The natural injection T of* Proposition 2.1 *induces a metric on $\mathscr{L}$ given by*

$$d(a,b)=\|T(a)-T(b)\| \quad (a,b\in\mathscr{L})$$

DEFINITION 2.3. Let $\mathscr{M}^+$ be the set of all bounded weakly countably additive X^*-valued measures on the Borel σ-algebra $\mathscr{B}$ of the real line. This is a real Banach space when equipped with the norm

$$\|\mu\|_1=\sup\{\|\mu(A)\|:A\in\mathscr{B}\} \quad (\mu\in\mathscr{M}^+)$$

For a proof see Dunford and Schwartz (1958). From now on $\|\cdot\|$ without a suffix will denote the norms of X or X^*. Finally we define

$$\mathscr{M}=\{\mu\in\mathscr{M}^+:\|\mu(A)\|\leqslant 1,\ \langle p,\mu(A)\rangle\geqslant 0 \quad (p\in\mathscr{S},\ A\in\mathscr{B}),$$
$$\|\mu(R)\|=1\}.$$

The next result is the main representation theorem for $\mathscr{O}$.

THEOREM 2.4. *There exists a one-one map $u\to\hat{u}$ on $\mathscr{O}$ into $\mathscr{M}$ such that*

$$\langle p,\hat{u}(A)\rangle=p(u(A)) \quad (p\in X,\ A\in\mathscr{B})$$

Proof: The required mapping $u\to\hat{u}$ is given by $\hat{u}=Tou$ where T is the injection in Proposition 2.1. The rest of the proof is straightforward using Proposition 2.1. ■

Remarks. Let $u\in\mathscr{O}$ and $\hat{u}\in\mathscr{M}$ be the corresponding vector-valued measure as in Theorem 2.4. Then for any proper state $p\in\mathscr{S}$ and $A\in\mathscr{B}$, $\langle p,\hat{u}(A)\rangle$ is the probability of finding the observable u in the Borel set A when the system is in the state p. Note also that $\|\hat{u}\|_1=1$. Compare the similarity of the above representation theorem with the corresponding one in quantum theory.

The next result is an important property of the set $\mathscr{M}$.

THEOREM 2.6. *The set $\mathscr{M}$ is a closed subset of* $(\mathscr{M}^+,\|\cdot\|_1)$.

Proof. Let (μ_n) be a sequence in $\mathscr{M}$ which converges to μ in $(\mathscr{M}^+,\|\cdot\|_1)$. Note that $\|\mu_n\|_1=1$ and for any $v\in\mathscr{M}^+$, $\|v(A)\|\leqslant\|v\|_1$ for all $A\in\mathscr{B}$. Thus

$$\begin{aligned}|[1-\|\mu(R)\|]|&=|[\|\mu_n(R)\|-\|\mu(R)\|]|\\&\leqslant\|\mu_n(R)-\mu(R)\|\\&\leqslant\|\mu_n-\mu\|_1\to 0 \quad (n\to\infty).\end{aligned}$$

Hence $\|\mu(R)\|=1$. Furthermore, for each $A\in\mathscr{B}$ the sequence $(\|\mu_n(A)\|)$ is bounded by 1 and $\mu_n\to\mu$ implying that $\|\mu(A)\|\leqslant 1$. Similarly $\langle p, \mu(A)\rangle\geqslant 0$ for all $p\in\mathscr{S}$ and $A\in\mathscr{B}$. Thus $\mu\in\mathscr{M}$ and $\mathscr{M}$ is closed. ■

III. TOTAL AND BOUNDED OBSERVABLES

We now come to the question of when can one regard each element of $\mathscr{M}$ as representing an observable. It is obvious that with each observable u there is a probability measure on R corresponding to each state of the system. It sounds reasonable to postulate the converse which roughly says that each map on $\mathscr{S}$ into the class of probability measures on R which satisfies certain properties defines an observable. No philosophical discussion of this is attempted here. We only give the mathematical formulation.

DEFINITION 3.1. Let $\mathscr{P}$ be the set of all probability measures on the real line and let Hom $(\mathscr{S}, \mathscr{P})$ be the set of all convex homomorphisms on $\mathscr{S}$ into $\mathscr{P}$. Then a set of observables $\mathscr{O}$ on the logic $\mathscr{L}$ is said to be *total* if to each element of Hom $(\mathscr{S}, \mathscr{P})$ corresponds a unique observable in $\mathscr{O}$.

THEOREM 3.2. *Let $\mathscr{O}$ be total. Then to each element of $\mathscr{M}$ corresponds an observable.*

Proof: Let $\mu\in\mathscr{M}$ and define $P_\mu(E)=\langle p, \mu(E)\rangle$ $(E\in\mathscr{B})$ for each $p\in\mathscr{S}$. By Definition 2.3, $\langle p, \mu(E)\rangle\geqslant 0$ when $p\in\mathscr{S}$. Furthermore,

$$P_\mu(E)=|\langle p, \mu(E)\rangle|\leqslant \|p\|\cdot\|\mu(E)\|\leqslant 1$$

since $p\in\mathscr{S}$ implies $\|p\|=1$. Also by definition of μ, $P_\mu(\emptyset)=0$, $P_\mu(R)=1$. Thus $P_\mu\in\mathscr{P}$. Now consider the map $J_\mu: p\to P_\mu$ of $\mathscr{S}$ into $\mathscr{P}$. It is obvious that $J_\mu\in\mathrm{Hom}(\mathscr{S}, \mathscr{P})$ and since $\mathscr{O}$ is total it defines a unique observable in $\mathscr{O}$. ■

Thus when $\mathscr{O}$ is total we can identify it with $\mathscr{M}$.

DEFINITION 3.3. An observable $u\in\mathscr{O}$ is said to be *bounded* if and only if its support, supp(u), is a compact subset of R. The set of all bounded observables will be denoted by $\mathscr{O}_0$.

THEOREM 3.4. *Let $u\to\hat{u}$ be the injection of $\mathscr{O}$ into $\mathscr{M}$ defined in Theorem 2.4.*

Then the map $u \to u^*$ given by

$$u^* = \int_R x\hat{u}(dx) \quad (u \in \mathcal{O}_0)$$

maps $\mathcal{O}_0$ into X^* and is such that $\langle p, u^* \rangle$ is the expectation value of u when the system is in the state $p \in \mathcal{S}$.

Proof: Since $\text{supp}(\hat{u})$ is compact, the integral $\int_R x\hat{u}(dx)$ exists and is single-valued in X^*. Thus $u \to u^*$ is a mapping of $\mathcal{O}_0$ into X^*. Now let $p \in X$, then $\langle p, u^* \rangle = \int_R x \langle p, \hat{u}(dx) \rangle$ since $\hat{u}$ is of compact support and hence a regular vector-valued Borel measure, see Dinculeanu (1967). By Theorem 2.4, $\langle p, \hat{u}(A) \rangle = p(u(A))$ $(A \in \mathcal{B})$ and hence for $p \in \mathcal{S}$ $\int_R x \langle p, \hat{u}(dx) \rangle = = \int_R xpou(dx)$ which is the expectation value of u when the system is in the state p. ■

Remarks. With the notation of Theorem 3.4 we have a mapping of $\mathcal{O}_0$ onto the subset

$$\mathcal{X} = \left\{ \int_R x\hat{u}(dx) : u \in \mathcal{O}_0 \right\}$$

of the space X^*. Physically this gives an important role to X^* since each $u^* \in \mathcal{X} \subset X^*$ is a bounded linear functional on X 'representing' a bounded observable u such that for any proper state p, $u^*(p)$ is the expectation value of u in this state.

Department of Mathematics, Birkbeck College,
London

BIBLIOGRAPHY

Dinculeanu, N., *Vector Measures*, Pergamon Press, Oxford, 1967.

Dunford, N. and Schwartz, J. T., *Linear Operators*, Part I, Interscience Publishers Inc., New York, 1958.

Kronfli, N. S., *International Journal of Theoretical Physics* **2** (1969), 345.

Kronfli, N. S., *International Journal of Theoretical Physics* **3** (1970), 191.

Varadarajan, V. S., *Geometry of Quantum Theory*, Vol. I. Van Nostrand Co. Inc., Princeton, N.J., 1968.

N. S. KRONFLI

PROBABILISTIC FORMULATION OF CLASSICAL MECHANICS

ABSTRACT. Starting axiomatically with a system of finite degrees of freedom whose logic $\mathscr{L}_c$ is an atomic Boolean σ-algebra, we prove the existence of phase space Ω_c, as a separable metric space, and a natural (weak) topology on the set of states $\mathscr{S}$ (all the probability measures on $\mathscr{L}_c$) such that Ω_c, the subspace of pure states $\mathscr{P}$, the set of atoms of $\mathscr{L}_c$ and the space $\mathscr{P}(\Omega_c)$ of all the atomic measures on Ω_c, are all homeomorphic. The only physically accessible states are the points of Ω_c. This probabilistic formulation is shown to be reducible to a purely deterministic theory.

I

This note treats the probabilistic theory of a system whose logic is a Boolean σ-algebra and shows its reduction to a completely deterministic one. The set of states $\mathscr{S}$ consists of all the probability measures on the logic, and the set of observables $\mathscr{O}$ consists of all the σ-homomorphisms on the real Borel σ-field into the logic. This is a special case of generalised quantum theory as defined by Kronfli (1970) which includes both quantum logics and σ-algebras as special cases. On the other hand, conventional quantum theory yields classical mechanics only as an approximation. The results obtained are not very surprising, although they are more detailed technically. The main point, however, is the conclusion that *a theory is deterministic if and only if its logic a Boolean σ-algebra.* It is hoped that this will lend support to the probabilistic point of view, adopted in the lattice-theoretic formulation of the generalised theory, as a fruitful approach to the mathematical analysis of fundamental physics.

From this axiomatic formulation follows the existence of phase space as a separable metric space which is topologically and set theoretically equivalent to the set of all the pure states $\mathscr{P}$ of $\mathscr{S}$. In any state in $\mathscr{P}$ each observable is sharply defined with zero variance such that the states $\mathscr{S}\backslash\mathscr{P}$ become inaccessible physically.

II

The definition and existence of phase space follow from

C. A. Hooker (ed.), The Logico-Algebraic Approach to Quantum Mechanics, 503–507.

THEOREM 1. *For a system with finite degrees of freedom whose logic* $\mathscr{L}_c$ *is a Boolean σ-algebra, there exist a separable metric space* Ω *and a σ-homomorphism* φ *on the Borel σ-field* $\mathscr{B}(\Omega)$ *of* Ω *onto* $\mathscr{L}_c$.

Proof: The finiteness of the degrees of freedom implies the existence of a finite set of observables which is complete. Let $\mathscr{L}_1$ be the Boolean sub-σ-algebra of $\mathscr{L}_c$ generated by the ranges of these observables. Then $\mathscr{L}_1$ is countably generated, since the range of each observable is countably generated. Completeness means that $\mathscr{L}_1$ is maximal, and hence equals $\mathscr{L}_c$. Thus $\mathscr{L}_c$ is countably generated.

By Loomis theorem (Loomis, 1947) there exists a set X, a σ-algebra $\mathscr{A}_1$ of subsets of X and a σ-homomorphism h_1 of $\mathscr{A}_1$ onto $\mathscr{L}_c$. Let $(a_n)\subset\mathscr{L}_c$ generate $\mathscr{L}_c$. Since h_1 is onto, there exists a countable set $(A_n)\subset\mathscr{A}_1$ such that $a_n=h_1(A_n)$ $(n\in N)$. Let $\mathscr{A}_2$ be the sub-σ-algebra of $\mathscr{A}_1$ generated by (A_n). Then $h_1(\mathscr{A}_1)$ is a sub-σ-algebra of $\mathscr{L}_c$ containing its generators (a_n), and hence equals $\mathscr{L}_c$. Since $\mathscr{A}_2$ is countably generated, there exists a separable metric space Ω and a σ-isomorphism h_2 on $\mathscr{B}(\Omega)$ onto $\mathscr{A}_2$ (see Parthasarathy, 1967, p. 133, Theorem 2.2). The proof is completed by putting $\varphi=h_1\circ h_2$. ■

From now on $\mathscr{L}_c$ denotes a countably generated Boolean σ-algebra, for instance when it is a Boolean σ-algebra and the system has finite degrees of freedom. The set of all states will be denoted by $\mathscr{S}$ and the pure ones by $\mathscr{P}$. The space Ω is as in Theorem 1. Let $\Omega_c=\{x\in\Omega:\varphi(\{x\})\neq\emptyset\}$. The separable metric space Ω_c will be called the *phase space* associated with $\mathscr{L}_c$. [The author is not aware if there exists a φ, as in Theorem 1, such that $\Omega_c\in\mathscr{B}(\Omega)$. In this case $\mathscr{B}(\Omega_c)=\{A\in\mathscr{B}(\Omega):A\subset\Omega_c\}$. This would then make $\mathscr{B}(\Omega_c)$ and $\mathscr{L}_c$ σ-isomorphic. No such assumption is made here.]

Recall that a Boolean algebra is *atomic* if every non-atomic element $(\neq\emptyset)$ dominates at least one atom. It is essential that $\Omega_c\neq\emptyset$. The next result shows that atomicity of $\mathscr{L}_c$ is a sufficient condition. In a forthcoming paper we shall show that it is also necessary in order to make a Boolean system deterministic.

THEOREM 2. *Let* $\mathscr{L}_c$ *be atomic. Then* $\Omega_c\neq\phi$. *Furthermore, the mapping* $\gamma:x\to\varphi(\{x\})$ *is a bijection on* Ω_c *onto the set* $\mathscr{A}$ *of the atoms of* $\mathscr{L}_c$.

Proof: Let $a_0\in\mathscr{A}$ and (a_n) generate $\mathscr{L}_c$. For each n, either $a_0<a_n$ or $a_0<a_n'$. If necessary replace a_n by a_n' so that $a_0<a_n$ for all n. Clearly (a_n) still generates $\mathscr{L}_c$. Since φ is onto let $A_n\in\mathscr{B}(\Omega)$ such that $\varphi(A_n)=a_n$. Put

$B=\bigcap_n A_n$. Then $\varphi(B)=\bigwedge_n a_n>a_0\neq\emptyset$, implying $B\neq\emptyset$. Note that (A_n) generates $\mathscr{B}(\Omega)$. Now $\mathscr{R}=\{E\in\mathscr{B}(\Omega):B\subset E$ or $B\cap E=\emptyset\}$ is a sub-σ-algebra of $\mathscr{B}(\Omega)$ containing its generators and hence equals $\mathscr{B}(\Omega)$. But since the latter contains all singletons, $\mathscr{R}=\mathscr{B}(\Omega)$ is possible only if $B=\{x\}$ for some $x\in\Omega$. But $\varphi(\{x\})\neq\emptyset$, thus $x\in\Omega_c$ and $\Omega_c\neq\emptyset$.

Now let x, $y\in\Omega_c$, $x\neq y$ and $\gamma(x)=\gamma(y)=a$ say. Then $\{x\}\subset\{y\}'$ implying $a<a'$, i.e. $a=\emptyset$, which is a contradiction. Hence $x=y$ and γ is one-one on Ω_c.

Let $x\in\Omega_c$, $a_0=\gamma(x)$, $a\in\mathscr{L}_c$ and $a<a_0$. Then there exists $A\in\mathscr{B}(\Omega)$ such that $\varphi(A)=a$. Now either $x\in A$ or $x\in A'$. Thus either $a_0<a$ implying $a=a_0$; or $a_0<a'$ implying $a<a'$ i.e. $a=\emptyset$. Hence a_0 is an atom. Thus γ is an injection on Ω_c into $\mathscr{A}$.

Let $a_0\in\mathscr{A}$. As before we can choose a generating sequence (a_n) in $\mathscr{L}_c$ such that $a_0<a_n$ for all n. Let A, $A_n\in\mathscr{B}(\Omega)$ such that $\varphi(A)=a_0$ and $\varphi(A_n)=a_n$. Let $B_n=A\cup A_n$. Then $\varphi(B_n)=a_n$ and, therefore, (B_n) generates $\mathscr{B}(\Omega)$. Put $B=\bigcap_n B_n$. Then $\varphi(B)\neq\emptyset$ and as in the first paragraph of the proof $B=\{x\}$ for some $x\in\Omega_c$. Clearly $x\in A$ and hence $A\cap\Omega_c\neq\emptyset$. Thus $\emptyset\neq\gamma(x)<a_0$ implying $\gamma(x)=a_0$. Take $y\in A\cap\Omega_c$ and $x\neq y$. Then $\emptyset\neq\gamma(y)<a_0$ implying $\gamma(x)=\gamma(y)=a_0$. But γ is one-one making $x=y$, a contradiction. Thus A is a singleton and γ is onto. ■

From now on $\mathscr{L}_c$ will denote a countably generated atomic Boolean σ-algebra.

Let $\mathscr{A}$ be the set of all atoms of $\mathscr{L}_c$. For each $a\in\mathscr{A}$, define q_a by

$$q_a(b)=\begin{cases}1 & a<b\\ 0 & a<b'\end{cases}\quad(b\in\mathscr{L}_c)$$

Clearly, $q_a\in\mathscr{S}$, since $\mathscr{L}_c$ is atomic. Let $\mathscr{M}(\Omega)$ be the set of all probability measures on $(\Omega, \mathscr{B}(\Omega))$. For each $x\in\Omega$ define δ_x to be the atomic measure $\in\mathscr{M}(\Omega)$ concentrated at x. Put $\mathscr{P}(\Omega_c)=\{\delta_x:x\in\Omega_c\}$. The next result shows the simple structure of the set of pure states $\mathscr{P}$.

THEOREM 3. *The set of pure states on $\mathscr{L}_c$ is precisely $\mathscr{P}=\{q_a:a\in\mathscr{A}\}$. Furthermore, φ induces a bijection $\hat{\varphi}$ of $\mathscr{P}$ onto $\mathscr{P}(\Omega_c)$.*

Proof: The proof of the first part is very much the same as that by Varadarajan (1968, Theorem 6.6). For the second part define $\hat{\varphi}:p\to p\circ\varphi$ $(p\in\mathscr{S})$. Clearly, $\hat{\varphi}$ is a convex homomorphism on $\mathscr{S}$ into $\mathscr{M}(\Omega)$ mapping the extreme points $\mathscr{P}$ of $\mathscr{S}$ into the extreme points $\mathscr{P}(\Omega)$ of $\mathscr{M}(\Omega)$. To prove that $\hat{\varphi}$ is a bijection of $\mathscr{P}$ onto $\mathscr{P}(\Omega_c)$, let $x \in \Omega_c$, $\delta_x \in \mathscr{P}(\Omega_c)$ as defined above, and $a_x = \gamma(x)$.

Define $p_x = q_{a_x}$. Clearly

$$P_x(\varphi(A)) = \begin{cases} 1 & x \in A \\ 0 & x \in A' \end{cases} \quad (A \in \mathscr{B}(\Omega))$$

Thus $\delta_x = p_x \circ \varphi = \hat{\varphi}(p_x)$, implying that $\hat{\varphi}$ maps $\mathscr{P}$ onto $\mathscr{P}(\Omega_c)$. Finally, let $q_a, q_b \in \mathscr{P}$ such that $\hat{\varphi}(q_a) = \hat{\varphi}(q_b) = \delta_x$, say. This implies that $\gamma^{-1}(a) = \gamma^{-1}(b) = x$. But γ is a bijection of Ω_c onto $\mathscr{A}$, and hence $a = b$ or, equivalently, $q_a = q_b$, proving that $\hat{\varphi}$ is one-to-one on $\mathscr{P}$. ■

So far no use was made of the topological properties of Ω. All the results obtained would work for Ω as an abstract set and $\mathscr{B}(\Omega)$ a countably generated σ-field of subsets of Ω containing all singletons.

Let $\mathscr{M}(\Omega)$ be equipped with its weak topology. We define the *weak topology* on $\mathscr{S}$ to be the weakest such that $\hat{\varphi}$ is continuous on $\mathscr{S}$. Since Ω is a separable metric space, then $\mathscr{M}(\Omega)$ is metrisable as a separable metric space and the spaces Ω_c and $\mathscr{P}(\Omega_c)$ are homeomorphic (see Parthasarathy, 1967, pp. 42–43). We have now both topological and set theoretical equivalence of all three spaces Ω_c, $\mathscr{P}(\Omega_c)$ and $\mathscr{P}$. This topological equivalence is important when considering continuous groups of convex automorphisms of $\mathscr{S}$ and their induced representation for motions in Ω_c, in particular the (one-parameter) dynamical group.

III

In this section is shown that the points of Ω_c are the only physically accessible states of the system such that at each point every observable is sharply defined with zero variance. This depends on an important theorem of Varadarajan. Let $\mathbf{B}(\Omega, R)$ be the set of equivalence classes $[f]$ of all real-valued Borel functions f on Ω, where $f_1, f_2 \in [f]$ if and only if $\{x \in \Omega : f_1(x) \neq f_2(x)\} \in \mathrm{Ker}(\varphi)$. Let $\mathscr{O}$ be the set of observables on $\mathscr{L}_c$.

THEOREM 4. *There exists a mapping $f: u \to [f_u]$ of $\mathscr{O}$ into $\mathbf{B}(\Omega, R)$ such that*
(i) $u(E) = \varphi(f_u^{-1}(E)) \quad (E \in \mathscr{B}(R))$,
(ii) $f_u(x) = 0$ *for all* $x \in \Omega_{c'}$.
Proof: See Varadarajan, 1968, Theorem 1.4. ■

COROLLARY. *Let* $x\in\Omega_c$, $u\in\mathcal{O}$ *and* $[f_u]$ *be the corresponding element in* $\mathbf{B}(\Omega, R)$ *as in Theorem 4. Then the expectation value of u in the state x is* $f_u(x)$, *and its variance is zero.*

Proof. Let $p_x=\hat{\varphi}^{-1}(\delta_x)$ and $\mu_x:E\to p_x\circ\varphi(f_u^{-1}(E))$ $(E\in\mathscr{B}(R))$. Then, clearly, μ_x is the atomic measure on R concentrated at $f_u(x)$. The expectation value of u in the state p_x is

$$\int_{-\infty}^{+\infty} tp_x(u(dt))=\int_{\infty}^{+\infty} t\mu_x(dt)=f_u(x)$$

The variance is

$$\int_{-\infty}^{+\infty} t^2p_x(u(dt))-f_u(x)^2=0 \blacksquare$$

Now any convex automorphism on $\mathscr{S}$ is a one-to-one mapping of $\mathscr{P}$ onto itself. This is the same for the dynamical group $\{U_t:t\in R\}$. Starting the system in a well defined state $p_x\in\mathscr{P}(x\in\Omega_c)$, its state will always remain in $\mathscr{P}$ for all time t.

Department of Mathematics, Birkbeck College,
London

BIBLIOGRAPHY

Kronfli, N. S., *International Journal of Theoretical Physics* **3** (1970), 199.

Loomis, L. H., *Bulletin of the American Mathematical Society* **53** (1947), 754.

Parthasarathy, K. R., *Probability Measures on Metric Spaces*, Academic Press, New York, 1967.

Varadarajan, V. S., *Geometry of Quantum Theory*, Vol. I. Van Nostrand Co. Inc., Princeton, N.J., 1968.

N. S. KRONFLI

ATOMICITY AND DETERMINISM IN BOOLEAN SYSTEMS

ABSTRACT. The logic of a Boolean system of finite degrees of freedom is shown to be atomic if and only if the system obeys a deterministic theory. This is, therefore, the physical meaning of atomicity. Furthermore, it is proved that nondeterminacy of such a system implies the nonexistence of phase space.

I

In this note classical mechanics is again considered as a special case of generalised physical theory defined by Kronfli (1970a). The same notation will be employed. We distinguish between a classical theory and a (classical) deterministic one. The former is one in which each pair of observables is compatible. Axiomatically, therefore, a classical system of finite degrees of freedom will necessarily have a logic $\mathscr{L}$ which is a countably generated Boolean σ-algebra.

We define a *deterministic* system to be a classical one such that there exists a non-empty subset $\mathscr{S}_0$ of the set of its states $\mathscr{S}$ where for each state in $\mathscr{S}_0$ at least one observable has zero variance and furthermore $\mathscr{S}_0$ is an invariant of the dynamical subgroup $\mathscr{D}$ of the group $\mathrm{Aut}(\mathscr{S})$ of convex automorphisms of $\mathscr{S}$. This apparently weak condition gives the full determinism of classical mechanics.

The meaning of lattice atomicity for the logic of a classical mechanical system has been obscured in the literature. This note shows that atomicity is both a necessary and sufficient condition for a Boolean system of finite degrees of freedom to be deterministic. In other words, the two are synonymous.

Unlike a countably generated σ-field of subsets of a set, an abstract countably generated Boolean σ-algebra need not be atomic. (Take, for example, the quotient of all the Borel subsets of the unit interval on the line modulo the sets of Lebesgue measure zero.) In a previous paper (Kronfli, 1970b), atomicity of the (Boolean) logic of a classical system of finite degrees of freedom, was shown to be a sufficient condition for the system to be deterministic. A Boolean system, although classical, need not

C. A. Hooker (ed.), The Logico-Algebraic Approach to Quantum Mechanics, 509–512.

be deterministic. In this paper it is shown that atomicity is also a necessary condition. It will also be shown that without the condition of determinism, classical phase space need not even exist. The reader may compare the above-mentioned results with the objections to atomicity stated by Birkhoff and von Neumann (1936).

II

From now on $\mathscr{L}$ is a countably generated Boolean σ-algebra (the logic of a classical system S of finite degrees of freedom), $\mathscr{S}$ is the set of all probability measures on $\mathscr{L}$ (the states of S) and $\mathscr{P}$ the set of all extreme points of the convex set $\mathscr{S}$ (the pure states of S). Our first result shows the equivalence of $\mathscr{P}\neq\emptyset$ to the atomicity of $\mathscr{L}$.

PROPOSITION 1. *The set $\mathscr{P}$ is not empty if and only if $\mathscr{L}$ is atomic. In this case $\mathscr{P}=\{q_a : a\in\mathscr{A}\}$, where $\mathscr{A}$ is the set of atoms of $\mathscr{L}$ and q_a is the atomic measure concentrated at the atom a.*

Proof. Assume $\mathscr{L}$ is atomic. Then clearly for each $a\in\mathscr{A}$, $q_a\in\mathscr{P}$. Thus $\mathscr{P}\neq\emptyset$. That $\{q_a : a\in\mathscr{A}\}$ equals $\mathscr{P}$ follows from Theorem 3 of Kronfli (1970b).

Conversely, assume $\mathscr{P}\neq\emptyset$ and $p\in\mathscr{P}$. First we assert that range $(p)=\{0, 1\}$. If this is not so, then there exists $a\in\mathscr{L}\setminus\{\emptyset, 1\}$ such that $0<p(a)<1$. Define $p_1, p_2\in\mathscr{S}$ by

$$\begin{aligned} p_1(x)&=(p(a))^{-1}p(x\wedge a)\\ p_2(x)&=(1-p(a))^{-1}p(x\wedge a') \end{aligned} \qquad (x\in\mathscr{L})$$

This gives

$$p=p(a).p_1+(1-p(a)).p_2$$

with $p_1\neq p_2$, since $p_1(a)=1$ and $p_2(a)=0$. This is a contradiction, since p is an extreme point of the convex set $\mathscr{S}$. Hence, range $(p)=\{0, 1\}$. Now let $(a_n)\subset\mathscr{L}$ generate $\mathscr{L}$. Since $p(x)$ is either 0 or 1 for each $x\in\mathscr{L}$, we choose $b_n=a_n$ or $b_n=a_n'$ such that $p(b_n)=1$ for all n. Clearly (b_n) generates $\mathscr{L}$. Put $b=\bigwedge_n b_n$. Since $p(b)=1$ then $b\neq\emptyset$. Let

$$\mathscr{R}=\{x\in\mathscr{L} : b<x \quad \text{or} \quad b<x'\}$$

Clearly, $\mathscr{R}$ is a Boolean sub-σ-algebra of $\mathscr{L}$ containing its generators (b_n)

and, therefore, $\mathscr{R}=\mathscr{L}$. But since $b\neq\emptyset$, the above result can not be possible unless b is an atom of $\mathscr{L}$. For, let $x\in\mathscr{L}$ and $x<b$. Since $\mathscr{R}=\mathscr{L}$, then either $b<x$ or $b<x'$. The first case gives $x=b$ and the second $x=\emptyset$. Hence b is an atom.

With the assumption $\mathscr{P}\neq\emptyset$ we proved that $\mathscr{A}\neq\emptyset$. It remains to show that $\mathscr{L}$ is atomic, i.e. each non-zero element of $\mathscr{L}$ dominates at least one atom. Let, therefore Ω, Ω_c, $\mathscr{B}(\Omega)$, φ and γ be as in Theorems 1 and 2 of Kronfli (1970b). Since φ maps $\mathscr{B}(\Omega)$ onto $\mathscr{L}$, then for each non-zero and non-atomic element $a\in\mathscr{L}$ there exists $A\in\mathscr{B}(\Omega)$, which is not a singleton, since γ is a bijection on Ω_c onto $\mathscr{A}$, with $A\cap\Omega_c\neq\emptyset$ and $\varphi(A)=a$. (It is easy to see that this corollary to Theorem 2 holds simply with the assumption $\mathscr{A}\neq\emptyset$.) Let $x\in A\cap\Omega_c$. Then $\gamma(x)$ is an atom and $\gamma(x)<a$. This completes the proof. ■

Using the definition of determinism given in the first section, we show that determinism and atomicity of the logic are equivalent. The obvious candidate for $\mathscr{S}_0$ is $\mathscr{P}$.

PROPOSITION 2. *The countably generated Boolean σ-algebra $\mathscr{L}$ is the logic of a deterministic system if and only if $\mathscr{L}$ is atomic.*

Proof. Let $p\in\mathscr{S}$ and u be an observable whose variance in p is zero. Let τ be its expectation value,

$$\int_R t\, p\circ u(dt)$$

which is necessarily finite. Then

$$\int_R (t-\tau)^2 p\circ u(dt)=0$$

Hence the function $t\to(t-\tau)^2$ on $R\to R$ is zero $(p\circ u)$ almost everywhere. This is possible only if $p\circ u$ is an atomic measure concentrated at τ. Put $a=u(\{\tau\})$. Then a is either $\emptyset$ or an atom of $\mathscr{L}$. But $p(a)=1$, and hence a is an atom and $p=q_a\in\mathscr{P}$. Thus $\mathscr{P}$ is not empty and by Proposition 1 $\mathscr{L}$ is atomic. Thus at least $\mathscr{S}_0\subset\mathscr{P}$. Also, each element of $\mathrm{Aut}(\mathscr{S})$, and hence of the dynamical group $\mathscr{D}$, is a bijection of $\mathscr{P}$ onto itself. Therefore determinism implies the atomicity of $\mathscr{L}$.

Conversely, atomicity of $\mathscr{L}$ implies that the theory of the system is deterministic as was shown in Kronfli (1970b). ■

COROLLARY. *Let $\mathscr{L}$ be the logic of a Boolean system S of finite degrees of freedom. Then the phase space of S does not exist if S is non-deterministic.*

Proof. Assume S is non-deterministic with non-empty phase space. Since phase space is equipotent to $\mathscr{P}$ (Kronfli, 1970b), then $\mathscr{P}$ is not empty. From Propositions 1 and 2, S is deterministic, which is a contradiction. ■

Department of Mathematics, Birkbeck College,
London

BIBLIOGRAPHY

Birkhoff, G. and Neumann, J. von, *Annals of Mathematics* **37** (1936), 823.
Kronfli, N. S., *International Journal of Theoretical Physics* **3** (1970a), 199.
Kronfli, N. S., *International Journal of Theoretical Physics* **3** (1970b), 395.

C. PIRON**

SURVEY OF GENERAL QUANTUM PHYSICS*

The abstract description of a physical system is developed, along lines originally suggested by Birkhoff and von Neumann, in terms of the complete lattice of propositions associated with that system, and the distinction between classical and quantum systems is made precise. With the help of the notion of state, a propositional system is defined: it is remarked that every irreducible propositional system (of more than three dimensions) is isomorphic to the lattice of all closed subspaces of a Hilbert space constructed on some division ring with involution. The propositional system consisting of a family of separable complex Hilbert spaces is treated as a particular case which is sufficiently general to include both classical and quantum mechanics. The theory of the Galilean particle without spin is given as an illustration. Finally the basis for the statistical interpretation of wave mechanics is developed with the help of Gleason's theorem in an appendix, a proof of essentially the first part of Gleason's theorem is given which is a little different (perhaps more geometric) from that originally given by Gleason.

I. INTRODUCTION

This paper is based on a series of seminars given at the University of Denver in the autumn of 1970, and consists of a review of work done by the author and collaborators at the Institute of Theoretical Physics, University of Geneva, during the past several years. It presents a formalism for the description of physical processes including both classical and quantum phenomena.

This formalism is obtained by taking seriously the realistic point of view of Einstein [4], and describing a physical system in terms of "elements of reality." Contrary to what one might expect, it is possible to justify the use of the linear structure of the Hilbert space without the use, at the outset, of the additivity which is characteristic of probability theory (see, for example, Ludwig [9]).

The statistical interpretation of quantum theory appears to provide a consistent basis for understanding the wave mechanical formalism. The additivity characteristic of probability theory is clearly associated with this interpretation. However, it is possible, and of considerable interest, to inquire on a more fundamental level for the source of this idea.

If the basis for the statistical interpretation of wave mechanics is to

C. A. Hooker (ed.), The Logico-Algebraic Approach to Quantum Mechanics, 513–543.

emerge in a mathematically well-defined way from the study of a set of more fundamental axioms, it is clear that these axioms should be nonstatistical in nature, and should lead as well to a description of classical systems as to quantal systems. An axiomatic framework of this kind can therefore provide a unified structure for physics, which we shall call general quantum physics.

It is also clear that a satisfactory axiomatic structure of the kind referred to above cannot be formulated *a priori* in terms of wave functions, since their use would imply a statistical interpretation at the outset. The role of the wave functions in general quantum physics must emerge from the analysis of the more fundamental theory. As said above, the linear structure of the Hilbert space does appear, without reference to any statistical notions, as the appropriate description of general quantal systems (a set consisting of a family of Hilbert spaces describes systems which are not purely quantal, and the purely classical limit is described by a family of trivial one-dimensional Hilbert spaces). It is in this way that the statistical interpretation for wave mechanics will emerge as a consequence of essentially nonstatistical axioms, and what is presupposed in classical physics is clearly brought into evidence.

It has been found that the mathematical language of formal logic is appropriate for the formulation at the axioms of general quantum physics. This language, first introduced formally into physics by Birkhoff and von Neumann, [2] provides, within the context of a single general mathematical scheme (lattices), a precise way to state the difference between classical and quantal systems. Ballentine [14] has remarked that, like von Neumann's original formulation of the hidden variable problem, this lattice theoretic formalism may be "too abstract." What I hope to show in the discussion to follow is that such a formalism is not "too abstract," but, on the contrary, provides a precise mathematical framework for the systematic examination of the consequences of the intuitive ideas which are commonly held to be the basis for physical theories. It is only by such systematic analysis that one can answer questions about the structure of the theory which would otherwise be lost in endless semantic controversy. The setting for the development of new theoretical ideas, furthermore, can best be established by stating the conceptual basis of the present quantum theory in a precise way.

In Section II, a physical interpretation for the axioms of the theory is given. The discussion is completely self-contained, and all definitions required for the mathematical treatment of systems of propositions are given. The complete lattice of propositions associated with a physical system is defined; the distinction between classical and quantal systems is then made precise with the second theorem of this section. In the classical theory, the lattice is distributive [2]. The axioms of general quantum theory are then presented (no restriction with respect to distributivity is made), and the notion of *state* is defined. The axiom of atomicity for the lattice is justified from this point of view, i.e., each state defines an atom (a minimal nontrivial proposition), and every atom defines a state. With these axioms, one may define a *propositional system.*

In Section III, it is remarked that every propositional system can be decomposed into irreducible ones and that every irreducible propositional system (of more than three dimensions) is isomorphic to the lattice of all closed subspaces of a Hilbert space constructed on some division ring with involution, and therefore the Hilbert space provides a concrete realization. The definition of observables and symmetries is given in the Section IV, and the propositional system consisting of a family of separable complex Hilbert spaces is treated as a particular case which is sufficiently general to contain both the classical theory and the usual quantum theory. The theory of the Galilean particle without spin is given as an illustration in Section V.

In Section VI, the basis for the statistical interpretation of wave mechanics is developed with the help of Gleason's theorem [5]. An appendix is given with a proof of what is essentially the first part of Gleason's theorem. The demonstration is a little different (perhaps more geometric) from that originally given by Gleason.

II. ABSTRACT DESCRIPTION OF A PHYSICAL SYSTEM

It is the goal of a physical theory to describe and predict the results of experiments on a physical system, defined as a part of reality existing in space-time external to the physicist. One can only hope to describe such a system if it is sufficiently isolated; the concept of "isolation" is a question of definition or idealization.

An affirmation made by a physicist concerning the actual properties of a particular physical system can be checked by experiment. Such a test consists, in general, in a measurement for which the result is expressed by "yes" or "no." If the answer is "yes," the physicist's affirmation is confirmed, but, of course, not proved. If the answer is "no," then the physicist's affirmation is mistaken. One cannot prove that an affirmation is correct on the basis of experiment, but one can only hope to disprove the affirmation. The possibility of disproving a theory is basic in physics.

DEFINITION. We define a question to be a measurement (or experiment) leading to an alternative of which the terms are "yes" and "no."

A question consists of a procedure to be carried out with the physical system under consideration, and a rule for interpreting the possible results in terms of "yes" or "no."

Examples. (E1) Take the chalk and bend in with all your strength; then see if it is broken.

(E2) Put Mr. X on the scale and ascertain that the pointer stops on the 200 lb mark.

It is important to take account of the enormous variety of possible questions and of the great generality of this concept.

If α is a question, we denote by $\alpha^{\sim}$ the question obtained by exchanging the terms of the alternative. For example, $(E1)^{\sim}$ corresponds to the following: Take the chalk and bend it with all your strength; then see if it is *not* broken.

If $\{\alpha_i\}$ is a family of questions, we denote by $\prod_i \alpha_i$ (if there are just two, we write $\alpha \cdot \beta$) the question defined in the following manner:

One measures an arbitrary one of the α_i and attributes to $\prod_i \alpha_i$ the answer thus obtained.

It is easy to verify the following rule:

$$(\prod_i \alpha_i)^{\sim} = \prod_i (\alpha_i^{\sim})$$

There exists a trivial question which we shall denote as I and which consists in doing anything (or nothing) and stating that the answer is "yes" each time.

DEFINITION. When the physical system has been prepared in such a way that the physicist may affirm that in the event of a measurement the result "yes" is certain, we shall say that the question is true.

If the outcome for the question β is not certain, the statement "β is true" is false, but we do not say "β is false."

For certain pairs of questions β, γ, one may have the following relation:

DEFINITION. If the physical system is prepared in such a way that β is "true," then one is sure that γ is "true." We denote it as $\beta<\gamma$ and read it as "β less than γ."

Such a relation expresses a physical law. It is transitive for if "β true" $\Rightarrow$ "γ true" and if "γ true" $\Rightarrow$ "δ true," then "β true" $\Rightarrow$ "δ true." Then this relation defines an equivalence:

$$\beta \sim \gamma \quad \text{iff} \quad \beta<\gamma \quad \text{and} \quad \gamma<\beta$$

DEFINITION. We define a proposition as an equivalence class of questions, and denote by b the equivalence class containing the question β; i.e.,

$$b=\{\gamma \mid \gamma \sim \beta, \quad \gamma \text{ a question}\}$$

It is easy to verify that if "β is true," then any $\gamma \sim \beta$ is "true." Hence we can say the proposition is "true" if and only if any and therefore all of its questions are "true."

Let us assume that it is possible to define the set of all propositions for a given system. It is, in fact, not *a priori* possible to define such a set, in general, but it is always possible to define one sufficiently large to contain all of the useful propositions.

Let $\mathscr{L}$, then, be the *set* of propositions defined for a given physical system. We then have the following theorem.

THEOREM. $\mathscr{L}$ is a complete lattice.

Proof. First a (partial) ordering relation is defined on $\mathscr{L}$:

$$b<d \quad \text{iff} \quad \beta<\delta \quad \text{with } \beta \in b \quad \text{and} \quad \delta \in d$$

Second, given any family of propositions b_i, there exists a greatest lower bound. For, let

$$\bigwedge_i b_i = \left\{\gamma \mid \gamma \sim \prod_i \beta_i, \quad \beta_i \in b_i\right\}$$

this proposition satisfies the relations

$$\bigwedge_i b_i < b_i \quad \forall_i \qquad \text{and} \qquad x < b_i \quad \forall_i \Rightarrow x < \bigwedge_i b_i$$

because "$\prod_i \beta_i$ is true" iff "β_i is true" $\forall i$, and if "α true"$\Rightarrow$"β_i true" $\forall i$, then "α true"$\Rightarrow$"$\prod_i \beta_i$ true."

But if the greatest lower bound exists for every family of propositions, then the least upper bound exists as well and is defined as

$$\bigvee_i b_i = \bigwedge_{x \in \mathscr{F}} x$$

where

$$\mathscr{F} = \{x \mid b_i < x \quad \forall i\}$$

This subset $\mathscr{F}$ is never void since it contains always I. ■

As apparent in the demonstration, the greatest lower bound of two propositions a and b has the following properties:

"$a \wedge b$ true"$\Rightarrow$"a true" and "b true"

which shows that $\wedge$ plays the same role as "and" in logic. However, for the least upper bound, we have only

"a true" or "b true"$\Rightarrow$"$a \vee b$ true."

In fact, one has the following:

THEOREM. If "$a \vee b$ true"$\Leftrightarrow$("a true" or "b true") for all $a, b \in \mathscr{L}$, then $\mathscr{L}$ is distributive: $a \wedge (b \vee c) = (a \wedge b) \vee (a \wedge c)$.

Proof. "$a \wedge (b \vee c)$ true"$\Leftrightarrow$"a true" and ("b true" or "c true")$\Leftrightarrow$("a true" and "b true") or ("a true" and "c true")$\Leftrightarrow$"$(a \wedge b) \vee (a \wedge c)$ true." ■

In classical theory, $\mathscr{L}$ is the set of subsets of phase space and is distributive. The implication to the right in the theorem stated above is the essential distinction between classical and quantum theory.

Examples. (1) The selection of eggs into small, medium, and large. We use a gauge which consists essentially of a hole such that an egg passing through the hole is called small; to this experiment there corresponds a question s. Another gauge with a larger hole classifies the egg as large if it does not pass through; to this experiment there corresponds a question l. Using $\{\alpha\}$ for the equivalence class of questions containing α, the lattice is as given in Figure 1. The lines correspond to order relations. Note that

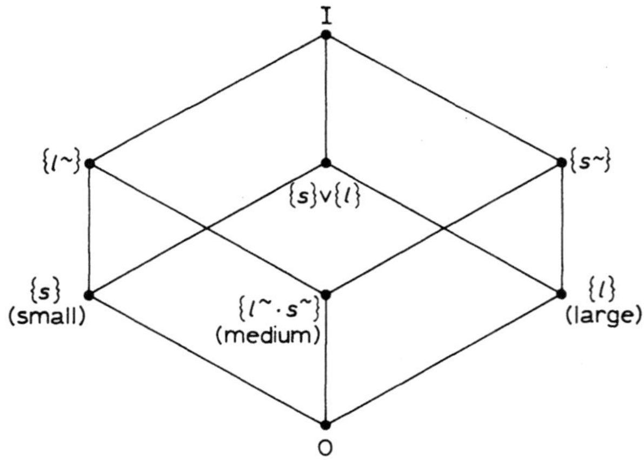

Fig. 1.

$\{s\} \vee \{l\}$ corresponds to a question, which consists of both measurements s and l; the answer is "yes" if one of the answers is "yes." We see that this lattice is distributive, and in fact "$\{s\} \vee \{l\}$ true" iff "$\{s\}$ true" or "$\{l\}$ true." We must make two operations to see if an egg is "small or large."

(2) The linear polarization of photons. The experiment consists in placing a polarizer in a beam of linearly polarized photons. In fact, it is possible to verify, by dispatching photons one by one, that this experiment leads to a plain alternative. Either the photon passes through or it is absorbed. We shall define the question a_ϕ by specifying the orientation of the polarizer (the angle ϕ) and interpreting the passage of a photon as a "yes." Experience shows that in order to obtain a photon prepared in such a way that a_ϕ is "true," it is sufficient to consider photons which have traversed a first polarizer oriented at this angle ϕ. But experiment also shows that it is impossible to prepare photons capable of traversing with complete certainty a polarizer oriented at angle ϕ as well as another oriented at an angle $\phi' \neq \phi$ (modulo π) i.e.,

$$a_\phi \cdot a_{\phi'} \sim 0$$

The corresponding lattice of propositions is then as given in Figure 2, for one cannot define a question which is "true" if and only if "a_ϕ is true" or "$a_{\phi'}$ is true." A photon is absorbed if it does not get through a polarizer; a second experiment on that photon cannot be performed. This lattice is not distributive; it therefore corresponds to a quantum system.

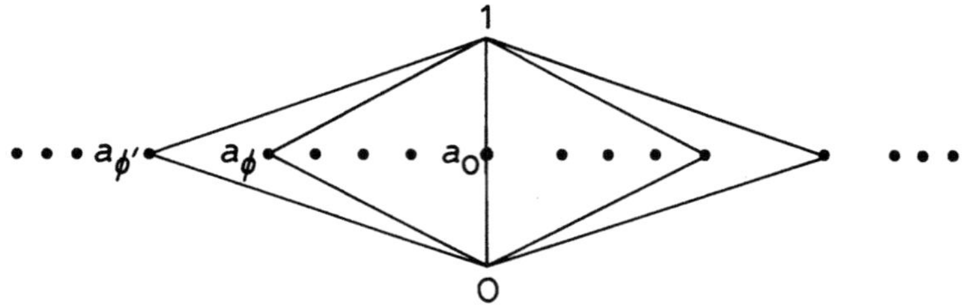

Fig. 2.

DEFINITION. We shall say that c is a *compatible complement* for b if it is a complement

$$b \wedge c = 0 \quad \text{and} \quad b \vee c = I$$

and if there exists a question β such that

$$\beta \in b \quad \text{and} \quad \beta^{\sim} \in c$$

In the case of example (1), $\{l^{\sim}\}$ and $\{l\}$ are compatible complements, and it is a little exercise to prove that $\{l^{\sim} \cdot s^{\sim}\}$ is a compatible complement for $\{l\} \vee \{s\}$. In the case of example (2), the proposition $a_{\phi+(\pi/2)}$ is a compatible complement for a_{ϕ}, but $a_{\phi'}$ with $\phi' \neq \phi (\text{mod}\, \pi/2)$ is only a complement of a_{ϕ}.

In the general case, we shall posit the following axiom:

AXIOM C. For each proposition, there exists at least one compatible complement.

Let us recall the definition of a sublattice generated by a family of propositions. A sublattice of $\mathscr{L}$ is a subset S of $\mathscr{L}$ such that ...

$$b, c \in S \Rightarrow b \vee c, b \wedge c \in S.$$

The sublattice generated by a family of propositions is the sublattice intersection of all sublattices containing this family.

AXIOM P. If $b < c$ are propositions of $\mathscr{L}$ and if b' and c' are compatible complements for b and c, respectively, then the sublattice generated by $\{b, b', c, c'\}$ is distributive.

From this axiom, it immediately follows that

$$b < c \Rightarrow c' < b'$$

and hence the compatible complement is unique. Thus the mapping

$$b \mapsto b'$$

is an orthocomplementation:

O_1 $(b')' = b$

O_2 $b \wedge b' = 0$ and $b \vee b' = I$

O_3 $b < c \Rightarrow c' < b'$

Furthermore, it is weakly modular:

O_4 $b < c \Rightarrow c \wedge (c' \vee b) = b$

THEOREM. Let $\mathscr{L}$ be a CROC, i.e., a complete, orthocomplemented, and weakly modular (O_1 to O_4) lattice. If one interprets the orthocomplement as a compatible complement, then $\mathscr{L}$ satisfies axioms C and P.

Proof. Let $b < c$ be elements of $\mathscr{L}$. We must show that the sublattice generated by $\{b, b', c, c'\}$ is distributive. Weak modularity implies

$$c \wedge (c' \vee b) = b \quad \text{and} \quad b' \wedge (b \vee c') = c'$$

These two relations are necessary and sufficient for the set

$$\{0, b, b' \wedge c, c, c', b \vee c', b', I\}$$

to form a distributive sublattice. ■

The axioms C and P permit us to define the very important concept of compatibility.

DEFINITION. In a CROC, two propositions b and c are said to be *compatible* if the sublattice generated by $\{b, b', c, c'\}$ is distributive. We shall denote this property by $b \leftrightarrow c$.

There are many characterizations of the compatibility of two propositions [8, 12]. Any one of these implies all of the others. The simplest is

$$b \leftrightarrow c \Leftrightarrow b \wedge (c \vee b') < c$$

which can be physically interpreted in the following way: b is compatible with c if, whenever "b is true" and "$c \vee b'$ is true," then c "is true."

The following properties of compatible propositions are useful for the "calculus of propositions":

$$b \leftrightarrow c \Rightarrow b \leftrightarrow c'$$

$$b \leftrightarrow c \text{ and } b \leftrightarrow d \Rightarrow b \leftrightarrow c \wedge d,\ b \leftrightarrow c \vee d$$

If, in the triplet $\{a, b, c\}$, one proposition is compatible with the two others, then we have the distributivity relation: $a \wedge (b \vee c) = (a \wedge b) \vee (a \wedge c)$.

The sublattice generated by a family of mutually compatible propositions is distributive.

DEFINITION. We define the notion of *state* as the family S of all propositions actually true for the given system.

To be consistent with our definition of the lattice of propositions $\mathscr{L}$, the family S must have the following properties:

(S_1) $O \notin S$ and $I \in S$.

(S_2) If $a \in S$ and $x \in \mathscr{L}$, with $a < x$, then $x \in S$.

(S_3) If $a_i \in S$, then $\bigwedge_i a_i \in S$.

In addition to these properties, S must describe the system completely; this imposes a fourth property:

(S_4) S is maximal; that is, if a subset of $\mathscr{L}$ satisfies S_1, S_2, S_3 and contains S, it is equal to S.

It is easy to see that with these properties, S possesses a minimal element $p = \bigwedge_{x \in S} x \in S$ (by S_3) and is not 0 (according to S_1), and is of the form

$$S = \{x \mid p < x \in \mathscr{L}\}$$

To satisfy the maximality condition S_4, this proposition p must be an atom, i.e., a proposition different from 0 and such that $0 < x < p \Rightarrow x = 0$ or $x = p$. We conclude that each state defines an atom, and every atom defines a state.

By definition, if a proposition is different from 0, then it must be true for some state of the system. This justifies the following axiom A_1:

AXIOM A_1. $\mathscr{L}$ is atomic: if a is different from 0, then there exists an atom $p < a$.

AXIOM A_2. If $a \in \mathscr{L}$ and p is an atom of $\mathscr{L}$ such that $a' \wedge p = 0$, then $(p \vee a') \wedge a$ is an atom.

To interpret this axiom A_2, we must give the following definitions:

DEFINITION. A question β is said to be an ideal question if "a is true" before the measurement and $a \leftrightarrow \{\beta\} \Rightarrow$ "a is true" after the measurement.

DEFINITION. A question β is said to be of the first kind if a "yes" answer to β implies that "$\{\beta\}$ is true" immediately after the measurement.

If we know the state p of the system and perform an ideal measurement α of the first kind, and if a "yes" answer is obtained, we shall see that we can predict only that $(p \vee a') \wedge a$ is true immediately after the measurement.

We first note that in an ideal measurement of a, $p_a = (p \vee a') \wedge (p \vee a)$ is true before the measurement because it contains p; but it is still true after the measurement because it is compatible with a. All other propositions x which are *a priori* true must contain p and be compatible with a and so are greater than p_a:

$$(p \vee a') \wedge (p \vee a) < (x \vee a') \wedge (x \vee a) = x$$

Since α is of the first kind, and if the answer is "yes", then all x greater than a are true after the measurement. Then, *a priori*, one knows only that all elements greater than

$$[(p \vee a') \wedge (p \vee a)] \wedge a = (p \vee a') \wedge a$$

are true. If this were not an atom, the state after the measurement would not be known, and information would be lost in spite of the fact that the measurement was to have been as nonperturbative as possible.

DEFINITION. A complete lattice satisfying the axioms C, P, A_1, and A_2 is called a *propositional system*.

The center $\mathscr{Z}$ of a propositional system $\mathscr{L}$ is the set of all propositions compatible with all others in $\mathscr{L}$.

THEOREM. The center $\mathscr{Z}$ of a propositional system is a Boolean CROC, i.e., a distributive CROC, which is isomorphic to the lattice of subsets of a set.

Proof. $\mathscr{Z}$ is a CROC because if z is in $\mathscr{Z}$, then z' is in $\mathscr{Z}$, and if z_i is in $\mathscr{Z}$ for all i, then $\bigvee_i z_i$ and $\bigwedge_i z_i$ are in $\mathscr{Z}$. It is obviously distributive. The rest of the proof is more technical; it can be found in [12].

This theorem allows one to distinguish the classical case from the quantum case and from intermediate cases. A proposition system is said to be *purely classical* if its center $\mathscr{Z} = \mathscr{L}$. A propositional system is said to be *purely quantal* if its center contains only 0 and I. In physics, there are a large number of intermediate cases for which the center contains non-trivial propositions. We shall say for such cases that the system possesses *superselection rules.*

III. REALIZATION OF A PROPOSITIONAL SYSTEM

In the preceding section, we have defined a propositional system and we have given its physical interpretation. We will construct, in what follows, a concrete realization in a structure which is more convenient for explicit calculation. To do this, it is necessary to define the notion of morphism:

DEFINITION. A morphism of a CROC $\mathscr{L}_1$ into a CROC $\mathscr{L}_2$ is a mapping μ from $\mathscr{L}_1$ into $\mathscr{L}_2$ such that:

(1) $\mu(\bigvee_i a_i) = \bigvee_i \mu a_i$.

(2) $a < b' \Rightarrow \mu a < (\mu b)'$.

We denote by 0_i the minimal element of $\mathscr{L}_i$ and by I_i the maximal element of $\mathscr{L}_i (i = 1, 2)$. The preceding definition then implies the following properties:

(3) $\mu 0_1 = 0_2$.

(4) $\mu(a') = (\mu a)' \wedge \mu I_1$.

(5) $\mu(\bigwedge_i a_i) = \bigwedge_i \mu a_i$.

Proof. (3) $0_1 < 0'_1$, then $\mu 0_1 < (\mu 0_1)' \Rightarrow \mu 0_1 = 0_2$.

(4) $\mu I_1 \wedge (\mu a)' = [\mu(a') \vee \mu a] \wedge (\mu a)'$. $\mu(a') < (\mu a)'$ by (2), and hence by weak modularity,

$$[\mu(a') \vee \mu a] \wedge (\mu a)' = \mu(a')$$

(5) $\mu(\bigwedge_i a_i) = \mu[(\bigvee_i a'_i)'] = [\mu(\bigvee_i a'_i)]' \wedge \mu I_1 = [\bigvee_i \mu(a'_i)]' \wedge \mu I_1 = \bigwedge_i ([\mu(a'_i)]' \wedge \mu I_1) = \bigwedge_i (\mu a_i)$. ■

Example. Let $[0, b]$ be a segment of a CROC $\mathscr{L}$, that is to say, the set

$$\{x \mid x < b, \quad x \in \mathscr{L}\}$$

with the relative orthocomplementation

$$x \mapsto x^r = x' \wedge b$$

Then $[0, b]$ is a sub-CROC; the canonical injection of $[0, b]$ into $\mathscr{L}$ is a morphism.

Proof. First, let us show that the mapping

$$x \mapsto x^r = x' \wedge b$$

is an orthocomplementation:

$$x \wedge x^r = x \wedge x' \wedge b = 0, \quad x \vee x^r = x \vee (x' \wedge b) = b$$

in view of weak modularity, and

$$x < y \Rightarrow y' \wedge b < x' \wedge b, \quad (x^r)^r = (x \vee b') \wedge b = x$$

Finally, it is trivial that the canonical injection is a morphism. ■

An isomorphism of two CROC's is just a morphism which is one to one and onto.

The structure of a general morphism μ is very simple. The kernel of μ, i.e., the set $\{x_1 \mid \mu x_1 = 0_2\}$, is a segment $[0, z]$ such that z is in the center of $\mathscr{L}_1$. To prove this, take z to be the least upper bound of all propositions of the kernal; then, it is in the kernel. It is easily verified that for all $x \in \mathscr{L}_1$, $(z \vee x') \wedge x$ is in the kernel; therefore $(z \vee x') \wedge x < z$, hence $z \leftrightarrow x$. If the center contains only 0_1 and I_1, then the morphism is injective or identically 0_2. When the center is not trivial, the CROC can be decomposed into sub-CROC's; to be precise, let us introduce a new concept as follows.

The direct union of a family of CROC's $\mathscr{L}_\alpha$ will be denoted by $\bigvee_\alpha \mathscr{L}_\alpha$ and is the CROC obtained as follows: It is the set of families $\{x_\alpha\}$, where $x_\alpha \in \mathscr{L}_\alpha$, with the order relation

$$\{x_\alpha\} < \{y_\alpha\} \Leftrightarrow x_\alpha < y_\alpha \quad \forall \alpha$$

and the orthocomplementation

$$\{x_\alpha\} \mapsto \{x'_\alpha\}$$

For each indexed β, we can define a projection π_β from $\bigvee_\alpha \mathscr{L}_\alpha$ onto $\mathscr{L}_\beta$ which is the morphism

$$\{x_\alpha\} \mapsto x_\beta$$

A CROC is said to be irreducible if it is not isomorphic to some direct union of two CROC's, each containing more than one element. The center of the direct union $\bigvee_\alpha \mathscr{L}_\alpha$ is isomorphic to the direct union of the centers of the $\mathscr{L}_\alpha$. A CROC is irreducible if and only if its center contains only 0 and I. Knowing that the center of a propositional system is isomorphic to the lattice of subsets of a set, it is not difficult to prove that every propositional system is the direct union of irreducible propositional systems.

With this result, to give a realization of a propositional system, it is sufficient to give such a realization in the irreducible case.

THEOREM. Every irreducible propositional system (with possible exceptions in two and three dimensions) is isomorphic to the lattice of all closed subspaces of a Hilbert space constructed on some division ring with involution.

We do not give here the proof of this fundamental theorem [12]. We only discuss the converse in a particular case. Consider a complex Hilbert space H; the set $\mathscr{P}(H)$ of all its closed subspaces ordered by inclusion defines a complete lattice because the intersection of any family of closed subspaces is closed. This lattice is obviously orthocomplemented, the orthocomplement of a subspace being defined by the orthogonal subspace. The easiest way to prove weak modularity is to utilize the one-to-one correspondence between closed subspaces and projectors. Let us recall that two projectors commute if and only if their product is a projector, and that in such a case this product is the projector onto their intersection subspace. Denote by P_b the projector corresponding to the closed subspace b; if $b<c$, P_b and P_c commute and we may write

$$\begin{aligned} P_{b\vee c'} &= I - P_{b'\wedge c} = I - (I - P_b)\, P_c \\ &= I - P_c + P_b \end{aligned}$$

thus

$$\begin{aligned} P_{b\vee (b'\wedge c)} &= I - P_{b'\wedge (b\vee c')} \\ &= I - (I - P_b)\,(I - P_c + P_b) = P_c \end{aligned}$$

To prove that $\mathscr{P}(H)$ is a propositional system, it remains to show that the axioms A_1 and A_2 are satisfied. The first is trivial, the rays of H corresponding to the atoms. To prove A_2, we must recall a classical result of analysis which affirms that the subspace linearly generated by a ray p and a closed subspace b' is closed, from which it follows that $(p \vee b') \wedge b$ is a ray.

To complete the interpretation of $\mathscr{P}(H)$, let us show that commutivity is completely equivalent to compatibility.

If P_b and P_c commute,

$$P_{b \vee c'} = I - (I - P_b)\, P_c$$
$$= I - P_c + P_b P_c$$

whence

$$P_{c \wedge (b \vee c')} = P_c(I - P_c + P_b P_c) = P_c P_b$$
$$= P_{c \wedge b}$$

Conversely, if $b \leftrightarrow c$, then

$$b = (b \vee c) \wedge (b \vee c')$$

and since $P_{b \vee c}$ commutes with $P_{b \vee c'} = I - P_{b' \wedge c}$, we can write

$$P_b = P_{(b \vee c) \wedge (b \vee c')} = P_{b \vee c} P_{b \vee c'}$$

Now, both $P_{b \vee c}$ and $P_{b \vee c'}$ commute with P_c, whence the conclusion.

IV. THE GENERAL FRAMEWORK OF THE QUANTUM THEORY

To give the framework of the quantum theory, it is not sufficient to describe the propositional system and the states; one must also define observables and symmetries. We have seen in the preceding section that every propositional system can be realized as the direct union of irreducible lattices of projectors of generalized Hilbert spaces. We will not consider here the most general case, but restrict our attention to a particular case which is sufficiently general to contain the classical theory and the usual quantum theory.

Let us take a family of separable complex Hilbert spaces H_α where $\alpha \in \Omega$ and Ω is some set of indices. We define the propositional system $\mathscr{L}$ as the direct union $\bigvee_\alpha \mathscr{P}(H_\alpha)$. Then a proposition is represented by a family of projectors $\{P_\alpha\}$. The ordering relation

$$\{P_\alpha\} < \{Q_\alpha\}$$

is equivalent to

$$P_\alpha Q_\alpha = P_\alpha \quad \forall \alpha \in \Omega$$

The orthocomplementation is given by the mapping

$$\{P_\alpha\} \mapsto \{I_\alpha - P_\alpha\}$$

Two propositions $\{P_\alpha\}$ and $\{Q_\alpha\}$ are compatible if and only if

$$[P_\alpha, Q_\alpha]=0 \quad \forall \alpha \in \Omega$$

The state of the system is completely determined by giving a ray (a subspace of dimension one) in one of the spaces H_α or, equivalently, by giving a family of projectors all of which are null except for one which is of rank one.

For a purely classical system (e.g., a particle), the observables map from the phase space into the values of the scale of the measuring apparatus. For a purely quantal system (e.g., a spin-1/2 system) the observables are self-adjoint operators whose eigenvalues are the values that are measured on the scale of the measuring apparatus. Our definition includes both of these notions. Roughly speaking, we call an observable a correspondence between the propositions defined by the measuring apparatus and certain propositions of the measured system:

DEFINITION. Every morphism ϕ of a Boolean CROC $\mathscr{B}$ into a proposition system $\mathscr{L}$ is called an *observable*:

$$\mathscr{B} \xrightarrow{\phi} \mathscr{L}$$

For example, the sub-CROC generated by a proposition and its orthocomplement defines an observable called a two-valued observable. More generally, every Boolean sub-CROC of a propositional system determines an observable.

The observables ϕ_i are said to be *compatible* with each other if one of the following three equivalent properties are satisfied:

(1) There exists an observable ϕ and morphisms μ_i such that the following diagram is commutative:

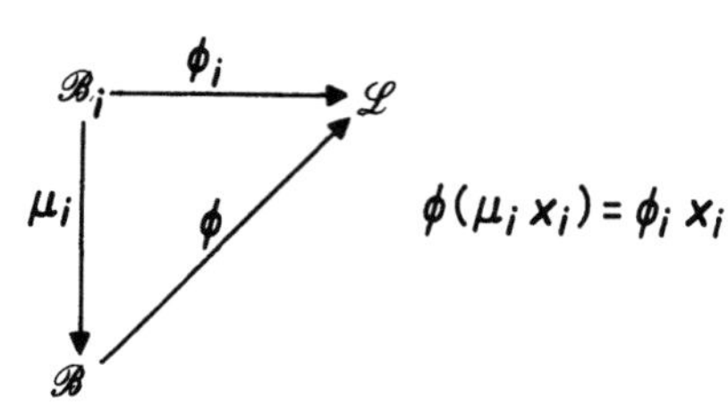

(2) The sub-CROC generated by the images of ϕ_i is Boolean.

(3) Every pair of propositions from the collection of the images of ϕ_i is compatible.

The structure of the observables is closely connected with the structure of a Boolean CROC. Let $\mathscr{B}$ be a boolean CROC and $\{x_i\}$ a family of elements of $\mathscr{B}$ such that

$$x_i \wedge x_j = 0 \quad \text{and} \quad \bigvee_i x_i = I$$

Then $\mathscr{B}$ is the direct union of the segments $[0, x_i]$. The restriction of the observable ϕ to one of the segments is a new observable which we denote by ϕ_{x_i}:

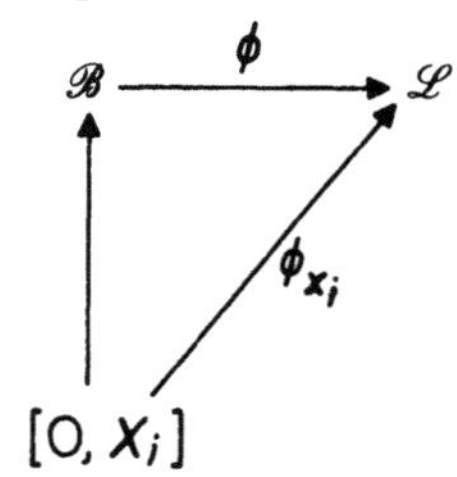

Let k be the least upper bound of Ker ϕ. Then ϕ_k is identically zero and $\phi_{k'}$ is injective. Let a be the least upper bound of the set of all atoms of $[0, k']$. If $a = k'$, the observable ϕ is said to have a *purely discrete spectrum.* On the other hand, if $a = 0$, ϕ is said to have a *purely continuous spectrum.* In general, an observable may be decomposed into a part ϕ_k which is identically zero, a part ϕ_a with purely discrete spectrum, and a part $\phi_{a' \wedge k'}$ with purely continuous spectrum.

An observable ϕ such that $\phi I = I$ is *compatible* with a state described by an atom p if p is compatible with all the propositions in the image of ϕ. In this case, for any $x \in \mathscr{B}$, either ϕx is true or $(\phi x)' = \phi x'$ is true, and the greatest lower bound of the inverse image of all ϕx which are true defines an atom of $\mathscr{B}$ which is the "*value*" of ϕ for the state p (in other words, p is an *eigenstate* of ϕ). This can never occur if ϕ has a purely continuous spectrum.

In the purely classical case, every space H_α is of dimension 1, and $\mathscr{L}$ is isomorphic to the lattice of the subsets of Ω. In this case, it can be proven [13] that an observable always has a purely discrete spectrum and that it is defined by a mapping of a part of Ω into the atoms of $\mathscr{B}$. In the purely quantal case, the set of indices Ω consists of a single element, and $\mathscr{L}$ is isomorphic to $\mathscr{P}(H)$. In this case, the image of ϕ in $\mathscr{P}(H)$ is a

Boolean sub-CROC which is the lattice of the projectors contained in an Abelian von Neumann algebra [1]. If H is separable, this algebra is generated by a self-adjoint operator [11].

In general, the morphisms π_{α_0} which project $\mathscr{L} = \bigvee_\alpha \mathscr{P}(H_\alpha)$ on the irreducible components $\mathscr{P}(H_{\alpha_0})$ define a decomposition of the observable ϕ:

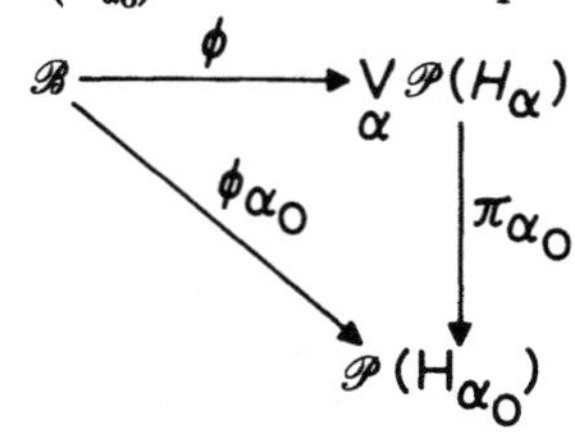

Conversely, ϕ is completely determined by giving all of the ϕ_{α_0}. In our particular case, to each observable there therefore corresponds a family of self-adjoint operators $\{A_\alpha\}$.

DEFINITION. Every automorphism S of the propositional system $\mathscr{L}$ is called a symmetry.

The structure of a symmetry is determined by the following result due to E. Wigner:

THEOREM. Every symmetry S of the propositional system $\bigvee_\alpha \mathscr{P}(H_\alpha)$ is given by a permutation f of the index set Ω and by a family of unitary or antiunitary transformations U_α mapping H_α onto $H_{f\alpha}$. Each U_α is defined only up to a phase.

With each model of a physical system, there is associated, in a natural way, a group of transformations G. This group acts not only on the system, but also on the measuring apparatus. The corresponding observable must be invariant under this double transformation. In other words, if $S(g)$ is a representation of G by symmetries of the propositional system $\mathscr{L}$, and $\sigma(g)$ a representation of G by automorphisms of the Boolean CROC $\mathscr{B}$, then the following diagram must be commutative:

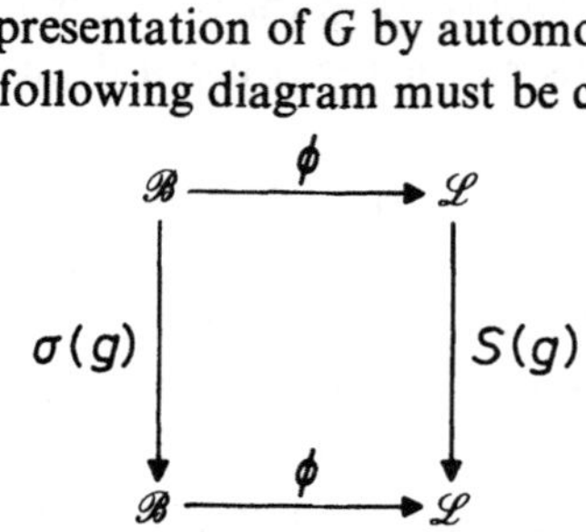

$$\phi \cdot \sigma(g) = S(g) \cdot \phi$$

Such a relation is called a system of imprimitivity [10].

V. THE GALILEAN PARTICLE WITHOUT SPIN

We define such a system by its observables and by the group which acts on them; only afterwards do we describe its dynamics in the presence of external interactions. A Galilean particle without spin is characterized by the observables momentum $\mathbf{p}$, position $\mathbf{q}$, and time t; all other observables are functions of these. The group that we have to consider is $G=\{\mathbf{v}, \mathbf{a}, R, \tau\}$, acting in the seven-dimensional space $(\mathbf{p}, \mathbf{q}, t)$, containing:

(1) The Galilei transformations $\mathbf{v}$:

$$\mathbf{p}\mapsto\mathbf{p}+m\mathbf{v}, \quad \mathbf{q}\mapsto\mathbf{q}, \quad t\mapsto t$$

(2) The space translations $\mathbf{a}$:

$$\mathbf{p}\mapsto\mathbf{p}, \quad \mathbf{q}\mapsto\mathbf{q}+\mathbf{a}, \quad t\mapsto t$$

(3) The rotations R:

$$\mathbf{p}\mapsto R\mathbf{p}, \quad \mathbf{q}\mapsto R\mathbf{q}, \quad t\mapsto t$$

(4) The time translations τ:

$$\mathbf{p}\mapsto\mathbf{p}, \quad \mathbf{q}\mapsto\mathbf{q}, \quad t\mapsto t+\tau$$

The Boolean CROC corresponding to the measuring apparatus is constructed from the subsets of R for the time and from the subsets of R^3 for the position and for momentum. In each of these cases, one has *a priori* two possibilities: Either $\mathscr{B}$ contains all subsets in question or $\mathscr{B}$ is the CROC of the Borel sets modulo the subsets of measure zero. The group G acts in a natural way on the elements $\Delta\in\mathscr{B}$ and one obtains the following systems of imprimitivity:

For the time t

$$t(\Delta)=S(\mathbf{v})\,t(\Delta), \qquad t(\Delta)=S(\mathbf{a})\,t(\Delta)$$
$$t(\Delta)=S(R)\,t(\Delta), \quad t(\Delta+\tau)=S(\tau)\,t(\Delta)$$

For the position $\mathbf{q}$

$$\mathbf{q}(\Delta)=S(\mathbf{v})\,\mathbf{q}(\Delta), \quad \mathbf{q}(\Delta+\mathbf{a})=S(\mathbf{a})\,\mathbf{q}(\Delta)$$
$$\mathbf{q}(R\Delta)=S(R)\,\mathbf{q}(\Delta), \qquad \mathbf{q}(\Delta)=S(\tau)\,\mathbf{q}(\Delta)$$

For the momentum $\mathbf{p}$

$$\mathbf{p}(\Delta+m\mathbf{v})=S(\mathbf{v})\,\mathbf{p}(\Delta),\quad \mathbf{p}(\Delta)=S(\mathbf{a})\,\mathbf{p}(\Delta)$$
$$\mathbf{p}(R\Delta)=S(R)\,\mathbf{p}(\Delta),\quad \mathbf{p}(\Delta)=S(\tau)\,\mathbf{p}(\Delta)$$

We suppose the dynamics of the Galilean particle to be reversible in the sense of thermodynamics, and postulate that the corresponding evolution is given by symmetries which induce a representation of the one-parameter translation group. The symmetry which describes the evolution of the system after a time τ must change the proposition $t(\Delta)$ into $t(\Delta+\tau)$. If this symmetry is represented by the family of unitary transformations

$$H_\alpha \xrightarrow{U_\alpha(\tau)} H_{\alpha_\tau}$$

then the group multiplication law imposes the relation

$$U_{\alpha_{\tau_1}}(\tau_2)\,U_\alpha(\tau_1)=\omega_\alpha(\tau_2,\tau_1)\,U_\alpha(\tau_1+\tau_2)$$

If one postulates some conditions of differentiability, we can define

$$i\,\partial_\tau\psi_{\alpha_\tau}=\lim_{\delta\tau\to 0}(i/\delta\tau)\,(\omega_\alpha^{-1}(\delta\tau,\tau)\,U_{\alpha_\tau}(\delta\tau)\,\psi_{\alpha_\tau}-\psi_{\alpha_\tau})$$
$$=\mathscr{H}_{\alpha_\tau}\psi_{\alpha_\tau}$$

and

$$\partial_\tau\alpha_\tau=\lim_{\delta\tau\to 0}(1/\delta\tau)\,(\alpha_{\tau+\delta\tau}-\alpha_\tau)$$
$$=X(\alpha_\tau)$$

which is the Schrödinger equation coupled with an ordinary differential equation.

Next, let us write all of the above results more explicitly for the cases of the purely classical Galilean particle and of the quantal Galilean particle.

A. *The Purely Classical Galilean Particle*

In this case, the proposition system is isomorphic to the subsets of the

other hand, since all observables have a purely discrete spectrum, one has to choose the first possibility for $\mathscr{B}$, i.e., one has to identify $\mathscr{B}$ with the subsets of R or R^3. It is easy to verify that the following observables satisfy the imprimitivity relations:

$$t(\Delta)=\{(\mathbf{p}, \mathbf{q}, t) \mid t\in\Delta\subset R\}$$
$$\mathbf{q}(\Delta)=\{(\mathbf{p}, \mathbf{q}, t) \mid \mathbf{q}\in\Delta\subset R^3\}$$
$$\mathbf{p}(\Delta)=\{(\mathbf{p}, \mathbf{q}, t) \mid \mathbf{p}\in\Delta\subset R^3\}$$

These observables may be defined as the inverse images of the following functions:

$$(\mathbf{p}, \mathbf{q}, t)\mapsto t$$
$$(\mathbf{p}, \mathbf{q}, t)\mapsto \mathbf{q}$$
$$(\mathbf{p}, \mathbf{q}, t)\mapsto \mathbf{p}$$

In the purely classical case, the equations of evolution reduce to

$$\partial_\tau \alpha_\tau = X(\alpha_\tau)$$

where here,

$$\alpha_\tau = \{\mathbf{p}_\tau, \mathbf{q}_\tau, t_\tau = t_0 + \tau\}$$

In classical mechanics, one imposes more, i.e., the canonical equations

$$\dot{\mathbf{p}}(t) = -\partial_{\mathbf{q}}\mathscr{H}(\mathbf{p}, \mathbf{q}, t), \quad \dot{\mathbf{q}}(t) = +\partial_{\mathbf{p}}\mathscr{H}(\mathbf{p}, \mathbf{q}, t)$$

The Hamiltonian $\mathscr{H}(\mathbf{p}, \mathbf{q}, t)$ must be such that the transformations of the group G act as canonical transformations, with a new Hamiltonian which is just the old one with the new variables. Because of this physical interpretation, the Galilei transformation $\mathbf{v}$ must change $\dot{\mathbf{q}}(t)$ into $\dot{\mathbf{q}}(t)+\mathbf{v}$. This condition can be written

$$\partial_{\mathbf{p}}\mathscr{H}(\mathbf{p}+m\mathbf{v}, \mathbf{q}, t) = \partial_{\mathbf{p}}\mathscr{H}(\mathbf{p}, \mathbf{q}, t) + \mathbf{v}$$

The general solution is

$$\partial_{\mathbf{p}}\mathscr{H}(\mathbf{p}, \mathbf{q}, t) = (1/m)\,[\mathbf{p} - \mathbf{A}(\mathbf{q}, t)]$$

By integration, we find

$$\mathscr{H}(\mathbf{p}, \mathbf{q}, t) = (1/2m)\,[\mathbf{p} - \mathbf{A}(\mathbf{q}, t)]^2 + V(\mathbf{q}, t)$$

which is the most general Galilean covariant Hamiltonian.

B. *The Quantal Galilean Particle*

One postulates that the observable t be compatible with all other observables. Then, as the system is represented by a family of Hilbert spaces $\{H_\alpha\}$, one is led to identify the index α with the points of R and to set $\alpha=t$. In these circumstances, the symmetry $S(\tau)$ defines for each value of t a unitary transformation $V_t(\tau)$ between the spaces H_t and $H_{t+\tau}$. Thus one can identify the spaces H_t in such a way as to have $V_t(\tau)=I$. In this way, the representation $S(g)$ of G is reduced in each H_t to a representation up to a phase of the subgroup $G_0=\{\mathbf{v}, \mathbf{a}, R\}$. In the space $L^2(R^3)$ of square integrable functions ϕ defined on R^3, this representation of G_0 may be written as follows:

$$[V(\mathbf{v})\,\phi]\,(\mathbf{x})=\exp(-i\mu\mathbf{v}\cdot\mathbf{x})\,\phi(\mathbf{x})$$
$$[V(\mathbf{a})\,\phi]\,(\mathbf{x})=\phi(\mathbf{x}-a)$$
$$[V(R)\,\phi]\,(\mathbf{x})=\phi(R^{-1}\mathbf{x})$$

It satisfies the commutation relations of H. Weyl:

$$V(\mathbf{v})\,V(\mathbf{a})=\exp(-i\mu\mathbf{v}\cdot\mathbf{a})\,V(\mathbf{a})\,V(\mathbf{v})$$

Finally, the following observables satisfy the imprimitivity relations:
For the time:

$$t(\Delta)=\{P_t\} \quad \text{with} \quad P_t=\begin{cases} I_t, & \text{if} \quad t\in\Delta \\ 0_t, & \text{if} \quad t\notin\Delta \end{cases}$$

The corresponding family for Hermitian operators is

$$\{t_t=tI_t\}$$

For the position:

$$\mathbf{q}(\Delta)=\{P_t=\chi_\Delta(x)\}$$

where χ_Δ is the characteristic function of Δ

$$\chi_\Delta(\mathbf{x})=\begin{cases} 1, & \text{if} \quad \mathbf{x}\in\Delta \\ 0, & \text{if} \quad \mathbf{x}\notin\Delta \end{cases}$$

The corresponding three families of Hermitian operators are:

$$\{q_t^i=x^i\}, \quad i=1, 2, 3$$

For the momentum: We give only the three families of Hermitian operators:

$$\{p_t^i = -i(m/\mu)\,\partial_{x^i}\}, \quad i=1,2,3, \quad m/\mu=\hbar$$

This model is identical with the usual model of spinless particles in nonrelativistic quantum mechanics in the Schrödinger picture. In this latter case, the time t is treated as a parameter and not as a variable of the L^2-space on the same footing as $\mathbf{x}$. This use of t agrees completely with the idea of time as a continuous superselection rule.

In this quantal case, the equation of evolution reduces to

$$i(\partial/\partial t)\,\psi_t = \mathscr{H}_t\psi_t$$

where we have put

$$t_\tau = t_0 + \tau = t$$

The Hamiltonian $\{\mathscr{H}_t\}$ must be such that the symmetry corresponding to the Galilean transformation $\mathbf{v}$ must change $\dot{\mathbf{q}}$ into $\dot{\mathbf{q}}+\mathbf{v}$, where $\dot{\mathbf{q}}$ is defined by the family $\{\dot{\mathbf{q}}_t\}$, and

$$\dot{\mathbf{q}}_t = i[\mathscr{H}_t, \mathbf{q}_t]$$

According to this requirement,

$$V(\mathbf{v})\,\dot{\mathbf{q}}_t V(\mathbf{v})^{-1} = \dot{\mathbf{q}}_t + \mathbf{v}$$

On the other hand,

$$V(\mathbf{v})\,p_t V(\mathbf{v})^{-1} = \mathbf{p}_t + m\mathbf{v}$$

and it then follows that

$$[V(\mathbf{v}), \mathbf{p}_t - m\dot{\mathbf{q}}_t] = 0$$

The operator $\mathbf{p}_t - m\dot{\mathbf{q}}_t$ is therefore a function of the operator $\mathbf{x}$ and, of course, of t; i.e.,

$$\mathbf{p}_t - m\dot{\mathbf{q}}_t = \mathbf{A}(\mathbf{x}, t)$$

The general solution for $\mathscr{H}_t$ is obtained by adding to a particular solution some operator which commutes with $\mathbf{q}_t$. For a particular solution, we take

$$\mathscr{H}_t^0 = \tfrac{1}{2}(1/m\hbar)\,[\mathbf{p}_t - \mathbf{A}(\mathbf{x}, t)]^2 = \tfrac{1}{2}(m/\hbar)\,\dot{\mathbf{q}}_t^2$$

The general solution for the quantal Galilean particle is therefore of the form [7]

$$\mathscr{H}_t = (1/\hbar)\,\{(1/2m)\,[\mathbf{p}_t - \mathbf{A}(\mathbf{x}, t)]^2 + V(\mathbf{x}, t)\}$$

The formal correspondence between the Hamiltonians for the purely classical and the quantal Galilean particles is the justification of the so-called correspondence principle.

VI. THE IDEAL MEASUREMENTS AND GLEASON'S THEOREM

Suppose the system to be given in a state p, and consider the measure of a proposition a. If a is compatible with p, "a is true" or "a' is true," and then the result of the measurement of a question α such that $\alpha \in a$ and $\alpha^{\sim} \in a'$ is certain. It is only in this case that one has really measured $\{a, a'\}$. If a is not compatible with p, it is impossible to predict the result; nevertheless, for the case in which α is ideal and of the first kind (see Section II), immediately after the measurement the state of the system is $(p \vee a') \wedge a = \phi_a p$ if the result is "yes." We will show in the following that it is possible to calculate the probability for the answer "yes" in such an ideal measurement of a on the basis of the following assumption: The probability for the answer "yes" depends only on the states before and after the ideal measurement.

In other words, if p is the state before and a the proposition considered, the probability $w_p(a)$ for the answer "yes" is only a function of p and $\phi_a p$.

THEOREM. The function $w_p(a)$ must have the following properties:

(1) $0 \leqslant w_p(a) \leqslant 1$.

(2) $w_p(a) = 1 \Leftrightarrow p < a$.

(3) $w_p(a) + w_p(a') = 1$.

(4) $w_p(a) + w_p(b) = w_p(a \vee b) a < b'$.

Proof. The three first properties are trivial. To prove (4), we must make a remark.

Let x and y be two compatible propositions. If we make an ideal measurement of x and obtain the answer "yes," and immediately afterward perform the ideal measurement of y, and obtain the answer "yes," this constitutes an ideal measurement of $x \wedge y$ because

$$\phi_y \phi_x = \phi_{y \wedge x} = \phi_x \phi_y$$

if $x \leftrightarrow y$, as it is easy to verify. Then, according to the rule of composition of probability,

$$w_p(x \wedge y) = w_p(x)\, w_{\phi_x p}(y)$$

If $a < b'$, this implies that

$$w_p(a) = w_p(a \vee b)\, w_q(a)$$
$$w_p(b) = w_p(a \vee b)\, w_q(b)$$

where $q = \phi_{a \vee b} p$. However, by our assumption,

$$w_q(a) = w_q(b')$$

since $\phi_a \phi_{a \vee b} = \phi_{b'} \phi_{a \vee b}$. Then

$$w_p(a) + w_p(b) = w_p(a \vee b)\, [w_q(b') + w_q(b)] = w_q(a \vee b). \blacksquare$$

Conversely, let w_p be a mapping from the propositional system $\mathscr{L}$ into the real interval $[0, 1]$ satisfying the four conditions of the theorem for a given atom p; we can interpret $w_p(a)$ as the probability to obtain the answer "yes" if one performs an ideal measurement of the proposition a on the system in the state p. It is an important result due to Gleason that given p, such a mapping is unique.

THEOREM. Given $\mathscr{L} = \bigvee_\alpha \mathscr{P}(H_\alpha)$, where the H_α are Hilbert spaces (of dimension $\neq 2$) on the reals, complexes, or quaternions and given p an atom of $\mathscr{L}$, there exists one and only one mapping w_p from $\mathscr{L}$ into the real interval $[0, 1]$ such that:

(1) $w_p(x) = 1 \Leftrightarrow p < x$.

(2) $w_p(a) + w_p(b) = w_p(a \vee b)$ if $a < b'$.

Proof. If $a = \{Q_\alpha\}$ and p is given by a projector of rank one P_{α_0} defined on H_{α_0}, it is easy to verify that the mapping

$$w_p(a) = \mathrm{tr}\,(P_{\alpha_0} Q_{\alpha_0})$$

satisfies all the conditions. If there exists another such mapping, it must take a different value for some proposition a and since by (1) and (2),

$$w_p(a) = w_p((p \vee a') \wedge a) + w_p(p' \wedge a)$$
$$= w_p((p \vee a') \wedge a)$$

it must be also different for some atom q. The ray corresponding to such an atom q must be in the Hilbert space H_{α_0}, otherwise $w_p(q)$ cannot be different from zero. Now, it is possible to choose two unit vectors **f** and **g**, respectively, in the rays corresponding to p and q, in such a way that their scalar product is real. Then, to complete the proof, it is sufficient to show the unicity for the lattice of projectors of the real Hilbert space generated by **f**, **g**, and an orthogonal vector; such unicity is a direct consequence of Gleason's theorem [5] (see appendix). ■

As a consequence of the theorem, it is important to remark that not only does w_p satisfy (2), but also the complete additivity

$$\sum_i w_p(a_i) = w_p\left(\bigvee_i a_i\right) \quad \text{if} \quad a_i < a_j' \quad \forall i \neq j$$

Then, for each observable ϕ, the mapping w_p defines a complete measured Boolean algebra:

$$\mathscr{B} \xrightarrow{\phi} \mathscr{L} \xrightarrow{w_p} [0, 1]$$

and we can apply the rules of probability in the usual way. If ϕ is represented by the family of self-adjoint operators $\{A_\alpha\}$, we can define the expectation value by

$$\bar{A} = \mathrm{tr}(P_{\alpha_0} A_{\alpha_0})$$

APPENDIX

In this appendix, we prove what is more or less the first part of Gleason's theorem. The demonstration is a little different, and perhaps more geometric.

THEOREM. There exists one and only one mapping w from the lattice of subspaces of the real three-dimensional Hilbert space R^3 into the real interval $[0, 1]$ such that:

(1) $w(x) = 1 \Leftrightarrow p < x$ for a given ray p.

(2) $w(a) + w(b) = w(a \vee b)$ for $a \perp b$.

Proof. The proof is given as the consequence of four lemmas. In the following, we represent the lattice of the subspaces of R^3 by the projective real plane in such a way that the line at infinity represents the two-dimensional subspace orthogonal to the given ray p. In this plane, every ray is represented by a point and the orthogonality relation between rays induces an orthogonality relation between points. The desired mapping can be defined as a function $w(q)$ of the points q of this space which takes on the value 1 for p and 0 for every point at infinity.

LEMMA 1. If $w(q)$ is continuous and if the value of $w(q)$ depends only on the angle ϕ between the rays p and q, then it is unique and is given by

$$w(q) = \cos^2\phi$$

Proof. In Figure 3, the position of a point q on the line is labeled by the square of the tangent of the angle between p and q; we have chosen $\lambda > 1$.

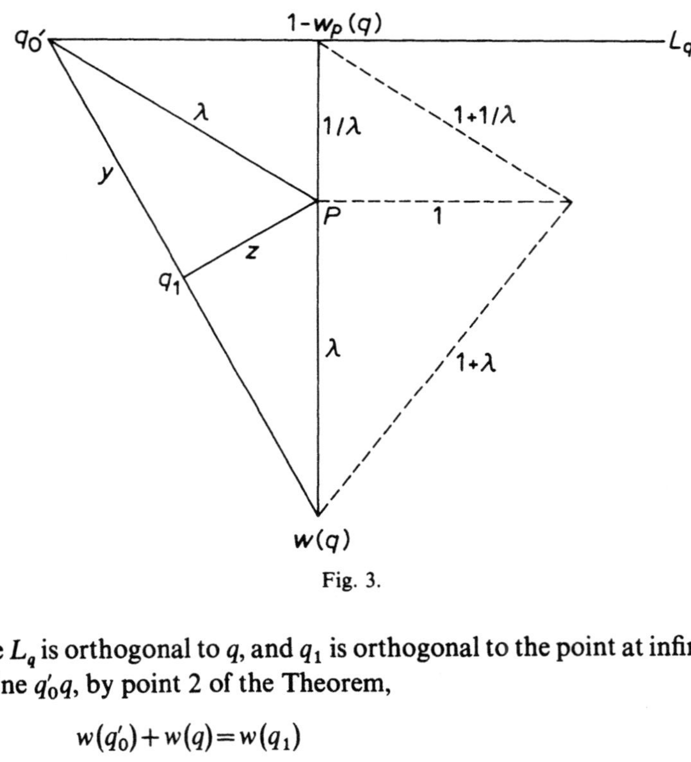

Fig. 3.

Since L_q is orthogonal to q, and q_1 is orthogonal to the point at infinity on the line $q_0'q$, by point 2 of the Theorem,

$$w(q_0') + w(q) = w(q_1)$$

Hence, by the hypothesis of Lemma 1,

$$w(q_1) = 2w(q)$$

Using Pythagoras' theorem,

$$4y=(1+\lambda)+[1+(1/\lambda)]+\lambda-(1/\lambda)=2(1+\lambda)$$

and therefore

$$z=\tfrac{1}{2}(\lambda-1)$$

The relation above is then

$$2w(\lambda)=w[\tfrac{1}{2}(\lambda-1)]$$

If we let $x=1/(1+\lambda)$ and $w[(1-x)/x]=f(x)$, this becomes

$$2f(x)=f(2x),\quad x<\tfrac{1}{2}$$

For $1/2\leqslant x\leqslant 1$, or $\lambda\leqslant 1$, we note that $w(1/\lambda)=1-w(\lambda)$, so that $f(x)=1-f(1-x)$.

The demonstration is completed by remarking that for $f(x)$ a continuous function, there exists one and only one solution,

$$f(x)=x$$

LEMMA 2. If $w(q)$ is continuous, then its value depends only on the angle ϕ between p and q.

Proof. Because of the continuity hypothesis, it is sufficient to prove that $w(q_1)\geqslant w(q_2)$ if $\lambda_2>\lambda_1$, where, as before, $\lambda=\tan^2\phi$.

From Figure 4, one sees that

$$\lambda_2=\lambda_1/\cos^2\psi>\lambda_1$$

and in this particular case,

$$w(q_1)=w(q_2)+w(q_2')\geqslant w(q_2)$$

where q_2' is the point on the line L orthogonal to q_2. On the other hand, for

$$0<\lambda_2-\lambda_1<\varepsilon$$

it is always possible, by dividing the angle ψ by n sufficiently large [such that $\psi^2/n<\varepsilon(1-\psi^2/n^2)$], to prove $w(q_1)\geqslant w(q_2)$ by repeating the argument given above.

LEMMA 3 (Gleason). If $w(q)$ is continuous at some point q_0, then it is continuous at every point.

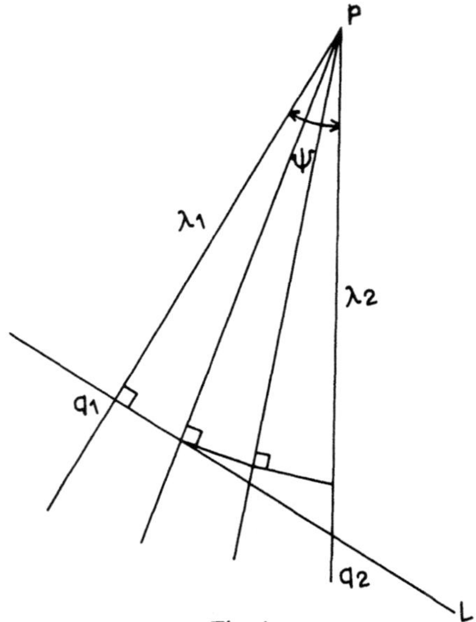

Fig. 4.

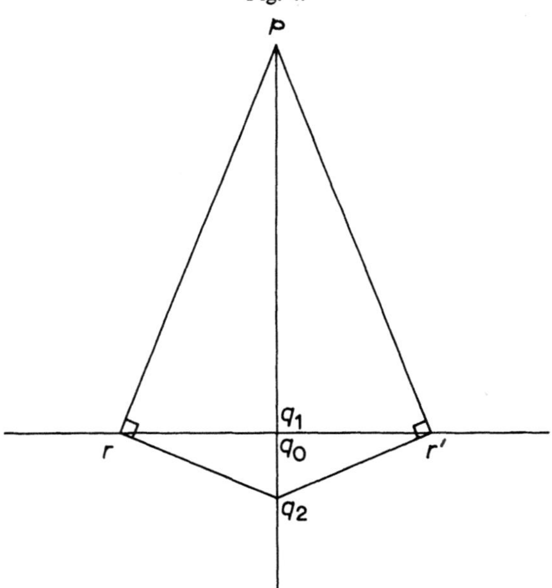

Fig. 5.

Proof. First, we will show that if $w(q)$ is continuous at q_0, it is continuous at every point q_1 orthogonal to q_0. Given $\varepsilon > 0$, let U be the corresponding neighborhood of q_0. Take the point $q \in q_0 q_1$ orthogonal to some point $q' \in U$ on the line $q_0 q_1$. For every point $r_0 \in U$, define by r_1 the point on the line $r_0 q$ orthogonal to r_0 and by r' the point on the same line $r_0 q$ orthogonal to q. The points r_1 for which the corresponding r' lies in U define a neighborhood of q_1. Now, $w(q_0) + w(q_1) = w(q') + w(q)$ and

$$w(r_0) + w(r_1) = w(r') + w(q).$$

Hence,

$$\begin{aligned} &|w(r_1) - w(q_1)| \\ &\quad = |w(q_0) - w(r_0) + w(r') - w(q')| \\ &\quad \leqslant |w(q_0) - w(r_0)| + |w(r') - w(q_0)| + |w(q_0) - w(q')| \leqslant 3\varepsilon \end{aligned}$$

To complete the proof, it is sufficient to remark that given two arbitrary points, there always exists a point orthogonal to both (as can easily be seen directly in R^3).

LEMMA 4. The function $w(q)$ is continuous at some point q_0.

Proof. The function $w(q)$ is a decreasing function of λ along a line passing through the given point p. Given q_1 and q_2, two points on such a line, suppose $\lambda_2 > \lambda_1$; then, there always exists a point r such that $q_2 r$ is perpendicular to pr and $q_1 r$ is perpendicular to pq_1. By the same reasoning used before,

$$w(q_1) > w(r) > w(q_2)$$

It is well known that a bounded decreasing function is continuous at some point. Then $w(q)$ is continuous at some point $q_0 \in pq$ along the line pq. Now,

$$\text{if} \quad w(q_1) - w(q_2) < \varepsilon, \quad \text{then} \quad |w(q) - w(q_0)| < \varepsilon$$

for every point of the triangle r, r', q_2 of Figure 5. ■

University of Denver

* Supported by the National Science Foundation under the auspices of Departmental Science Development Grant GU 2635.

** On leave of absence from Institute of Theoretical Physics, University of Geneva, Geneva, Switzerland.

BIBLIOGRAPHY

[1] Bade, W. G., *Pacific J. Math.* **4** (1954), 393.
[2] Birkhoff, G. and Neumann, J. von, *Ann. Math.* **37** (1936), 823.
[3] Dixmier, J., *Les algèbres d'opérateurs dans l'espace Hilbertien*, 2nd ed. (Gauthier-Villars, Paris, 1969) exercise 3f, Chapter I, paragraph 7, p. 120.
[4] Einstein, A., Podolsky, P., and Rosen, N., *Phys. Rev.* **47** (1935), 777.
[5] Gleason, A. M., *J. Rat. Mech. Anal.* **6** (1957), 885.
[6] Halmos, P. R., *Lectures on Boolean Algebra* (1963), Chapter 9, p. 35.
[7] Jauch, J. M., *Helv. Phys. Acta* **37** (1964), 284.
[8] Loomis, L. W., *Mem. Amer. Math. Soc.* (1955), No. 18.
[9] Ludwig, G., *Lecture Notes in Physics* **4**, Springer-Verlag, Berlin, 1970.
[10] Mackey, G. W., *Induced Representations of Groups and Quantum Mechanics*, Benjamin, New York, 1968.
[11] Neumann, J. von, *Math. Ann.* **102** (1929), 370.
[12] Piron, C., *Helv. Phys. Acta* **37** (1964), 439.
[13] Piron, C., 'Observables in General Quantum Theory', delivered at International School of Physics, "Enrico Fermi," Foundations of Quantum Mechanics, 29 June–11 July, 1970, to be published.
[14] Ballentine, D., *Rev. Mod. Phys.* **42** (1970), 358.

R. J. GREECHIE AND STANLEY P. GUDDER

QUANTUM LOGICS

PREFACE

Although the authors have great interest in physics and philosophy we are not 'experts' in these fields; moreover our primary interest (and training) is in mathematics. For these reasons our presentation is essentially mathematical in nature. We realize that, when one presents an approach which purports to deal with physical situations, physical justifications should be given for one's assumptions. If further this approach has philosophical ramifications then one ought to discuss these ramifications. In this paper we shall attempt to motivate the assumptions made; however, we shall minimize discussion of the philosophical import of these assumptions.

Our main aim is to present some mathematical tools necessary for the study of axiomatic quantum mechanics, in particular the quantum logic approach. We feel that the mathematical tools, techniques and theorems must be first understood before physical and philosophical discussions about the subject can take place. It is our opinion that the quantum logic approach can give one, if nothing else, deeper insight into the understanding of quantum phenomena.

I. INTRODUCTION

There are many advantages of axiomatic formulations for physical theories. First of all, by stating one's axioms carefully one is fully aware of the assumptions made in the theory and it is clear what hypotheses must be physically justified. These hypotheses may then be tested in the laboratory as a check of the theory. Of course, a physical theory can never be proved to be correct; the conclusions from the theory can only be compared with experimental results to test whether it is an approximate description of some small isolated portion of 'reality'. If the theory fails to compare favorably with the experimental results it must be abandoned as a theoretical description for that portion of reality. Otherwise it may be retained

C. A. Hooker (ed.), The Logico-Algebraic Approach to Quantum Mechanics, 545–575

until a better theory is discovered. Secondly, an axiomatic approach gives a common 'universe of discourse' in which ideas may be discussed and conjectures formulated. Many of the great controversies in physics seem to result from difficulties in semantics. For example, it is our feeling that the controversy over hidden variables in quantum mechanics is caused to a certain extent by a failure in laying out the 'ground rules' for the game. Many arguments seem to result from the fact that the debators have different underlying physical formulations in mind and it is never clearly stated exactly what assumptions one is making. If at the beginning a common universe of discourse in terms of an axiomatic formulation were established these types of problems might vanish. Thirdly, if one operates under a consistent axiomatic model for a physical theory one is assured that no mathematical contradictions will be encountered. For example, the difficulties in quantum electrodynamics stemming from the occurrence of divergences and infinities might be avoided if a consistent axiomatic model were constructed. Indeed this is one of the reasons for the introduction of axiomatic quantum field theory by Wightman and his co-workers.

There are several axiomatic approaches to the foundations of quantum mechanics available in the physical and mathematical literature. One of these is the quantum logic approach initiated by Birkhoff and von Neumann in 1936 [4]. This study has been continued by Mackey [54, 55], and Varadarajan [80], and has been refined and altered by Jauch, Piron [38, 39, 40, 45, 46, 64] and others [1, 5, 14, 18, 28, 30, 31, 32, 42, 56, 58, 62, 65, 67, 68, 69, 76, 78, 79, 82, 84]. Another is the algebraic approach first conceived by Jordan *et al.* [47] in 1934, developed further by Segal [71, 72] and others [50, 73, 74] culminating in the elegant theories of Haag [37], Wightman [77], and their co-workers. Another formulation has been proposed by Ludwig and his collaborators [51, 52] and has been recently refined by Mielnik [59, 60], Davies and Lewis [11, 12] and others [15, 16, 36]. These models present different approaches to what appears to be essentially the same underlying theory. In fact there have been studies made comparing these different approaches [27, 35, 63]. Now it may seem, at first sight, to be wasteful and redundant to proliferate the literature with different approaches to the same subject. However, it has turned out that each approach adds new insights and different viewpoints which have led to fruitful results and contributed to a deeper understanding of quantum theory.

Of course, most working physicists do not use and are probably even unaware of the above formulations. The majority of physicists rely upon the von Neumann [81] and/or Dirac [13] formulations of quantum mechanics and in so doing have achieved many extraordinary successes. We have no quarrel with these researchers. We only contend that a knowledge of some of the basic mathematical tools involved in a consistent axiomatic model grounded upon physically justified assumptions may prove both useful and rewarding.

One of the aims of the present paper is to present one of these models, namely the quantum logic approach, in some detail. This approach has evolved in several slightly different directions. One of these directions we attribute to Mackey and another to Jauch-Piron. We single out these researches only for expediency and, although they have had profound influence on the subject, it must be realized that many others have made equally important contributions.

Having treated these two major approaches to quantum logic in some detail we discuss several attempts to define a conditional, $a \supset b$, in quantum logic. The impact of our discussion is that there has been no successful attempt to define a conditional, $a \supset b$, which is an element of the logic and behaves in a fashion similar to that of classical logic.

We conclude with a discussion of combinatorial quantum logic. Orthogonality spaces are introduced in order to present a model which distinguishes between the two possible orderings induced by states. Other applications of this promising approach are given.

II. OBSERVABLES, STATES AND QUESTIONS [55]

Although, as we have stated in the introduction, there are several axiomatic formulations for the foundations of quantum mechanics, they all seem to involve in some degree the three basic, primitive notions of observable, state, and question. In the laboratory, the experimental physicist makes different measurements. He measures physical observables which have traditional names such as energy, momentum, position, spin, charge, magnetic moment, etc. He makes these measurements by subjecting a physical system to a measuring apparatus, and he is concerned with the outcomes of these interactions when the system is in some specified state or condition.

Let us suppose that we have some fixed physical system which may exist in any one of a collection of states $S=\{s, s_1, s_2, \ldots\}$ and that in this system one may measure the observables $\mathcal{O}=\{x, y, z, \ldots\}$. Now the result of measuring an observable x can usually be formulated as a number; e.g., the spin of the particle was $+2$, the energy of the electron was 3 erg. Of course, in practice, one repeats the experiment many times (keeping the state as fixed as possible) and obtains only a statistical distribution for the values of x. Thus, given an observable x and a state s, the experimentalist obtains a probability distribution $p(x, s)(\cdot)$. By this we mean that given any set E of real numbers (mathematicians usually only consider Borel sets but we will not concern ourselves with such technicalities now) $p(x, s)(E)$ is a number between 0 and 1 representing the probability that the observable x has a value in the set E when the system is in the state s. If we denote the set of all probability distributions by M we may now formulate our first axiom.

AXIOM 1. There is a map $p:\mathcal{O}\times S \rightarrow M$ denoted by $p(x, s)(\cdot)$. (Of course $\mathcal{O}\times S$ denotes the set of all ordered pairs (x, s) where $x\in\mathcal{O}$ and $s\in S$.)

If two observables have the same probability distribution in every state then there is no experimental way to distinguish them so they must be the same observable. Similarly if two states give the same probability distributions for all observables they must be equal. We are thus led to our next axiom.

AXIOM 2. If $p(x, s)(E)=p(y, s)(E)$ for all $s\in S$ and all sets E of real numbers then $x=y$. If $p(x, s_1)(E)=p(x, s_2)(E)$ for all $x\in\mathcal{O}$ and all sets E of real numbers then $s_1=s_2$.

If we can measure an observable x then it is just as easy to measure the observable x^2; simply take the measured values of x and square them. Now the probability that x^2 has a value λ is the same as the probability that x has the values $\pm\sqrt{\lambda}$, and more generally the probability that x^2 has a value in a set E of real numbers is the probability that x has a value in the set $\pm\sqrt{E}=\{\lambda:\lambda^2\in E\}$. In the same way, if f is a real valued function and x is an observable then $f(x)$ is an observable and the probability that $f(x)$ has a value in the set E is the probability that x has a value in the set $f^{-1}(E)=\{\lambda:f(\lambda)\in E\}$. Usually mathematicians consider only Borel functions but again we omit the technicality. We are now ready for our next axiom.

AXIOM 3. If $x \in \mathcal{O}$ and f is a real valued function then there is a $y \in \mathcal{O}$ such that $p(y, s)(E) = p(x, s)(f^{-1}(E))$ for every $s \in S$ and every set of real numbers E.

It follows from Axiom 2 that the observable y in Axiom 3 is unique. We denote this observable by $y = f(x)$.

Now there is a particular type of observable which is extremely simple. These are the observables with only two possible values, say 0 and 1. We call such observables 'questions'. For example, a counter is a question since it gives a measurement with only two possible outcomes: unactivated (or 0) and activated (or 1). We thus define a *question* to be any observable x that satisfies $p(x, s)(\{0, 1\}) = 1$ for all $s \in S$; that is, x has the value 0 or 1 with certainty in every state. It is easy to show that $x \in \mathcal{O}$ is a question if and only if $x^2 = x$. There is another convenient way to describe questions. If $E \subseteq R$ (R denotes the real line) then the *characteristic function*

$$\chi_E(\lambda) = \begin{cases} 1 \text{ if } \lambda \in E. \\ 0 \text{ if } \lambda \notin E. \end{cases}$$

Now it is easy to show that $x \in \mathcal{O}$ is a question if and only if $x = \chi_E(y)$ for some $E \subseteq R$, $y \in \mathcal{O}$. In particular if $x \in \mathcal{O}$ then we can associate with each $E \subseteq R$ a question $\chi_E(x)$. This question has the value 1 if x has a value in E and the value 0 if x has a value not in E. Notice if $\chi_E(x) = \chi_E(y)$ for all $E \subseteq R$ then

$$\begin{aligned} p(x, s)(E) &= p(x, s)(\chi_E^{-1}\{1\}) = p(\chi_E(x), s)(\{1\}) \\ &= p(\chi_E(y), s)(\{1\}) = p(y, s)(E) \end{aligned}$$

for all $s \in S$, $E \subseteq R$ and hence $x = y$. We thus see that not only can we associate with any observable x a collection of questions $\{\chi_E(x) : E \subseteq R\}$ but that this associated collection of questions determines x.

Denote the set of questions by Q. We have seen that the system (Q, S) contains all the information given in $(\mathcal{O}, S)$. Since questions are far simpler than general observables it appears that we can make a more fundamental study by considering (Q, S) instead of $(\mathcal{O}, S)$. Let us now try to discover some of the mathematical properties of the system (Q, S). If $\alpha \in Q$ and $s \in S$ we define $s(\alpha) = p(\alpha, s)(\{1\})$. Now $s(\alpha)$ may be interpreted as the probability that α has the value 1 (or α has the answer 'yes') in the state s. Notice if $\alpha, \beta \in Q$ and $s(\alpha) = s(\beta)$ for all $s \in S$ then by Axiom 2, $\alpha = \beta$. It also follows from Axiom 2 that if $s_1(\alpha) = s_2(\alpha)$ for all $\alpha \in Q$ then $s_1 = s_2$ so there are sufficiently many questions to determine the state. If $\alpha_1, \alpha_2 \in Q$ we define $\alpha_1 \leqslant \alpha_2$ if $s(\alpha_1) \leqslant s(\alpha_2)$ for all $s \in S$. Thus $\alpha_1 \leqslant \alpha_2$ if α_1 has a smaller probability of having an answer 'yes' than α_2 in every state. It is easy to check that $\leqslant$ is a partial order relation on Q; that is $\alpha \leqslant \alpha$ for all $\alpha \in Q$, $\alpha \leqslant \beta$ and $\beta \leqslant \gamma$ implies $\alpha \leqslant \gamma$, $\alpha \leqslant \beta$ and $\beta \leqslant \alpha$ implies $\alpha = \beta$. Thus $(Q, \leqslant)$ is a *partially ordered set* or *poset*. Let f be the function $f(\lambda) = 1 - \lambda$. If $\alpha \in Q$ we define the observable α' by $\alpha' = f(\alpha)$. Notice

$$p(\alpha', s)(\{0, 1\}) = p(\alpha, s)(f^{-1}\{0, 1\}) = p(\alpha, s)(\{0, 1\}) = 1$$

for every $s \in S$ so $\alpha' \in Q$. Also

$$s(\alpha') = p(\alpha', s)(\{1\}) = p(\alpha, s)(f^{-1}\{1\}) = p(\alpha, s)(\{0\})$$
$$= 1 - p(\alpha, s)(\{1\}) = 1 - s(\alpha).$$

Thus α' corresponds to the negation of the question α. If f_0 and f_1 are the functions that are identically zero and one respectively and $x \in \mathcal{O}$ we define the observables 0 and 1 by $0 = f_0(x)$ and $1 = f_1(x)$ respectively. Notice

$$p(0, s)(\{0, 1\}) = p(x, s)(f_0^{-1}\{0, 1\}) = p(x, s)(R) = 1.$$

Hence $0 \in Q$ and also

$$s(0) = p(0, s)(\{1\}) = p(x, s)(f_0^{-1}\{1\}) = p(x, s)(\varphi) = 0$$

for all $s \in S$. Similarly $1 \in Q$ and $s(1) = 1$ for all $s \in S$. Hence $0 \leqslant \alpha \leqslant 1$ for all $\alpha \in Q$. We may interpret 0 and 1 as the questions whose answers are always 'no' and 'yes' respectively. In the poset $(Q, \leqslant)$ we say that γ is the *least upper bound* of $\alpha, \beta \in Q$ if $\alpha, \beta \leqslant \gamma$ and whenever $\alpha, \beta \leqslant \delta$ we have $\gamma \leqslant \delta$. The least upper bound need not exist, but when it does it is unique. We denote the least upper bound (or sup) of α and β by $\alpha \vee \beta$ when it exists. We define the greatest lower bound (or inf) of α and β dually and denote it by $\alpha \wedge \beta$ when it exists. The following lemma is easily proved.

LEMMA 2.1. The operation $\alpha \to \alpha'$ is an orthocomplementation on Q. That is, $\alpha'' = \alpha$ for all $\alpha \in Q$, if $\alpha \leqslant \beta$ then $\beta' \leqslant \alpha'$, and $\alpha \vee \alpha'$ always exists and equals 1.

Thus $(\alpha, \leqslant, ')$ is an *orthocomplemented poset*. This section has served only as an introduction to the physical and mathematical notions involved in an axiomatic approach to quantum mechanics. The theory has not been carried out far enough to give a mathematical model for a physical system. In the next section we will start anew on a slightly different tack and present a more detailed and complete model. The framework that we have developed in the present section can be extended further [55] to obtain a model equivalent to that of the next section. However it is more common and possibly more instructive to begin with the questions as the axiomation elements as is done in the succeeding section.

III. THE BIRKHOFF-VON NEUMANN-MACKEY APPROACH [80]

In the last section we formulated an axiomatic theory based on the observables and states of a physical system. We then derived the notion of questions which turned out to be more elementary than the observables. For this reason we now formulate an axiomatic model in which the questions are the sole primitive axiomatic elements. We then derive the concepts of states and observables in terms of these primitive elements.

Let $Q=\{\alpha,\beta,\gamma,\ldots\}$ be the set of questions for a quantum system. A question may be interpreted as corresponding to a measurement or experiment leading to two alternatives which we call 'yes' and 'no'. The above measurement or experiment consists of a procedure to be carried out with the physical system under consideration, and a rule for interpreting the possible results in terms of 'yes' and 'no'. It is well-known that there are measurements in quantum mechanics that interfere with each other such as position and momentum measurements. Suppose α and β are non-interfering questions (i.e. performing the experiment α does not change the answers of β and vice-versa). If whenever the answer to α is 'yes' it follows that the answer to β is 'yes' we write $\alpha \leqslant \beta$. Notice the relation $\leqslant$ has no apparent connection with the order relation in Section II. We cannot use that order relation here since our only axiomatic elements are questions and we want to derive the states from these. The relation $\leqslant$ should satisfy:

(Q1) $\alpha \leqslant \alpha$, for all $\alpha \in Q$;

(Q2) if $\alpha \leqslant \beta$, $\beta \leqslant \alpha$ then $\alpha=\beta$;

(Q3) if $\alpha \leqslant \beta$, $\beta \leqslant \gamma$ then $\alpha \leqslant \gamma$;

(Q4) there are questions 0,1 such that $0 \leqslant \alpha \leqslant 1$ for all $\alpha \in Q$.

Thus Q is a poset with universal bounds 0,1. A fundamental problem is whether Q is a *lattice*, i.e. does $\alpha \vee \beta$ and $\alpha \wedge \beta$ exist for all α, $\beta \in Q$? For example if $\alpha \wedge \beta$ exists it would (in this setting) be interpreted as the question whose answer is 'yes' if and only if α and β both have the answer 'yes'. If α and β are interfering questions it is, to some researchers, doubtful that an experimental apparatus can be constructed corresponding to such a question except under special conditions. For this reason we do not assume Q is a lattice. In Figure 1 we diagram an example of a poset which is not a lattice. In reading such diagrams, a rising line from α to β means $\alpha \leqslant \beta$ and there is no $\delta \neq \alpha, \beta$ such that $\alpha \leqslant \delta \leqslant \beta$.

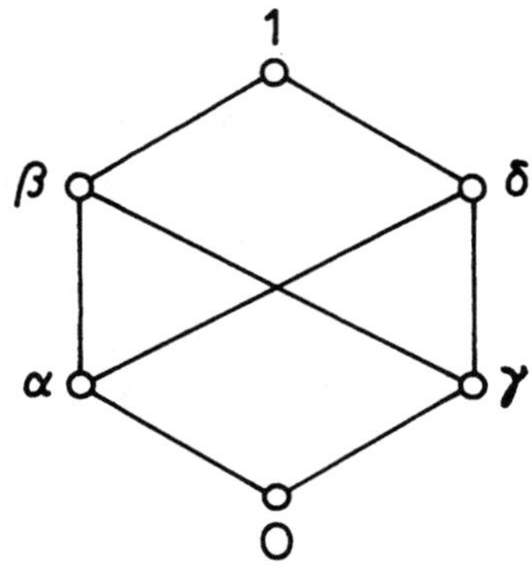

Fig. 1

Given $\alpha \in Q$ we define α' as the question whose alternatives are reversed; that is, we consider α' to have a 'yes' answer if and only if the result of α is 'no'. It is clear that the following conditions are satisfied:

(Q5) $\alpha'' = \alpha$ for all $\alpha \in Q$;

(Q6) if $\alpha \leqslant \beta$ then $\beta' \leqslant \alpha'$;

(Q7) $\alpha \vee \alpha'$ exists and equals 1 for all $\alpha \in Q$.

Thus $(Q, \leqslant, ')$ forms an orthocomplemented poset. If $\alpha \leqslant \beta'$ we say α and β are *orthogonal* are write $\alpha \perp \beta$. Notice $\alpha \perp \beta$ if and only if $\beta \perp \alpha$. We may interpret $\alpha \perp \beta$ to mean that α and β are non-interfering and that β has a 'no' result if α has a 'yes' result. In this case it *is regarded as* physically resonable to assume that $\alpha \vee \beta$ exists. We extend this conclusion, mainly for mathematical convenience, to a countable number of questions.

(Q8) If α_i is a sequence of mutually orthogonal questions then $\bigvee \alpha_i$ exists, i.e. Q is σ-orthocomplete.

We now make this into a statistical theory by introducing states. A state should completely describe the system as far as is physically possible. Experimentally the most we can determine about the questions of a physical system is the probabilities of getting 'yes' (and hence also 'no') results.

Thus given a question α, the state should determine the probability that α has a 'yes' result. We therefore define a *state* s as a probability measure on Q; that is, s is a map from Q to the unit interval $[0,1]\subset R$ such that (S1) $s(1)=1$; (S2) $s(\bigvee\alpha_i)=\sum s(\alpha_i)$ if $\alpha_i\perp\alpha_j$, $i\neq j=1, 2,\ldots$.

A set of states S on Q is full if $s(\alpha)\leqslant s(\beta)$ for all $s\in S$ implies $\alpha\leqslant\beta$. One can give examples of systems satisfying (Q1)–(Q7) which have no states at all [22]. Physically, one would expect to have enough states to capture the ordering on the questions. We therefore postulate:

(Q9) There is a full set of states S on Q.

If (Q, S) satisfies (Q1)–(Q9) we call it a *quantum logic*. An orthocomplemented poset is called an *orthomodular poset* if $\alpha\leqslant\beta$ implies $\alpha\vee(\beta\wedge\alpha')$ exists and equals β. The next lemma has a straightforward proof.

LEMMA 3.1. If (Q, S) is a quantum logic then Q is an orthomodular poset.

Now there are quantum logics which are not lattices. Let $\Omega=\{1, 2, 3, 4, 5, 6\}$ and let Q be the collection of subsets of Ω with an even number of elements. Order Q by inclusion and let $'$ be the usual set complementation. For $A\in Q$, $i=1,\ldots, 6$ define $s_i(A)=\begin{cases}1 \text{ if } i\in A\\ 0 \text{ if } i\notin A\end{cases}$. Then $S=\{s_i: i=1,\ldots, 6\}$ is a full set of states and (Q, S) is a quantum logic. However Q is not a lattice since, for example, $\{1, 2, 3, 4\}\wedge\{2, 3, 4, 5\}$ does not exist.

We say that two questions α, β are *compatible* (written $\alpha C\beta$) if there are mutually orthogonal questions α_1, β_1, γ such that $\alpha=\alpha_1\vee\gamma$, $\beta=\beta_1\vee\gamma$. We shall see that compatible questions are ones that can be answered simultaneously; that is, questions which do not interfere. In fact, the words compatible, simultaneously answerable, and non-interfering are frequently used synonymously. Notice if $\alpha\perp\beta$ then $\alpha C\beta$ and $0C\alpha$, $1C\alpha$ for all $\alpha\in$Q. Physically, our interpretation of $\alpha\leqslant\beta$ demands that $\alpha C\beta$ if $\alpha\leqslant\beta$. This need not happen if there is no full set of states which is another reason we insist upon (Q9).

LEMMA 3.2. If (Q, S) is a quantum logic then $\alpha C\beta$ whenever $\alpha\leqslant\beta$.

Dynamical variables are very important in classical physics and they are equally important in quantum mechanics. In classical mechanics dynamical variables are defined as functions on phase space. Since we have no phase space in our present setting, we must define them differently. To distinguish these objects from their classical counterpart (our definition will actually be a generalization of dynamical variables) we shall call them ‘observables’. An observable should be an object associated with our physical system which can be measured. That is, it determines a set of real numbers, the values of the observable. On the other hand, given any set of real numbers E an observable x gives us the question: ‘Does x have a value in E?’ Now it would be cumbersome mathematically to consider all subsets of R. For this reason mathematicians usually consider a class of subsets of R which is at the same time large enough to contain all the physically important subsets of R and yet small enough to make the theory manageable. Now open intervals are certainly important subsets of R (these intervals correspond to inexact results of measurements such as ‘the result is between 2.03 and 2.05 degrees centigrade’) and we would surely like to be able to take set complements and countable unions of sets and still remain within our class of subsets. The class of *Borel subsets* $B(R)$ of R is defined to be the smallest collection of subsets of R that contains the open intervals and that is closed under set complementation and countable unions [38, 48]. It is easy to show that open sets and closed sets are Borel sets and that $B(R)$ is closed under countable intersections. Now if x is an observable and $E \in B(R)$ we have the corresponding question $x(E)$: ‘Does x have a value in the set E?’. We thus define an observable x as a map from $B(R)$ to Q that satisfies the following physically plausible conditions:

($\mathcal{O}$1) $x(R) = 1$;

($\mathcal{O}$2) if $E \cap F = \emptyset$, then $x(E) \perp x(F)$;

($\mathcal{O}$3) $x(\bigcup E_i) = \bigvee x(E_i)$ if $E_i \cap E_j = \emptyset$, $i \neq j = 1, 2, \ldots$.

It follows that $x(\emptyset) = 0$ and denoting the complement of $E \in B(R)$ by E' we have $x(E') = x(E)'$. To give an example of an observable, let α_i be a sequence of mutually disjoint questions such that $\bigvee \alpha_i = 1$ and let λ_i be a sequence of distinct real numbers. Defining the map x by $x(E) = \bigvee \{\alpha_j : \lambda_j \in E\}$, $E \in B(R)$, it is easily checked that x is an observable.

Two observables x, y are *compatible* (written $x \mathrm{C} y$) if $x(E) \mathrm{C} y(F)$ for every E, $F \in B(R)$. We shall show later that observables which are compatible may be thought of physically as being observables which are simultaneously measurable. It can also be shown that a collection of compatible observables may be identified with a collection of dynamical variables.

The reader should notice that we have constructed a generalized probability theory. Instead of being a Boolean σ-algebra of subsets of a set, our events (questions) which are more general, form a logic with less structure than a Boolean σ-algebra. The probability measures are replaced by states and the random variables by observables. Notice if x is an observable and s a state then the probability that x has a value in $E \in B(R)$ is $s[x(E)]$. Thus $s[x(E)]$ corresponds to $p(x, s)(E)$ in Section II. Before proceeding further, let us consider two examples of quantum logics.

Example 1. Let Ω be a phase space and let $B(\Omega)$ be the Borel subsets of Ω (defined in a similar way as $B(R)$ above). $B(\Omega)$ may be thought of as the set of mechanical events. Now $B(\Omega)$ satisfies (Q1)–(Q8). A state is now a probability measure on $B(\Omega)$ and from the existence of measures concentrated at points we see that $B(\Omega)$ has a full set of states and is thus a quantum logic. If x is an observable it follows from a theorem of Sikorski-Varadarajan [80] that there exists a (measurable) function $f: \Omega \to R$ such that $x(E) = f^{-1}(E)$ for every $E \in B(R)$. Thus observables are just inverses of dynamical variables. We thus see that the quantum logic generalizes classical mechanics and also the conventional Kolmogorov formulation of probability theory [48]. It is easily checked that all events (questions) and observables are compatible in this example.

Example 2. Let H be a separable complex Hilbert space and let P be the collection of all closed subspaces of H. Ordering P by inclusion and defining the complement of a subspace as its orthocomplement it is easily seen that P satisfies (Q1)–(Q8). If $\alpha \in P$ we denote the unique orthogonal projection on α by P_α. Now if $\varphi \in H$ and $\|\varphi\| = 1$ then the map $\alpha \to \langle \varphi, P_\alpha \varphi \rangle$ is a state. If $\alpha \nleq \beta$ choose a unit vector φ_0 in α which is not in β. Then $\langle \varphi_0, P_\alpha \varphi_0 \rangle = 1$ and $\langle \varphi_0, P_\beta \varphi_0 \rangle \neq 1$ so P has a full set of states and is thus a quantum logic. It is an interesting and important fact that every state is a convex combination of states of the above form. Indeed, Gleason [21] has shown that any state s on P has the form $s(\alpha) = \sum_1^\infty \lambda_i \langle \varphi_i, P_\alpha \varphi_i \rangle$, $\lambda_i \geqslant 0$, $\sum \lambda_i = 1$, φ_i is an ortho-normal set of unit vectors. Identifying closed subspaces with their orthogonal projections, an observable may be thought of as a projection-valued measure. Since, using the spectral theorem [38], there is a one-one correspondence between projection-valued measures and self-adjoint operators, we may identify observables with self-adjoint operators. It is straightforward to show that $\alpha, \beta \in P$ are compatible if and only if P_α and P_β commute. It follows that two observables are compatible if and only if they commute. Of course, the present example gives the usual framework of conventional quantum mechanics. We thus see that the quantum logic is a generalization of conventional quantum mechanics.

Let us now return to general quantum logics. If x is an observable we call $\{x(E):E\in B(R)\}$ the *range* of x.

LEMMA 3.3 [80]. Two questions α, β are compatible if and only if they are in the range of a single observable.

This last lemma justifies the fact that compatible questions are physically non-interfering questions since to measure two compatible questions we need measure only a single observable.

A function $f:R\to R$ is said to be a *Borel function* if $f^{-1}(E)\in B(R)$ for every $E\in B(R)$. Again we consider Borel functions instead of arbitrary functions for mathematical manageability. It can be shown that the Borel functions form the smallest class of functions which contains the continuous functions and which is closed under pointwise convergence. Now if x is an observable and u a Borel function on R then there is an operational significance for $u(x)$. That is, if x has the value $\lambda\in R$ then $u(x)$ has the value $u(\lambda)$. This is equivalent to saying that the question 'Does $u(x)$ have a value in $E\in B(R)$?' is the same as the question 'Does x have a value in $u^{-1}(E)$?'. Motivated by this we define $u(x)$ as $u(x)(E)=x(u^{-1}(E))$ for all $E\in B(R)$. It is easily checked that $u(x)$ is an observable and that $u(x)Cx$.

THEOREM 3.4 [80]. Two observables x, y are compatible if and only if there is an observable z and Borel functions u, v such that $x=u(z)$ and $y=v(z)$.

This last theorem shows that, physically, compatible observables are measurements that can be performed simultaneously (i.e., non-interfering) since to measure compatible observables one need only measure a single observable.

One can continue this approach to a considerable extent and introduce such notions as time evolution, spectral theory, symmetry, superposition principle, superselection rules, scattering theory, and many others [18, 27, 30, 31, 32, 40, 41, 45]. However we refer the reader, interested in further study, to the literature and hope that we have conveyed some of the flavor of this subject.

IV. THE JAUCH-PIRON APPROACH [42, 65]

In the last section we gave an approach to quantum logics and in this section we present a slightly different approach. Each approach has its advantages and disadvantages. One of the advantages of the present approach is that it gives a much richer (and hence more specific) mathematical structure than that of Section III. For example one is able to derive the existence of sups and infs in a reasonable physical manner and thus show that in this case Q is a lattice. A disadvantage of this approach is that the probabilistic interpretation seems to disappear (although it can be partly recovered later under certain conditions); this is at the same time an advantage since states can be defined without recourse to probabilistic statements and therefore no difficulties arise attributing a state to an individual system.

In this approach the questions $Q=\{\alpha, \beta, \gamma, \ldots\}$ are again taken as the primitive axiomatic elements where the questions are interpreted exactly the same as in Section III. If $\alpha \in Q$ the question $\alpha^{\sim}$ is the question obtained by interchanging the alternatives of α. We have changed notation because we want to use $'$ for something else later. If $\{\alpha_i\}$ is a collection of questions (not necessarily countable) we denoted by $\pi\alpha_i$ ($\alpha \cdot \beta$ if there are two) the question defined in the following manner: Measure an arbitrary one of the α_i's and attribute to $\pi\ a_i$ the answer thus obtained. Clearly $(\pi\alpha_i)^{\sim}=\pi\alpha_i{}^{\sim}$. There exists a trivial question 1 which consists in doing anything (or nothing!) and stating that the answer is 'yes'. Let $0=1^{\sim}$.

Definitions. When the physical system is prepared in such a way that the result of a measurement of α is certain to be 'yes' then α is *true*. If whenever the physical system is prepared so that α is true we have β true also then we write $\alpha \leqslant \beta$.

This last relation expresses a physical law. This order relation is weaker than the order given in Section III and in fact this is the essential difference between the two formulations. We shall see that this difference enables one to develop a much richer structure. Clearly $\alpha \leqslant \alpha$ for all $\alpha \in Q$ and $\alpha \leqslant \beta$, $\beta \leqslant \gamma$ implies $\alpha \leqslant \gamma$. We define an equivalence relation $\alpha \sim \beta$ if $\alpha \leqslant \beta$ and $\beta \leqslant \alpha$. A *proposition* is an equivalence class of questions. We denote the equivalence class containing α by $a=\{\beta \in Q : \beta \sim \alpha\}$. We say a is *true* if any (and hence all) questions in a are true.

At this point we see that the probabilistic interpretation of propositions is lost. For example let α and β be questions.Define the question γ in the following way: Flip a fair coin; if the coin comes up heads measure α, if the coin comes up tails measure β. Define the question δ as follows: Flip a weighted coin in which the probability of heads is $\frac{2}{3}$ and the probability of tails is $\frac{1}{3}$; if the coin comes up heads measure α, if the coin comes up tails measure β. Now the two questions γ and δ are equivalent since the only way γ or δ can be true is if α and β are both true. Suppose we had a notion of the probability λ that α has a 'yes' answer and the probability μ that β has a 'yes' answer. Then the probability that γ has a 'yes' answer would be $\frac{1}{2}\lambda+\frac{1}{2}\mu$ while the probability δ has a 'yes' answer would be $\frac{2}{3}\lambda+\frac{1}{3}\mu$ which is, in general, different from $\frac{1}{2}\lambda+\frac{1}{2}\mu$. Thus there would be no unique way to associate a probability to the proposition containing γ and δ.

Let L be the set of propositions defined for a given physical system. If $a, b\in L$ define $a \leqslant b$ if $\alpha \leqslant \beta$ for all $\alpha\in a$, $\beta\in b$. It is easy to see that if a_i is a collection of propositions then $\bigwedge a_i=\{\beta:\beta\sim\pi\alpha_i, \alpha_i\in a_i\}$ and that $\bigvee a_i = \bigwedge\{x\in L:x \geqslant a_i \text{ for all } i\}$. We thus see that L is a *complete lattice*. It is clear that '$a\wedge b$ true'$\Leftrightarrow$'a true' and 'b true' so $\wedge$ plays the same role as 'and' in ordinary logic. This formulation overcomes the difficulties one has in the formulation of Section III where one gives the interpretation to $a\wedge b$ as the question which has answer 'yes' when a and b have a 'yes' answer in which case a and b must be measured simultaneously. In this formulation a and b are not measured simultaneously but one at a time.

For the sup in this formulation, however, we only have 'a true' or 'b true'$\Rightarrow$'$a\vee b$ true'. In fact we have the following lemma.

LEMMA 4.1 [63]. If '$a\vee b$ true'$\Leftrightarrow$('a true' or 'b true') for every $a, b\in L$ then L is distributive; i.e., $a\wedge(b\vee c)=(a\wedge b)\vee(a\wedge c)$ for all $a, b, c\in L$.

The implication '$a\vee b$ true'$\Rightarrow$'a true' or 'b true' is an essential distinction between classical and quantum theory. This implication holds in classical theory but in general not in quantum mechanics. Thus in classical theory L is distributive and it can then be shown [3] that L is isomorphic to the set of subsets of some phase space.

Example. The linear polarization of photons. The experiment consists of placing a polarizer in a beam of linearly polarized photons. By dispatching photons one by one, this experiment leads to an alternative: Either the photon passes through or is absorbed. We define the question α_φ by specifying the orientation of the polarizer (the angle φ) and interpreting the passage of a photon as a 'yes'. By experiment one can show that in order to obtain a photon prepared so that α_φ is 'true' it is sufficient to consider photons which have passed a first polarizer oriented at this angle φ. Also experiments show that it is impossible to prepare photons capable of traversing with complete certainty a polarizer oriented at angle φ as well as another oriented at angle $\varphi' \neq \varphi (\text{mod } \pi)$ i.e. $\alpha_\varphi \cdot \alpha_{\varphi'} \sim 0$. The corresponding lattice of propositions is given in Figure 2.

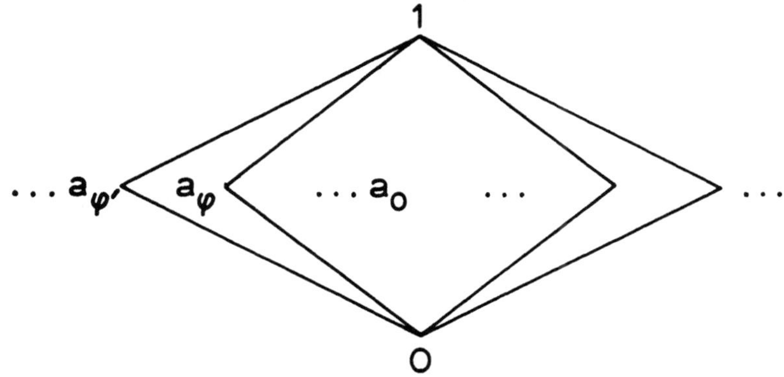

Fig. 2

One cannot define a question which is 'true' if and only if 'a_φ is true or $a_{\varphi'}$ is true' since a photon which is absorbed does not get through the polarizer so a second experiment cannot be performed. Notice in this case '$a_\varphi \vee a_{\varphi'}$ is true' $\not\Rightarrow$ 'a_φ is true or $a_{\varphi'}$ is true'. The lattice is not distributive and therefore corresponds to a quantum system.

We say that b is a *compatible complement* of a if it is a complement of a (i.e., $a \wedge b = 0$, $a \vee b = 1$) and if there is a question $\alpha \in a$ such that $\alpha^\sim \in b$. For example in the polarization experiment $a_{\varphi+\pi/2}$ and a_φ are compatible complements but no other $a_{\varphi'}$ is a compatible complement for a_φ.

Note. Every proposition has at least one compatible complement. This can be seen as follows. If $a \in L$, let $\alpha \in a$ and let b be the equivalence class containing $\alpha^\sim$. Then b is a compatible complement of a.

AXIOM P. If $a \leqslant b$ and a', b' are compatible complements of a, b respectively, then the sub-lattice generated by $\{a, b, a', b'\}$ is distributive.

We can justify Axiom P as follows. If $a \leqslant b$ then a and b are non-interfering so they may be considered part of a classical subsystem and we have seen that classical systems are distributive.

Suppose a_1, a_2 are both compatible complements of a. Then since $a \leqslant a$ by Axiom P, a, a_1, a_2 are in a distributive sub-lattice. But in a distributive lattice complements are unique so $a_1 = a_2$. Hence compatible complements are unique. We denote the compatible complement of a by a'. It follows from Axiom P that $a''=a$, $a \leqslant b$ implies $b' \leqslant a'$ and $a \leqslant b$ implies $b=a \vee (b \wedge a')$. Thus L is a complete orthomodular lattice or CROC. We say that two propositions a, b are *compatible* if the sub-lattice generated by $\{a, b, a', b'\}$ is distributive. This definition is equivalent to the definition of compatibility given in Section III.

A *JP-state* of L is a subset $S \subset L$ satisfying:

(S_1) $0 \notin S$;

(S_2) if $a \in S$ and $a \leqslant x$ then $x \in S$;

(S_3) if $a_i \in S$ then $\bigwedge a_i \in S$;

(S_4) S is maximal, i.e., if a subset of L satisfies (S_1), (S_2), (S_3) and contains S, it equals S.

A JP-state corresponds physically to the set of propositions that are true for some preparation of the system. Since a preparation determines and is determined by the set of propositions that are true for that preparation, a JP-state may be thought of as a preparation of the system.

Let S be a JP-state and let $p= \bigwedge\{a:a \in S\}$. Then by ($S_3$), $p \in S$. By (S_2), $S=\{a \in L:p \leqslant a\}$. Now p is an atom (i.e. $0 \leqslant p_1 \leqslant p$ implies $p_1=p$ or 0). Indeed, if there exists $0 \leqslant p_1 \leqslant p$ and $p_1 \neq 0, p$ then $\{a \in L:p_1 \leqslant a\}$ satisfies (S_1), (S_2), (S_3) and properly contains S which contradicts (S_4). Thus every JP-state defines an atom and every atom defines a JP-state.

By definition if a proposition is different from 0, then it must be true for some preparation. We have therefore justified the next axiom.

Axiom A_1. L is atomic (i.e. if $a \neq 0$ there is an atom $p \leqslant a$).

Axiom A_1 is equivalent to the axiom: Every proposition is contained in a JP-state.

Our last axiom is the following:

Axiom A_2. If $a \in L$ and p is an atom of L such that $p \nleqslant a'$ then $(p \vee a') \wedge a$ is an atom.

Roughly speaking, the justification for A_2 is that if p is the JP-state of the system and if an 'ideal measurement of the first kind' of a is made [42, 65] with the resulting answer 'yes' then the smallest proposition that is 'true' after the measurement is $(p \vee a') \wedge a$. Hence the resulting JP-state should correspond to $(p \vee a') \wedge a$ so this proposition must be an atom. A similar interpretation for this axiom may be found in [67, 68].

Axiom A_2 is called the *semimodular* or *covering law*. A complete lattice satisfying Axioms P, A_1, and A_2, i.e., a complete, atomic, semimodular, orthomodular lattice is called a *propositional system*. A propositional system L is *irreducible* if the only elements of L compatible with all other elements of L are 0 and 1.

THEOREM 4.2. Any propositional system is isomorphic to a direct product of irreducible propositional systems.

One of the great achievements of this theory is the following.

THEOREM 4.3 (Piron) [64]. Every irreducible propositional system of dimension $\geqslant 4$ is isomorphic to the lattice of all closed subspaces of a Hilbert space over a division ring with involution.

We thus see that a propositional system gives a structure which is very close to the conventional quantum mechanical formalism in terms of a complex Hilbert space. This theory can then be considered as a 'derivation' of the Hilbert space that mysteriously occupies such an important place in conventional quantum mechanics. Of course an important problem is to give more physical information so that the division ring is determined (hopefully to be the complex numbers). One result along these lines has been obtained by Gudder and Piron [34]. It states that if L admits an observable that is maximal in a certain sense then the division ring must contain the reals as a subfield. Examples of maximal observables are the position, momentum, and magnetic moment observables in conventional quantum mechanics. If further the division ring is a finite extension of R it follows from a theorem of Frobenius [66] that it must be the reals, complexes, or quaternions.

It follows from Theorem 4.3 that a propositional system L with dim $L \geqslant 4$ has a full set of states $\mathscr{S}$, although, as we have pointed out earlier, these states cannot be defined in terms of questions. In this way $(L, \mathscr{S})$ becomes a quantum logic which enjoys a still richer structure.

In summary, we define an *abstract quantum logic* $(\mathscr{L}, \mathscr{S})$ to be a σ-orthocomplete orthomodular poset $\mathscr{L}$ together with a full set of states $\mathscr{S}$. This structure seems to be the underlying framework for (almost) all quantum logical studies. In particular the models presented in Sections III and IV contain abstract quantum logics in which the axiomatic elements are given specific physical interpretations.

V. HIDDEN VARIABLE THEORIES

As was mentioned in the introduction, one of the advantages gained by an axiomatic formulation is that it presents a common universe of discourse in which to study deep quantum mechanical concepts. We also stated in Section III that many of the important concepts of quantum mechanics can be formulated within the quantum logic framework. Since we do not have the space to consider all these concepts, we shall attempt to illustrate the utility of the quantum logic approach by considering one concept which is not only important to this conference but has played a recurrent role throughout the history of the development of quantum theory. This is the concept of hidden variables.

One of the problems in hidden variable discussions is that they have occurred in different frameworks simultaneously and therefore investigators in many cases have been writing (and talking) about different subjects but have called them the same thing. Also it is very common that what one researcher calls hidden variables is entirely different from what another calls them so it is not surprising that some investigators are able to prove they do not exist while others prove they do. We would like to consider two hidden variable proofs in the quantum logic framework, one a proof that they do not exist and one a proof that they do.

Jauch and Piron's interpretation of hidden variables is that if hidden variables exist then there would be states for which every proposition is either true or false; that is, any question would have the answer 'yes' with certainty or the answer 'no' with certainty. They call such JP-states *dispersion-free*. Thus, in their approach to quantum logics, if an atom p corresponds to a dispersion-free state we must have $p \leqslant a$ or $p \leqslant a'$ for every proposition $a \in L$. It follows that p is compatible with every proposition. Jauch-Piron go even further than this. Precisely, they say that a propositional system L *admits hidden variables* if every JP-state is dispersion-free [44]. Thus if L admits hidden variables then each atom is compatible with every atom. It follows that each proposition is compatible with every proposition (in particular L is distributive) and hence there are no interfering experiments and L is a classical propositional system. This is a contradiction since there are noncompatible propositions for quantum mechanical systems. These concepts can also be phrased in the BVM (Birkhoff-von Neumann-Mackey) framework for quantum logic in which case we say a state is *dispersion-free* if its values are just 0 and 1, and a similar proof goes through.

We now consider another interpretation of hidden variables phrased in the BVM quantum logic context. We feel that the main difficulties in hidden variables discussions is that the investigators giving impossibility proofs [29, 44, 46, 49, 61, 81, 84] are not referring to the same thing as the hidden variable proponents. They are proving something is impossible but these things are not what hidden variable researchers such as Einstein [17], Bohm and Bub [6, 7, 8, 9], and others [2] are referring to when they speak of hidden variables. We will give a general definition in the BVM quantum logic context of what we feel an HV (hidden variables) theory is as described by HV proponents and prove that such a theory is *always* possible and is, in fact, unique in a certain sense.

We first give an English-language version of what we feel the HV researchers mean by an HV theory.

The state s of a quantum mechanical system is not complete in the sense that another variable ω can be adjoined to s so that the pair (s, ω) completely determines the system. That is, a knowledge of (s, ω) enables one to predict precisely the outcome of any single measurement. Furthermore, an average of (s, ω) over the values of ω gives the usual quantum state s.

We now attempt to translate the above version of an HV theory into a mathematical-language version on a quantum logic (Q, S). First a single measurement corresponds to a Boolean sub σ-algebra of Q. This is because in a single measurement there is no possibility of interference so the measurement corresponds to a distributive subsystem. To say that the results of a measurement (corresponding to a Boolean sub σ-algebra $B \subset Q$) are completely determined means that one has a dispersion-free state s_0 defined on B (*not* on Q). We denote the set of dispersion-free states on B by S_B. Recall that in probability theory a *probability space* is a triple (Ω, F, μ) where Ω is the set of elementary outcomes, F is the Boolean σ-algebra of events, and μ is a probability measure on F.

DEFINITIONS. A quantum logic (Q, S) *admits an* HV *theory* if there is a probability space (Ω, F, μ) with the following property: For any maximal Boolean sub σ-algebra $B \subset Q$ there is a map H_B from $S \times \Omega$ onto S_B such that (i) $\omega \to H_B(s, \omega)(\alpha)$ is measurable for every $s \in S$, $\alpha \in B$; (ii) $\int_\Omega H_B(s, \omega)(\alpha) d\mu(\omega) = s(\omega)$ for every $s \in S$, $\alpha \in B$. Denote the set of maximal Boolean sub σ-algebras of Q by $\mathscr{B}$. We call $((\Omega, F, \mu), \{H_B : B \in \mathscr{B}\})$ an HV *theory for* (Q, S). An HV theory $((\Omega, F, \mu), \{H_B : B \in \mathscr{B}\})$ is *minimal* if $H_B(s, \omega_1) = H_B(s, \omega_2)$ for every $s \in S$, $B \in \mathscr{B}$ implies $\omega_1 = \omega_2$.

The definition merely says that for each 'completed' state (s, ω) there is a dispersion-free state $H_B(s, \omega)$ for any single measurement B and (i), (ii) say that the average of these dispersion-free states over ω give back the quantum state s. We consider only maximal Boolean sub σ-algebras so that the theory does not become too cumbersome. This is really only a technicality since any Boolean sub σ-algebra is contained in a maximal one. The probability space (Ω, F, μ) may be thought of as the space of hidden variables. If an HV theory is minimal, there exists a minimal number of hidden variables – just enough to give all the dispersion-free states. We now state our main theorem [26].

THEOREM 5.1. Any quantum logic (Q, S) admits a minimal HV theory $((\Omega, F, \mu), \{H_B : B \in \mathscr{B}\})$. Furthermore, $((\Omega, F, \mu), \{H_B : B \in \mathscr{B}\})$ is the unique minimal HV theory in the sense that if $((\Omega', F', \mu'), \{H_B' : B \in \mathscr{B}\})$ is another HV theory, there exists a measurable map τ from Ω' into Ω such that $H_B(s, \tau\omega')(\alpha) = H_B'(s, \omega')(\alpha)$ for all $B \in \mathscr{B}$, $\omega' \in \Omega'$, $s \in S$, $\alpha \in B$, $\mu'(\tau^{-1}(\Lambda)) = \mu(\lambda)$ for all $\lambda \in F$ and if $((\Omega', F', \mu'), \{H_B' : B \in \mathscr{B}\})$ is minimal then τ is one-one.

VI. THE CONDITIONAL

We begin this section with a brief sketch of the treatment of the conditional in classical logic. Let B be a Boolean algebra and $\mathfrak{F}$ a (lattice) filter (i.e., (I) $a, b \in \mathfrak{F}$ implies $a \wedge b \in \mathfrak{F}$; (II) $a \in \mathfrak{F}$, $x \in B$ with $x \geqslant a$ implies $x \in \mathfrak{F}$) in B. We regard B as a logic and $\mathfrak{F}$ as the set of true propositions of B. For any two propositions a, b in B the conditional $a \supset b$ (read 'a implies b') is the proposition $a' \cup b$. Thus $a \supset b$ is true in case $a' \cup b \in \mathfrak{F}$. By 'dividing out' the filter $\mathfrak{F}$ (or the dual filter of all false propositions) we obtain another Boolean algebra called the reduced algebra, $B_1 = B/\mathfrak{F}$. Let the elements of B_1 be denoted by $\bar{a}$ where $a \in B$. Formally $\bar{a} = \{b \in B \mid b \vee c = a \vee c$ for some $c \in B$ with $c' \in \mathfrak{F}\}$. It turns out that the following statements are equivalent: (1) $a \supset b \in \mathfrak{F}$, (2) $\overline{a' \vee b} = \bar{1}$, (3) $\bar{a} \leqslant \bar{b}$. Thus in a Boolean algebra the statement '$a \supset b$ is true' may be translated into the statement '$\bar{a} \leqslant \bar{b}$'.

More than anything else the fact which allows us to regard the Boolean algebra B as a logic is the existence of a valid inference scheme (modus ponens): If $a \in \mathfrak{F}$ and $a \supset b \in \mathfrak{F}$ then $b \in \mathfrak{F}$. The argument runs as follows: $a \in \mathfrak{F}$, $a' \cup b \in \mathfrak{F}$ so $a \wedge (a' \cup b) \in \mathfrak{F}$ by (I); but since B is a Boolean algebra

$$(*) \qquad a \wedge (a' \cup b) = a \wedge b \in \mathfrak{F};$$

moreover $a \wedge b \leqslant b$ so that $b \in \mathfrak{F}$ by (II).

(*) is the key to the above computations. The equation given in (*) is equivalent to saying that $a\mathrm{C}b$.

The question of whether there exists in quantum logic a conditional which is a logical proposition allowing a valid inference scheme has received some attention. So far the results are one-sided: If, for every $a, b \in \mathscr{L}$ there is a proposition $a \supset b$ behaving enough like the conditional of classical logic then $\mathscr{L}$ is classical (a Boolean algebra). Here are four examples:

(1) Fay [19] has proved the following: If a relation R is defined on an orthomodular lattice L by $(a, b) \in R$ if and only if $a' \vee b = 1$, then R is transitive if and only if L is a Boolean algebra.

(2) Skolem [75] defines an *implicative lattice* to be a lattice L with the property that for every a, b in L there exists an element $a \supset b$ in L such that

$$\text{(i)} \qquad a \wedge (a \supset b) \leqslant b$$

and

$$\text{(ii)} \qquad \text{if } a \wedge c \leqslant b \text{ then } c \leqslant (a \supset b).$$

He then proves that any implicative lattice is distributive.

(3) Working in an orthomodular lattice L, Catlin [10] defines a relation I on L by $(a, b) \in I$ if there exists an element $a \supset b$ satisfying (i) and (ii) above. (If such an element exists it is unique; in fact, $a \supset b = a' \vee b$.) Catlin proves (for example) that $(a, b) \in I$ and $(b', a') \in I$ both hold if and only if $a\mathrm{C}b$ and $(a' \vee b)\mathrm{C}x$ for all $x \in L$. Thus (again) if $a' \vee b$ behaves as the conditional for every a, b in L, then L is a Boolean algebra.

(4) Łukasiewicz [53] has developed a (classical) system in which a proposition P is assigned a truth value $[P] \in \mathbb{R}$ satisfying $0 \leqslant [P] \leqslant 1$. He defines the truth values of the conditional $P \to Q$ and negation $\bar{P}$ as follows:

$$[P \to Q] = \begin{cases} 1 & \text{if } [P] \leqslant [Q] \\ 1 - [P] + [Q] & \text{if } [Q] < [P] \end{cases}$$

and

$$[\bar{P}] = 1 - [P].$$

$[P] = 1$ is interpreted as 'P is true'. Intermediate values of $[P]$ stand for various degrees of certainty. Note that

(i) $[P] = 1$ and $[P \to Q] = 1$ implies $[Q] = 1$

and

(ii) $[P \to Q] = 1$ and $[Q \to R] = 1$ implies $[P \to R] = 1$.

Jauch and Piron [43] considered an adaptation of the infinite-valued logic as developed by Łukasiewicz [53]. They argued as follows. Reichenbach [70] has proposed that the

elementary propositions about quantum mechanical systems should admit three truth values: True, false, and undetermined. In view of the fact, however, that the state of a system attributes to each yes-no experiment a probability function $p(a)$ with $0 \leqslant p(a) \leqslant 1$, it seems more natural, once one has passed beyond the ordinary double-valued logic, to consider 'quantum-logic' as an infinite-valued logic.

The various degrees of certainty should depend on the state of the physical system. Thus they sought the existence in a quantum logic $(L, \mathscr{S})$ of a conditional $p \to q$ for each pair p, q in L. It must satisfy, *for each* $m \in \mathscr{S}$,

$$(*) \qquad m(p \to q) = \begin{cases} 1 & \text{if } m(p) \leqslant m(q) \\ 1 - m(p) + m(q) & \text{if } m(q) < m(p) \end{cases}.$$

They showed that in a standard quantum logic there exist p and q that admit no conditional.

Greechie and Gudder [24] generalized this result to an arbitrary quantum logic. In fact, they did more. An outline of the technical results is given below.

Let $(L, \mathscr{S})$ be an abstract quantum logic and assume that $\mathscr{S}$ is *closed under the formation of midpoints*, i.e.

$$m_1, m_2 \in \mathscr{S} \text{ implies } \tfrac{1}{2}m_1 + \tfrac{1}{2}m_2 \in \mathscr{S}.$$

Call a pair $(a, b) \in L \times L$ conditional if there exists $c \in L$ such that for all $m \in \mathscr{S}$

$$(**) \qquad m(c) = \min\{1, m(a') + m(b)\}.$$

If c exists it is unique, write $c = a \to b$. (Note that (*) and (**) are, in fact, the same condition.) Call $(L, \mathscr{S})$ *conditional* if $a \to b$ exists for all pairs (a, b). Then $(L, \mathscr{S})$ is conditional if and only if $L = \{0,1\}$. Moreover, if

$\mathscr{S}$ is strongly order determining, then $a \to b$ exists if and only if $a \leqslant b$ or $b \leqslant a$.

From a different point of view, Piron has argued that a conditional $a \supset b$ does not exist because 'no experimental arrangement is possible which measures the proposition $a \supset b$'.

Thus the indications are that the conditional in quantum logic must be treated as a relation, perhaps nothing more than the relation $\leqslant$. In this case the required valid inference scheme would be: a is true and $a \leqslant b$, therefore b is true. Here 'a is true' is apparently interpreted as $a \in \mathfrak{F}$ where $\mathfrak{F}$ is the filter of all propositions true in some (fixed) state.

E. L. Marsden has made an interesting observation (oral communication) on implication in a quantum logic. His approach has the advantage (or disadvantage?) of ignoring the states, i.e. of working completely within the orthomodular poset.

Let P be an orthomodular poset, S a subset of P with $1 \in S$. Modifying a notion of classical logic (cf. A. Church, *Mathematical Logic*, Princeton University Press, 1956) Marsden defines an element $d \in P$ to be *a theorem based on S*, written $S \vdash d$ in case there exist $a_1, \ldots, a_n \in P$ such that $a_n = d$ and, for all i, either $a_i \in S$ or there exist j, $k < i$ with $\{a_i, a_j, a_k\}$ a commuting set and $a_j = a'_k \vee a_i$. He also defines a *C-filter* in P to be a non-empty subset $\mathfrak{F}$ of P such that (1) if $x \in \mathfrak{F}$ and $x \leqslant y$ then $y \in \mathfrak{F}$, and (2) if $x, y \in \mathfrak{F}$ and $x \mathrm{C} y$ then $x \wedge y \in \mathfrak{F}$.

Let $\bar{S}$ denote the C-filter generated by S. Theorem: $d \in \bar{S}$ if and only if $S \vdash d$. The difficulty with this is that $\bar{S}$ is not associated with any congruence relation on P. Thus $\bar{S}$ cannot be used to form a reduced logic.

VII. COMBINATORIAL QUANTUM LOGIC

In this section we discuss some aspects of the combinatorial approach to quantum logic.

We restrict our considerations to finite structures. Let $(P, \leqslant, ')$ be a finite orthomodular poset and A the set of atoms in P. Make A into a graph by defining, for $a, b \in A$, $a \perp b$ to mean that $a \leqslant b'$ in P. For $M \subseteq A$, let $M^{\perp} = \{x \in A \mid x \perp m \text{ for all } m \in M\}$ and $M^{\perp\perp} = (M^{\perp})^{\perp}$. Let $\mathscr{L} = \{D^{\perp\perp} \mid D$ is an orthogonal subset of $A\}$. Then $(\mathscr{L}, \subseteq, {}^{\perp})$ is isomorphic to $(P, \leqslant, ')$. Thus we may recapture $(P, \leqslant, ')$ from the *orthogonality graph* $(A, \perp)$. We now pass to the *orthogonality space* $(A, \mathscr{E})$ by defining $\mathscr{E}$ to be the set of all maximal orthogonal sets in $(A, \perp)$, that is, the set of maximal complete subgraphs (or cliques) of the graph $(A, \perp)$. It is easy to see that $(A, \mathscr{E})$ determines the graph $(A, \perp)$; thus we can recapture $(P, \leqslant, ')$ from $(A, \mathscr{E})$. The main reason for passing from the poset to the associated orthogonality space is psychological: The diagramatic representation for $(A, \mathscr{E})$ is most perspicuous.

We illustrate this process for the well-known orthomodular lattice D_{16}, the Hasse diagram of which is given in Figure 3.

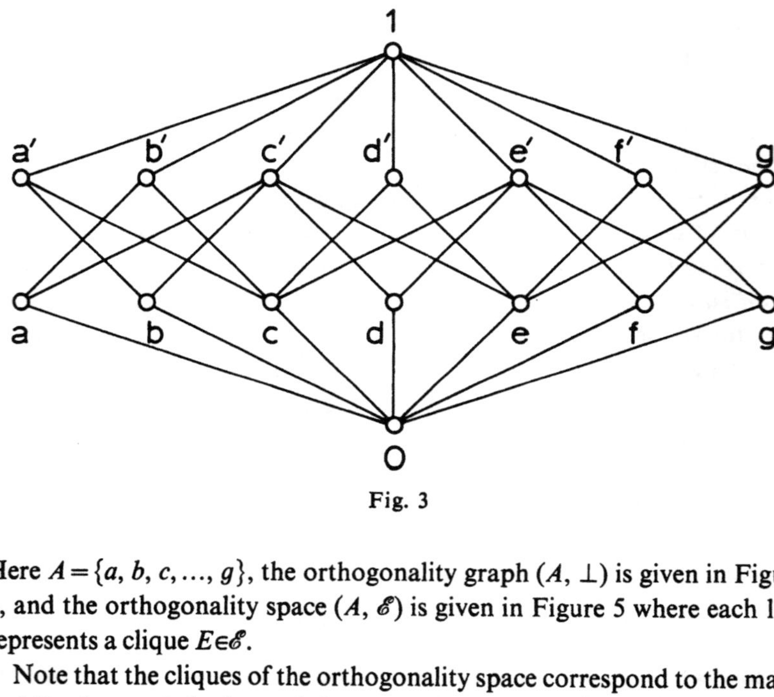

Fig. 3

Here $A = \{a, b, c, \ldots, g\}$, the orthogonality graph $(A, \perp)$ is given in Figure 4, and the orthogonality space $(A, \mathscr{E})$ is given in Figure 5 where each line represents a clique $E \in \mathscr{E}$.

Note that the cliques of the orthogonality space correspond to the maximal Boolean subalgebras of the poset, for example $\{c, d, e\}$ of Figure 5 corresponds to $\{0, c, d, e, c', d', e', 1\}$ of Figure 3.

By focussing on the intertwining of the maximal subalgebras of an orthomodular poset and then translating into the orthogonality space we are able, in certain instances, to create structures tailored to predescribed criteria. We illustrate this with an example of a quantum logic $(\mathscr{L}, \mathscr{S})$ in which $\mathscr{L}$ is an orthomodular poset and $\mathscr{S}$ is a full but not strongly order determining set of states; moreover $\mathscr{S}$ is sufficient. First, recall the

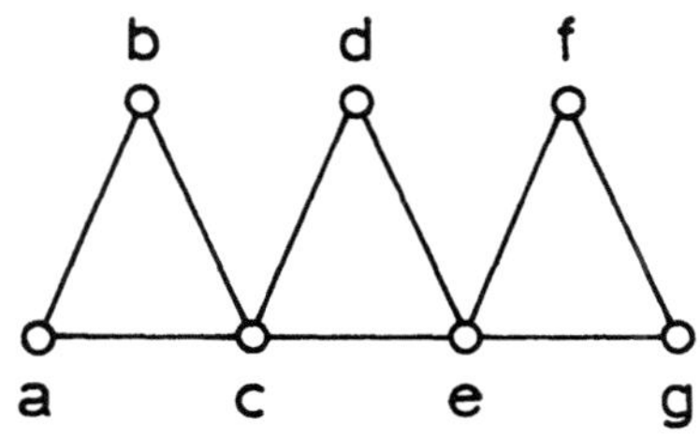

Fig. 4

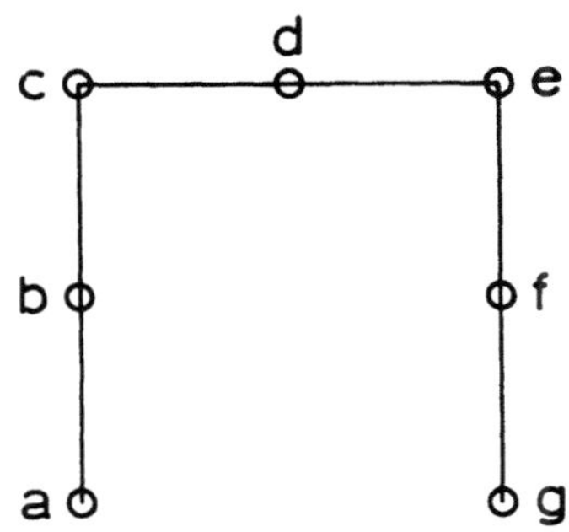

Fig. 5

meaning of the terms full, strongly order determining, and sufficient. A set of states $\mathscr{S}$ on an orthomodular poset $\mathscr{L}$ is *full* (or *order determining*) in case $m(x) \leqslant m(y)$ for all $m \in \mathscr{S}$ implies $x \leqslant y$; $\mathscr{S}$ is *strongly order determining in case* $\{m \in \mathscr{S} | m(x) = 1\} \subseteq \{m \in \mathscr{S} | m(y) = 1\}$ implies $x \leqslant y$; $\mathscr{S}$ is *sufficient* (or satisfies the *projection postulate*) in case for all non-zero $x \in \mathscr{L}$ there exists a (not necessarily unique) state $m \in \mathscr{S}$ with $m(x) = 1$. (Note that in Section II we defined $\leqslant$ in such a way that $\mathscr{S}$ was full. Also, in Section III, we eventually assume ($Q9$) that the set of states on Q is full. Because of Theorem 4.3 there is, in fact, a strongly order determining set of states on the resulting structure of Section IV.)

The example is given in Figure 6. This is a diagram of an orthogonality space $(X, \mathscr{E})$. The corresponding orthomodular poset may be obtained by the construction outlined above, by passing through the associated orthogonality graph $(X, \perp)$ and on to the poset $\mathscr{L} = \mathscr{L}(X, \perp)$. The details of the argument that the structure has the required property appear in [23].

It is easy to see that an orthomodular poset with a strongly order deter-

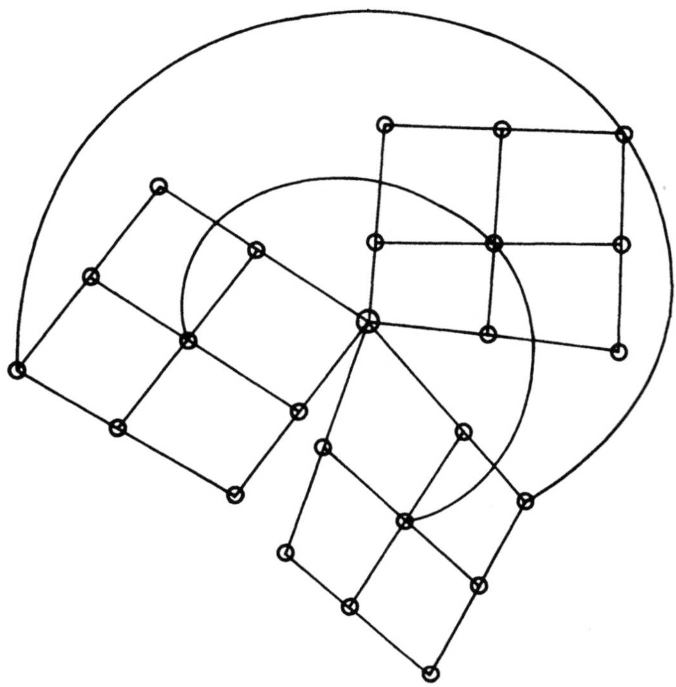

Fig. 6

mining set of states $\mathscr{S}$ is in fact a quantum logic, i.e. $\mathscr{S}$ is full. We have just proven that the reverse implication fails: There exists a quantum logic with a set of states which is (sufficient but) not strongly order determining. B. Collings has given an example of an orthomodular lattice with a state space having these same properties. The example, however, is too complicated to be given here.

There exists [25] an orthomodular poset $P_{j,k}$ with a distinguished element $x \in P_{j,k}$ such that $m(x) < j/k\,(0 < j < k)$ for every state m on $P_{j,k}$; moreover $P_{j,k}$ admits a full set of states if and only if $\frac{1}{2} < j/k$. Thus we see in a rather dramatic fashion that the axiom of sufficiency does not follow from the other axioms of a quantum logic.

Combinatorial quantum logic is useful for reasons other than axiomatics. For example, one can prove the following theorem using the combinatorial methods alluded to above and expounded in [22, 25].

THEOREM (Schrag). For any finite group G there exists a quantum logic $(\mathscr{L}, \mathscr{S})$ such that the full automorphism group of $\mathscr{L}$ is G.

The combinatorial approach has also yielded results of a negative character. For example, consider Finkelstein's [20] very readable presentation of what he calls a model quantum logic. Beginning with a finite set of entities he constructs a structure which is mathematically equivalent to a finite orthocomplemented projective geometry large enough to contain at least one plane, which, in turn, must be orthocomplemented. But there exists a combinatorial argument, due to R. Baer, that no such plane exists and therefore no such model quantum logic exists. This comment is not intended to be a criticism of Finkelstein's carefully presented argument but serves only to point out, once again, that combinatorics can play an important role in quantum logic.

VIII. CONCLUSION

In conclusion we wish to emphasize that quantum logic is not a closed subject. Although penetrating investigations have been made there exists no universally accepted theory. The basic framework has not been established to the satisfaction of even a majority of researchers. We feel that Jauch and Piron's axiom A_2 has not been sufficiently justified; the problem appears intricately interwoven with the philosophical problem of what is meant by 'truth with certainty' in an empirical setting, with axiom (S_3) in the definition of state, and with the 'ideal measurements of the first kind'. While in the Mackey-von Neumann presentation the axioms presented herein may be more easily accepted by some, in order to complete this development it appears necessary to make the *ad hoc* axiom that the logic, in fact, is the usual Hilbert space structure.

There are numerous lines of investigation which may provide insight into the sought-for underlying theory. We take this opportunity to list four directions which we consider of prime importance:

(1) Provide a physically meaningful interpretation for existing infima in a quantum logic. This, of course, is one of Birkhoff and von Neumann's original suggestions.

(2) Provide a connection between the structure of the state space and that of the underlying logic.

(3) Develop a general theory of group representations on an abstract quantum logic.

(4) Explain the meaning of the word 'logic' in the title of this paper.

Department of Mathematics,
Kansas State University
and
Department of Mathematics,
University of Denver

BIBLIOGRAPHY

[1] Amrien, W. O., 'Localizability for Particles of Mass Zero', *Helvetia Physica Acta* **42** (1969), 149–90.
[2] Bell, J. S., 'On the Hidden Variable Problem in Quantum Mechanics', *Reviews of Modern Physics* **38** (1966), 447–52.
[3] Birkhoff, G., 'Lattice Theory', *American Mathematical Society Colloquium* Publ. 25 (1967).
[4] Birkhoff, G. and von Neumann, J., 'The Logic of Quantum Mechanics', *Annals of Mathematics* **37** (1936), 823–43.
[5] Bodiou, G., *Theorie Dialectique des Probabilities*, Gauthier-Villars, Paris, 1964.
[6] Bohm, D. and Bub, J., 'A Proposed Solution of the Measurement Problem in Quantum Mechanics by Hidden Variables', *Reviews of Modern Physics* **38**, (1966) 453–69.
[7] Bohm, D. and Bub, J., 'A Refutation of the Proof by Jauch and Piron that Hidden Variables can be Excluded in Quantum Mechanics', *Reviews of Modern Physics* **38** (1966), 470–5.
[8] Bub, J., 'Hidden Variables and the Copenhagen Interpretation – a Reconcilliation', *The British Journal for the Philosophy of Science* **19** (1968), 185–210.
[9] Bub, J., 'What is a Hidden Variable Theory of Quantum Phenomena?', *International Journal of Theoretical Physics* **2** (1969), 101-23.
[10] Catlin, D., 'Implicative Pairs in Orthomodular Lattices', *Carribbean Journal of Science and Mathematics* **1** (1969), 69–79.
[11] Davies, E. B., 'On the Repeated Measurements of Continuous Observables in Quantum Mechanics', *Journal of Functional Analysis* **6** (1970), 318–46.
[12] Davies, E. B. and Lewis, J. T., 'An Operational Approach to Quantum Probability', *Communications in Mathematical Physics* **17** (1970), 239–60.
[13] Dirac, P. A. M., *The Principles of Quantum Mechanics*, Oxford University Press, London, 1958.
[14] Eckmann, J. P. and Zabey, Ph., 'Impossibility of Quantum Mechanics in a Hilbert Space over a Finite Field', *Helvetia Physica Acta* **42** (1969), 420–4.
[15] Edwards, C. M., 'The Operational Approach to Algebraic Quantum Theory, I', *Communications in Mathematical Physics* **16** (1970), 207–30.
[16] Edwards, C. M., 'Classes of Operations in Quantum Theory', *Communications in Mathematical Physics* **20** (1971), 26–56.
[17] Einstein, A, Rosen, N., and Podolsky, B., 'Can Quantum-Mechanical Description of Physical Reality be Considered Complete?', *Physical Review* **47** (1935), 777–80.
[18] Emch, G. and Piron, C., 'Symmetry on Quantum Theory', *Journal of Mathematical Physics* **4** (1963), 469–73.

[19] Fay, Gy., 'Transitivity of Implication in Orthomodular Lattices', *Acta Scientiarum Mathematicarum* **28** (1967), 267–70.
[20] Finkelstein, D., '*The Logic of Quantum Physics*', *Transactions New York Academy of Science* Ser. 2, **28** (1962/63) 621–37.
[21] Gleason, A. M., 'Measures on Closed Subspaces of a Hilbert Space', *Journal of Rational Mechanical Analysis* **6** (1957), 885–93.
[22] Greechie, R. J., 'Orthomodular Lattices Admitting no States', *Journal of Combinatorial Theory* **10** (1971), 119–32.
[23] Greechie, R. J., 'An Orthomodular Poset With a Full Set of States not Embeddable in Hilbert Space', *Carribbean Journal of Science and Mathematics* **1** (1969), 15–26.
[24] Greechie, R. J., and Gudder, S. P., 'Is Quantum Logic a Logic?', *Helvetia Physica Acta* **44** (1971), 238–40.
[25] Greechie, R. J. and Miller, F. R., 'On Structures Related to States on an Empirical Logic: I. Weights on Finite Spaces', *K.S.U. Department of Mathematics Technical Report* No. 14, April (1970).
[26] Gudder, S. P., 'On Hidden Variable Theories', *Journal of Mathematical Physics* **11** (1970), 431–6.
[27] Gudder, S. P., *Spectral Methods for a Generalized Probability Theory*, Transl. *Amer. Math. Soc.* **119** (1965), 428–42.
[28] Gudder, S. P., 'Hilbert Space, Independence, and Generalized Probability', *Journal Math. Anal. Appl.* **20** (1967), 48–61.
[29] Gudder, S. P., 'Dispersion-Free States and the Existence of Hidden Variables', *Proceeding of the American Mathematical Society* **19** (1968) 319–24.
[30] Gudder, S. P., 'Uniqueness and Existence Properties of Bounded Observables', *Pacific Journal of Mathematics* **19** (1964), 81–93; **19** (1966) 588–9.
[31] Gudder, S. P., 'Coordinate and Momentum Observables in Axiomatic Quantum Mechanics', *Journal of Mathematical Physics* **8** (1967), 1848–58.
[32] Gudder, S. P., 'Systems of Observables in Axiomatic Quantum Mechanics', *Journal of Mathematical Physics* **8** (1967), 2109–13.
[33] Gudder, S. P. and Boyce, S., 'A Comparison of the Mackey and Segal Models for Quantum Mechanics', *International Journal of Theoretical Physics* **3** (1970), 7–21.
[34] Gudder, S. P. and Piron, C., 'Observables and the Field in Quantum Mechanics', *Journal of Mathematical Physics* **12** (1971), 1583–8.
[35] Guenin, M., 'Axiomatic Formulations of Quantum Theories', *Journal of Mathematical Physics* **7** (1966), 271-82.
[36] Gunson, J., 'On the Algebraic Structure of Quantum Mechanics', *Communications in Mathematical Physics* **6** (1967), 262–85.
[37] Haag, R. and Kastler, D., 'An Algebraic Approach to Quantum Field Theory', *Journal of Mathematical Physics* **5** (1964), 848–61.
[38] Jauch, J. M., *Foundations of Quantum Mechanics*, Addison-Wesley, Reading, Mass., 1968.
[39] Jauch, J. M., 'The Problem of Measurement in Quantum Mechanics', *Helvetia Physica Acta* **37** (1964), 293–316.
[40] Jauch, J. M., 'Systems of Observables in Quantum Mechanics', *Helvetia Physica Acta* **33** (1960), 711–26.
[41] Jauch, J. M. and Misra, B., 'Supersymmetry and Essential Observables', *Helvetia Physica Acta* **34** (1961), 699–710.
[42] Jauch, J. M. and Piron, C., 'On the Structure of Quantal Proposition Systems', *Helvetia Physica Acta* **42** (1969), 842–8.

[43] Jauch, J. M., 'What is Quantum-Logic?', *Quanta*, University of Chicago Press, Chicago and London, 1970.
[44] Jauch, J. M., 'Can Hidden Variables be Excluded from Quantum Mechanics?', *Helvetia Physica Acta* **36** (1963), 827–37.
[45] Jauch, J. M., 'Generalized Localizability', *Helvetia Physica Acta* **40** (1967), 559–70.
[46] Jauch, J. M., 'Hidden Variables Revisited', *Reviews of Modern Physics* **40** (1968), 228–9.
[47] Jordan, P., von Neumann, J., and Wigner, E., 'On an Algebraic Generalization of the Quantum Mechanical Formalism', *Annals of Mathematics* **35** (1934), 29–64.
[48] Kolmogorov, A. N., *Foundations of the Theory of Probability*, Chelsea, New York, 1956.
[49] Kochen, S. and Specker, E. P., 'The Problem of Hidden Variables is Quantum Mechanics, *Journal of Mathematics and Mechanics* **17** (1967), 331–48.
[50] Lowdenslager, D. B., 'On Postulates for General Quantum Mechanics', *Proceedings of the American Mathematical Society* **8** (1957) 88–91.
[51] Ludwig, G., 'Versuch einer axiomatischen Grundlegung der Quantemechanik und allgemeinerer physikalischer Theorien', *Zeitschrift für Physik* **181** (1964), 233–60.
[52] Ludwig, G., 'Attempt of an Axiomatic Foundation of Quantum Mechanics and More General Theories, II, III', *Communications in Mathematical Physics* **4** (1967), 331–48; **9** (1968), 1–12.
[53] Łukasiewicz, J., *Aristotelic Syllogistic from the Standpoint of Modern Formal Logic*, Oxford University Press, London, 1957.
[54] Mackey, G. W., 'Quantum Mechanics and Hilbert Space', *American Mathematical Monthly* **64** (1957), 45–57.
[55] Mackey, G. W., *Mathematical Foundations of Quantum Mechanics*, W. A. Benjamin, New York, 1963.
[56] MacLaren, M. D., 'Notes on Axioms for Quantum Mechanics', *Atomic Energy Commission Research and Development Report ANL*-7065, Argonne National Laboratory (1965).
[57] MacLaren, M. D., 'Nearly Modular Orthocomplemented Lattices', Mathematical Note No. 358, Mathematics Research Laborary, Boeing Scientific Research Laboratories, (1964).
[58] Maczynsky, M. J., 'A Remark on Mackey's Axiom System for Quantum Mechanics', *Académie Polonaise des Sciences Bulletin* **15** (1967), 583–7.
[59] Mielnik, B., 'Geometry of Quantum States', *Communications in Mathematical Physics* **9** (1968) 55–80.
[60] Mielnik, B., 'Theory of Filters', *Communications in Mathematical Physics* **15** (1969), 1–46.
[61] Misra, B., 'When can Hidden Variables be Excluded in Quantum Mechanics?', *Il Nuovo Cimento* **47** (1967), 841–59.
[62] Plyman, R. J., 'A Modification of Piron's Axioms', *Helvetia Physica Acta* **41** (1968), 69–74.
[63] Plymen, R. J., 'C^*-Algebras and Mackey's Axioms', *Communications in Mathematical Physics* **8** (1968), 132–46.
[64] Piron, C., 'Axiomatique' Quantique, *Helvetia Physica Acta* **37** (1964), 439–68.
[65] Piron, C., 'Survey of General Quantum Physics', *University of Denver Report* (1970).
[66] Pontrjagin, L., *Topological Groups*, Gordon and Breach, Inc., New York, 1966.
[67] Pool, J. C. T., 'Baer *-semigroups and the Logic of Quantum Mechanics', *Communications in Mathematical Physics* **9** (1968), 118–41.

[68] Pool, J. C. T., 'Semimodularity and the Logic of Quantum Mechanics', *Communications in Mathematical Physics* **9** (1968), 212–28.
[69] Ramsey, A., 'A Theorem on Two Commuting Observables', *Journal of Mathematics and Mechanics* **15** (1966), 227–34.
[70] Reichenbach, H., *Philosophical Foundations of Quantum Mechanics*, California Press, 1941.
[71] Segal, I., 'Postulates of General Quantum Mechanics', *Annals of Mathematics* **48** (1947), 930–48.
[72] Segal, I., *Mathematical Problems of Relativistic Physics*, American Mathematical Society, Providence, R.I., 1963.
[73] Sherman, S., 'On Segal's Postulates for General Quantum Mechanics', *Annals of Mathematics* **64** (1956), 593–601.
[74] Sherman, S., 'Non-Negative Observables are Squares', *Proceedings of the American Mathematical Society* **2** (1951), 31–3.
[75] Skolem, Th., 'Untersuchungen über die Axiome des Klassenkalküls und über Produktions-und Summations Probleme, welche gewisse Klassen von Aussagen Betreffen', *Videnskapsselskapets Skrifter, I Mat.-Nat. Klasse* (1919) No. 3, p. 37.
[76] Sourian, J. M., 'Quantification Geometrique', *Communications in Mathematical Physics* **1** (1966), 374–98.
[77] Streater, R. F. and Wightman, A. S., *PCT, Spin and Statistics and All That*, W. A. Benjamin, New York, 1964.
[78] Suppes, P., 'Probability Concepts in Quantum Mechanics', *Philosophy of Science* **22** (1961), 378–89.
[79] Suppes, P., 'The Probabilistic Argument for a Non-Classical Logic of Quantum Mechanics', *Philosophy of Science* **33** (1966), 14–21.
[80] Varadarajan, V. S., *Geometry of Quantum Theory*, Vols. I, II, Van Nostrand, Princeton, N. J., 1968, 1970.
[81] von Neumann, J., *Mathematical Foundations of Quantum Mechanics*, Princeton University Press, 1955.
[82] Zierler, N., 'Axioms for Non-Relativistic Quantum Mechanics', *Pacific Journal of Mathematics* **11** (1961), 1151–69.
[83] Zierler, N., 'Order Properties of Bounded Observables', *Proceedings of the American Mathematical Society* **10** (1963), 346–51.
[84] Zierler, N. and Schlessinger, M., 'Boolean Embeddings of Orthomodular Sets and Quantum Logic', *Duke Mathematical Journal* **32** (1965), 251–62.

BAS C. VAN FRAASSEN

THE LABYRINTH OF QUANTUM LOGICS*

1. The Development of Quantum Logic

The conceptual structure of the new quantum theory is in some respects so different from that of classical physics that it has from the very beginning suggested radical departures in philosophy and logic. Specifically, a number of writers have considered non-standard systems of logic in connection with quantum mechanics (see bibliography). Two main directions may be discerned, initiated by Reichenbach, and by Birkhoff and von Neumann. The aim of the present paper is *first* to present a unified exposition of the main systems found in the literature, and *second* to discuss and evaluate the main logical and philosophical theses and arguments which have concerned the subject of quantum logic.

The starting point for our exposition is Beth's semantic analysis of physical theories. This will make possible a semantic analysis of each of the systems of quantum logic to be discussed. Since the original presentation of these logical systems was in most cases rather imprecise, it would perhaps be better to speak of a *reconstruction* than of an analysis. But we do not think that our treatment does more violence to the original intent of the authors than does, say, the current semantic analysis of modal logic to the intent of the originators of the standard modal systems.

2. The Formal Structure of a Physical Theory

We begin by outlining a model for a certain kind of physical theory, derived from Beth's ideas concerning the application of formal semantic concepts to the study of scientific theories. In the next section we shall apply this to the specific case of the elementary quantum theory, and show how this leads to the conception of a 'logic of quantum mechanics'.

Beth addressed himself specifically to non-relativistic theories which use a mathematical model to represent the behavior of a certain kind of

C. A. Hooker (ed.), The Logico-Algebraic Approach to Quantum Mechanics, 577–607

physical system [1], [3], [4], [5]. A physical system is capable of a certain set of *states*, and these states are represented by the elements of a certain mathematical space, the *state-space*. Specific examples are the use of Euclidean n-space in classical mechanics and Hilbert space in quantum mechanics.[1]

Besides the state-space, the theory uses a certain set of *measurable physical magnitudes* to characterize the physical system. This yields the set of *elementary statements* of the theory: each elementary statement U formulates a proposition to the effect that a certain such physical magnitude m has a certain value r at a certain time t. (Thus we write $U=U(m, r, t)$; or $U=U(m, r)$ when abstracting from variation with time.) Whether or not the magnitude m has the value r depends on the state of the system; in some states m has the value r and in others it does not. This relation between states on the one hand and the values of physical magnitudes on the other may also be expressed as a relation between the state-space and the elementary statements. Thus, there corresponds to an elementary statement $U=U(m, r)$ a certain subset $h(U)$ of the state-space H: m has the value r if and only if the state of the system is represented by an element of H which belongs to $h(U)$. (We also say that $h(U)$ is the set of elements of H which *satisfy* U.) The mapping h is the third characteristic feature of the theory; it connects the state-space with the elementary statements, and hence, the mathematical model of the theory with the measurement-results.[2]

This mapping h induces the semantic relations among the elementary statements:

(1) U is *true* if and only if the state of the system is represented by an element of $h(U)$.

(2) U is *valid* if and only if $h(U)=H$.

(3) U is *semantically entailed* by V if and only if $h(V)\subseteq h(U)$.[3]

That this is as yet only a rough and preliminary characterization of these semantic notions will be clear especially from the informal nature of (1). But one can already see how this may lead to a consideration of various logical systems: it has traditionally been the task of logic to give a systematic account of validity and entailment.

3. The Elementary Statements of Quantum Mechanics

In the case of quantum mechanics, the states of a system are represented by the elements (*state-vectors*) of a Hilbert space.[4] For each measurable physical magnitude ('observable') m there is a Hermitean operator M on Hilbert space, with the following significance: m has the value r in state x if and only if $Mx=rx$. The following terminology is used here: when $Mx=rx$, then r is called an *eigenvalue* of M, and x an *eigenvector* of M corresponding to the eigenvalue r. So the vectors which satisfy the elementary statement $U=U(m, r)$ are given by:

(4) $\quad h(U)=\{x: Mx=rx\}.$

This abstracts from the possibility of variation with time. Quantum mechanics was originally developed in two forms: Heisenberg's matrix mechanics, in which the operator M is a function of the time, and Schrödinger's wave mechanics, in which the state-vector is a function of the time.[5] Thus, depending on which of these specific formulations we choose, we have either

(4a) $\quad h(U(m, r, t))=\{x: M_t x=rx\},$ or
(4b) $\quad h(U(m, r, t))=\{x(t): Mx(t)=rx(t)\}.$

But for our purpose the shallower analysis given by (4) will do in most cases.

An operator transforms a vector into another vector, and in general the new vector is not merely a scalar multiple of the first one. So if the state of the system is represented by the state-vector x, we have three possibilities:

(5a) $\quad Mx=rx,$
(5b) $\quad Mx=r'x$ for some $r'\neq r,$
(5c) $\quad Mx\neq r'x$ for any value r'.

If the third possibility obtains, then the magnitude m does *not have any* of its possible values $r, r', r'', \ldots$. This sounds strange to the classical ear; the Bohr-Heisenberg explanation is that a measurable magnitude has a (sharp) value only in a certain kind of experimental situation.[6] But this feature is quite independent of the Copenhagen interpretation; for example, Feyerabend sees a central difference between classical and quantum theory exactly in the breakdown of the principle that "each entity possesses *always* one property out of each category" [13], p. 51, and pp. 52–53.

Of special interest here is the question of *compatibility or incompatibility* of two physical magnitudes m and m'. This corresponds to the question whether or not two operators M and M' commute, i.e. whether $MM' = M'M$. In particular, it is found that

$$QPx - PQx = i\hbar x$$

where Q and P are the operators corresponding to the X-coordinate of position and momentum respectively. Thus Q and P do not commute. It means specifically that if x is an eigenvector of the one, it is necessarily not an eigenvector of the other; for if it were, say corresponding to the eigenvalues r and r', we would have

$$QPx = Qr'x = r'Qx = r'rx = rr'x = rPx = Prx = PQx\,.$$

in which case the difference between QPx and PQx would not be $i\hbar x$ but the zero-vector (since x cannot be the zero-vector, which represents no possible physical state). Therefore when the X-coordinate of position has a value r, then the X-coordinate of momentum does not have any of its possible values r'.

4. Alternatives in Quantum Logic

A *logic* is a system of axioms and/or rules which characterizes the set of valid sentences and the set of valid arguments for a certain language. Thus a logic of quantum mechanics must do this for a certain language pertaining to quantum mechanics. We shall take this language to be the language of the elementary statements (and perhaps complex sentences built up out of these).

At this point we must present a schematic formalization of this language. This formalization is schematic in two ways. First, we intend to remain on the level of abstraction assumed by the writers whose work we discuss below. (There can be no doubt that quantum logic represents, in each case, a relatively shallow analysis of the language of quantum theory. But then, so does standard logic with respect to mathematical discourse; the term 'logic' is traditionally associated with such a high level of abstraction.) Second, the various writers (implicitly) differ on certain points in this construction, and we mean to accommodate all their approaches. The general structure of this language L is then given by:

(6)(a) *Syntax of L*: The set of sentences of L comprises (at least) a set of elementary statements $U(m, r)$.

(b) *Semantics of L*:

(i) Associated with L is a *state-space H*.

(ii) Associated with each elementary statement $U(m, r)$ is a Hermitean operator M on H; r denotes an eigenvalue of M.

(iii) A *model* for L is a couple $K=\langle X, f\rangle$ where X is an element (the system) to which the function f assigns a location $f(X)$ in H. (Here $f(X)$ is the state-vector for the system X.)

(iv) The elementary statement $U(m, r)$ is *true* in the model $K=\langle X, f\rangle$ if and only if the vector $x=f(X)$ is such that $Mx=rx$.

The definitions of validity and semantic entailment can be given in the form

(7) U is a *valid* sentence of L if and only if U is true in every model for L;

(8) U is *semantically entailed by V* in L if and only if U is true in every model in which V is true;

which are equivalent to those given by (1)–(4) for elementary statements U.[7]

We may note about L, *first* that it is what we have elsewhere called a *semi-interpreted language* [32], and *second* that its syntax and semantics have been left under-specified in a number of respects. Specifically, in the syntax we have left open the possibility that there are sentences of L which are not elementary statements. About the elementary statements, we have said when they are true in a model, but not when they are false. For both points there are two main alternatives, and various writers have chosen differently among these alternatives.

Let us first consider the second question, the question when an elementary statement is false. We have already noted that $U(m, r)$ may fail to be true for either of two reasons: because for some values r' other than r, $U(m, r')$ is true, or because for no value r' whatsoever $U(m, r')$ is true. The latter possibility represents the divergence from the classical case.

One possible reaction is to say that $U(m, r)$ is false whenever it is not true – for whatever reason. However, this is not the only possible alternative; the history of logic provides ample precedent for distinguishing *false* from *not true*. This distinction is usually made via the principle:

(9) A sentence is *false* if and only if its denial is true.

If we construe the denial in such a way that it is true whenever the original sentence fails to be true, then this reduces to the previous case. When denial is construed in this way, we speak of *exclusion negation*.[8] But, as has been pointed out many times, a denial is usually understood in the context of a definite set of alternatives. And then the denial is construed to *assert* that one of the *other* alternatives obtains. In this case we speak of *choice negation*. Specifically, we may take the set of eigenvalues of the operator M as the set of alternatives providing the context for $U(m, r)$; its choice negation would then assert that the statevector x is such that $Mx=r'x$ for a value $r'\neq r$. If we now uphold principle (9). for choice negation rather than exclusion negation, *false* no longer coincides with *not true*. This yields the first set of basic alternatives for quantum logic:

(IA) $U(m, r)$ is *false* in the model $K=\langle X,f\rangle$ if and only if $Mx\neq rx$ for $x=f(x)$.

(IB) $U(m, r)$ is *false* in the model $K=\langle X,f\rangle$ if and only if $Mx=r'x$ for $x=f(X)$ and some value $r'\neq r$. .

In the first case, the principle of bivalence holds for elementary statements: $U(m, r)$ is always either true or false. In the second case, the truth-or-falsity of $U(m, r)$ presupposes that the system is in an eigenstate of the operator M.[9]

The second question, whether L comprises sentences which are not elementary statements, must be understood in the context of our intention that L be a language of elementary statements. That means that any sentence of L must either be an elementary statement, or have been constructed out of elementary statements. We shall call an n-ary function f of sentences into sentences an *n-ary connective* provided the truth-value (or lack of truth-value) of $f(U_1,\ldots, U_n)$ in a model K depends entirely on $K, f, U_1,\ldots, U_n$. So for any sentence V of L there must be elementary statements $U_1,\ldots, U_n$ of L such that $V=f(U_1,\ldots, U_n)$ for a certain connective f.

Connectives are usually associated with certain symbols of the language; for example, the conjunction of two atomic sentences in the language of the propositional calculus is a non-atomic sentence produced by infixing a dot or ampersand. If we follow this course of action then connectives will map elementary statements into sentences which are not elementary statements. It might be thought that the question of the logic of L will become trivial if we do not introduce non-elementary statements by means of such connective symbols. But this is not so, due to the existence of various semantic relations among the elementary statements (in the language of the standard propositional logic, the atomic sentences are semantically independent of each other). So one might ask the question: is there an elementary statement V which qualifies as the exclusion negation of $U(m, r)$, or as the choice negation of $U(m, r)$, or as the conjunction of $U(m, r)$ and $U(m', r')$ – and so on. This yields our second set of basic alternatives for quantum logic:

(IIA) The language L has a set of connectives, each of an integral degree $n<0$ and defined for a certain class of n-tuples of elementary statements; the values of the connectives are again elementary statements.

(IIB) The language L has a set of connectives, each of an integral degree $n<0$ and defined for a certain class of n-tuples of sentences of L; the values of the connectives are not elementary statements.

Only when the structure of L has been further specified, by a choice among these alternatives and a characterization of the set of connectives, can the question of the logic adequate with respect to L be broached.

In the following sections we shall present some main approaches to quantum logic, explicating their intent to be that of formulating a logic adequate with respect to L, given a certain choice from the pairs of alternatives (I) and (II). That a certain amount of extrapolation will be involved here, is unavoidable. Our purpose will be served if under our explication, each of these approaches to quantum logic is intelligible, in the sense of having a well-defined objective and well-defined criteria for success. The criteria for success are those which are standard in metalogical appraisal: the system must be at least sound and preferably complete with respect to validity and semantic entailment. (In the present paper, questions of completeness will mostly be disregarded; most of the writers discussed did not raise these questions.)

5. Reichenbach's Approach to Quantum Logic

Reichenbach chose alternatives (IB) and (IIB); this choice is what we shall mean by 'Reichenbach's approach'. (His discussions in [26], Sections 30–34, are always in terms of measurement, but reference to his theoretical discussion of this subject in Section 21 shows how to represent his choice in terms of the mathematical model of operators and eigenvectors.) We shall first discuss Reichenbach's use of *three-valued matrices*, to implement this choice, then Lambert's use of *supervaluations* for the same purpose, and finally, consider the limitations of this approach.

5.1. *Reichenbach's Use of Matrices*

Reichenbach introduced his choice of alternative (IB) as follows:

> Ordinary logic is two-valued; it is constructed in terms of the truth-values *truth* and *falsehood*. It is possible to introduce an intermediate truth-value which may be called *indeterminacy*, and to coordinate this truth-value to the group of statements which in the Bohr-Heisenberg interpretation are called *meaningless*. Several reasons can be adduced for such an interpretation. If an entity which can be measured under certain conditions cannot be measured under other conditions, it appears natural to consider its value under the latter conditions as indeterminate. It is not necessary to cross out statements about this entity from the domain of meaningful statements; all we need is a direction that such statements can be dealt with neither as true nor as false statements. This is achieved with the introduction of a third truth-value of indeterminacy. ([26], p. 145)

Much philosophical puzzlement has been occasioned by the use of the term 'third truth value'. This term is unfortunate; the case can be stated much more perspicuously by saying that certain statement are neither true nor false, rather than that they have a third truth value, the value *indeterminacy*. A value assignment may assign T, F, or I to a sentence; or perhaps 1, 0, or $\frac{1}{2}$; but this assignment is only a marker for the corresponding class of sentences.

The connectives introduced by Reichenbach are all defined by three-valued truth-tables. These connectives are given by syncategorematic expressions in the language L; thus we see that Reichenbach chose alternative (IIB). It must be noted that Reichenbach lists only a few of the connectives definable by means of a three-valued matrix, and does not use even all of these in his subsequent discussions. It is not clear whether he meant all such connectives to appear in L, or only those which he lists.

The main use made of this machinery is to express the statement

(10) The elementary statements U and V correspond to incompatible magnitudes (are incompatible)

in L by means of the sentence

(11) $(U \vee \sim U) \rightarrow \sim\sim V$

where the connectives $\vee$, $\sim$, $\rightarrow$ are given by the tables

$\vee$	T	I	F
T	T	T	T
I	T	I	I
F	T	I	F

$\rightarrow$	T	I	F
T	T	F	F
I	T	T	T
F	T	T	T

$\sim$	
T	I
I	F
F	T

It is easy to see that (11) was constructed to satisfy the criterion that it be true if and only if U and V are such that if U is true, or U is false, then V is indeterminate. This criterion is also satisfied by any connective ⫰ defined by placing T or F or I in the blank spaces of the following table

⫰	T	F	I
T	F	F	T
F	F	F	T
I	–	–	–

There are 27 such connectives, and the one given implicitly by (11) is one defined by placing a T in each blank space. In any case, a single connective could be used to express (10) in L. Certainly this connective is one which is not an obvious generalization of any two-valued truth-functional connective; but neither is Reichenbach's $\sim$, which does not correspond to either choice negation or exclusion negation.

From another point of view, however, no such connective can be used to express the notion of incompatibility adequately. For when m and m′ are incompatible magnitudes, then (on choice (IB)) $U(m', r')$ fails to be true or false in *any* model in which $U(m, r)$ is true or false. But of course it could happen that U is true in a model, and V does not have a truth-value in that model, although U and V do not correspond to incompatible magnitudes. For example, let Q and Q' be the operators corresponding to the X and Y coordinates of position respectively and P the operator corresponding to the X coordinate of momentum. Then suppose that in

$K=\langle X,f\rangle$, $x=f(X)$ is such that $Qx=rx$ and $Q'x=r'x$. The former entails that x is not an eigenvector of P. Hence $U=U(q', r')$ is true in U and $V=U(p, r'')$ is indeterminate in K; by the above table we see that $U \overset{\circ}{\mid} V$ is true in K. Yet U and V do not correspond to incompatible magnitudes.

In this sense, the semantic relation of incompatibility is not adequately expressed by any three-valued truth-functional connective, but at most by a three-valued *modal* connective. Reichenbach's sentence (11) stands to (10) approximately as does material implication to semantic entailment. Instead of choosing an object-language equivalent for (10), however, Reichenbach might have treated incompatibility as a relation among statements (to which one could appeal in arguments) and this relation he might have defined by:

(12) $U \vee \sim U$ semantically entails $\sim\sim V$

The logic adequate to Reichenbach's formulation of L must rely both on the systems of three-valued logic developed by Post, Lucasiewicz, and Tarski (to which he refers in [26], p. 147) and special rules concerning this relation of incompatibility. An example of such a rule would be one validating the following argument form:

(13) $A \vee B$
C (C incompatible with B)
Therefore, A

– a generalized form of the disjunctive syllogism. We may understand Reichenbach's discussion of (11) as implicitly conveying the conjecture that a complete system need only have the special rule:

(14) If U is incompatible with V, infer (11)[10]

The inferable sentences of form (11) are not three-valued tautologies of course; they are rather what Carnap would have called *meaning postulates*.

We may point out here that if it were decided not to use a connective corresponding to incompatibility, then one could also choose the three-valued connectives for the language L in such a way that the classical propositional logic would govern them. One way to do this is to define all the usual connectives in terms of disjunction and negation in the usual manner, and then to let these two basic connectives be *exclusion negation* and *exclusion disjunction* defined by:

+	T	I	F	–
T	T	T	T	F
I	T	F	F	T
F	T	F	F	T

The result would be that bivalence holds for complex sentences, classical propositional logic is sound, and for example, the conjunction of two incompatible sentences is an always-false sentence. In that case alternatives (IB) and (IIB) would be implemented by a procedure which only adds certain special rules to the standard logic (and which would require no reschooling of our logical intuitions).

Two objections might be raised against this by Reichenbach: first, he appears to feel that (10) ought to have an object-language counterpart, and second, the non-standard connectives play a role in the 'suppression of causal anomalies' in [26], Section 33. We have already suggested a sense in which no three-valued truth-functional connective can be an adequate object-language counterpart to the semantic relations of incompatibility. But the further alternative of using a three-valued modal connective for this has its drawbacks also: it still leaves us with the tasks of giving a sense to the iteration of this connective, in such sentences as 'That U and V are incompatible, is incompatible with W'. This use of 'incompatible' is so far from its original intent that it seems to me to raise serious doubts concerning the introduction of any such connective.

Secondly, if an anomaly can be shown to be avoidable by an analysis of its reformulation in L, this must be so due to the semantic structure of L. We can make this point clear by showing that the mere choice of alternative (IB) makes Reichenbach's argument possible in one of his main examples. The following passage contains Reichenbach's use of the properties of his three-valued disjunction in his analysis of the well known n-slit interference experiment.

> Let B_1 be the statement: 'The particle passes through slit B_i'. After a particle has been observed on the screen we know that... if the particle did not go through $n-1$ of the slits, it went through the nth slit, and that if it went through one of the slits, it did not go through the others.... But since from these relations the disjunction $[B_1 \vee \ldots \vee B_n]$ is not derivable, we cannot maintain that this disjunction is *true*; all we can say is that it is *not false*. It can be indeterminate. This will be the case if no observation of the particle at one of the slits has been made. [26], pp. 162–163

Were the disjunction true, one could deduce that the particle must have gone through one of the slits, and this deduction would lead to the well-known anomaly that opening a new slit affects the probability that a particle going through one of the old slits will reach a given point on the screen (as discussed by Reichenbach in [26], Section 7). Hence the importance of showing that the disjunction need not be true.

But the analysis quoted above can be paralleled by a semantic analysis as follows. Let $r_1, \ldots, r_n$ be the positions of the slits; let q be the measurable physical magnitude which is the position of the particle at the time of its passage through the diaphragm, corresponding to the operator Q. Since q has $r_1, \ldots, r_n$ as its only possible values, these are the eigenvalues of Q; and we have:

(15) At most one of $U(q, r_1), \ldots, U(q, r_n)$ is true; if all but $U(q, r_i)$ are false, then $U(q, r_i)$ is true,

provided 'false' is construed in accordance with alternative (IB). The assumption of bivalence leads from (15) to: one of $U(q, r_1) \ldots, U(q, r_n)$ must be true – and hence to anomaly. But since 'false' in (15) is construed in accordance with (IB), this assumption of bivalence does not hold. This recognition of the failure of bivalence as a necessary condition for the correctness of (15) blocks the anomalous inference in just the same way as Reichenbach's analysis.

Moreover, since '$U(q, r_i)$ is true' can here be translated into '$Qx = r_i x$', we see that this semantic analysis can in turn be paralleled by a purely quantum-theoretic analysis of the experimental situation. Thus the suppression of the anomalies is also independent of the choice between alternatives (IA) and (IB).

5.2. *Lambert's Use of Supervaluations*

Karel Lambert has recently sought to implement Reichenbach's approach (choice of (IB) and (IIB) in [19] by using not matrices but *supervaluations*, a device introduced by the present author in [35] and [33]. The advantages of doing so are that the connectives are automatically the classical ones, and classical logic remains sound for them. The semantic relation of incompatibility has no object-language counterpart, but as we have argued in the preceding section, this cannot be considered a weakness.

Just as did Reichenbach, Lambert discusses the semantics in terms of measurement. We shall again reformulate this work as pertaining to a completion of the language L in accordance with alternatives (IB) and (IIB). Since supervaluations are a much less well-known device than matrices, we shall do so in some more detail. To the syntax of L we add:

(16a) If A and B are sentences of L then $\sim A$ and $(A \vee B)$ are complex sentences of L. Every sentence of L is either an elementary statement of L or is a complex sentence constructed out of elementary statements of L in this manner.

And to the semantics we add:

(16b) (i) A *classical valuation* over a model $K = \langle X, f \rangle$ is a function v which assigns to each sentence of L the value T or F, subject to the conditions:

1. if $Mf(X) = rf(X)$ then $v(U(m, r)) = T$
2. if $Mf(X) = r'f(X)$ and $r' \neq r$, then $v(U(m, r)) = F$
3. $v(\sim A) = T$ if and only if $v(A) = F$
4. $v(A \vee B) = T$ if and only if $v(A) = T$ or $v(B) = T$

(ii) The *supervaluation* induced by the model K is the function s such that

1. $s(A) = T$ if and only if $v(A) = T$ for every classical valuation v over K
2. $s(A) = F$ if and only if $v(A) = F$ for every classical valuation v over K
3. $s(A)$ is not defined otherwise.

(iii) The sentence A is *true in the model* K if it is assigned T by the supervaluation induced by K, and is *false in the model* K if it is assigned F by that supervaluation, and otherwise is neither true nor false in K.

It must be noted that clause (16b) (iii) does not contradict clause (6b) (iv). The search for the correct logical system is made simpler by the following useful result:

(17) A sentence A is valid in L (as completed by (16)) if and only if A is assigned T by all classical valuations over all models.

This means that the semantic relation of incompatibility plays *no* role with respect to the validity of sentences; also that classical propositional logic is sound with respect to L.

The relation of incompatibility now plays a role only with respect to the validity of arguments (semantic entailment). All classically valid arguments are still valid, but there are further arguments which are valid due to the relation of incompatibility. We shall here discuss one example of this, which leads to a semantic characterization of incompatibility for *L*.

Consider again the operators *P* and *Q* governed by the Heisenberg exchange relation

(18) $QPx - PQx = i\hbar x$

It is clear that if $Qf(X) = rf(X)$ in the model $K = \langle X, f \rangle$, then $f(X)$ is not an eigenvector of *P*, and vice versa. From this it follows that some classical valuation *v* over *K* will assign *F* to the conjunction

$U(q, r)$ & $U(p, r')$

where & is defined in terms of $\vee$ and $\sim$ as usual. Hence the supervaluation induced by *K* cannot assign *T* to this conjunction. In other words, this conjunction is not true in any model. From this it follows that it semantically entails any other sentence of *L*. Thus the argument

$U(q, r)$ & $U(p, r')$
hence $\sim (U(q, r)$ & $U(p, r'))$

is valid in *L*, though the corresponding conditional is not a valid sentence.

It is interesting to note that, conversely, any sentence which semantically entails its own negation is not true in any model. But it will not be false in every model, unless the corresponding conditional is valid.[11]
So the following is a reasonable generalization of the notion of incompatibility to all the sentences of *L*:

(19) *A* and *B* are incompatible with each other if and only (*A* & *B*) semantically entails $\sim$(*A* & *B*), but (*A* & *B*)$\supset \sim$(*A* & *B*) is not a valid sentence of *L*

– where $\supset$ is defined as usual in terms of $\vee$ and $\sim$.

5.3. *The Limitations of Reichenbach's Approach*

As we have seen, the choice of alternatives (IB) and (IIB) can be implemented in various ways (of which Lambert's seems the most direct). However what we have considered so far represents a rather shallow analysis of the semantic structure of the set of elementary statements. And these are surely more interesting in themselves than any complex sentences which may be constructed out of them by means of extraneous linguistic devices. Let me hasten to add that the bare presentation of Reichenbach's and Lambert's work, reformulated as pertaining to our language L, cannot hope to do justice to the philosophical discussions which accompanied their formalization. But it is only to be expected that the choice of alternative (IIA) will lead to a closer, deeper, analysis of the semantic relations among elementary statements.

6. Von Neumann's Approach to Quantum Logic

John von Neumann chose alternatives (IA) and (IIA) in the earliest work in quantum logic in the section called 'Projections as Propositions' in [37].[12] This initiative was pursued in different ways by Birkhoff and von Neumann [7] and Strauss [27], as well as some later writers. We shall discuss these in turn, but first we wish to explain the basic idea of 'Projections as Propositions'.

Let M be a Hermitean operator with the eigenvalues $r_1, \ldots, r_k, \ldots$.[13] Then there is a special set of Hermitean operators $P_1, \ldots, P_k, \ldots$ each of which has only the eigenvalues 0 and 1, and $Mx = r_i x$ if and only if $P_i x = x$. These form a disjoint family of *projective operators* in the sense that

(20)(a) $P_i P_i x = P_i x$

(b) $P_i P_j x = 0$ when $i \neq j$

(c) $x = P_1 x + P_2 x + \cdots + P_k x + \cdots$

From this it follows that the elementary statements $U(m, r_i)$ and $U(p_i, 1)$ semantically entail each other (given only the conditions on L given by (6)).[14] When alternative (IA) (bivalence) is adopted, that means that $U(m, r_i)$ and $U(p_i, 1)$ are semantically entirely equivalent. Thus we can *without loss* regard every elementary statement as concerning the value of a quantity corresponding to a projective operator.

One point is in order to avoid confusion in the discussion of negation below. The statements

$$Mx = r_i x \quad \text{and} \quad P_i x = x$$

are equivalent (and hence so are $Mx \neq r_i x$ and $P_i x \neq x$). But the statements

$$Mx = r'x \quad \text{for some} \quad r' \neq r_i$$
$$P_i x = r'x \quad \text{for some} \quad r' \neq 1 \quad (\text{i.e. } P_i x = 0)$$

are not equivalent.

There are several ways to give some intuitive content to this formal possibility. First, we may see the transition from the Hermitean operator M to the projective (i.e. idempotent Hermitean) operators P_i as a transition from

quantity m has the value r in system X

to

system X has the *property* that the quantity m has the value r in it. (this property being a measurable quantity with possible values 1 and 0)

A second point concerns the calculation of probabilities. Since $PPx = Px$, it follows that Px is always an eigenvector of P corresponding to the eigenvalue 1, or the zero vector. Thus the expansion (20c) can be given in the form

(21) $$x = c_1 x_1 + c_2 x_2 + \cdots + c_k x_k \ldots$$

where $P_i x = c_i x_i$, and x_i is a unit eigenvector of M corresponding to the value r_i. A basic postulate of quantum mechanics is that (after normalization) the *probability* that a measurement for m upon a system with state vector x will (would) yield the value r_i, equals $c_i^* c_i$. It is easily shown that this probability equals 1 (resp. 0) if and only if $c_i^* c_i = c_i = 1$ (resp. 0). Remembering that $P_i x = c_i x_i$, we see that $U(p_i, 1)$ – respectively $U(p_i, 0)$ – may be read as: the probability that a measurement for m will (would) yield the value r_i equals 1 – respectively 0.[15] (We may just note that this makes $U(p_i 0)$ equivalent also to 'the probability that a measurement for m will (would) yield a value other than r_i equals 1' – which does *not* entail that $U(m, r')$ is true for some value $r' \neq r_j$.)

In the section 'Projections as Propositions' von Neumann explicitly considered the sentential connectives of negation, conjunction, and disjunction, suggesting two ways to implement the choice of (IA) and (IIA). These two ways were worked out in some more detail by Strauss and by Birkhoff and von Neumann respectively.

6.1. *Strauss' Use of Projection Operators*

Martin Strauss formulated the elementary statements of his language Q in [27] in such a way that they contain an explicit reference to a system. This means that one can refer to several systems in the same language (though no single sentence of Q refers to more than one system). We shall continue to consider only the simple case in which elementary statements concern the same system.

To emphasize that all the sentences of Q are elementary statements (choice (IIA)), Strauss considers not sentential connectives but what he calls *predicational connections*; that is, he uses the form 'Y is (F and G)' rather than 'Y is F and Y is G', for example. This constitutes no essential difference; in view of the practice of the other writers, we present his work as if he had considered sentential connectives.

But finally, Strauss' elementary statements correspond *only* to those which may be given the form $U(p, 1)$ – where p corresponds to a projective operator P. Now it is easily seen that $U(p, 0)$ is the *choice negation* of $U(p, 1)$, since

$$Px = 0x = 0$$
$$Px = rx \quad \text{for some} \quad r \neq 1$$

are equivalent for the projective operator P. Strauss points out (following von Neumann) that the set of elementary statement of form $U(p, 1)$ is still closed under choice negation. For the operator P defined by

(22) $\bar{P}x = x - Px \quad (\text{or: } \bar{P} = I - P)$

is again a projective operator, and is such that

(23) $\bar{P}x = x \quad \text{if and only if} \quad Px = 0$
$\bar{P}x = 0 \quad \text{if and only if} \quad Px = x$

Hence the statement $U(\bar{p}, 1)$ is equivalent to $U(p, 0)$ and is the choice negation of $U(p, 1)$. Henceforth writing '$U(p)$' for '$U(p, 1)$', we see that every elementary statement can be given the form '$U(p)$', and:

(24) Every elementary statement $U(p)$ has a *choice negation* $\neg U(p)$ namely the elementary statement $U(\bar{p})$.

Next we may consider conjunction. The conjunction of $U(p)$ and $U(q)$ must be true if and only if both of these are true. So we must find projective operator R such that $Rx=x$ if and only if $Px=Qx=x$. Here the situation is that the calculus of operators defines such an operator R for the operators P and Q, namely their product PQ, but only if they commute. So the set of elementary statements is not closed under conjunction on this approach, though conjunction is defined for an important subset:

(25) The pair of elementary statements $U(p)$ and $U(q)$ has a *conjunction* $(U(p)\ \&\ U(q))$, namely $U(r)$ where r is the quantity corresponding to the operator PQ, if and only if P and Q commute.

Disjunction can be introduced definitionally as $\neg(\neg U(p)\ \&\neg U(q))$ and is defined only if P and Q commute: similarly, of course, any other classical connective can be introduced by definition. It must be noted that choice negation is the only significant familiar connective under which the set of elementary statements is closed for Strauss' approach.

6.2. *Birkhoff and Von Neumann's Use of Subspaces*

The well-known joint paper of Birkhoff and von Neumann [7] provides a way to adopt alternatives (IA) and (IIA), and yet have the language closed under the usual statement connectives. This is not done by broadening the set of elementary statements. Rather it is based on the observation that the projective operator stand in a one-to-one correspondence to the subspaces (closed linear manifolds) of the Hilbert space, but that the calculus of projections is not isomorphic to the calculus of subspaces.

To make this clear, we have to say more about the geometry of Hilbert space. For any vectors x, y of the Hilbert space, the following are defined:

the vector sum $x+y$, a vector
the inner product (x, y), a scalar

and the vectors x and y are *orthogonal* to each other if and only if $(x, y)=0$. A *linear manifold* is a set of vectors which is closed under addition of vectors and multiplication by scalars (if x and y belong, so does $rx+r'y$). If a linear manifold is topologically closed, then it is a *subspace*; a subspace of a Hilbert space is again a Hilbert space. Each subspace contains the *null-vector* 0; the smallest subspace is the *null-space* 0 which contains only the null-vector. The subspaces of a Hilbert space play a role analogous to the lines and planes through the origin in ordinary Euclidean 3-space. This analogy provides a good intuitive guide to many of the points made below.

If we perform the usual set-theoretic operations of intersection, union, and complementation on subspaces, we find that only the first of these produces new subspaces. But similar to union is *linear union*: the linear union $S \oplus T$ of S and T is the least subspace containing both. Similar to complement is *orthogonal complement*: the orthogonal complement $S^{\perp}$ of S is the set of vectors which are orthogonal to all the elements of S (again a subspace). The subspaces of a Hilbert space ordered by set inclusion, and with the operations of intersection, linear union, and orthogonal complementation form an *orthocomplemented lattice*.

A basic theorem concerning subspaces is that if S is a subspace, then any vector z is equal to a sum $x+y$, where x belongs to S and y to $S^{\perp}$. The *projection on the subspace* S is then defined by

$$P(x+y)=x \quad \text{where} \quad x \in S, \quad y \in S^{\perp}$$

Above we defined a projective operator as an idempotent Hermitean operator. It can now be shown that every projection on a subspace is a projective operator and vice versa. Specifically:

(26) P is the projection on S if and only if P is the projective operator such that $S=\{x : Px=x\}$.

The operator $\bar{P}$ defined by $\bar{P}x=x-Px$ is clearly the projection on $S^{\perp}$ in this case; equivalently, $S^{\perp}=\{x : Px=0\}$.

In the preceding section we remarked that (given alternative (IA)) every elementary statement $U(m, r)$ may be given the form $U(p)=U(p, 1)$ with p corresponding to a certain projective operator P. Thus the Birkhoff-von Neumann completion of L can be given by adding to the syntax and semantics:

(27)(a) Every sentence is an elementary statement.

(b) $U(p)$ is true in $K=\langle X, f\rangle$ if and only if $f(X)$ belongs to $h(U(p))=\{x : \mathrm{P}x=x\}$, and false otherwise.

Note that (27b) does not contradict (6b) (iv) but only adds the principle of bivalence. We now have the result that every sentence U of L corresponds to a subspace $h(U)$, the set of vectors which satisfy U, in the sense of Section 2. The connectives will be defined in terms of this mapping h.[16]

The complement $\bar{P}$ of a projective operator P corresponds to the orthogonal complement $S^{\perp}$ of a subspace S, as we remarked above, so we have choice negation as before:

(28) Every elementary statement U has a *choice negation* $\neg U$, namely an elementary statement V such that $h(V) = h(U)^{\perp}$.

Considering now conjunction, recall that if P and Q commute, then PQ is again a projective operator with the property:

$$PQx = x \quad \text{if and only if} \quad Px = Qx = x.$$

But that means that PQ is the projection on the *intersection* of the subspaces on which P and Q project. Accordingly, Birkhoff and von Neumann adopt the following generalization:

(29) Every pair of elementary statements U and V has a *conjunction* $(U \& V)$, namely the elementary statement W such that $h(W) = h(U) \cap h(V)$.

The question is now what the conjunction means if P and Q do not commute. We still have: $(U \& V)$ is true if and only if U and V are both true, for all cases. Hence it is entirely correct to call this connective *conjunction.* For a pair of incompatible propositions, recall our discussion of the operators corresponding to the X-coordinates of momentum and position respectively. These have *no* eigenvectors in common. Hence for two projective operators which correspond to specific values of position and momentum, the intersection of the two subspaces is as small as it possibly can be: it is the null-space. Thus the conjunction

electron X has position r and momentum r'

which was meaningless for Bohr and Heisenberg, *always indeterminate-or-false* for Reichenbach, and *not well-formed* for Strauss, is *always false* for Birkhoff and von Neumann. This assimilates it to such other necessarily false propositions as

the table is red and green all over;
The bar is warmer than itself;
The cupola is both round and not round.

This is the obvious move if one accepts bivalence and also wishes the language to be closed under conjunction. For there is no eigenstate of both position and momentum – hence the conjunction in question cannot be true. Bivalence entails then that it is false.

If we introduce disjunction $\vee^*$ again definitionally through de Morgan's law, the principle

$$(S^{\perp} \cap T^{\perp})^{\perp} = S \oplus T$$

for subspaces S and T shows that

$$h(U \vee^* V) = h(U) \oplus h(V).$$

This is like ordinary disjunction in that each disjunct semantically entails the disjunction.

Nevertheless, this is by no means ordinary disjunction, as is shown by the fact that $U \vee^* V$ can be true even if neither U nor V is true ($S \cup T$ is in general a proper subset of $S \oplus T$). Reminiscent of Aristotle's discussion of future contingents is the fact that the law of excluded middle ($U \vee^* \neg U$) holds, although neither U nor $\neg U$ may be true in a given model.

The character of the logic adequate with respect to the language L thus construed is of course determined by the character of the lattice of subspaces. This is in all cases a complete orthocomplemented lattice. If the dimension of the Hilbert space H is 0 or 1, the lattice satisfies the *distributive law:*

$$(30) \qquad A \cap (B \oplus C) = (A \cap B) \oplus (A \cap C)$$

and hence is a Boolean algebra. If H is finite-dimensional but of dimension greater than 1, the lattice is not distributive but still satisfies the weaker *modular law:*

$$(31) \qquad A \oplus (B \cap C) = (A \oplus B) \cap C \quad \text{provided} \quad A \subseteq C.$$

However, in quantum mechanics infinite-dimensional Hilbert spaces are used, and here even the modular law breaks down ([16], p. 22) Birkhoff and von Neumann were quite content to have distributivity fail, but they considered changes in quantum theory to preserve modularity (see especially [6]).

6.3. *Extensions of Birkhoff and Von Neumann's Approach*

The work of Birkhoff and von Neumann certainly influenced every subsequent writer on the subject. Besides the ones we have already discussed there has been important further work by Février [11] and [12], Fuchs [14], Emch and Jauch [10], Varadarayan [36], Mackey [21], Kochen and Specker [18], and Suppes [28] and [29]. Of these, we shall briefly discuss the contributions of Février and Fuchs.

Paulette Février first attempted to develop a quantum logic by means of three-valued matrices; however, the values were *true, contingently false*, and *necessarily false* (or: false by virtue of incompatibility). She soon linked the truth-values with the subspaces of Hilbert space, and Birkhoff regarded her work as a continuation of von Neumann's and his own ([6], p. 157). Most interesting from our point of view is that she eventually attempted to combine alternatives (IIA) and (IIB) (following, she said, a suggestion by Beth – see [11], p. 382). Thus one would have one kind of connective which leads only from elementary statements to elementary statements, and another kind of connective which leads to sentences in general not even logically equivalent to elementary statements (such as exclusion negations). Other such combinations might be interesting – for example the choice of (IB) and (IIA).

W. Fuchs' short article [14] is notable mainly for his attempt to introduce a kind of implication, which is not merely the definitional analogue of the material conditional. Using our notation, he introduces his arrow by:

(32) $A \rightarrow B$ is the proposition which says that $h(A) \subseteq h(B)$.

He then discusses restrictions on the introduction rules for implication to avoid the theorem $A \rightarrow (B \rightarrow A)$. Such iteration of arrows is well-formed only if the arrow is a sentential connective of L (and is not simply a symbol for the metalinguistic relation of semantic entailment). But then (32) is not satisfactory; since it does not define $h(A \rightarrow B)$, it does not tell us how to understand such iteration. A possible definition of $h(A \rightarrow B)$ is given by

$$(33) \qquad h(A \rightarrow B) = \begin{cases} H & \text{if } h(A) \subseteq h(B) \\ O & \text{otherwise} \end{cases}$$

which means that $A \to B$ is again an elementary statement if A and B are. This would yield an analogue to Lewis S_5 strict implication (and $A \to (B \to A)$ is not valid for it).[17]

7. Arguments concerning the significance of quantum logic

The work discussed above has occasioned a number of reviews and discussions; many of these rather critical. Many of the criticisms I consider to have been cogent with respect to the original formulation. But in this paper we have presented a reconstruction of the subject of quantum logic: from our point of view, each attempt at quantum logic has been an attempt to elucidate and exhibit semantic relations among the elementary statements. While I believe the reconstruction to be entirely faithful to the *intent* of the original work, it nevertheless avoids the difficulties noted by the above-mentioned criticisms – as I shall attempt to show.

On the other hand, it has been suggested at various times that the subject of quantum logic throws a radical new light on the foundations of logic. If our reconstruction is correct, the work in quantum logic falls entirely within the scope of the semantic analysis of logic in general, so that this suggestion can hardly be true. But I would maintain that the semantic analysis of logic is of major importance to the foundations of logic, *and* that the attempts to formulate a logic of quantum mechanics constitute pioneering work in this field.

7.1. *Quantum Logic and Quantum Theory*

We may begin with a look at McKinsey and Suppes' very critical review of P. Février's *La structure des théories physiques* [22]. They understood her aim to be a formalization of quantum theory *based on* a non-standard logic. Implicit in the work of Carnap and many of his contemporaries in philosophy of science is the following picture of a physical theory: it is ideally constructed by adding axioms with empirical content to a formalized system of logic and mathematics, say *Principia Mathematica*. The latter consists of standard logic plus axioms for sets. In that sense one might say that the current picture of a physical theory implied that its (eventual) formalization must be 'based on' standard logic. McKinsey and Suppes understood Février to be challenging this picture of a physical theory.

Perhaps that is what she had in mind; in that case it is indeed fair to say that she "does not appear to appreciate the difficulties that would be involved in the heroic task of developing quantum mechanics in the framework of a non-classical logic" ([22], p. 52), and that the mathematical part of this project "would be somewhat analogous to the writing of *Principia Mathematica*, though vastly more onerous" ([22], p. 54).

But from our point of view a logic of quantum mechanics is simply an attempt to give a systematic account of the semantic relations among the elementary statements of that theory. And these semantic relations are to be *deduced from* the quantum theory – *that* is the sense in which this logic is a quantum logic. It is not meant to be the basis for a formalization of the theory, or for a new, non-standard *Principia*.

On the other hand, this does not mean that the research in quantum logic has no value for foundational research in quantum theory. An exposition of the Birkhoff and von Neumann system is part of Ludwig's *Die Grundlagen der Quantenmechanik* [20]. Elegant recent work in this area by Mackey [21] and Zierler [42, 43], is clearly indebted to (or inspired by) that of Birkhoff and von Neumann. But none of these attempts involve the proposal to develop the theory in the framework of a non-standard mathematics.

The point that the relation between quantum mechanics and quantum logic is that the former provides the semantics for the latter, is also relevant to Feyerabend's critique of Reichenbach.

First, Feyerabend sees as a central assumption in the classical point of view the principle that if m is a physical magnitude, then $U(m, r)$ must be true for *some* value of r at any given time. He correctly points out the semantic feature of the elementary statements of quantum mechanics which is the violation of this assumption – and that this is sufficient to dissolve the anomalies exhibited by Reichenbach ([13], pp. 51–53). But of course, Reichenbach's three-valued logic is intended to reflect exactly this semantic feature. Thus it is hardly *à propos* to say that "Reichenbach is one of those thinkers who are prepared to give up... even classical logic because they cannot adjust themselves to" the inadequacy of the classical point of view ([13], p. 53).

In Section 4 of [13], Feyerabend argues that Reichenbach's system violates his own criteria of adequacy. Reichenbach wrote:

When we wish to incorporate all quantum mechanical statements into three-valued logic, it will be the leading idea to put into the true-false class those statements which we call quantum mechanical laws. ([26], pp. 159–160)

Now Feyerabend argues, and Reichenbach admits, that the law of conservation of energy gets the middle value. Reichenbach writes:

The principle requiring that the sum of kinetic and potential energy be constant connects simultaneous values of momentum and position.... It follows that the principle of conservation of energy is eliminated... from the domain of true statements, without being transformed into a false statement; it is an indeterminate statement. ([26], p. 166)

Feyerabend adds:

The same results if we use the statement in the form in which it appears in quantum mechanics. In this form the statement asserts that the sum of various operators, not all of them commuting, will disappear....

The last argument admits of generalization; every quantum-mechanical statement containing non-commuting operators can only possess the value 'indeterminate'. This implies that *the commutation rules* which are among the basic laws of quantum mechanics *as well as the equations of motion*... will be indeterminate.... ([13], p. 54)

These would indeed be disastrous consequences. But from the point of view of our present reconstruction, the discussion went astray at its very beginning.

First, we have sharply distinguished between the language L of elementary statements, and the general language of quantum theory. In the latter we encounter such statements as:

Postulate I. A system with n degrees of freedom is completely specified by the normed state vector $\psi_t(q_1, \ldots, q_n)$....
Postulate II. To every observable corresponds a Hermitean operator....

which are found in many texts and discussions of quantum theory (here quoted from Mandl [23]). The above statements are in fact examples of what philosophers of science have generally called *correspondence rules* (assuming that 'system' and 'observable' have counterparts in the 'observation language'); for us they are part of the semantics of L; and for the physicist they are basic principles of the quantum theory.

There is an obvious and excellent reason for distinguishing between these two languages: for it is easy to construct an artificial language satisfying the conditions we have placed on L, but no one has as yet shown how to formalize a language adequate to the complete development of quantum theory. Were the latter formalized, we could join the two languages (or perhaps the former would already be part of the latter). However, this should not affect the semantic structure of either. Since the development of quantum theory is within the framework of classical mathematics, one would find that

$$Mx = rx$$

is always true or false, even if

$$U(m, r)$$

is sometimes neither true nor false. (This is perfectly compatible with the fact that the one statement is true if and only if the other is true.) In Reichenbach's formulation, these two statements *cannot* be identified, for this very reason.

As we have construed it, Reichenbach's proposal generates indeterminacies *only* among elementary statements. Both Reichenbach and Feyerabend begin by considering the law of conservation of energy stated in its classical form and construed, one is led to assume, in terms of elementary statements and numerical quantifiers. This is not *à propos*, and Feyerabend very rightly turns to the quantum theoretic formalism, in which the law has the form of an assertion that a certain sum of operators equals zero. Some of these operators do not commute. But the important point is that this statement is *not* a sentence of L, but of the language of the theory of linear operators on Hilbert space. Therefore this statement must be true or false. The correct identification of the operators makes it true. Mutatis mutandis for the exchange relations and the equations of motion.

To sum up, the laws of quantum mechanics concern the semantics of the language of elementary statements; they do not belong to it. (This shows equally why these laws are not indeterminate in Reichenbach's approach, and not necessarily false in the Birkhoff-von Neumann approach.)

7.2. *Quantum Logic and the Foundations of Logic*

Turning now to the significance of quantum logic for the philosophy of logic, we find that Février, and Emch and Jauch, see in its development concrete support for Gonseth's point of view that logic is a 'physique de l'objet quelconque'. But once we accept that it is a logic only of the set of elementary statements for a certain physical theory, we must fully concur in Jordan's judgment on this matter:

> Man kann [es] nach Birkhoff-Neumann so ausdrücken, dass man von einer *Quantenlogik* im Gegensatz zu einer *klassischen Logik* spricht. Natürlich ist es Geschmacksache, ob man diese Bezeichnung anerkennen will; jedoch ist sie jedenfalls dann naturgemäss, wenn man unter 'Logik' die Gesetze der möglichen Verknüpfungen von Aussagen... über den Zustand eines physikalischen Systems verstehen will – in *dieser* Auffassungsweise ist auch die Logik eine empirische Wissenschaft.... ([17], p. 368)

Only from such a general point of view does a multiplicity of logics make sense; and then the discovery of a non-standard logical structure does not constitute a conceptual revolution.

Gonseth's characterization of logic is somewhat reminiscent of the views of Quine, according to whom logical principles, no more than physical laws, are immune to revision in the face of new experimental evidence:

> Revision even of the logical law of the excluded middle has been proposed as a means of simplifying quantum mechanics; and what difference is there in principle between such a shift and the shift whereby Kepler superseded Ptolemy, or Einstein Newton, or Darwin Aristotle? ([25], p. 43)

But this is an entirely misleading way of stating the point. No law is contradicted by the proposal of a non-standard quantum logic. Rather, what is shown is that we can construct languages for which the familiar laws do not hold; hence, what is shown is that standard logic has a limited domain of application. This should come as no surprise; as purely formal possibilities, these non-standard languages have been investigated by logicians since the second decade of this century. The (pleasant) surprise is rather to see that these ideas concerning non-standard logical systems have a natural application in the set of elementary statements of a physical theory. This is well argued by Putnam:

> perhaps this is what is meant when it is said that a three-valued logic does not constitute a real alternative to the standard variety: it exists as a calculus, and perhaps as a non-standard way of using logical words, but there is no *point* to this use.
>
> ... three-valued logic and other non-standard logics had first to be shown to exist as consistent formal structures ... The only remaining question is whether one can describe a physical situation in which this use of logical words would have a point.
>
> Such a physical situation (in the microcosm) has indeed been described by Reichenbach. ([24], pp. 76–77)

That is, the question is whether there is a philosophically or scientifically interesting language of which we can say: this non-standard logical system is the logic of that language.

Finally, I should like to emphasize what I find philosophically the most significant feature of these attempts to formulate a 'logic of quantum mechanics'. These very diverse attempts can all be understood, as pursuing different but related aims, from the point of view of a semantic analysis of physical theories. I see in this point of view a powerful rival for the concept of a physical theory as a syntactic system with formation, transformation, and correspondence rules, and its emendations by means of coordinative definitions, reduction sentences, and meaning postulates. This strongly syntactic approach certainly led to a great deal of insight into the structure of scientific theories; perhaps the semantic approach, so fruitful in foundational work in logic and mathematics, will prove equally successful in further philosophical inquiry in science.

University of Toronto

NOTES

* Presented at the Philosophy of Science Association First Biennial Meeting, Pittsburgh, October 1968. The research for this paper was supported by NSF grant GS-1566.

1 Birkhoff and von Neuman [7] generalized the term phase space' to cover both cases; Weyl [40] used 'system space'. The term 'state-space' is adopted from systems theory; cf. [41]. Similar conceptions of the structure of physical theories can be found in Destouches [8], [9], Strauss [27] and in Weyl's interesting article [39].

2 While $U(m, r, t)$ may be read as 'A measurement of m at t will (would) find the value r', this must not be understood as an operationalist identification of meaning. The exact relation between $U(m, r, t)$ and the outcomes of actual experiments is the subject of an auxiliary theory of measurement, which is only caricatured by the notion of a set of 'correspondence rules'; cf. [28].

3 Here we can let V be a set of elementary statements in which case $h(V) = \cap \{h(A): A \in \in V\}$.

4 In this simplified presentation we only consider pure cases, and not mixtures; this appears to be sufficient for a discussion of quantum logic.

5 Cf. [15], pp. 35–36 or [23], pp. 100–101.

6 We add the word 'sharp' since for each state (hence also those which are not eigenstates) there is a determinate *probability* that the value r will be found upon a measurement for m.

7 The principles of the theory should govern the choice of the state-space and of the operators, and/or limit the set of models; superselection rules may set further limitations

8 The terms 'exclusion negation' and 'choice negation' are taken from Mannoury's discussions of the foundations of mathematics, which played an important role in the development of Intuitionism (cf. [2], pp. 20–22). Similar distinctions go back at least to the Arab logicians of the Middle Ages.

9 For a semantic discussion of this notion of presupposition, see [33], Sections I and II.

10 Concerning incompatibility; also needed are axioms to the effect that if $r \neq r'$ and $U(m, r)$ then not $U(m, r')$.

[11] In terms of [33], $U(m, r)$ and $\sim U(m, r)$ presuppose that the state-vector is an eigenvector of M (hence $\sim$ is choice negation), and a conjunction of incompatible sentences presupposes its own denial. We may note in addition that if we are content with the present, relatively shallow analysis of semantic relations among the elementary statements, L can be reformulated as a radical presuppositional language [34], for which metalogical theorems are very straightforward.
[12] The question of bivalence is not raised explicitly; in the historical context this can be taken to signify its acceptance.
[13] We consider only a simple case: M is assumed to have a discrete spectrum and to be non-degenerate. For the extension to other cases we refer to von Neumann's work.
[14] It should also be noted that the statement 'the value of m lies in the interval I' is also expressed by an elementary statement of the form $U(p, 1)$, where p corresponds to a projective operator P; see [37], p. 252. A presentation of quantum theory which follows this approach closely is that of Temple [31].
[15] This suggests, of course, that the set of elementary statements ought really to be broadened to include *all* probability statements concerning what a measurement for m will (would) find. J. von Neumann sketched such an extension in [38].
[16] This will show more perspicuously how L is here a semi-interpreted language in the sense of [32]. The presentation of the subject most directly along these lines is that of Ludwig [20], pp. 54–59.
[17] Cf. my discussion of the relation between S_5 and semi-interpreted languages in [32], Section 5.

BIBLIOGRAPHY

[1] Beth, E. W., 'Analyse Sémantique des Théories Physiques', *Synthese* **7** (1948–49) 206–207.

[2] Beth, E. W., *Mathematical Thought*, D. Reidel Publ. Co., Dordrecht-Holland, 1966.

[3] Beth, E. W., *Natuurphilosophie*, Noorduijn, Gorinchem, 1948.

[4] Beth, E. W., 'Semantics of Physical Theories', *Synthese* **12** (1960), 172–175.

[5] Beth, E. W., 'Towards an Up-to-date Philosophy of the Natural Sciences', *Methodos* **1** (1949), 178–185.

[6] Birkhoff, G., 'Lattices in Applied Mathematics', *American Mathematical Society, Proc. of Symposia in Pure Mathematics* **2** (1961), 155–184.

[7] Birkhoff, G. and von Neumann, J., 'The Logic of Quantum Mechanics', *Annals of Mathematics* **37** (1936), 823–843.

[8] Destouches, J-L., 'Les Principes de la Mécanique Générale', *Actualités Scientifiques et Industrielles* **140** (1934).

[9] Destouches, J.-L., 'Le Role des Espaces Abstraits en Physique Nouvelle', *Actualités Scientifiques et Industrielles* **223** (1935).

[10] Emch, G. and Jauch, J. M., 'Structures logiques et mathematiques en physique quantique', *Dialectica* **19** (1965), 259–279.

[11] Février, P., 'Logical Structure of Physical Theories' in L. Henkin *et al.* (eds.), *The Axiomatic Method*, North-Holland Publ. Co., Amsterdam, Holland, 1959, pp. 376–389.

[12] Février, P., 'Les relations d'incertitude de Heisenberg et la logique', *Comptes Rendus de l Académie des Sciences* **204** (1937), 481–483.

[13] Feyerabend, P., 'Reichenbach's Interpretation of Quantum Mechanics', *Philosophical Studies* **9** (1958), 49–59.

[14] Fuchs, W. R., 'Ansatze zu einer Quantenlogik', *Theoria* **30** (1964), 137–140.
[15] Green, N. S., *Matrix Mechanics*, Noordhoff, Groningen, 1965.
[16] Halmos, P. R., *Introduction to Hilbert Space*, Chelsea, New York, 1957.
[17] Jordan, P., 'Quantenlogik und das Kommutative Gesetz', in L. Henkin *et al.* (eds.), *The Axiomatic Method*, North-Holland Publ. Co., Amsterdam, Holland, 1959, pp. 365–375.
[18] Kochen, S. and Specker, E. P., 'Logical Structures Arising in Quantum Theory. in J. W. Addison *et al.* (eds.), *The Theory of Models*, North-Holland Publ. Co., Amsterdam, Holland 1965, pp. 177–189.
[19] Lambert, K., 'Logic and Microphysics', in K. Lambert (ed.), *The Logical Way of Doing Things*, Yale University Press, New Haven, 1969, pp. 93–117.
[20] Ludwig, G., *Die Grundlagen der Quantenmechanik*, Springer-Verlag, Berlin, 1954.
[21] Mackey, G. W., *The Mathematical Foundations of Quantum Mechanics*, Benjamin, New York, 1963.
[22] McKinsey, J. C. C. and Suppes, P., *Review* of P. Destouches-Février, 'La Structure des Theories Physiques', *Journal of Symbolic Logic* **19** (1954), 52–55.
[23] Mandl, F., *Quantum Mechanics*, Academic Press, New York, 1957.
[24] Putman, H., 'Three-Valued Logic', *Philosophical Studies* **8** (1957), 73–80.
[25] Quine, W. V. O., *From a Logical Point of View*, Harper and Row, New York, 1963.
[26] Reichenbach, H., *Philosophic Foundations of Quantum Mechanics*, University of California Press, Berkeley, 1944.
[27] Strauss, M., 'Mathematics as Logical Syntax – a Method to Formalize the Language of a Physical Theory', *Erkenntnis* **7** (1937–38), 147–153.
[28] Suppes, P., 'Logics Appropriate to Empirical Theories', in J. W. Addison *et al.* (eds.), *The Theory of Models*, North-Holland Publ. Co., Amsterdam, Holland, 1965, pp. 364–375.
[28] Suppes, P., 'The Probabilistic Argument for a Non-Classical Logic in Quantum Mechanics', *Philosophy of Science* **33** (1966), 14–21.
[30] Suppes, M., 'Probability Concepts in Quantum Mechanics', *Philosophy of Science* **28** (1961), 378–389.
[31] Temple, C., *The General Principles of Quantum Theory*, Methuen, London, 1961.
[32] van Fraassen, B. C., 'Meaning Relations Among Predicates', *Nous* **1** (1967), 161–179.
[33] van Fraassen, B. C., 'Presupposition, Implication, and Self-Reference', *Journal of Philosophy* **65** (1968), 136–152.
[34] van Fraassen B. C., 'Presuppositions, Supervaluations, and Free Logic', in K. Lambert (ed.), *The Logical Way of Doing Things*, Yale University Press, New Haven 1969, pp. 67–91.
[35] van Fraassen, B. C., 'Singular Terms, Truth-Value Gaps, and Free Logic', *Journal of Philosophy* **63** (1966), 481–495.
[36] Varadarajan, V. S., 'Probability in Physics and a Theorem on Simultaneous Observability', *Comm. Pure Appl. Math.* **15** (1962), 189–217.
[37] von Neumann, J., *Mathematical Foundations of Quantum Mechanics* (tr. by R. T. Beyer), Princeton University Press, Princeton, 1955.
[38] von Neumann, J., 'Quantum Logics', unpublished; reviewed by A. H. Taub in *John von Neumann, Collected Works* (vol. **4**, pp. 195–197), Macmillan, New York, 1962.
[39] Weyl, H., 'The Ghost of Modality', in M. Farber (ed.), *Philosophical Essays in Memory of Edmund Husserl*, Harvard University Press, Cambridge, Mass., 1940, pp. 278–303.

[40] Weyl, H., *The Theory of Groups and Quantum Mechanics*, Dover, New York, 1950.
[41] Zadeh, L. A. and Desoer, C. A., *Linear System Theory: The State Space Approach*, McGraw-Hill, New York, 1963.
[42] Zierler, N., 'Axioms for Non-Relativistic Quantum Mechanics', *Pacific Jounal of Mathematics* **11** (1961), 1151–1169.
[43] Zierler, N. and Schlesinger, M., 'Boolean Embeddings of Orthomodular Sets and Quantum Logic', *Duke Math. Journal* **32** (1965), 251–262.

THE UNIVERSITY OF WESTERN ONTARIO SERIES IN PHILOSOPHY OF SCIENCE

A Series of Books on Philosophy of Science, Methodology, and Epistemology published in connection with the University of Western Ontario Philosophy of Science Programme

1. J. LEACH, R. BUTTS, and G. PEARCE (eds.), *Science, Decision and Value*. Proceedings of the Fifth University of Western Ontario Philosophy Colloquium, 1969. 1973, vii + 213 pp.
2. C. A. HOOKER (ed.), *Contemporary Research in the Foundations and Philosophy of Quantum Theory*. Proceedings of a Conference held at the University of Western Ontario, London, Canada, 1973, xx + 385 pp.
3. J. BUB, *The Interpretation of Quantum Mechanics*. 1974, ix + 155 pp.
4. D. HOCKNEY, W. HARPER, and B. FREED (eds.), *Contemporary Research in Philosophical Logic and Linguistic Semantics*. Proceedings of a Conference held at the University of Western Ontario, London, Canada, 1975, viii + 334 pp.

GPSR Compliance

The European Union's (EU) General Product Safety Regulation (GPSR) is a set of rules that requires consumer products to be safe and our obligations to ensure this.

If you have any concerns about our products, you can contact us on ProductSafety@springernature.com

In case Publisher is established outside the EU, the EU authorized representative is:

Springer Nature Customer Service Center GmbH
Europaplatz 3
69115 Heidelberg, Germany

Batch number: 10607216

Printed by Printforce, the Netherlands